I,II,III,A
metals

Indicates element essential for health

	IIIa	IVa	Va	VIa	VIIa	He 2
						4.0026

5 boron **B** 2 3 10.81	6 carbon **C** 2 4 12.011	7 nitrogen **N** 2 5 14.0067	8 oxygen **O** 2 6 15.9994	9 fluorine **F** 2 7 18.9984	10 neon **Ne** 2 8 20.179
13 aluminum **Al** 2 8 3 26.9815	14 silicon **Si** 2 8 4 28.086	15 phos-phorus **P** 2 8 5 30.9738	16 sulfur **S** 2 8 6 32.06	17 chlorine **Cl** 2 8 7 35.453	18 argon **Ar** 2 8 8 39.948

	Ib	IIb					

28 nickel **Ni** 2 8 16 2 58.70	29 copper **Cu** 2 8 18 1 63.546	30 zinc **Zn** 2 8 18 2 65.38	31 gallium **Ga** 2 8 18 3 69.72	32 germa-nium **Ge** 2 8 18 4 72.59	33 arsenic **As** 2 8 18 5 74.9216	34 selenium **Se** 2 8 18 6 78.96	35 bromine **Br** 2 8 18 7 79.904	36 krypton **Kr** 2 8 18 8 83.80
46 palladium **Pd** 2 8 18 18 0 106.4	47 silver **Ag** 2 8 18 18 1 107.868	48 cadmium **Cd** 2 8 18 18 2 112.41	49 indium **In** 2 8 18 18 3 114.82	50 tin **Sn** 2 8 18 18 4 118.69	51 antimony **Sb** 2 8 18 18 5 121.75	52 tellurium **Te** 2 8 18 18 6 127.60	53 iodine **I** 2 8 18 18 7 126.9045	54 xenon **Xe** 2 8 18 18 8 131.30
78 platinum **Pt** 2 8 18 32 17 1 195.09	79 gold **Au** 2 8 18 32 18 1 196.9665	80 mercury **Hg** 2 8 18 32 18 2 200.59	81 thallium **Tl** 2 8 18 32 18 3 204.37	82 lead **Pb** 2 8 18 32 18 4 207.2	83 bismuth **Bi** 2 8 18 32 18 5 208.9804	84 polonium **Po** 2 8 18 32 18 6 (209)	85 astatine **At** 2 8 18 32 18 7 (210)	86 radon **Rn** 2 8 18 32 18 8 (222)

63 europium **Eu** 2 8 18 25 2 151.96	64 gado-linium **Gd** 2 8 18 25 9 2 157.25	65 terbium **Tb** 2 8 18 27 2 158.9254	66 dyspro-sium **Dy** 2 8 18 28 2 162.50	67 holmium **Ho** 2 8 18 29 8 2 164.9304	68 erbium **Er** 2 8 18 30 8 2 167.26	69 thulium **Tm** 2 8 18 31 8 2 168.9342	70 ytterbium **Yb** 2 8 18 32 8 2 173.04	71 lutetium **Lu** 2 8 18 32 9 2 174.97
95 americium **Am** 2 8 18 32 25 2 (243)	96 curium **Cm** 2 8 18 32 25 9 2 (247)	97 berkelium **Bk** 2 8 18 32 26 9 2 (247)	98 califor-nium **Cf** 2 8 18 32 28 8 2 (251)	99 einstein-ium **Es** 2 8 18 32 29 8 2 (254)	100 fermium **Fm** 2 8 18 32 30 8 2 (257)	101 mende-levium **Md** 2 8 18 32 31 8 2 (258)	102 nobelium **No** 2 8 18 32 32 8 2 (255)	103 lawren-cium **Lr** 2 8 18 32 32 9 2 (260)

LIVING CHEMISTRY

LIVING CHEMISTRY

DAVID A. UCKO

Antioch College
formerly of Hostos Community College
City University of New York

ACADEMIC PRESS   New York   San Francisco   London
A Subsidiary of Harcourt Brace Jovanovich, Publishers

ACADEMIC PRESS, INC.
111 Fifth Avenue, New York, New York 10003

United Kingdom Edition published by
ACADEMIC PRESS, INC. (LONDON) LTD.
24/28 Oval Road, London NW1

Library of Congress Cataloging in Publication Data

Ucko, David A
    Living chemistry.

    Includes index.
    1.   Chemistry.   2.   Biological chemistry.   I.   Title.
QD31.2.U24   540'.2'461   76-13951
ISBN 0–12–705950–4

PREFACE

"Living Chemistry" was written to serve the needs of students who are concerned with health—both their own and that of others. The book had its origins in my experience teaching chemistry for allied health students at Hostos Community College of the City University of New York. Based on an "open-door" admissions policy, Hostos encouraged the enrollment of students who might not otherwise have the opportunity to obtain a career education. Although the background, ability, and preparation of my students varied widely, they had one thing in common. On entering the course they viewed chemistry as an abstract, difficult subject, bearing little relationship to their own lives, the world around them, or to their future health science careers. Existing textbooks either confirmed my students' fear of chemistry or, if easier to read and understand, provided insufficient preparation for subsequent courses and possible certification exams.

"Living Chemistry" was designed to avoid both these pitfalls. It provides thorough, systematic coverage of the chemical information related to health, and it does so in an understandable and easily readable way. Chemical principles are reinforced throughout with examples and applications drawn from medicine, nursing, dentistry, biology, and nutrition. These applications, along with over 300 line drawings and photographs, play a significant role in developing student interest while emphasizing the practical importance of the chemical topic discussed.

I have found that this approach is also well suited for teaching chemistry to liberal arts students. "Living Chemistry" is presently the basis for a health-related introductory course at Antioch College called Chemistry of Life which is taken by students majoring in the humanities and social sciences.

The students using this book will probably have had little previous exposure to science. Accordingly, no assumptions have been made about the students' high school science preparation. Every new term is carefully explained, while basic calculations are carried out at a simple arithmetical level. The language has been kept intentionally simple and direct: "big" words are avoided wherever possible. Key words are emphasized in boldface type, and important concepts appear in italics. Each chapter concludes with a comprehensive summary reinforcing the major topics covered. As a further study aid, exercises at the end of each chapter are keyed to the individual sections within the chapter. Throughout the book, important and useful information is summarized in tabular form. The nearly 150 tables serve a dual role in providing a convenient format for review and reference, while at the same time offering at-a-glance illustrations of interesting comparative data.

The first part of the text develops the basic concepts required for understanding the "language" and principles of chemistry. SI units are introduced, but the units stressed are those that the student will use professionally or see in popular articles. The study of chemistry begins with the atom and the elements; the molecule and formula unit are mentioned for the first time in Chapter 4, only after a discussion of chemical bonding. The introduction to "general" chemistry in Chapters 1 through 9 is followed by a brief study of carbon compounds based on functional groups, with no mention of reaction mechanisms. Only those concepts are presented that are essential to the second half of the text, the study of biologically important molecules. In this part, the chemistry of carbohydrates, lipids, and proteins is discussed; details of individual reaction steps for important complex metabolic pathways are presented in an appendix. Sections from the final chapters—Vitamins and Hormones, Chemistry of the Body Fluids, Drugs and Poisons—can be studied separately at different times throughout the course if desired.

Appendixes at the end of the book provide coverage of optional topics, including a mathematics review, scientific notation, the unit-factor and proportion methods, metric conversion with practice problems, atomic orbitals, hybridization, metabolic pathways, and the cell. Answers to all numerical problems have also been included.

"Living Chemistry" can be used in either a traditional format or a mastery-type approach. It is part of a complete learning package keyed to the text on a chapter by chapter basis. The other package components include:

Student Guide for Living Chemistry—Contains learning objectives for each chapter, self-tests to check student understanding, lists of important terms, and a complete glossary.

Experiments for Living Chemistry—Designed to illustrate the concepts and applications covered in the text. Each chapter contains three to four short experiments, allowing flexibility in designing a laboratory program.

Instructor's Guide for Living Chemistry—Includes four examinations for each chapter, references to demonstrations, suggested supplementary materials, and a guide to mastery learning.

Masters for key figures and tables will be available on request for use in preparing slides and transparencies.

I wish to acknowledge the many people who have contributed to "Living Chemistry." Thanks go to Professor Donald Carter, Milwaukee Area Technical College, for reviewing the entire manuscript and suggesting the use of flow charts in several chapters. I am grateful for the many helpful comments provided by Dr. Evangelos Gizis and Dr. Robert Dreyfuss in their reviews, and for the encouragement given by Dr. Clara Watnick, all of Hostos Community College. I am indebted to all the individuals who helped provide or arrange for the many photographs in the text and to Ms. Lydia Vabre who typed the manuscript. The staff of Academic Press deserves thanks for their kind assistance. Above all, I wish to thank my wife, Barbara, for her valuable assistance and for putting up with seeing just the back of my head for such a long time.

DAVID A. UCKO
Yellow Springs, Ohio

Adaptability

"Living Chemistry" can be used with equal success in both one-semester and one-year courses. Below is a suggested outline for a one-semester course. This selection of topics is only one possible alternative; changes should be made based on the needs and interests of the students.

Outline for a One-Semester Course

Chapter	Suggested sections of text[a]
1	May be used as introductory material
2	2.1–2.10, (2.11)
3	3.1–3.13
4	4.1, 4.3–4.5, 4.7, (4.9–4.13)
5	5.1–5.13
6, 7	6.1, 6.2, 6.6, (6.12)
	7.1–7.12
8	8.1–8.7, 8.9–8.12
9	9.1, 9.2, 9.5, (9.7), 9.8–9.13
10	10.1–10.12
11, 12	11.1–11.13
	12.1–12.7
13, 17	13.1–13.7, (13.8)
	17.1, 17.3, 17.4, 17.8, 17.10, 17.11
14, 18	14.1–14.7, 14.9, 14.10
	18.4–18.6, (18.7–18.10)
15, 16, 19	15.1–15.5, (15.6), 15.7–15.9, (15.10)
	16.1–16.5, (16.6)
	19.1, 19.6–19.9
20	20.1–20.5, 20.7–20.11

[a] Sections in parentheses are optional.

Special interest topics

Many topics covered in "Living Chemistry" will be of particular interest to students in health-related career programs. These topics are listed below under the appropriate career headings:

Medicine/Nursing
acidosis/alkalosis, 171–173
alcohol rub, 237
anemia, 510
antacids, 162
antibiotics, 528–530
antiseptics, 362
aspirin and analgesics, 531–535
autoclave, 90, 362
basal metabolic rate, 390
Benedict's solution, 302–304
blood groups and Rh, 510–511
blood sugar level, 401–402, 486–489
body temperature, 14, 517
children's doses, 140–141
corticosteroids, 491–493
disinfectants, 239, 272
drug concentrations, 136–140
enzymes for therapy, 376–377
fluid pressure (IV, blood), 116, 418
glucose tolerance, 402
hemodialysis, 148
hormonal disorders, 481–502
ice packs, 110
medical diagnosis, 1
medical use of organic compounds,
    237–282
narcotics, 532–535
oxygen administration, 97–99
phenylketonuria, 462
poisons, 547–548
sedative–hypnotics, 536–538
specific gravity, 111–113
steam burns, 110
stimulants, 541–543
sulfa drugs, 276, 375
urine tests, 522

Dental Hygiene/Assisting
acrylics, 287
amalgam, 128
anesthetics, 535–536
calculus, 364
composition of teeth, 363–365
dental use of organic compounds, 240,
    243, 271
dental caries and decalcification, 364
dental wax, 326–327
fluoride treatment, 365
hormonal calcium regulation, 485
hydrocolloid impression material, 311
plaster, 118–119
polysulfide impression material, 287
"quat" solutions, 271–272
radiation safety, 198–201
X-rays and photography, 187–188

Respiratory Therapy
breathing, 87–88
diffusion, 84
ethers, 242–245
evaporation, 106–108
gas law calculations, 86–92
gaseous anesthetics, 536
halogenated hydrocarbon anesthetics,
    535–536
Henry's law, 133
humidity therapy, 122
hyperbaric chamber, 135
intermittent partial pressure breathing
    apparatus, 94
kinetic theory of gases, 82–83
nebulization, 123
oxygen therapy, 97–99
respiration, 94–97, 171, 394

CONTENTS

14   Lipids   320

15   Proteins   344

19 Metabolism of proteins 424

20 Heredity and protein synthesis 444

21 Vitamins and hormones 470

Matter and measurement

Most of the functions of your body, including those taking place right now, depend on chemical principles. Health care, whose role is maintaining the body functions, also depends on chemistry. Studying chemistry will therefore help you understand how the body works in times of health and how it can be treated when disease is present.

Chemistry is a science, a systematic and logical organization of facts that describe our world. Modern chemistry began in the eighteenth century with the development of experiments: observations and measurements carried out under controlled conditions, like tests performed on blood or urine. Scientific observations are summarized in **laws**, which are statements about the way that nature behaves, such as the law of gravity. As you will see, many scientific laws have important applications to health.

1.1   The scientific method

To explain their observations, scientists propose a **model** or **hypothesis**. Testing of this possible explanation by further experimentation and observation is the basis of the **scientific method**. If the hypothesis does not agree with the new results, it must be changed or replaced. A **theory** is an explanation that has been tested and confirmed many times.

In several ways, the scientific method is similar to medical diagnosis. First, observations are made of the state of the patient's health. This step may involve examining the patient and performing measurements like taking the temperature or analyzing a blood sample. Next, a possible explanation or hypothesis is proposed, such as the presence of a specific disease. Testing this hypothesis involves treatment for the condition—for example, by the administration of a drug. The patient is observed again to determine whether the proposed explanation, and the therapy based on it, were correct. If the symptoms disappear, the diagnosis (hypothesis) is confirmed; if not, another explanation may be necessary and, with it, further treatment.

The health professional must work under certain limitations that may not apply to the scientist. Time is a critical factor; an initial diagnosis must often be decided rapidly, without making all possible tests or observations. Furthermore, in the health area, many problems may be involved at once, while the scientist tries to limit the investigation to one part of a single problem.

1.2 The metric system

Observations and measurement are fundamental both to the health area and to chemistry. A nurse, for example, may check the vital signs of each patient several times a day. In chemistry, you will be concerned with **matter**, the "stuff" all around you. To describe matter, a uniform method of measurement must be used.

The system of measurement used throughout the world is the **metric system.** (The present official version is called the International System of Units, abbreviated SI.) Scientific and medical fields use this system almost exclusively. The United States is one of the few countries that still uses the English system of measurement, but plans are being made to switch to the metric system.

Because it is based on the number 10, the metric system is easy to learn. *Each unit or standard of measurement is related to other units by some multiple of 10.* Therefore, you do not have to memorize numbers like 5280 (the number of feet in a mile) to convert between units in this system. Every unit is either 10 times, 100 times, 1000 times, and so on, larger or smaller than another unit.

Table 1-1 lists the prefixes commonly used in the metric system. The most important are "kilo-," meaning one thousand (1000) times larger than another unit, and "milli-," meaning one thousand times smaller (1/1000 or 0.001). Other common prefixes are "centi-," a hundred times smaller (1/100 or 0.01), and "micro-," a million times smaller (1/1,000,000 or 0.000001). Note that each prefix has a one-letter symbol, the first letter of its name. Since both milli- and micro- begin with m, the Greek letter for m, which is μ (pronounced mū), is used for micro-.

Table 1-1   Common Metric Prefixes

Prefix	Symbol	Meaning
kilo-	k	one thousand times larger (1000)
centi-	c	one hundred times smaller (1/100 or 0.01)
milli-	m	one thousand times smaller (1/1000 or 0.001)
micro-	μ	one million times smaller (1/1,000,000 or 0.000001)

1.3 Mass

The **mass** of an object is the amount of matter it contains. Everything around you—this book, your clothing, the lamp—consists of matter and has mass. Mass and weight are often used to mean the same thing, but **weight is the earth's attraction for matter because of gravity.** The weight of an object changes with its location, but the mass stays the same. The weight of an astronaut on the moon, for example, is only about one-sixth the weight measured on earth because the force of gravity on the moon is smaller, but the mass remains constant. Even on earth, the weight of an object varies slightly depending on its geographical location.

A **balance** is used to measure mass. It works like a seesaw, comparing masses on either side. Figure 1-1 shows a beam balanced on a sharp edge. In part (b), the object whose mass is being measured is placed on one side; the beam is no longer balanced. As shown in part (c), known masses are added to the other side until the beam balances again. The mass of the original object is now equal to the sum of the known masses, since the mass on both sides must

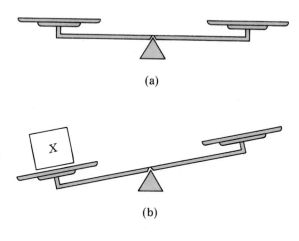

(a)

(b)

Figure 1-1. The operation of a balance. (a) The beam is initially balanced. (b) Placing a mass on one side makes the beam unbalanced. (c) Masses placed on the other side restore the balance. The masses on both sides are now equal. The unknown object (*X*) has a mass of 5 grams + 1 gram + 1 gram, or 7 grams.

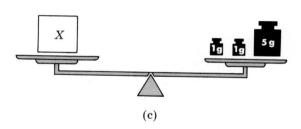

(c)

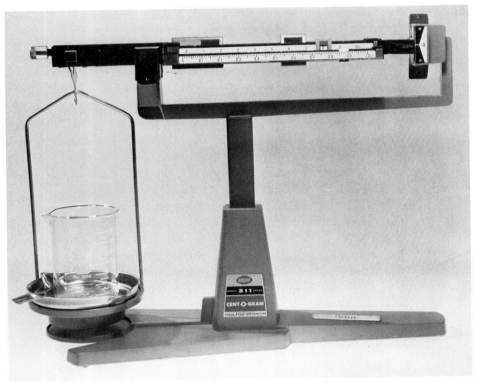

Figure 1-2. A commercial balance. The beaker shown has a mass of 68.87 grams. (Photo by Al Green.)

be the same for the beam to balance. Figure 1-2 illustrates a commercial balance based on this principle.

Mass is measured in the metric system in terms of the **gram**. Its symbol is g, but you may also see it written as gm. The gram represents a small amount of matter, since it takes 454 grams of mass to weigh 1 pound (lb) on earth.

Larger quantities of mass are measured in kilograms. Since kilo- means one thousand times larger, a kilogram is 1000 grams:

$$\begin{array}{c|c} \text{kilo} & \text{gram} \\ 1000 \text{ times} & \text{unit of mass} \end{array}$$

The symbol is kg, made by combining the symbol for kilo- with the symbol for gram. A kilogram of mass has a weight of about $2\frac{1}{5}$ (2.2) pounds. Therefore, a woman weighing 120 pounds has a mass of 55 kilograms. The masses of the parts of her body are given in Table 1-2.

Smaller units of mass are needed for dispensing medications to patients. Milli- means one thousand times smaller, so milligram, abbreviated mg, is

Table 1-2 **Mass of Body Components of a 120-Pound Female**

Body component	Mass (kilograms)	Weight (pounds)
muscle	24	52
fat	8	17
skeleton	8	17
skin	5	11
blood	4	10
miscellaneous	6	13
	55	120

one-thousandth of a gram:

milli	gram
1/1000 times	unit of mass

There are 1000 milligrams in 1 gram. Similarly, the microgram, abbreviated μg, is 1 million times smaller than the gram. A gram contains 1 million micrograms. These relationships are summarized in Table 1-3; illustrative problems are worked out in Appendix B.1 and B.2. (Appendix A contains a review of basic mathematics.)

Table 1-3 **Metric Units for Mass**

Unit	Symbol	Meaning
kilogram	kg	1000 grams
gram	g	basic unit; 454 grams = 1 pound
milligram	mg	1/1000 (or 0.001) gram
microgram	μg	1/1,000,000 (or 0.000001) gram

Drug dosages are sometimes determined on the basis of a person's mass. For a particular medication, the recommended dose may be 2 milligrams of drug per kilogram of body mass (2 mg/kg). Therefore, a patient with a mass of 55 kilograms would need 55 kilograms times 2 milligrams/kilogram or 110 milligrams of the drug.

1.4 Length

Another important property of matter is its size. You measure **length** by the distance between two points, two lines, or two surfaces. In the English system, with which you are familiar, length is expressed in inches (in.), feet

Table 1-4 **Lengths of Objects in Meters**

Object	Length (meters)
distance to sun	150,000,000,000
diameter of earth	13,000,000
Mt. Everest	10,000
Empire State Building	400
human height	2
cockroach	0.04
grain of sand	0.000 1
red blood cell	0.000 006 5
poliovirus	0.000 000 025

(ft), yards (yd), and miles (mi). In the metric system, the basic unit is the **meter**, abbreviated m. A meter is slightly longer than a yard (3 feet). Table 1-4 lists a sampling of representative lengths in terms of the meter.

In the medical field, units smaller than the meter are needed. The centimeter, abbreviated cm, is 1/100 or 0.01 meter, since centi- means one hundred times smaller:

centi | meter
1/100 times | unit of length

There are 100 centimeters in a meter. A centimeter is less than half an inch (Figure 1-3).

Just as a milligram is one-thousandth of a gram, a millimeter, abbreviated mm, is one thousand times smaller than a meter:

milli | meter
1/1000 times | unit of length

One meter contains 1000 millimeters. The millimeter is 10 times smaller than the centimeter; there are 10 millimeters in a centimeter.

Figure 1-4 compares a yardstick to a meter stick, which is a "ruler" exactly 1

Figure 1-3. Copyright 1972, United Feature Syndicate, Inc.

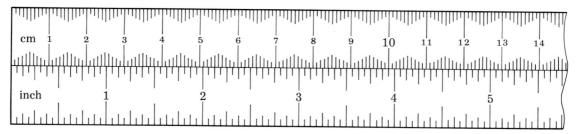

Figure 1-4.   A meter stick compared to a yardstick. The large divisions of the meter stick are centimeters ($\frac{1}{100}$ meter); the smallest divisions are millimeters ($\frac{1}{1000}$ meter). There are 10 millimeters in a centimeter. Notice that 1 inch corresponds to about 2.5 centimeters.

meter long used to measure lengths in the metric system. The meter stick is divided into 100 smaller divisions, the centimeters, each of which is further divided into 10 divisions, the millimeters. Also notice that 1 inch is approximately equal to 2.5 centimeters.

Table 1-5 summarizes the relationships between units of length in the metric system. Detailed examples illustrating the conversion between units are presented in Appendix B.1 and B.2.

Table 1-5   Metric Units for Length

Unit	Symbol	Meaning
meter	m	basic unit; 1 meter = 39 inches
centimeter	cm	1/100 (or 0.01) meter
millimeter	mm	1/1000 (or 0.001) meter

1.5   Volume

Matter takes up space. **Volume** is a measure of the amount of space occupied. The volume of a solid rectangular object, like a box, is found by multiplying the length times the height times the width or depth. If you measure length in centimeters, the units of volume will be cm × cm × cm or cm³, **cubic centimeters**. One cubic centimeter, abbreviated cm³ or cc, is shown in Figure 1-5.

Another unit of volume, useful for liquids, is the **liter**, abbreviated l. The liter was formerly defined as the volume of 1 kilogram of water, but it is now defined as the volume of a cube (a box with equal sides) having a length of 10 centimeters on each side. A liter, therefore, has a volume of 10 centimeters × 10 centimeters × 10 centimeters, or 1000 cubic centimeters. This quantity

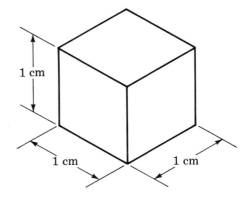

Figure 1-5. The volume of 1 cubic centimeter. Its volume is equal to 1 centimeter × 1 centimeter × 1 centimeter = 1 cubic centimeter (cm³, or cc).

is slightly larger than a quart (qt), the unit of volume you use in the English system.

Again, you will need to measure volumes smaller than a liter, so another unit will be more useful. The milliliter, abbreviated ml, is one thousand times smaller than the liter:

milli	liter
1/1000 times	unit of volume

Thus, the liter contains 1000 milliliters, as shown in Figure 1-6.

Because a liter also contains 1000 cubic centimeters, the milliliter and the cubic centimeter must have the same volume,

$$\text{milliliter} = \text{cubic centimeter}$$

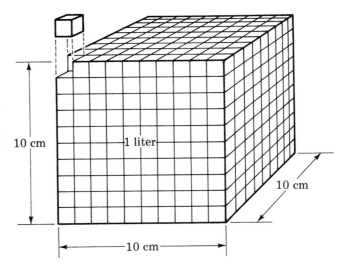

Figure 1-6. The relationship between a milliliter and a liter. One liter contains 1000 milliliters, or 1000 cubic centimeters. The cubic centimeter and the milliliter have the same volume.

Table 1-6 **Metric Units for Volume**

Unit	Symbol	Meaning
liter	l	basic unit; 1 liter = 1.06 quarts
milliliter	ml	1/1000 (or 0.001) liter
cubic centimeter	cm³ (cc)	same as milliliter

These two units are therefore used interchangeably. An injection of 5 cubic centimeters of insulin is the same as one of 5 milliliters. The relationship between these units is summarized in Table 1-6; conversion problems are illustrated in Appendix B.1 and B.2.

Volumes of liquids can be measured in various ways, as shown in Figure 1-7. A graduated cylinder is used to measure approximate volumes. A buret adds small amounts of liquid of known volume (the stopcock at its bottom controls the flow). A pipet transfers a measured volume from one container to another. A syringe is similar to a pipet, but it is used to inject a known volume.

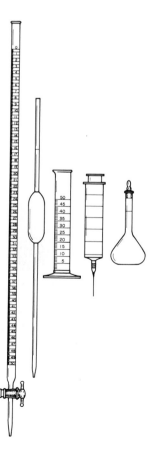

Figure 1-7. Equipment used to measure volume, from left to right: buret, pipet, graduated cylinder, syringe, volumetric flask; each measures a volume of 50 milliliters.

Volumetric flasks hold a fixed volume when filled up to the mark on the neck. Each of these pieces of glassware comes in different sizes depending on its use.

1.6 Density

An important property of matter is the relationship between its mass and volume. You know that iron is "heavier" than wood. This difference means that a certain amount of iron has a greater mass than an equal volume of wood. Another way to say the same thing is that you need a larger volume of wood to get the same mass as the iron, as shown in Figure 1-8.

The ratio of the mass of an object to its volume is its **density**. When mass is measured in grams and volume is measured in cubic centimeters (or milliliters), density has the units of grams per cubic centimeter, g/cm³ (g/ml):

$$\text{density} = \frac{\text{mass (g)}}{\text{volume (cm}^3)}$$

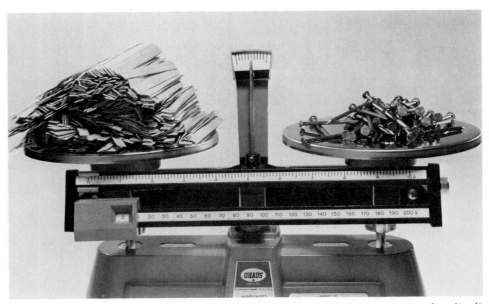

Figure 1-8. Equal masses of wood and iron. Because iron has a greater density, its volume is smaller than a quantity of wood having the same mass. (Photo by Al Green.)

Table 1-7 **Densities of Common Substances**

Substance	Density (grams/cubic centimeter)
gold	19.3
mercury	13.6
lead	11.3
iron	7.9
limestone	3.2
aluminum	2.7
seawater	1.03
pure water	1.00
ice	0.92
gasoline	0.70
wood	0.50
air	0.0013

Thus, iron is denser than wood. The numerical values of the density of wood, iron, and other common substances or types of matter are presented in Table 1-7.

1.7 Temperature

Temperature measures how "hot" or "cold" a substance is. Body temperature is related to health; fever, an abnormally high temperature, is usually a sign of illness. To measure temperature, **thermometers** are used; several types are illustrated in Figure 1-9.

The mercury thermometer has a glass bulb containing mercury connected to a thin (capillary) tube inside a thicker tube for support. An increase in the temperature of the glass bulb when placed in the patient's mouth or rectum causes the mercury to expand in volume and rise in the tube. The highest point reached by the mercury, representing the patient's temperature, is read on a scale etched on the outside of the thermometer. A narrowing or constriction just above the bulb stops the mercury from falling as the thermometer cools when you read it. The thermometers used in a chemical laboratory do not have such a constriction and can break if you try to shake them down.

Electronic thermometers are based on the flow of electricity through a device called a thermistor. The amount of electric current it conducts depends on the temperature. As the temperature goes up, a greater current flows and is registered on a dial. These thermometers are generally faster and safer than the mercury type. New disposable thermometers are based on a change in color with temperature.

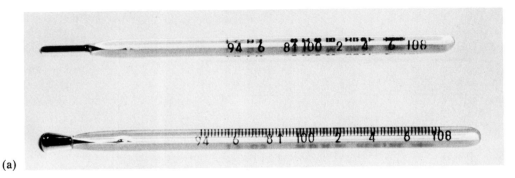

(a)

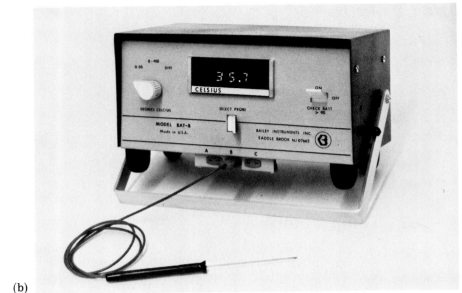

(b)

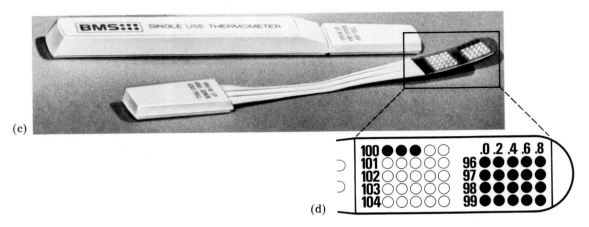

(c)

(d)

Figure 1-9. Thermometers: (a) mercury, oral and rectal, (b) electronic, (c) disposable, (d) close-up of disposable (reading 100.4° F). [Photo (a) by Al Green; (b) courtesy of Bailey Instruments Co., Inc.; (c) and (d) courtesy of Bio-Medical Sciences, Inc.]

Three different scales exist to measure temperature: **Celsius** or centigrade (°C), **Fahrenheit** (°F), and **Kelvin** or absolute (K). The size of each Celsius degree is the same as the size of the Kelvin degree, but the Fahrenheit degree is smaller, only five-ninths as large. Thus, there are 100 degrees or divisions on the Celsius and Kelvin scales between the temperatures of ice and boiling water but 180 degrees on the Fahrenheit thermometer. Also, each scale has a different starting point, as shown in Figure 1-10. The temperature of ice can be given as either 0°C, 32°F, or 273 K, while the boiling temperature of water is 100°C, 212°F, or 373 K.

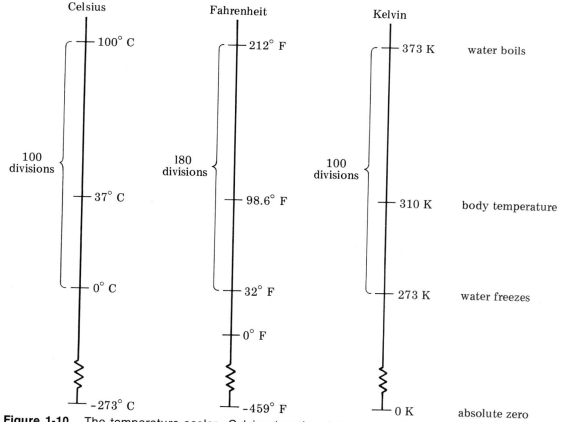

Figure 1-10. The temperature scales: Celsius (centigrade), Fahrenheit, and Kelvin (absolute). Note the different location of zero degrees on each scale and the different sizes of 1 degree.

Mathematically, the Celsius and Fahrenheit temperature scales are related by simple formulas:

$$°C = \tfrac{5}{9}(°F - 32)$$

$$°F = \tfrac{9}{5}(°C) + 32$$

You can understand these formulas by noting that the fraction $\tfrac{5}{9}$ comes from the ratio between the number of divisions between the temperatures of boiling water and ice on the Celsius scale, 100, to the number of divisions on the Fahrenheit scale, 180 ($100/180 = \tfrac{5}{9}$). The 32 term is just a correction for the different location of zero degrees on the two scales. The Kelvin scale is related to the Celsius scale by the following formula:

$$K = °C + 273$$

The 273 term corrects for the different starting points of these scales. Examples of temperature conversion using these formulas are worked out in Appendix B.3.

Table 1-8 shows the conversion between Celsius and Fahrenheit temperatures in the clinical region. The oral temperature range for a healthy adult is from 97.0°F (36.1°C) to 99.1°F (37.3°C), with an average of 98.6°F (37.0°C). Temperatures measured rectally are higher by about 1°F (0.6°C). Your body temperature is controlled by the hypothalamus, a gland that regulates blood circulation through the skin, sweating, and muscle activity. Setting of this body "thermostat" to a higher level produces fever, or oral temperature above 98.6°F (37.0°C) for a person confined to bed or above 99.0°F (37.2°C) for a moderately active person.

Table 1-8   Clinical Temperature Conversion

Temperature (°C)	Temperature (°F)
35.5	95.9
36.0	96.8
36.5	97.7
37.0 (normal oral)	98.6
37.5 (normal rectal)	99.5
38.0	100.4
38.5	101.3
39.0	102.2
39.5	103.1
40.0	104.0
40.5	104.9

1.8   Three states of matter

Given what you now know, **matter**, the "stuff" from which everything is made, can be defined as anything that has mass and therefore occupies space. Matter commonly exists in three possible forms or states: solid, liquid, and gas (Figure 1-11). (A fourth state, plasma, exists only under unusual conditions.)

Solids have a definite volume. They generally have a rigid shape and resist changes in their structure. Solids are thus incompressible (cannot be compressed). Their density is usually high, and they expand only very slightly when heated. Examples of solids include such substances as bone, gold, glass, wood, and salt.

Liquids also have a constant volume but not a fixed shape. They take the shape of their container, filling it from the bottom up. Liquids are nearly incompressible and expand a small amount if the temperature rises. Their density is generally lower than that of solids. Examples include blood, urine, milk, and alcohol.

solids

liquid

gas

Figure 1-11.   Examples of the three states of matter: solid, liquid, and gas.

Table 1-9 **The Three States of Matter**

Property	Solid	Liquid	Gas
shape	rigid	not definite	not definite
volume	definite	definite	expands without limit
mass	definite	definite	definite
density	high	medium	low
compressibility	not compressible	not compressible	can be compressed
effect of heat	very slight expansion	small expansion	large expansion

Gases do not have a definite volume or shape. They expand without limit to fill the space they are in. Gases can also be compressed into smaller volumes, such as in the oxygen tanks of a health center. Under normal conditions, their density is very low. Examples of gases are air, steam, oxygen, and neon.

Table 1-9 summarizes the most important properties of these three states of matter.

SUMMARY

Modern chemistry began in the eighteenth century with the development of experiments: observations and measurements carried out under controlled conditions. Chemistry is a science, a systematic and logical organization of facts that describe our world. Scientific observations are summarized in laws, which are statements about the way nature behaves.

To explain their observations, scientists propose a model or hypothesis. Testing of this possible explanation by further experimentation and observation is the basis of the scientific method. If the hypothesis does not agree with the new results, it must be changed or replaced. A theory is an explanation that has been tested and confirmed many times.

Measurements in chemistry are made in the metric system, which is based on the number 10. Each unit or standard of measurement is related to other units by some multiple of 10. The most important metric prefixes are "kilo-," a thousand times larger than another unit, and "milli-," a thousand times smaller.

The mass of an object is the amount of matter or "stuff" that it contains. Every substance around you consists of matter and has mass. Mass and weight are often used to mean the same thing, but weight is the earth's attraction for matter because of gravity. A balance, which works like a seesaw, is used to measure mass.

Mass in the metric system is measured in terms of the gram, whose symbol is g. The gram represents a small amount of matter, since 454 grams of mass correspond to a weight of 1 pound. Larger quantities of mass are measured in kilograms; a kilogram (kg) is 1000 grams. The milligram (mg), one-thousandth (1/1000) the size of a gram, is used to measure small amounts of matter.

You measure length by the distance between two points, two lines, or two surfaces. In the metric system, the basic unit is the meter (m), which is slightly larger than a yard (3 feet). The centimeter (cm) is 1/100 (0.01) meter, and the millimeter (mm) is 1/1000 (0.001) meter.

Volume is a measure of the amount of space occupied by matter. The volume can be measured in cubic centimeters (cm³) or liters (l). A liter has a volume of 1000 cubic centimeters, which is a little larger than a quart. The milliliter (ml), one thousand times smaller than a liter, has the same volume as a cubic centimeter.

The density of an object is the ratio between its mass and volume. When mass is measured in grams and volume is measured in cubic centimeters, density has the units grams per cubic centimeter (g/cm³).

Temperature measures how "hot" or "cold" a substance is. Three different scales exist: Celsius or centigrade (°C), Fahrenheit (°F), and Kelvin or absolute (K). The size of each Celsius degree is the same as the size of the Kelvin degree, but the starting point of each scale is different. The Fahrenheit degree is only five-ninths as large as the Celsius or Kelvin degree. The scales are related by the following formulas: $°C = \frac{5}{9}(°F - 32)$, $°F = \frac{9}{5}(°C) + 32$, $K = °C + 273$.

Matter can be defined as anything that has mass and therefore occupies space. It commonly exists in three possible forms or states. Solids have a definite volume and a rigid shape. Liquids have a constant volume but not a fixed shape. Gases have neither a definite volume or a definite shape; they expand without limit to fill the space they are in.

Exercises

Note: The number in parentheses after the question number indicates the section to which you should refer if you cannot answer the question. Answers to numerical problems are given in Appendix F.

1. (Intro.) Why is chemistry a science?

2. (1.1) Define law, hypothesis, scientific method.

3. (1.1) How is the scientific method related to medical diagnosis?

4. (1.2) Describe the basis of the metric system.

5. (1.3) How does mass differ from weight?

6. (1.3) Explain how a balance operates.

7. (1.3) How are the kilogram, milligram, and microgram related to the gram?

8. (1.4) Describe the relationships between the units of length in the metric system.

9. (1.4) Arrange the following in order of decreasing length: 1000 centimeters, 0.01 meter, 100 millimeters, 10,000 meters.

10. (1.5) What is a liter? a milliliter? a cubic centimeter?

11. (1.5) Describe three ways of measuring volume.

12. (1.6) Define density.

13. (1.6) Which has a greater mass: 1 cubic centimeter of gold or 10 cubic centimeters of wood? (Refer to Table 1-7.)

14. (1.7) Describe the operation of a mercury thermometer and an electronic thermometer.

16. (1.7) What is the normal oral and normal rectal body temperature in the Fahrenheit scale? in the Celsius scale?

17. (1.8) Describe the three states of matter.

18. (1.8) Identify the following as solid, liquid, or gas (at room temperature): oxygen, water, aspirin tablet, rubbing alcohol, silver, nylon.

The composition
of matter

All matter, whether solid, liquid, or gas, is composed of **atoms**, the basic building blocks of nature. There are presently 106 different "kinds" of these fundamental particles known. Just as the 26 letters of the alphabet create the entire English language, this limited number of types of atoms forms all the matter around you. *Chemistry is the study of these basic units, the atoms, how they combine, and how substances made of atoms are changed into other substances.*

2.1 The atom

The atom is extremely small. It is too little to be seen with the sharpest eye and too light to be weighed on the finest balance. To get a few grams of atoms, that is, enough to weigh a fraction of a pound, you would need about a million times a million times a million of them.

Yet this incredibly tiny atom is composed of even smaller pieces, the **subatomic particles**. Atoms consist of a certain combination of these three particles: **protons** (symbol p), **neutrons** (symbol n), and **electrons** (symbol e or e⁻). The smallest atoms are made from only a few subatomic particles; the largest atoms consist of more than 100.

Despite its small size, most of the atom consists of empty space. Its protons and neutrons are packed into a central core 1/10,000 the size of the atom called the **nucleus.** The electrons are very far away, outside the nucleus. If you imagined the period at the end of this sentence to be the nucleus of an atom, the electrons would be at the walls of the room you are in.

The electrons are located in definite **shells** or **energy levels** which surround the nucleus. In this sense, an atom can be compared to our solar system: The sun represents the nucleus, and the revolving planets represent the electrons.

(a)

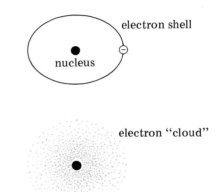

Figure 2-1. Models of the atom. (a) In the electron shell model, the electron is pictured as orbiting around the nucleus. (b) In the electron "cloud" model, the shading represents the chance of finding the electron in regions outside the nucleus.

(b)

But, in fact, because the electrons are moving so quickly, their location cannot be pinpointed exactly. Instead, you can only say that there are definite regions around the nucleus having the greatest chance of containing an electron. Figure 2-1 compares these two models of the atom.

Each energy level can hold only a limited number of electrons. The first and closest one to the nucleus (K shell) contains up to two electrons. The second level (L shell) holds up to eight electrons. The third (M shell) may also hold eight (but in some atoms up to eighteen). For the simplest atoms, each level must be filled before electrons can be placed in the next furthest shell. Thus, the K shell must contain its maximum of two electrons before any electrons can appear in the next energy level, the L shell.

2.2 Charge

An important property of the subatomic particles is their electrical **charge**. It is charge that builds up when you walk across a rug on a cold, dry day. (You feel a mild shock when the "static electricity" is released.) This fundamental property of nature exists in two forms called positive, symbolized by the plus sign (+), and negative, symbolized by the minus sign (−). As shown in Figure 2-2, when electrical charges are brought together, they affect each other. Charges that are the *same*, either both positive or both negative, push each other apart (repulsion). *Opposite* charges, positive and negative, pull themselves together (attraction).

In addition, unlike charges can cancel out each other's effect. For example, if you could put one positive and one negative charge together in a container, it would be neutral, having no charge at all when observed from outside. Ad-

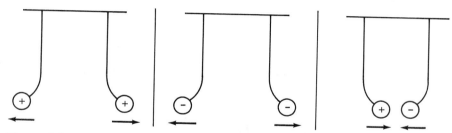

Figure 2-2. Electrical charges. Like charges (both positive or both negative) repel each other; opposite charges (positive and negative) attract.

ding two opposite charges is like taking one step forward and then one step backward, ending up at your starting point.

The proton has a positive (+) charge, while the electron has a negative (−) charge. As its name tells you, the neutron is neutral. It has neither a positive nor a negative charge. One of the most important properties of the atom is that *the number of protons is exactly the same as the number of electrons.* Therefore, the number of positive charges must be equal to the number of negative charges. *Every atom is thus neutral,* even though it contains charged particles.

2.3 Atomic number

The **atomic number** of an atom is equal to the number of protons in its nucleus. Since every atom has the same number of electrons as protons, the atomic number is also equal to the number of electrons surrounding the nucleus. Each of the 106 (chemically) different types of atoms has its own atomic number, its own "signature." *The atomic number is the most important identification of an atom.*

The smallest atom possible, hydrogen, has an atomic number of 1. It has only one proton and also one electron. The next largest atom has an atomic number of 2, meaning that it has two protons and two electrons. Every succeeding type of atom has one more proton (in addition to a certain number of neutrons) in its nucleus and one more electron outside of it. The atomic numbers of the 106 "kinds" of atoms thus go from 1 up to 106. The largest known atom, having an atomic number of 106, has 106 protons and 106 electrons.

2.4 Atomic mass

The mass of the proton and the mass of the neutron are about the same, as shown by the number 1 in Table 2-1. The approximate actual mass of each is

0.000 000 000 000 000 000 000 001 66 g

The electron is even lighter; its mass is almost 2000 times smaller. Thus, *the mass of an atom is determined almost completely by the mass of the protons and neutrons concentrated in its nucleus.* You can think of the atom as an elephant with fleas: The elephant represents the nucleus and the fleas represent its electrons. The electrons have so little mass compared to the protons and neutrons that, when the mass of an atom is given, the contribution of the electrons is usually ignored.

Table 2-1   The Subatomic Particles

Particle	Charge	Relative mass	Symbol
proton	+1	1	p
neutron	0	1	n
electron	−1	1/1836	e, e⁻

To avoid very small numbers, chemists use a relative scale of masses, based on comparison with another small mass taken as a standard. The reference mass is a particular atom, carbon, which has six protons and six neutrons in its nucleus. (Of course, it also has six electrons, but their mass can be neglected.) A new unit of mass, the **atomic mass unit**, abbreviated amu, is then defined as one-twelfth the mass of this atom. On this basis, the masses of the subatomic particles have the values shown in Table 2-2. With the mass of the proton and the mass of the neutron taken as 1 amu, you can now calculate the approximate mass of each type of atom.

The **mass number** of an atom is found by adding the number of protons and the number of neutrons in its nucleus:

mass number = number of protons + number of neutrons

Table 2-2   Masses of the Subatomic Particles

Particle	Exact mass (amu)	Approximate mass (amu)
proton	1.00728	1
neutron	1.00867	1
electron	0.000549	0

Table 2-3 **Examples of Atomic Composition**

Atomic number	Mass number	Number of protons	Number of neutrons	Number of electrons
4	9	4	5	4
11	23	11	12	11
17	35	17	18	17
26	56	26	30	26
78	195	78	117	78

For example, the mass number of carbon, with its six protons and six neutrons, is equal to 6 + 6, or 12. This number represents the approximate mass of a carbon atom in atomic mass units.

Using the same relationship, you can find the number of neutrons in an atom if you know its mass number and atomic number. You merely subtract the atomic number, that is, the number of protons, from the mass number:

number of neutrons = mass number − atomic number (number of protons)

Thus, if you knew that the mass number of carbon was 12 and its atomic number was 6, its number of neutrons would be equal to 12 − 6, or 6.

Table 2-3 gives examples of some atoms, their atomic number, mass number, and composition in terms of protons, neutrons, and electrons. Try covering up one column at a time and see if you can fill in the missing piece of information.

2.5 The structure of atoms

You can now draw a simple diagram to represent the structure of an atom. The nucleus, shown as a circle, contains the protons and neutrons. The electrons are then placed in shells outside the nucleus as illustrated for hydrogen, the simplest atom. Note that it has no neutrons at all. Ordinary hydrogen has just one proton in its nucleus and one electron in the first (K) shell. Its atomic number is 1 and its mass number is 1.

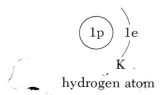

hydrogen atom

The next atom, helium, has two protons and two neutrons in the nucleus and two electrons in the first shell. Its atomic number is 2 and its mass number is 4. The K shell is now completely filled; new electrons must go into the next level.

helium atom

The atom with atomic number 3 is lithium. It has three protons, four neutrons, and three electrons. The mass number is 7. Notice that the third electron must go into the second shell.

lithium atom

Atomic number 4 corresponds to beryllium, with four protons, five neutrons, and four electrons. The mass number is 9. There are now two electrons each in the K and L shells.

beryllium atom

The next type of atom is boron, with an atomic number of 5 and a mass number of 11. It has five protons, six neutrons, and five electrons. The fifth electron can still be added to the second level because this shell can hold up to eight electrons, unlike the first shell, which holds only two.

boron atom

For the next five atoms, with the atomic number increasing by 1 each time, electrons continue to be added to the second shell. Thus, carbon (atomic number 6, mass number 12) has two electrons in the K shell and four in the L shell, in addition to the six protons and six neutrons in its nucleus:

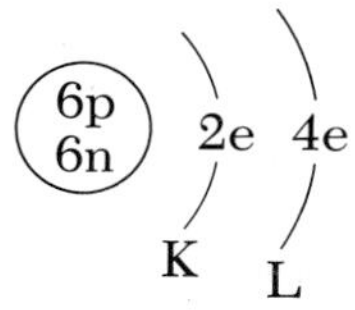

carbon atom

For nitrogen (atomic number 7, mass number 14), the last shell contains five electrons:

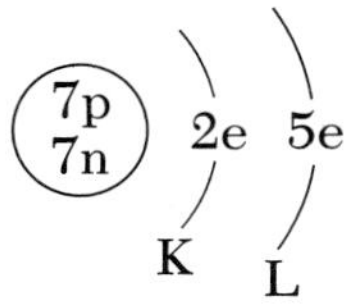

nitrogen atom

Oxygen (atomic number 8, mass number 16) has six electrons in the second level:

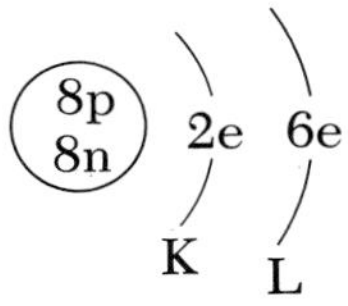

oxygen atom

There are nine electrons in fluorine (atomic number 9, mass number 19) with seven in the outermost shell:

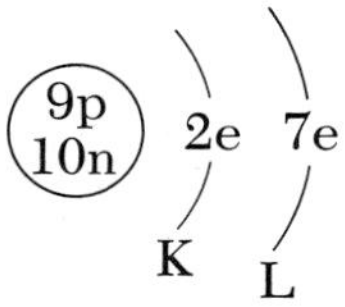

fluorine atom

Finally, with neon (atomic number 10, mass number 20), the second shell is complete with eight electrons:

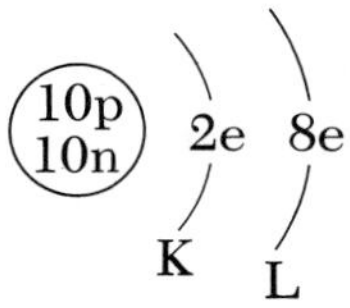

neon atom

Sodium (atomic number 11, mass number 23) has one electron in the third level, the M shell, because the first and second levels can hold no more electrons:

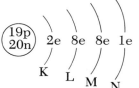

sodium atom

The next seven larger atoms, magnesium (atomic number 12), aluminum (13), silicon (14), phosphorus (15), sulfur (16), chlorine (17), and argon (18), have additional electrons added to the M shell. This level can actually hold up to eighteen electrons but stops at eight temporarily to allow the next two atoms, potassium (19) and calcium (20), to have one and two electrons, respectively, in the fourth shell. Thus, potassium, with a mass number of 39, can be represented as

(19p 20n) 2e 8e 8e 1e

K   L   M   N

potassium atom

After the next atom, calcium, electrons continue to add to the third shell, bringing it up to eighteen. See Appendix C.1 for a more detailed description of the electronic structure of atoms (orbitals).

2.6   Isotopes

All of the atoms, up to atomic number 106, are built up by this method of successively adding one more proton to the nucleus and one more electron to an unfilled energy level. The situation, however, is slightly more complicated. It turns out that atoms of a given type, say hydrogen atoms all having the atomic number 1, can exist in different forms. They all have the *same* number of protons (and therefore electrons) because they are hydrogen atoms. But they can have *different* mass numbers because these related atoms do not have the same number of neutrons. The three known forms of hydrogen are shown in Figure 2-3.

Atoms such as these, which have the same atomic number but different mass numbers, are **isotopes**. Because they have the same atomic number and therefore the same number of protons and the same number and arrangement of electrons, they are chemically alike. The sole difference is the number of

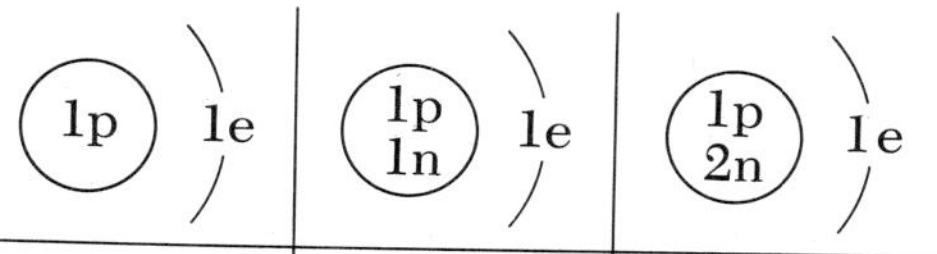

	1p 1e	1p 1n 1e	1p 2n 1e
atomic number	1	1	1
mass number	1	2	3

Figure 2-3. Isotopes of the hydrogen atom. Note that each has the same atomic number, although the mass numbers differ. The isotope of mass 2 is known as deuterium (D) and the isotope of mass 3 is called tritium (T).

neutrons, which contributes only to the mass of the atom. *The identity of an atom is fixed by its atomic number and not by its mass.* It is as if you had two beads that were completely alike in color, size, and all other properties, except that one was slightly heavier than the other.

2.7 Elements

In nature, isotopes for each kind of atom are found together. A large collection of such atoms having the same atomic number is called an **element**. You can see elements like silver, iron, or mercury, although you cannot see the individual atoms. The atom is the smallest part of an element that still has the properties of the element. That is, if you could take a piece of silver and cut it into smaller and smaller pieces, the tiniest possible piece that still behaved like silver would be one atom. Because of the possibility of isotopes, not all the atoms of an element must be identical. Some may have different masses, but they all must have the *same atomic number,* the same number of protons and therefore electrons. Examples of elements are shown in Figure 2-4.

As shown in Figure 2-5, the element chlorine contains two kinds of chlorine atoms: one with a mass number of 35 and the other with a mass number of 37. The element as it is found in nature contains 77.5% of the isotope of mass 35 and 22.5% of the isotope of mass 37.

Since atoms exist with atomic numbers from 1 to 106, there are 106 different elements. Of these, 90 are found in nature; the others are made artificially. As investigations go on, new elements with even higher atomic numbers may be added to the present 106.

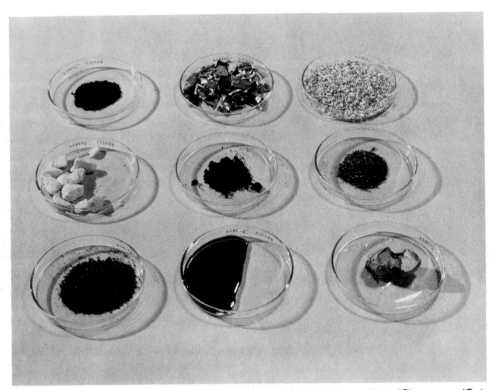

Figure 2-4. Examples of elements. From left to right, top row: carbon (C), copper (Cu), magnesium (Mg); middle row: sulfur (S), phosphorus (P), iodine (I); bottom row: iron (Fe), mercury (Hg), potassium (K). (Photo by Al Green.)

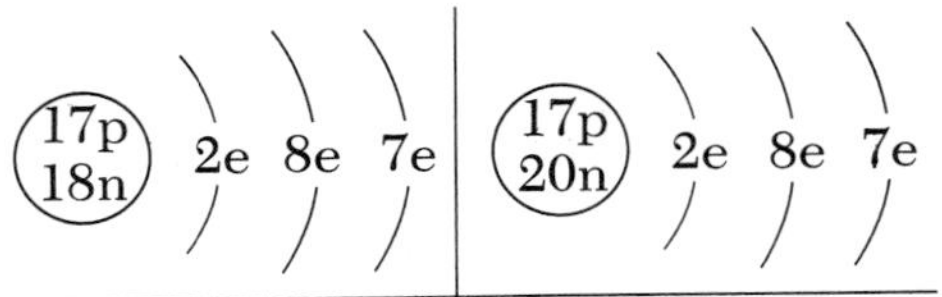

(17p 18n) 2e 8e 7e		(17p 20n) 2e 8e 7e

atomic number	17	17
mass number	35	37
natural abundance	77.5%	22.5%

Figure 2-5. Isotopes of the chlorine atom. The only difference between isotopes is the number of neutrons. The atomic weight of chlorine is 35.45 amu, which is closer to the more abundant isotope of mass 35.

2.8   Atomic weight

Since many elements contain atoms having different masses, you can no longer use the term mass number. You are dealing with a collection of atoms and must describe their average mass. The **atomic weight** of an element is the average of the atomic masses of its isotopes as they are found in nature. (Unfortunately, this term does not follow the distinction between mass and weight but is established by common usage anyway.) Just like atomic masses, these weights are relative. They are based on comparison with the most common isotope of carbon (the one with six protons and six neutrons), which is assigned a mass of exactly 12 atomic mass units (amu).

As an example, look again at the isotopes of chlorine in Figure 2-5. The atomic weight of the element must be an average of the masses of the two kinds of chlorine atoms that are found in nature. It must reflect the fact that more atoms have a mass of 35 than 37, so the average should be closer to 35. In fact, the atomic weight of chlorine is 35.45 amu. Because most atoms have isotopes, few of the atomic weights of the elements are simple whole numbers.

2.9   The periodic table

All of the 106 known chemical elements can be arranged in a special way, called the **periodic table** (Table 2-4). It was created during the nineteenth century, when scientists had gathered tremendous quantities of information about the properties of elements. In order to use this great store of knowledge effectively, some sort of organization was necessary.

The table is made by writing the elements in the order of *increasing atomic number*. Therefore, the table starts with hydrogen (atomic number 1); the next element is helium (atomic number 2), and so on. When the atoms are arranged in this manner, certain chemical properties repeat themselves in a regular way. So instead of just listing the 106 elements in one long row, a new row starts every time the same properties start coming up again. Those elements whose chemistry is similar become part of the same vertical columns. The periodic table is thus based on what is known as the **periodic law**: The properties of the elements repeat in a regular (or periodic) way when the elements are arranged by their atomic numbers.

The table contains **symbols**, the "shorthand" ways of abbreviating the names of the elements. Each consists of one or two letters with the first always capitalized and the second always lowercase. This distinction is very important as you can see by comparing the following symbols:

Co   cobalt, an element
CO   carbon monoxide, formed from two elements, carbon and oxygen

Table 2-4 The Periodic Table

Handwritten annotations: Groups — same # of electrons in valence shell (outermost) ⟶ similar chemical properties; nonmetals →; metals ←; period

Period	Ia	IIa	IIIb	IVb	Vb	VIb	VIIb	VIII	VIII	VIII	Ib	IIb	IIIa	IVa	Va	VIa	VIIa	Noble gases
1	1 **H** 1.0079																	2 **He** 4.0026
2	3 **Li** 6.941	4 **Be** 9.0122											5 **B** 10.81	6 **C** 12.011	7 **N** 14.0067	8 **O** 15.9994	9 **F** 18.9984	10 **Ne** 20.179
3	11 **Na** 22.9898	12 **Mg** 24.305											13 **Al** 26.9815	14 **Si** 28.086	15 **P** 30.9738	16 **S** 32.06	17 **Cl** 35.453	18 **Ar** 39.948
4	19 **K** 39.098	20 **Ca** 40.08	21 **Sc** 44.9559	22 **Ti** 47.90	23 **V** 50.9414	24 **Cr** 51.996	25 **Mn** 54.9380	26 **Fe** 55.847	27 **Co** 58.9332	28 **Ni** 58.70	29 **Cu** 63.546	30 **Zn** 65.38	31 **Ga** 69.72	32 **Ge** 72.59	33 **As** 74.9216	34 **Se** 78.96	35 **Br** 79.904	36 **Kr** 83.80
5	37 **Rb** 85.4678	38 **Sr** 87.62	39 **Y** 88.9059	40 **Zr** 91.22	41 **Nb** 92.9064	42 **Mo** 95.94	43 **Tc** (97)	44 **Ru** 101.07	45 **Rh** 102.9055	46 **Pd** 106.4	47 **Ag** 107.868	48 **Cd** 112.41	49 **In** 114.82	50 **Sn** 118.69	51 **Sb** 121.75	52 **Te** 127.60	53 **I** 126.9045	54 **Xe** 131.30
6	55 **Cs** 132.9054	56 **Ba** 137.33	57 **La*** 138.905	72 **Hf** 178.49	73 **Ta** 180.9479	74 **W** 183.85	75 **Re** 186.207	76 **Os** 190.2	77 **Ir** 192.22	78 **Pt** 195.09	79 **Au** 196.9665	80 **Hg** 200.59	81 **Tl** 204.37	82 **Pb** 207.2	83 **Bi** 208.9804	84 **Po** (209)	85 **At** (210)	86 **Rn** (222)
7	87 **Fr** (223)	88 **Ra** 226.0254	89 **Ac**** (227)	104 **Ku** (261)	105 **Ha** (262)	106 **?**												

Transition elements

* Lanthanide series	58 **Ce** 140.12	59 **Pr** 140.9077	60 **Nd** 144.24	61 **Pm** (145)	62 **Sm** 150.4	63 **Eu** 151.96	64 **Gd** 157.25	65 **Tb** 158.9254	66 **Dy** 162.50	67 **Ho** 164.9304	68 **Er** 167.26	69 **Tm** 168.9342	70 **Yb** 173.04	71 **Lu** 174.97
**Actinide series	90 **Th** 232.038	91 **Pa** 231.0359	92 **U** 238.029	93 **Np** 237.0482	94 **Pu** (244)	95 **Am** (243)	96 **Cm** (247)	97 **Bk** (247)	98 **Cf** (251)	99 **Es** (254)	100 **Fm** (257)	101 **Md** (258)	102 **No** (255)	103 **Lr** (260)

Table 2-5   Latin Origins for the Names of Elements

Symbol	Latin name	Present name
Au	aurum	gold
Cu	cuprum	copper
Hg	hydrargyrum	mercury
Na	natrium	sodium
Pb	plumbum	lead
Sn	stannum	tin

Many of the symbols come from the Latin name of the elements as listed in Table 2-5. Others come from different languages such as Arabic (K, kalium—potassium) or German (W, wolfram—tungsten). Do not try to memorize all the symbols. You will need to know only the first 20 elements and a few others such as iron, copper, bromine, silver, iodine, cesium, gold, mercury, and uranium.

Each element has its own box in the periodic table. Sometimes both the atomic number and atomic weight are given; look at the table on the inside cover of this book. The number above the symbol is the atomic number, the most important identification of the element. The number below the symbol, given to several decimal places, is the atomic weight, the average of the masses of the isotopes for that element. The numbers on the right side of the box show the arrangement of electrons in shells outside the nucleus. For example, sodium (atomic number 11 and atomic weight 22.9898) is shown in Figure 2-6.

The horizontal rows running across the table called **periods** are numbered 1 through 7. Each row represents the filling of one shell of electrons. Periods 1, 2, and 3 are especially short, while Periods 6 and 7 are extra long. (The long lanthanide series and actinide series are written separately at the bottom to make the table more compact.) The vertical columns are called **groups** or families and are headed by roman numerals. Some have special family names such as alkali (Ia), alkaline earth (IIa), and halogen (VIIa); the last group, the noble gases, is sometimes designated Group 0 (or VIIIa).

Figure 2-6. Interpretation of the periodic table. Each box indicates the name, symbol, atomic number, atomic weight, and electron arrangement for an element. See table on inside front cover of book.

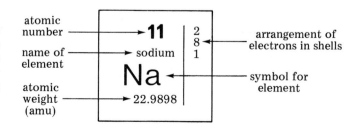

2.10 Properties of the elements

Elements can be labeled in another way. Notice the zigzag line that looks like a staircase running diagonally on the right side of the periodic table. The elements on the left side of this line, in other words, the majority, are **metals**, while those on the right side are **nonmetals**. Elements that touch the line sometimes act like metals and sometimes like nonmetals, so they are often called "semimetals" or "metalloids." Table 2-6 compares the important properties of metals and nonmetals. Pick out the metals and nonmetals in Figure 2-4.

Elements become more metallic, that is, have the properties of metals, as you move down a group from top to bottom and across a period from right to left. Thus, the most metallic elements are found at the lower left-hand side of the periodic table; cesium is the most metallic naturally occurring element. In the opposite way, nonmetallic properties increase as you move up a group from bottom to top and across a period from left to right. Thus, fluorine is the most nonmetallic element. Note that the noble gases are special elements in that they are usually considered to be neither metals nor nonmetals. Also, although hydrogen appears in Group Ia, it does *not* really act like a metal. It is an exception because of its very small size and should really be in a group by itself. Hydrogen is generally classified as a nonmetal.

The elements in a group have similar properties. But why should all the elements in Group Ia, for example, be very reactive metals? You can find the reason by looking at the electron arrangement for the elements of that group, particularly the outermost energy level, called the **valence shell.** In Group Ia (alkali metals), notice that the valence shell for each element contains one electron. Similarly, in Group IIa (alkaline earth metals), all the elements have two electrons in the valence shell. In fact, if you look at all the "a" group elements (elements that have the letter "a" after the roman numeral) you will find the same thing. *Every element in the same group has the same number of*

Table 2-6   **Properties of Metals and Nonmetals**

Metals	Nonmetals
include most elements	only 17 elements
all solids at room temperature (except Hg)	solids, liquid (Br only), and gases
shiny appearance	dull appearance
easily shaped	brittle (for solids)
good heat conductor	poor heat conductor
good electrical conductor	poor electrical conductor

electrons in the valence shell, and this number is given by the group number. Therefore, you should know immediately that all the elements in Group VIIa (halogens) must have seven electrons in their outermost shell. *It is because they have the same number of valence electrons that the chemical properties of the members of a group are so similar.*

The properties of the elements repeat themselves periodically because as one shell becomes filled, a new one begins which repeats the number of valence electrons. You will see in the next chapter the importance of this outermost shell of electrons in chemistry.

2.11 Elements important to health

In terms of health, only a relatively small number of elements are essential. They are shown by shading in the periodic table on the inside front cover of this book. The important elements of the human body are presented in Table 2-7.

Hydrogen, as a gas, is the lightest of all elements. Hydrogen atoms are found in acids, water, and most organic (carbon-containing) substances in the body. Chapter 8 (Acids, Bases, and Salts) is devoted to this important element.

Oxygen is the most abundant element on the earth. As a gas, it makes up about 21% of our atmosphere and, combined with hydrogen, makes up 89% of our water. It is essential for respiration and necessary for the processing of food by your body. Note that oxygen is the only element that can be used by the body in its elemental form. All other elements can be used only if they have first been combined with other elements.

Table 2-7 Elements in the Human Body

Element	Symbol	Percentage of atoms[a]	Percentage of mass
hydrogen	H	63	9.5
oxygen	O	25.5	65.0
carbon	C	9.5	18.5
nitrogen	N	1.4	3.3
calcium	Ca	0.31	1.5
phosphorus	P	0.22	1.0
potassium	K	0.06	0.35
sulfur	S	0.05	0.25
chlorine	Cl	0.03	0.20
sodium	Na	0.03	0.15
trace elements		0.01	0.25

[a] The total is more than 100% because of round-off errors.

Carbon is probably the most important element because all plant and animal life is based on it. Elemental carbon is found in many forms: charcoal, coke, coal, graphite, and diamond. In the form of carbon dioxide, it is a waste product of the body from foods containing carbon and, through photosynthesis, serves as a basic raw material for plants.

Nitrogen is a gas that makes up about four-fifths of our air. It is present in all living matter. As gaseous nitrogen goes through what is called the nitrogen cycle, it is converted to forms essential for life in all living cells. It is found in protein, the building material of tissue, and in enzymes, substances that assist chemical changes in your body.

Sodium and **potassium** are both very active metals. When combined with other elements (such as chlorine), they are needed for proper functioning of the nervous system and maintenance of the volume of the body fluids. **Calcium**, a metal, forms an important part of the bones and teeth; it is contained mainly in milk products. **Phosphorus**, a solid nonmetal, is also essential for the formation of bones and teeth, as well as brain and nervous tissue. A yellow solid, **sulfur** is found in many proteins and is used to make important medications, such as the sulfa drugs. Elemental **chlorine** is a greenish yellow gas that is very irritating and poisonous. But when combined with sodium, it is also vital for maintaining a proper fluid balance in the body.

Certain elements needed only in extremely small amounts, the **trace elements**, are listed in Table 2-8.

Table 2-8 **Trace Elements Essential for Health**

Element	Symbol	Role in human body
fluorine	F	dental health
magnesium	Mg	enzyme activity
silicon	Si	unknown
vanadium	V	unknown
chromium	Cr	sugar breakdown
manganese	Mn	enzyme activity
iron	Fe	oxygen transport; enzyme activity
cobalt	Co	enzyme activity
nickel	Ni	unknown
copper	Cu	enzyme activity
zinc	Zn	enzyme activity
selenium	Se	liver function
molybdenum	Mo	enzyme activity
tin	Sn	unknown
iodine	I	thyroid gland function

SUMMARY

All matter is composed of atoms, the basic building blocks of nature. There are 106 different "kinds" of these fundamental particles presently known. Chemistry is the study of these basic units, the atoms, how they combine, and how substances made of atoms are changed into other substances.

The atom is extremely small, yet it is composed of even tinier pieces, the subatomic particles. Atoms consist of a certain combination of protons (p), neutrons (n), and electrons (e or e⁻). The protons and neutrons are packed into a dense central core called the nucleus.

The electrons are far away, outside the nucleus. They are located in definite shells or energy levels which surround the nucleus. Each level can hold only a limited number of electrons: two in the first (K shell), eight in the second (L shell), and eight (or eighteen) in the third (M shell).

A fundamental property of nature called charge exists in two forms: positive (+) and negative (−). Charges that are the same, either both positive or both negative, repel each other; opposite charges attract each other. The proton has a positive charge, the electron has a negative charge, and the neutron has no charge (neutral). Every atom is neutral because it contains the same number of positive protons and negative electrons.

The atomic number of an atom is equal to the number of protons in its nucleus. Since every atom has the same number of electrons as protons, the atomic number is also equal to the number of electrons. Each different kind of atom has its own atomic number, beginning with 1 for the smallest atom, hydrogen, and ending with 106 for the largest known atom.

The masses of the proton and neutron are about the same; the mass of an electron is about 2000 times smaller. Based on comparison with a reference (carbon), the masses of the proton and neutron are each taken to be 1 amu (atomic mass unit). The mass number of an atom, its approximate atomic mass, is found by adding the number of protons and neutrons in its nucleus.

Simple pictures of atoms can be drawn using a circle to represent the nucleus with "shells" outside it for the electrons. Atoms are built up, starting with hydrogen (one proton, one electron), by successively adding one proton and one electron (and a certain number of neutrons) to the previous atom. Electrons are generally placed in the next shell only when the closer ones have been filled (for the simplest atoms).

Isotopes are atoms with the same atomic number but different mass numbers; they differ only in the number of neutrons. Because the identity of an atom is fixed by its atomic number, isotopes are chemically alike.

An element is a large collection of atoms having the same atomic number. The atom is the smallest part of an element that still has the properties of that element. Because of the possibility of isotopes, not all atoms of an element must be identical. Some may have different masses, but they all must have the same atomic number. There are 106 different elements, of which 90 are found in nature.

The atomic weight of an element is the average of the atomic masses of its isotopes as they are found in nature. Just like atomic mass, it is based on comparison with an isotope of carbon that is assigned a mass of exactly 12 amu. Because most atoms have isotopes, few atomic weights are simple whole numbers.

All of the 106 known chemical elements can be arranged in a special way, called the periodic table. It is based on the periodic law: The properties of the elements repeat in a regular (periodic) way when the elements are arranged by their atomic numbers. The table contains symbols, the one- or two-letter shorthand notation for the name of the element. The horizontal rows are called periods, and the vertical columns are called groups or families.

Most of the elements are metals—they are nearly all solids, conduct heat and electricity well, and are shiny and easily shaped. Others are either non-metals or semimetals (metalloids). The elements in a group have similar properties because they all have the same number of electrons in their outermost, or valence, shell.

In terms of health, only a relatively small number of elements are essential. They include hydrogen, oxygen, carbon, nitrogen, sodium, potassium, calcium, phosphorus, sulfur, and chlorine. Certain elements, known as the trace elements, are needed only in extremely small amounts.

Exercises

1. (Intro.) Define chemistry in terms of the atom.
2. (2.1) Where are the protons, neutrons, and electrons located in the atom?
3. (2.1) What are energy levels (shells)?
4. (2.2) Describe how like charges interact; how unlike charges interact.
5. (2.2) If an atom has 23 protons, how many electrons does it have? Explain.
6. (2.3) What is the atomic number?
7. (2.4) Compare the masses of the proton, neutron, and electron.
8. (2.4) Explain the atomic mass unit (amu) scale.
9. (2.4) How do you find the mass number of an atom?
10. (2.4) Fill in the following table:

	Atomic number	Mass number	Number of protons	Number of neutrons	Number of electrons
(a)	15	31	?	?	?
(b)	?	?	9	10	?
(c)	?	45	?	?	21

11. (2.5) Draw a simple "picture" of the following atoms: (a) hydrogen (atomic number 1, mass number 1); (b) carbon (atomic number 6, mass number 12); (c) magnesium (atomic number 12, mass number 24); (d) chlorine (atomic number 17, mass number 35); (e) fluorine (atomic number 9, mass number 19); (f) sulfur (atomic number 16, mass number 32); (g) lithium (atomic number 3, mass number 7); (h) phosphorus (atomic number 15, mass number 31).

12. (2.6) Which of the following atoms are isotopes of each other? Explain.

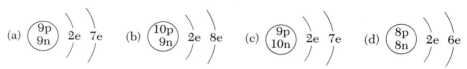

13. (2.6) How are isotopes similar? How do they differ?

14. (2.7) What is an element? What do all the atoms of an element have in common?

15. (2.8) What factors determine the atomic weight of an element?

16. (2.8) An element has two isotopes occurring in nature in equal amounts. One has a mass of 15 amu and the other a mass of 17 amu. What is the atomic weight of this element?

17. (2.9) How is the periodic table constructed?

18. (2.9) What is the periodic law?

19. (2.9) Write the symbols for the following elements: sulfur, sodium, carbon, chlorine, iron, potassium, silver, phosphorus, nitrogen, magnesium.

20. (2.9) Write the names of the elements with the following symbols: O, Al, He, Ca, Cu, Hg, Li, Ne, Au, Br.

21. (2.9) Give an example of (a) an alkaline earth element; (b) a halogen; (c) an element in the third row; (d) a noble gas; (e) an alkali element.

22. (2.10) How do metals differ from nonmetals? Give an example of each.

23. (2.10) Identify the following as metal, semimetal, or nonmetal: Cl, Na, Si, He, Zn, H, I, Al, K, N.

24. (2.10) Why do elements in the same group have similar chemical properties?

25. (2.11) Which elements are essential to health?

26. (2.11) What are trace elements? Give three examples.

Chemical bonding

Look at the periodic table and find the noble gases; they appear at the extreme right side. These are the final elements in each row because they have the maximum number of electrons in their valence shell, the outermost energy level that determines the chemical properties of an atom. *Each atom "wants" a completely filled valence shell as found in a stable noble gas element.*

3.1 The chemical bond

An atom gets a complete outer shell by joining with another atom to either share, gain, or lose electrons. Any of these processes results in a **chemical bond**, a force or sort of "chemical glue" that holds the atoms together. Since a complete valence shell contains eight electrons for the lighter elements, this basis for chemical bonding is often called the **octet rule** ("octet" means a group of eight). It can be stated in the following way:

> Atoms combine to permit each to have the electron arrangement of a noble gas, a filled valence shell of (usually) eight electrons.

Because atoms can interact to form chemical bonds, there is a tremendous variety of substances in the world. Almost everything in your body comes from the chemical combination of atoms. When different atoms join through bonding, they form a **chemical compound.** *All matter consists either of elements, which contain one type of atom, or of compounds, which contain different atoms chemically combined, or of mixtures of these two basic substances—elements and compounds.*

3.2 Covalent bonding

Hydrogen atoms form the simplest bond. An H atom consists of one proton and one electron. When two such hydrogen atoms are brought together, the

38

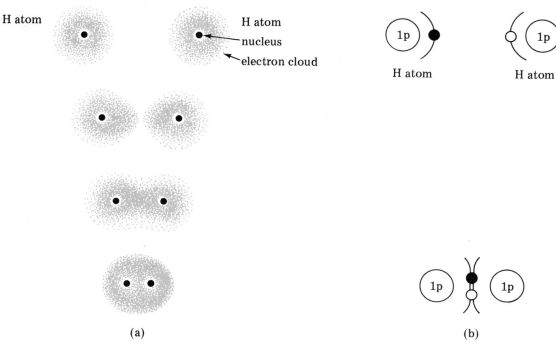

Figure 3-1. The bonding of two hydrogen atoms. (a) Electron "cloud" model; (b) electron shell model.

negative electron of each atom is attracted by the positive nucleus of the other atom, as illustrated in Figure 3-1a. Each electron is now attracted by the positive charge of two nuclei instead of just one.

By combining their valence shells, as shown in Figure 3-1b, the two atoms *share* a pair of electrons between them. Each hydrogen atom now gets to "use" an electron from the other atom, giving it a filled valence shell. (Remember that the first energy level is complete when it holds two electrons, not eight.) Both hydrogen atoms now have the electron arrangement of the noble gas helium. This type of attraction, formed by the sharing of a pair of electrons between the nuclei of two atoms, is called a **covalent bond.**

You can think of the formation of a covalent bond in the following manner. Two people are each reading books under lamps at opposite ends of a room. To get more light, they decide to move closer together, putting their lamps between them. In this way, both readers receive more light than if they stayed apart.

After the lamps are placed next to each other, the light overlaps and the readers cannot tell which light is coming from which lamp. In the same way, both electrons become equivalent when shared in a covalent bond. All electrons are alike, even though they may have come originally from different atoms.

3.3 The molecule

The unit formed by the covalent bonding of two or more atoms is a **molecule**. Thus, the combination of two atoms of hydrogen results in one molecule of hydrogen. Molecules range in size from just two atoms to many thousands of atoms bonded to each other. These very large units, called macromolecules ("macro" means long or great), play major roles in your body.

The number and kind of atoms in a molecule are given by its **formula**. The formula for a hydrogen molecule is H_2. The number 2 at the lower right side is called a subscript. It tells you how many atoms of a given kind, hydrogen in this case, are present in the molecule. You cannot write 2H, which would mean two separate atoms of hydrogen. You must always use a subscript when showing that atoms are chemically combined (except when only one atom of a particular type is present. The formula of a hydrogen molecule, H_2, is read out loud as H-two.)

3.4 Lewis symbols

Drawing pictures for covalent bonds can be simplified if you use what are known as **Lewis symbols** for atoms. Also called electron-dot notation, the Lewis symbols consist of the symbol for the element with only the *valence electrons* written around it. A dot (·) or cross (×) represents one electron. Thus, the hydrogen atom would be written as H· in this system. Lewis symbols for the first ten elements are presented in Table 3-1.

Electrons are placed around the atom on four sides. Thus, neon, with a complete valence shell, has four pairs of electrons around it. Notice that the number of electrons in the valence shell and therefore *the number of dots in the Lewis symbol is the same as the group number of the atom in the periodic table* (except for the noble gases). In addition, elements in the same group have the same number of electron dots in their Lewis symbols. Thus, chlorine, which appears under fluorine in Group VIIa of the table, has the Lewis symbol

$$\cdot \overset{\cdot\cdot}{\underset{\cdot\cdot}{Cl}} \colon$$

Table 3-1   **The Lewis Structures of Atoms**

Atom	Number of valence electrons	Lewis structure
hydrogen	1	H ·
helium	2	He :
lithium	1	Li ·
beryllium	2	Be :
boron	3	Ḃ :
carbon	4	·Ċ:
nitrogen	5	·Ṅ:
oxygen	6	·Ö:
fluorine	7	·F̈:
neon	8	:Ne:

3.5   Diatomic molecules

Using Lewis symbols, you can represent the hydrogen molecule, H_2, in the following way:

H:H   or   H×H

This is known as a **Lewis structure**. Crosses and dots are often used together to show that the electrons in the bond came from different atoms. This molecule is called **diatomic** because it consists of only two atoms bonded together ("di" means two). It is the simplest covalent molecule possible.

The element hydrogen is normally found in nature as the diatomic molecule, H_2, and not as single atoms. As shown in Table 3-2, there are six other elements that also normally exist as a pair of atoms covalently bonded. Their formulas all consist of the symbol for the element with the subscript 2 at the lower right-hand side.

Look at the structures of the diatomic elements. Like hydrogen, the halogens reach the noble gas arrangement by sharing one pair of electrons, one from each atom. Each of these covalent bonds is called a **single bond** because

Table 3-2 The Diatomic Elements

Element	Formula	Structure	Type of bond
hydrogen	H_2	H ⠶ H	single
fluorine	F_2	⠶F̈ ⠶ F̈ ⠶	single
chlorine	Cl_2	⠶C̈l ⠶ C̈l ⠶	single
bromine	Br_2	⠶B̈r ⠶ B̈r ⠶	single
iodine	I_2	⠶Ï ⠶ Ï ⠶	single
oxygen	O_2	·Ö ⠶ Ö ×	double
nitrogen	N_2	⠶N ⠶⠶ N ⠶	triple

it consists of a *single pair of electrons* being shared between the nuclei of the atoms. The single bond is represented by a long dash (—).

$$H—H \quad F—F \quad Cl—Cl \quad Br—Br \quad I—I \quad \text{single bonds}$$

Oxygen, in Group VIa, has six valence electrons and needs two more to fill its valence shell up to eight. It must share two pairs of electrons with a second oxygen atom, with two electrons coming from each, as shown in Table 3-2. Check for yourself that each oxygen originally had six valence electrons but now has a total of eight surrounding it. This covalent bond, formed by the sharing of *two pairs of electrons*, is a **double bond**, represented by a long double dash.

$$O{=}O \quad \text{double bond}$$

Nitrogen, in Group Va, with five valence electrons requires three more electrons to fill its valence shell. It obtains them through a triple bond to a second nitrogen atom. **A triple bond** consists of the sharing of *three pairs of electrons* between the nuclei of two atoms. Again, look at the table and check that each nitrogen has eight electrons, a complete octet, in the nitrogen molecule. The triple bond is represented by three long dashes.

$$N{\equiv}N \quad \text{triple bond}$$

3.6 Polar covalent bonds

In the diatomic elements, electrons are shared *equally* between two identical atoms. This type of bonding is known as a **nonpolar covalent bond.** But molecules can also contain different atoms. For example, hydrogen and chlorine atoms combine to form a molecule with a covalent bond, hydrogen chloride.

$$H : \overset{..}{\underset{..}{Cl}} :$$

hydrogen chloride, HCl

In this case, the electrons are not shared equally. The chlorine atom has a greater attraction for the electrons in the covalent bond largely because it has a greater number of positive protons in its nucleus.

The attraction of an atom for the electrons in a chemical bond is its **electronegativity.** Chlorine thus has a greater electronegativity than hydrogen. In general, the more nonmetallic an element, the stronger is its attraction for electrons and the greater is its electronegativity. It is as if two people were sharing blankets; the "stronger" person would be able to pull the blankets closer to one side of the bed. This type of *unequal* sharing of electrons is called a **polar covalent bond.**

In a diatomic molecule with a polar covalent bond, there is an uneven distribution of electrons. Because one atom pulls the electrons closer to its side, one part or "pole" of the molecule has more negative charge than the other "pole." In the example of people sharing blankets unequally, the person "hogging" the blankets would be warmer, while the other one would be colder.

The chlorine end of the HCl molecule has more of the electronic charge and is thus partially negative, symbolized as $\delta-$, where the Greek letter δ (delta) means "partial." The hydrogen end becomes partially positive, $\delta+$, because its negative electrons are being pulled away. The polar covalent molecule

$$\overset{\delta+ \quad \delta-}{H-Cl}$$

a polar molecule

stays neutral as a whole, but the chlorine has a slight negative charge and the hydrogen has a slight positive charge. As you will learn later, these slight charges can play important chemical roles.

3.7   Valence

Most covalent molecules consist of more than two atoms. For example, instead of nitrogen forming one triple bond with another nitrogen atom to get the noble gas arrangement in N_2, it can form three single bonds to other atoms as in ammonia. Its formula, NH_3, shows that the molecule contains one nitrogen atom bonded to three hydrogen atoms. When no subscript is written at the right side of the symbol for an atom, as with nitrogen, the subscript 1 is understood. Thus, the formula NH_3 really means N_1H_3.

$$H \overset{\cdot\cdot}{\underset{\overset{\times}{H}}{\overset{\times}{\underset{\times}{N}}}} H$$

ammonia, NH_3

Carbon, the most important element from the viewpoint of health, forms four covalent bonds to get the four more electrons it needs to complete its valence shell. The simplest example, methane, CH_4, consists of a carbon with single bonds to four hydrogen atoms. Note that the location of the dots (electrons), but not their number, can be changed from the way they are shown in Table 3-1.

$$\begin{array}{c} H \\ \overset{\times\times}{} \\ H \overset{\cdot}{\underset{\overset{\cdot\times}{}}{\overset{\times}{C}}} H \\ H \end{array}$$

methane, CH_4

Oxygen, in Group VIa, needs two more electrons. It can form two single bonds as in water, H_2O, or one double bond as in carbon dioxide, CO_2.

$$H \overset{\cdot\cdot}{\underset{\overset{\times\times}{H}}{\overset{\times}{O}}} \cdot$$

water, H_2O

$$\times \overset{\times\times}{\underset{\times}{O}} \overset{\cdot\cdot}{\underset{\times}{:}} C \overset{\cdot\cdot}{\underset{\times}{:}} \overset{\times\times}{\underset{\times}{O}} \times$$

carbon dioxide, CO_2

Depending on their number of valence electrons, different atoms form different numbers of bonds. This "combining capacity" of an atom is known as its **valence.** Since carbon must form four bonds for a complete outer shell, its valence is 4. Similarly, nitrogen usually has a valence of 3, oxygen a valence of 2, and fluorine (along with the other halogens) a valence of 1.

3.8 Naming covalent compounds

Covalent compounds containing two elements (binary compounds) are named by writing the name of one element in full followed by a space and the root of the second element with the ending "-ide." The roots of common ele-

Table 3-3 **Roots of Common Elements**

Element	Root	Element	Root
hydrogen	hydr-	oxygen	ox-
boron	bor-	sulfur	sulf-
carbon	carb-	fluorine	fluor-
nitrogen	nitr-	chlorine	chlor-
phosphorus	phosph-	bromine	brom-

Table 3-4   Prefixes for Numbers of Atoms

Prefix	Meaning
mono- (optional)	one
di-	two
tri-	three
tetra-	four
penta-	five
hexa-	six

ments are listed in Table 3-3. The first element of the name is the one that appears earlier in this sequence: B, Si, C, Sb, As, P, N, H, Se, S, I, Br, Cl, O, F. In addition, a Greek or Latin prefix, given in Table 3-4, which shows the number of atoms, is placed before each part of the name.

As an example, consider CO_2, which consists of one carbon atom and two oxygen atoms. Carbon appears before oxygen in the list given above and is written out in full. The root of oxygen is "ox-." The prefix "di-" is used to indicate two oxygens and the ending "-ide" is added. The complete name is carbon dioxide:

carbon	di	ox	ide
name of first element	"di" means 2	"ox" is root of oxygen	ending "ide"

More examples are given in Table 3-5.

Table 3-5   Examples of Names of Covalent Compounds

Formula	Name
CO	carbon monoxide
NO_2	nitrogen dioxide
SO_3	sulfur trioxide
CCl_4	carbon tetrachloride
PF_5	phosphorus pentafluoride
N_2O_5	dinitrogen pentoxide

3.9 Ions

Another way in which atoms combine is by the complete *transfer* of one or more electrons from one atom to another. This case represents polar covalent bonding carried to an extreme in which the situation can no longer be called "sharing" at all. Some atoms give away all their valence electrons to others which accept them and use these same electrons to fill up their own outer shells. Both the "givers" and the "takers" of electrons are then satisfied, having the noble gas arrangement of complete valence shells.

For example, lithium (atomic number 3) is a metal with three electrons; fluorine (atomic number 9) is a nonmetal with nine electrons. The simplest way for each to have complete outer shells is for lithium to give up its one valence electron to fluorine (as opposed to fluorine transferring seven electrons to lithium). In this manner, the outermost shells of both lithium and fluorine are filled. Each has reached a noble gas arrangement. Lithium has the same number of electrons as helium, and fluorine the same number as neon. (Remember that the first shell is complete with only two electrons, not eight.)

$$\begin{array}{c} \boxed{\begin{array}{c}3p\\4n\end{array}}\ 2e\ \ 1e \quad \text{or} \quad \text{Li}\cdot \qquad\qquad \boxed{\begin{array}{c}9p\\10n\end{array}}\ 2e\ \ 7e \quad \text{or} \quad \cdot\ddot{\text{F}}\colon \end{array}$$

$$\qquad\qquad\text{lithium atom} \qquad\qquad\qquad\qquad\qquad \text{fluorine atom}$$

After the transfer, the situation can be represented in the following way:

$$\left[\boxed{\begin{array}{c}3p\\4n\end{array}}\ 2e\right]^{+} \quad \text{or} \quad \text{Li}^{+} \qquad\qquad \left[\boxed{\begin{array}{c}9p\\10n\end{array}}\ 2e\ \ 8e\right]^{-} \quad \text{or} \quad :\ddot{\text{F}}\colon^{-}$$

$$\qquad\text{lithium ion} \qquad\qquad\qquad\qquad\qquad\qquad \text{fluoride ion}$$

You no longer have neutral atoms because in each case the number of positive protons does *not* equal the number of negative electrons. By the transfer of an electron, **ions**, electrically charged atoms, are formed. Lithium now has a charge of $+1$ because it has one more proton $(3+)$ than it has electrons $(2-)$; fluorine has a charge of -1 since it has one more electron $(10-)$ than it has protons $(9+)$.

A positively charged ion is a **cation** (pronounced cat'-eye-on), while a negative ion is an **anion**. The symbol for an ion is simply the symbol for the atom with its charge placed at the top right. Thus, Li^{+} (or Li^{1+}) and F^{-} (or F^{1-}) are formed. Cations have the same name as the atom they came from: Li^{+} is called lithium ion or lithium cation. A simple anion (one made from a single atom) is named by replacing the ending of the element by the suffix "-ide"; F^{-} is called fluoride ion or fluoride anion. Table 3-6 lists the most common simple ions and their names.

Table 3-6   Names of Simple Ions

Cation		Anion	
Symbol	Name	Symbol	Name
H^+	hydrogen ion	F^-	fluoride ion
Li^+	lithium ion	Cl^-	chloride ion
Na^+	sodium ion	Br^-	bromide ion
K^+	potassium ion	I^-	iodide ion
Mg^{2+}	magnesium ion	O^{2-}	oxide ion
Ca^{2+}	calcium ion	S^{2-}	sulfide ion
Al^{3+}	aluminum ion	N^{3-}	nitride ion

3.10 Charges of ions

It is easy to predict the charges of ions. When a neutral atom loses one or more electrons, a process called **oxidation**, a cation always forms. The atoms that can most easily lose electrons are those that have either one, two, or at most three valence electrons. These elements are the metals on the left side of the periodic table. The alkali metals in Group Ia, for example, each have one valence electron. By losing this electron, the metal atom forms a 1+ ion. Similarly, those metals in Groups IIa and IIIa form ions with charges of 2+ and 3+ by losing their two or three valence electrons, respectively.

On the other hand, when an atom gains electrons, a process called **reduction**, an anion forms. The elements that readily accept electrons are those that need only a few more to fill the valence shell—the nonmetals of Groups VIIa and VIa, and sometimes Va or even IVa, on the right side of the periodic table. The halogens of Group VIIa have seven valence electrons; by gaining one more, they form 1− ions. In the same way, the nonmetals of groups VIa and Va, which have six and five valence electrons, respectively, gain two $(2 + 6 = 8)$ or three $(3 + 5 = 8)$ more to form ions with charges of 2− or 3−. Table 3-7 summarizes this relationship. Notice that the elements in Group IVa can either gain or lose electrons since they have a half-filled valence shell.

Table 3-7   Relation between Charge of Ion and Group Number

	Group Number							
	Ia	IIa	IIIa	IVa	Va	VIa	VIIa	Noble gases
number of electrons	1	2	3	4	5	6	7	8
charge of ion	1+	2+	3+	4+ or 4−	3−	2−	1−	0

Table 3-8 **Atoms That Form Several Ions**

Element	Ion	Formal name	Common name
iron	Fe^{2+}	iron(II) ion	ferrous ion
	Fe^{3+}	iron(III) ion	ferric ion
copper	Cu^+	copper(I) ion	cuprous ion
	Cu^{2+}	copper(II) ion	cupric ion
tin	Sn^{2+}	tin(II) ion	stannous ion
	Sn^{4+}	tin(IV) ion	stannic ion
mercury	$Hg^{+\ a}$	mercury(I) ion	mercurous ion
	Hg^{2+}	mercury(II) ion	mercuric ion

[a] Occurs as Hg_2^{2+}.

For some elements, it is not always obvious what ion will be formed. In fact, the elements in the center of the periodic table, the transition metals (Groups Ib to VIIb and VIII), as well as some of the "a" group elements, can form ions with different charges. Table 3-8 lists several of these ions. Notice the two different ways of writing their names. In the formal name, the charge is given in roman numerals in parentheses after the name of the element. In the common name, the original name of the element is used followed by the ending "-ous" or "-ic"; the ion with lower charge is always given the "-ous" suffix. Other important ions that are not in "a" groups include: manganese ion, Mn^{2+}; zinc ion, Zn^{2+}; and silver ion, Ag^+.

3.11 Ionic bonds

Now that you know something about ions, look back to the original reaction between lithium and fluorine (Section 3.9). Lithium was oxidized, forming Li^+, by transferring its valence electron to fluorine, producing F^- by reduction. Remember the way in which particles with opposite charges interact—they attract each other. The cation and anion have a force holding them together. This force is an **ionic** (or electrovalent) **bond**. An ionic bond is formed by the attraction between oppositely charged ions.

The substance made from lithium and fluorine is therefore an ionic compound. It contains an ion with a 1+ charge and one with a 1− charge. But the compound is neutral because the positive charge of the cation is exactly balanced by the negative charge of the anion. Thus, *all compounds, like all atoms, are electrically neutral.*

You write the formula of this compound as LiF. The cation is always written first, followed by the anion with no spaces in between. The charges of ions are left out when the formula is written since the resulting compound is neutral. In writing LiF, it is understood that the formula really is Li_1F_1: one lithium ion and one fluoride ion. As with covalent formulas, the numbers at the lower right-hand side of each symbol, the subscripts, are always omitted when they have the value 1.

Look at some other examples. The formation of an ionic bond between sodium and chlorine can be represented in the following way:

$$\left(\begin{matrix}11p \\ 12n\end{matrix}\right) \quad 2e \quad 8e \quad 1e \rightarrow 7e \quad 8e \quad 2e \quad \left(\begin{matrix}17p \\ 18n\end{matrix}\right) \qquad \text{or} \qquad Na^{\times} \curvearrowright \cdot \ddot{C}l \colon$$

$$\underbrace{\qquad\qquad\qquad}_{\text{sodium atom}} \qquad \underbrace{\qquad\qquad\qquad}_{\text{chlorine atom}}$$

Sodium metal transfers its one valence electron to the nonmetal chlorine to produce two ions:

$$\left[\left(\begin{matrix}11p \\ 12n\end{matrix}\right) \quad 2e \quad 8e\right]^{+} \quad \left[8e \quad 8e \quad 2e \left(\begin{matrix}17p \\ 18n\end{matrix}\right)\right]^{-} \qquad \text{or} \qquad Na^{+} \quad {}^{-}{\times}\ddot{C}l \colon$$

$$\underbrace{\qquad\qquad\qquad}_{\text{sodium ion}} \qquad \underbrace{\qquad\qquad\qquad}_{\text{chloride ion}}$$

(Chlorine is drawn "backward" to make it easier to show the electron transfer. Obviously, it makes no difference since electrons are all around the nucleus.) The transfer results in a sodium cation, Na^+, and a chloride anion, Cl^-. The compound is therefore NaCl, sodium chloride, or table salt. Notice that the name is written in the same order as the formula—first the cation and then the anion.

The actual structure of sodium chloride consists of an orderly arrangement of Na^+ ions and Cl^- ions in which each ion is surrounded by six others of opposite charge. This solid is illustrated in Figure 3-2. There are no distinct pairs of ions, no "molecules" of sodium chloride. One sodium cation and one chloride anion, however, together make one **formula unit** of sodium chloride. It is the smallest part of the ionic compound that still has the properties of that compound.

It is important for you to understand the change that has taken place here. Sodium is a reactive shiny metal which burns spontaneously when put into water. Chlorine is a green poisonous gas that was used in World War I. But when sodium loses one electron and chlorine gains it, each forming an ion, the resulting compound has completely different properties. After all, sodium chloride is something you sprinkle on food. The point is that *the properties of each ion differ greatly from those of the atom from which it came* (Figure 3-3).

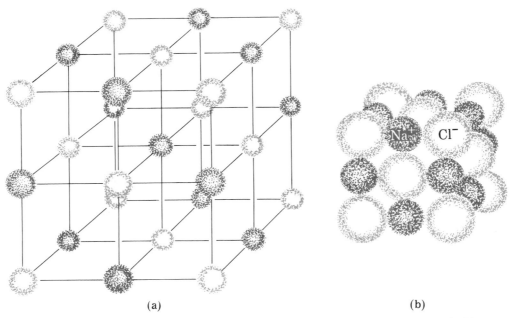

(a)   (b)

Figure 3-2. The structure of sodium chloride, two views. Each ion is surrounded by six ions of opposite charge.

Figure 3-3. Comparison of sodium metal (left) and sodium ion as sodium chloride (right). Although they differ by only one electron, the atom and the ion have properties that differ greatly. (Photo by Al Green.)

In the examples you have studied so far, only one electron has been transferred. But what happens, for example, when magnesium reacts with chlorine? Magnesium must lose two electrons, but chlorine can accept only one.

magnesium atom   chlorine atom

You can solve this problem by bringing in a second atom of chlorine. The magnesium donates one of its two valence electrons to each of the two chlorine atoms, as shown in Figure 3-4 on page 52. The resulting compound contains two chloride ions for each magnesium ion and therefore has the formula $MgCl_2$ and the name magnesium chloride. Using the same kind of thinking, you should be able to show why sodium and oxygen form a compound with the formula Na_2O.

3.12   Writing formulas of ionic compounds

A shortcut to writing the formulas of ionic compounds is based on the charges of ions. For example, consider the reaction between sodium and chlorine. You know that the cation formed from sodium is Na^+ and the anion from chlorine is Cl^-. Since compounds are neutral, *one formula unit must contain an equal number of positive and negative charges.* Therefore, one Na^+ and one Cl^- are needed, making NaCl.

Similarly, for magnesium and chlorine, magnesium forms Mg^{2+} (since it is in Group IIa) and chlorine again forms Cl^-. To make a neutral combination, two negative charges are needed to balance the two positive charges of magnesium ion. Thus, the formula unit must consist of Mg^{2+} and two Cl^- ions, $MgCl_2$.

The procedure for writing the formula of an ionic compound consists of the following steps:

1. Write the symbols of the atoms in the compound.
2. Using the group number from the periodic table, write the charge for each ion.
3. Multiply one of the ions (or both if necessary) by a whole number to get the same number of positive and negative charges.
4. Write these numbers, in lowest terms, as the subscripts in the formula.

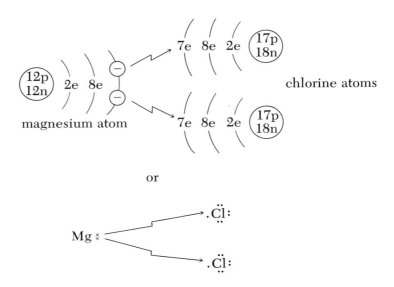

or

Mg ⁚ ⟨ → .C̈l:
       → .C̈l:

forming

formula: $MgCl_2$

name: magnesium chloride

Figure 3-4. The reaction of magnesium and chlorine to form magnesium chloride, $MgCl_2$. Note that the formula unit contains one magnesium ion and two chloride ions.

Example: aluminum and iodine

Step 1. Al $\qquad\qquad$ I

Step 2. Al^{3+} (Group IIIa) $\qquad$ I^- (Group VIIa)

Step 3. $1(Al^{3+})$, one aluminum ion $\qquad$ $3(I^-)$, three iodide ions

three positive charges $\quad$ balance $\quad$ three negative charges

Step 4. AlI_3, aluminum iodide

Example: calcium and nitrogen

Step 1. Ca $\qquad\qquad$ N

Step 2. Ca^{2+} (Group IIa) $\qquad$ N^{3-} (Group Va)

Step 3. You must multiply both ions
in order to balance the
charges; the lowest common
multiple of 2 and 3 is 6

six positive charges $\quad$ balance $\quad$ six negative charges

Step 4. Ca_3N_2, calcium nitride

By making sure that the number of positive charges is equal to the number of negative charges you are really saying that *the number of electrons lost by one atom is the same as the number of electrons gained by the other atom.* This principle is the basis for all reactions involving the simultaneous processes of oxidation and reduction.

3.13 Polyatomic ions

Certain covalent molecules can become charged and therefore act as ions. Because they do not consist of a single atom but a group of atoms which together have either a positive or negative charge, they are called **polyatomic ions** ("poly" means many). These covalently bonded atoms form a combination which acts like a single ion. The common polyatomic ions are listed in Table 3-9. The charge, written in the upper right-hand corner, just as in a simple ion, is the charge of the *entire group* of atoms. Notice in the names that the ending

Table 3-9 Common Polyatomic Ions

Name	Symbol	Name	Symbol
hydroxide ion	OH^-	peroxide ion	O_2^{2-}
nitrate ion	NO_3^-	carbonate ion	CO_3^{2-}
nitrite ion	NO_2^-	sulfate ion	SO_4^{2-}
acetate ion	$C_2H_3O_2^-$	sulfite ion	SO_3^{2-}
permanganate ion	MnO_4^-	phosphate ion	PO_4^{3-}
cyanide ion	CN^-	ammonium ion	NH_4^+
bicarbonate ion	HCO_3^-	hydronium ion	H_3O^+

"-ate" indicates that oxygen is present in the anion; "-ite" means that the ion contains less oxygen than the "-ate" form. Compounds containing polyatomic ions are named in the same way as simple ionic compounds—*first the name of the cation, then the name of the anion,* as illustrated for many of the compounds listed in Table 3-10. The Lewis structures of several such ions are shown in Figure 3-5. These groups of atoms are charged because of an "extra" electron (shown by an open dot) or proton (hydrogen ion in bold type) or sometimes one less electron. Note that in the case of the NH_4^+ ion, both electrons in the bond to the fourth hydrogen come from the nitrogen atom (an example of coordinate covalence).

When you write the formula of a compound containing more than one com-

Table 3-10   Examples of Ionic Compounds

Name	Formula	Medical use
aluminum acetate	$Al(C_2H_3O_2)_3$	wet dressing for skin, astringent (Burrow's solution)
ammonium carbonate	$(NH_4)_2CO_3$	smelling salts
ammonium chloride	NH_4Cl	diuretic (helps urine form), expectorant (liquefies bronchial secretion)
barium sulfate	$BaSO_4$	x-ray examination of internal organs
calcium carbonate	$CaCO_3$	antacid
calcium sulfate	$CaSO_4$	plaster
iron sulfate	$FeSO_4$	treatment of iron-deficiency anemia
magnesium sulfate (Epsom salt)	$MgSO_4$	laxative
potassium iodide	KI	expectorant, antifungal agent
potassium nitrate (saltpeter)	KNO_3	diuretic, antiseptic
potassium permanganate	$KMnO_4$	oxidizing agent, anti-infectant
silver nitrate	$AgNO_3$	antiseptic, prevents eye infection in newborn
sodium bicarbonate (baking soda)	$NaHCO_3$	antacid
sodium bromide	$NaBr$	sedative
sodium chloride	$NaCl$	normal saline solution
sodium fluoride	NaF	dental application
sodium iodide	NaI	treatment of iodine deficiency in thyroid conditions
zinc oxide	ZnO	astringent and protective (in calamine lotion)

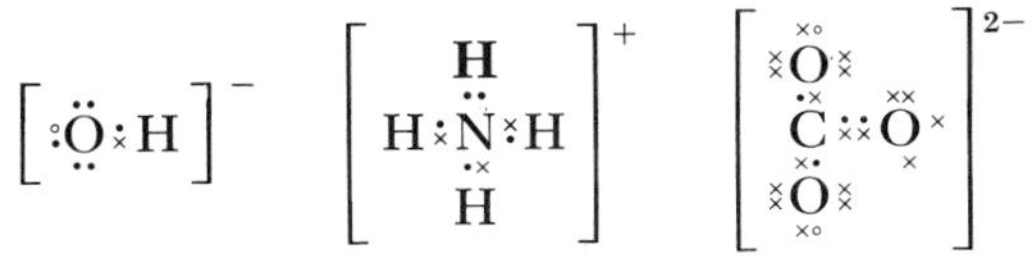

Figure 3-5. Lewis structures of polyatomic ions. (a) Hydroxide ion; (b) ammonium ion; (c) carbonate ion.

$$\left[\,\ddot{:}\ddot{O}\!:\!H\,\right]^{-} \qquad \left[\begin{array}{c} H \\ H\!:\!\ddot{N}\!:\!H \\ H \end{array}\right]^{+} \qquad \left[\begin{array}{c} \ddot{O} \\ C\!::\!\ddot{O} \\ \ddot{O} \end{array}\right]^{2-}$$

(a) OH^- (b) NH_4^+ (c) CO_3^{2-}

plex ion, you must often place the ion in parentheses with a subscript at its lower right. For example, calcium hydroxide must be written $Ca(OH)_2$, meaning one calcium ion and two hydroxide ions. You *cannot* write $CaOH_2$ because the subscript 2 would apply only to the hydrogen and not to the oxygen. The parentheses around the hydroxide ion make the subscript apply to the entire group, OH^-, not just part of it. Sometimes, there are subscripts within the parentheses. Thus, aluminum nitrate is written $Al(NO_3)_3$. The formula unit of this compound consists of one aluminum ion, Al^{3+}, and three nitrate ions, NO_3^-.

You have come a long way in this chapter. You now know that all matter is made from atoms or from groups of atoms joined by chemical bonds. These groups are either molecules with covalent bonds or ions of opposite charge held together by ionic bonds. It is also possible for covalently bonded atoms to carry charges as in the polyatomic ions.

Most often, the bonding in a compound is a blend, somewhere between the extremes of equal sharing and complete transfer of electrons. The exact nature of the bonding that exists determines to a large degree the chemical and physical properties of the compound. This fact is important because throughout this text you will be learning about many different compounds and how they are related to health.

SUMMARY

A chemical bond is a force, or sort of chemical "glue," that holds atoms together. Atoms combine to permit each to have the electron arrangement of a noble gas element, a filled valence shell of usually eight electrons (the octet rule). When two or more atoms combine, they form a chemical compound.

Covalent bonding results from the sharing of electrons between the nuclei of two atoms. The negative electrons are attracted by the positive charge of two nuclei instead of just one. A molecule is the unit formed by the covalent bonding of two or more atoms. It is the smallest part of a covalent compound that still has the properties of that compound. The formula gives the number and kind of atoms present in a molecule.

Lewis symbols are simplified ways of representing the valence shell of an atom. The number of dots, or electrons, around the symbol for the atom is the same as the group number in the periodic table (for "a" group elements).

Certain elements, known as the diatomic elements, are found in nature as pairs of atoms covalently bonded. They are hydrogen (H_2), nitrogen (N_2), oxygen (O_2), fluorine (F_2), chlorine (Cl_2), bromine (Br_2), and iodine (I_2). The bonds of hydrogen, fluorine, chlorine, bromine, and iodine molecules are single bonds—one pair of electrons is shared. The oxygen molecule has a double bond, two pairs of electrons shared, and the nitrogen molecule a triple bond, three pairs of electrons shared.

The diatomic elements all have nonpolar covalent bonds—the electrons are shared equally. In other molecules, electrons are pulled closer to one of the atoms because it has a greater ability to attract them, called electronegativity. The resulting bond is polar: There is an uneven distribution of electrons, creating a slightly positive and slightly negative "pole" or part in the molecule.

Most covalent molecules consist of more than two atoms. The "combining capacity" of an atom, the number of bonds it must form to have the noble gas arrangement, is called its valence. The valence for carbon is 4, that for nitrogen is 3, that for oxygen is 2, and that for fluorine is 1. Covalent compounds containing two elements are named by writing the name of one element in full followed by a space and the root of the second element with the ending "-ide."

Another way in which atoms combine is by the complete transfer of electrons from one atom to another. By giving away or receiving electrons, the atoms reach a noble gas arrangement. Since the number of protons and the number of electrons are no longer equal, the atom becomes electrically charged and is called an ion. A cation is a positive ion; an anion is a negative ion.

Metals tend to lose one or more electrons, a process known as oxidation, to form cations. Nonmetals can gain these electrons, a process of reduction, to form anions. The charge of the resulting ion can be predicted in many cases from the group number of the atom in the periodic table.

 An ionic bond forms by the attraction between oppositely charged ions. The smallest part of an ionic compound that still has the properties of that compound is called its formula unit. Ionic compounds generally exist as solids with an orderly repeating arrangement of ions. An ion has very different properties than the atom from which it forms.

Since all compounds are neutral, one formula unit must contain an equal number of positive and negative charges. This principle is the basis for writing the formulas of ionic compounds. They are named by first writing the cation and then the anion.

Polyatomic ions consist of groups of atoms which together have a positive or negative charge. When you write the formula of a compound containing more than one complex ion, you must place the ion in parentheses with a subscript at its lower right side.

Exercises

1. (3.1) Explain the basis for chemical bonding.
2. (3.1) What is a compound?
3. (3.2) How does a covalent bond form?
4. (3.3) What is a molecule? What does its formula represent?
5. (3.4) Write the Lewis symbol for the following atoms: (a) phosphorus; (b) calcium; (c) argon; (d) bromine; (e) potassium.
6. (3.5) Draw the Lewis structure for each of the seven diatomic elements. Identify the single, double, or triple bond.
7. (3.6) How does a polar covalent bond differ from a nonpolar covalent bond? Give an example of each kind.
8. (3.6) What is electronegativity?
9. (3.7) Draw the Lewis structure of (a) CCl_4; (b) PF_3; (c) H_2S. Identify the valence of each atom.
10. (3.8) Name the following covalent compounds: (a) SiF_4; (b) SO_2; (c) PCl_3; (d) CS_2.
11. (3.8) Write the formula for the following molecules: (a) nitrogen triiodide; (b) diarsenic pentasulfide; (c) phosphorus tribromide; (d) carbon tetrachloride.
12. (3.9) How does an ion differ from an atom?
13. (3.9) Write the formulas of the following ions: (a) oxide ion; (b) sodium ion; (c) chloride ion; (d) aluminum ion. Identify the cations and anions.
14. (3.9) Name the following: (a) K^+; (b) F^-; (c) H^+; (d) Al^{3+}; (e) N^{3-}.
15. (3.10) Compare oxidation and reduction.
16. (3.10) Predict the charge of the ion formed by (a) barium; (b) cesium; (c) selenium; (d) strontium.
17. (3.10) Write the symbol for (a) mercuric ion; (b) ferrous ion; (c) cupric ion; (d) stannous ion.
18. (3.11) What is an ionic bond? How is it different from a covalent bond?
19. (3.11) Draw a diagram to show the formation of an ionic bond between (a) calcium and oxygen; (b) sodium and sulfur; (c) magnesium and iodine.
20. (3.11) Define formula unit. Does it mean the same thing as molecule? Explain.
21. (3.12) Write the formula of the ionic compound formed from (a) potassium and oxygen; (b) magnesium and bromine; (c) aluminum and fluorine; (d) aluminum and oxygen; (e) sodium and iodine.
22. (3.13) Write the symbol for the following ions: (a) sulfate ion; (b) ammonium ion; (c) hydroxide ion; (d) hydronium ion; (e) phosphate ion.
23. (3.13) Name the following: (a) NO_3^-; (b) HCO_3^-; (c) $C_2H_3O_2^-$; (d) CO_3^{2-}; (e) CN^-.
24. (3.13) Write the formula of (a) sodium acetate; (b) ammonium carbonate; (c) aluminum nitrate; (d) potassium bicarbonate; (e) calcium phosphate; (f) magnesium hydroxide; (g) sodium sulfate; (h) barium nitrate.
25. (3.13) List three ionic compounds and their medical uses.

Compounds and chemical change

Nearly everything that takes place in your body involves changes in chemical compounds. To understand these processes, you must know more about the properties of compounds.

4.1 Formula or molecular weight of compounds

The smallest part of a compound still having the properties of that compound is the molecule or formula unit, depending on whether the bonding is mainly covalent or ionic. You will find it very useful to be able to calculate the mass of these smallest units of the compound, known as the **molecular weight** or **formula weight**. (Just as in atomic weight, the difference between mass and weight is ignored here.) Even though all compounds do not exist as molecules, the term "molecular weight" is sometimes incorrectly used for both molecules and formula units.

When you measure things, it is often helpful to know the mass of the smallest unit. For example, if someone asked you for 1000 marbles, you could sit down and count them out one by one. A much simpler way is to calculate the mass of the 1000 marbles and then pour them into a container on a balance until you reach that mass. Of course, you first need the mass of a single marble to figure out the mass of the larger amount. The same principle holds true for chemical compounds.

The molecular or formula weight is simply the *sum of the atomic weights of all the atoms (or ions) present in the molecule or formula unit*. Being able to read a chemical formula correctly is the most important part of finding these weights. Table 4-1 illustrates how you can interpret the formula of a com-

Table 4-1 **Composition of Compounds Based on Formulas**

Formula	Composition
H_2O	2 hydrogen atoms, 1 oxygen atom
$C_6H_{12}O_6$	6 carbon atoms, 12 hydrogen atoms, 6 oxygen atoms
Na_3PO_4	3 sodium atoms, 1 phosphorus atom, 4 oxygen atoms
$Al_2(SO_4)_3$	2 aluminum atoms, 3 sulfur atoms, 12 oxygen atoms
$(NH_4)_2CO_3$	2 nitrogen atoms, 8 hydrogen atoms, 1 carbon atom, 3 oxygen atoms

pound in terms of atoms. The subscript at the lower right side of the symbol for each element gives the number of that type of atom present. When written next to a group of atoms in parentheses, all atoms inside the parentheses are multiplied by that subscript. Thus, in aluminum sulfate, $Al_2(SO_4)_3$, there are 3 × 1 sulfur, or 3 sulfur atoms, and 3 × 4 oxygens, or 12 oxygen atoms, present in addition to the 2 aluminum atoms. When finding formula weights, you do not have to distinguish between atoms and ions.

Once you know how to read a chemical formula, it is easy to find the molecular or formula weight. You simply add up the atomic weights of each of the atoms present in that one unit of the compound. Their values can be taken from either the periodic table or a table of atomic weights; they are usually rounded off before the calculation is done. For example, the molecular weight of water, H_2O, is found by adding the atomic weights of two hydrogen atoms and one oxygen atom (see following listing):

H_2O	Number of atoms		Atomic weight (amu)		Total weight (amu)
2 hydrogen atoms	2	×	1	=	2
1 oxygen atom	1	×	16	=	16
molecular weight of H_2O					18

The same procedure can be followed for a more complicated compound, as shown in Table 4-2.

4.2 Percentage composition

One of the most important properties of a compound is that it has a *definite composition*. Everywhere in the world, water has exactly the same makeup: two atoms of hydrogen and one atom of oxygen. The molecular weight has a fixed value (18 amu) as do the contributions of the two hydrogen atoms (2

Table 4-2 Calculation of Formula Weight: $Al(C_2H_3O_2)_3$, Aluminum Acetate[a]

Composition	Number of atoms		Atomic weight (amu)		Total weight (amu)
1 aluminum atom	1	$\times$	27	$=$	27
6 carbon atoms	6	$\times$	12	$=$	72
9 hydrogen atoms	9	$\times$	1	$=$	9
6 oxygen atoms	6	$\times$	16	$=$	96
formula weight of $Al(C_2H_3O_2)_3$					204

[a] Used in Burrow's solution, an astringent.

amu) and the one oxygen atom (16 amu) to that weight. To put it another way, the percentage of hydrogen and the percentage of oxygen of the total weight of water have definite values.

The percentage of each element by weight in a compound is its **percentage composition**. It is found by dividing the weight of all the atoms of each element by the formula or molecular weight and multiplying by 100 to convert this fraction to a percentage, parts per hundred (see Appendix A.1). This procedure is illustrated in Table 4-3 for water, H_2O, and hydrogen peroxide, H_2O_2. A molecule of water consists of 2 parts of hydrogen, 11% of the weight, to 1 part of oxygen, 89% of the weight. Hydrogen peroxide, a poisonous compound used in disinfectants and hair bleach, has 2 parts hydrogen, 6% of the weight, to 2 parts oxygen, 94% of the weight. Thus, there are two ways of looking at

Table 4-3 Percentage Composition of H_2O and H_2O_2

H_2O (water)	weight of 2 H atoms $= 2 \times 1$ amu $= 2$ amu weight of 1 O atom $= 1 \times 16$ amu $= 16$ amu weight of H_2O $= 18$ amu
% H $=$	$\dfrac{2 \text{ amu}}{18 \text{ amu}} \times 100 = 0.11 \times 100 = 11\%$
% O $=$	$\dfrac{16 \text{ amu}}{18 \text{ amu}} \times 100 = 0.89 \times 100 = 89\%$
H_2O_2 (hydrogen peroxide)	weight of 2 H atoms $= 2 \times 1$ amu $= 2$ amu weight of 2 O atoms $= 2 \times 16$ amu $= 32$ amu weight of H_2O_2 $= 34$ amu
% H $=$	$\dfrac{2 \text{ amu}}{34 \text{ amu}} \times 100 = 0.06 \times 100 = 6\%$
% O $=$	$\dfrac{32 \text{ amu}}{34 \text{ amu}} \times 100 = 0.94 \times 100 = 94\%$

the makeup of a molecule or formula unit: by the number of atoms of each element or by the percentage composition based on the molecular or formula weight.

4.3 The mole

The mass of a single molecule or formula unit cannot be measured on a balance, since the mass of each atom is so small. Ordinary laboratory balances can only measure quantities heavier than several milligrams. To weigh atoms or groups of atoms, you must therefore take an extremely large number of them and measure them together. It is as if you were weighing feathers; they are so light that many would be needed before the balance would begin to register. Chemists therefore work with large collections of atoms or molecules.

A special unit, the **mole** (abbreviated mol), is defined for this big collection. A mole always contains the *same number of particles*, just as a "dozen" always means 12 of something. A mole of anything, whether atoms, ions, or molecules, consists of the following number of particles: 602,000,000,000,000,000,000,000, abbreviated 6.02×10^{23} (see Appendix A.2). This huge number of particles in a mole is called **Avogadro's number** (Figure 4-1).

By taking such a huge number of atoms or molecules, you have an amount that can easily be weighed. Its weight is simply the *atomic or formula weight written in grams*, as will be explained. For instance, since the molecular weight of water is 18 amu, 1 mole of water weighs 18 g. Table 4-4 gives more examples. Figure 4-2 compares 1 mole of carbon with 1 mole of glucose (dextrose).

One mole always contains Avogadro's number of units of the substance, but its total weight varies, depending on the weight of the individual unit. Think of a mole of feathers and a mole of marbles. You have the same number of each (Avogadro's number), but the mole of marbles weighs much more because each marble is heavier than each feather. In the same way, a mole of oxygen atoms and a mole of hydrogen atoms each contains Avogadro's number of atoms, but the mole of oxygen weighs 16 g while the mole of hydrogen weighs only 1 g. The reason is, of course, that each atom of oxygen (16 amu) weighs 16 times as much as each hydrogen atom (1 amu) because it contains more protons and neutrons in its nucleus. In summary, 1 mole of a substance

1. contains Avogadro's number (6.02×10^{23}) of atoms, formula units, or molecules, and
2. has a weight in grams equal to the atomic, formula, or molecular weight.

The mole is very useful for calculations involving quantities of chemical substances. But the amounts you use are seldom exactly 1 mole; instead you must be able to measure some multiple or some fraction of a mole. For example, since 1 mole of sodium chloride weighs 58 g, 2 moles weigh 2 moles × 58 g/mole = 116 g, 3 moles weigh 3 moles × 58 g/mole = 174 g, and so on.

Figure 4-1.   A representation of the tremendous number of particles in 1 mole, in this case, a mole of something you can see—feathers. (Based on an original drawing by Wendy Lindboe.)

Table 4-4   **Mass of 1 Mole of Various Substances**

Atom	Atomic weight (amu)	Mass of 1 mole (g)
C	12	12
Fe	56	56
Hg	201	201

Molecule	Molecular weight (amu)	Mass of 1 mole (g)
NH_3	17	17
CO_2	44	44
$C_6H_{12}O_6$	180	180

Formula unit	Formula weight (amu)	Mass of 1 mole (g)
NaCl	58	58
$KMnO_4$	158	158
$AgNO_3$	170	170

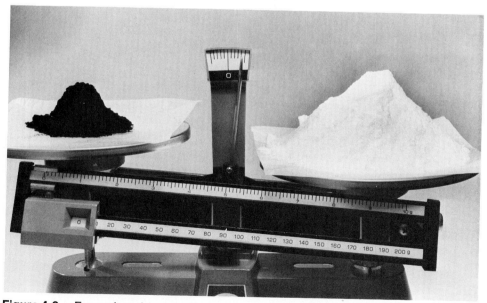

Figure 4-2.   Examples of 1 mole. One mole of glucose (dextrose), on the right, weighs much more (180 g) than 1 mole of carbon (12 g). The number of glucose molecules, however, is exactly the same as the number of carbon atoms—Avogadro's number. (Photo by Al Green.)

Similarly, $\frac{1}{2}$ (0.5) mole weighs 0.5 mole × 58 g/mole = 29 g, $\frac{1}{4}$ (0.25) mole weighs 0.25 mole × 58 g/mole = 15 g, and so on. In each case, to find the weight needed, you simply multiply the number of moles by the weight of 1 mole:

$$\text{number of grams} = \text{number of moles} \times \text{weight of 1 mole}$$

Example: Find the weight in grams of 0.10 mole of glucose, $C_6H_{12}O_6$.

C: $6 \times 12 = 72$
H: $12 \times 1 = 12$
O: $6 \times 16 = \underline{96}$
                180 amu

1 mole = 180 g
number of grams = (0.10 mole) × 180 g/mole
                = 18 g

4.4   Compounds vs mixtures

A chemical compound consists of atoms joined by chemical bonding. It is very different from just a mixture. When substances are simply mixed, the resulting composition can vary and the components or parts of the mixture can be physically separated. For example, iron and sulfur can be mixed together in any amounts. The iron can be easily removed from the mixture, as shown in Figure 4-3a, using a magnet, leaving only the sulfur behind. In addition, the properties of each part of the mixture remain the same; the iron still acts like iron and the sulfur like sulfur.

If the mixture of iron and sulfur is heated, a change takes place. A *new substance,* a compound, is formed. It looks different from either iron or sulfur and has different properties. A magnet no longer can separate the iron from the sulfur (Figure 4-3b). There is now one kind of substance present instead of two. A chemical bond forms between the iron and sulfur atoms to make iron sulfide. The differences between a compound and mixture are summarized in Table 4-5.

4.5   Chemical reactions

By heating iron with sulfur, you cause a **chemical reaction** to take place. The substances you start with, the **reactants**, undergo a chemical change to new substances, the **products**, having *different arrangements of atoms.* Thus, the compositions of the substances at the end of a chemical reaction are different from those at the beginning. New chemical bonds form, and perhaps old ones have been broken.

(a)   (b)

Figure 4-3.   Iron and sulfur (a) as a mixture and (b) chemically combined as iron sulfide. The magnet can separate the iron from the mixture but not from the compound.
(Photo by Al Green.)

Examples of chemical reactions you are familiar with are the rusting of iron
and the burning of paper or wood. All of the chemical processes in your body
involve reactions, changing certain compounds into other compounds. You
will learn more about these reactions later in the book.

A chemical reaction is summarized by writing a **chemical equation** that describes the change taking place. For example, the equation representing the
formation of iron sulfide from iron and sulfur is written

$$Fe + S \longrightarrow FeS$$

Table 4-5 Comparison of Compounds and Mixtures

Compound	Mixture
one kind of substance	two or more kinds of substances
definite composition	composition varies
individual properties of components lost	individual properties of components kept
components cannot be physically separated	components can be physically separated

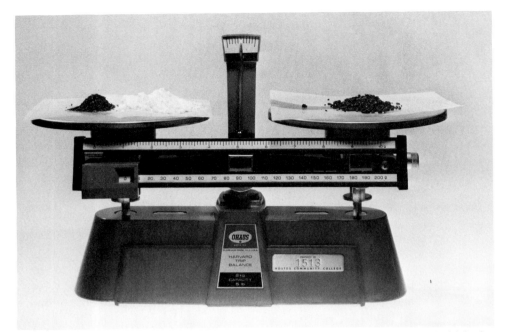

Figure 4-4.   Conservation of mass. The mass of iron and sulfur (left-hand side), the reactants, is equal to the mass of iron sulfide (right-hand side), the product of the reaction. (Photo by Al Green.)

An arrow points from the reactants, iron and sulfur, to the product of the reaction, iron sulfide.

In every chemical reaction, *the total mass of the reactants must equal the total mass of the products.* Thus, the mass of the iron added to the mass of the sulfur that reacts is the same as the mass of the iron sulfide produced, as shown in Figure 4-4. A chemical reaction took place, but matter was neither created nor destroyed. This relationship is known as the **law of conservation of mass.**

4.6   Balancing chemical equations

Because of the law of conservation of mass, *the number of atoms of each element must be the same before and after the reaction.* One kind of atom cannot be changed to another type of atom by a chemical process. The total number of atoms of each kind in the reactants, therefore, is equal to the total number of atoms of each kind in the products. This principle is the basis for "balancing" equations.

The equation Fe + S $\longrightarrow$ FeS is balanced already. There is one iron atom in the reactants and one in the product: the same holds for sulfur. But consider the following equation, which represents the reaction of glucose, $C_6H_{12}O_6$, with oxygen, an important process that takes place in your body:

$$C_6H_{12}O_6 + O_2 \longrightarrow CO_2 + H_2O$$

The reactants are glucose and oxygen; the products of the reaction are carbon dioxide and water. Count the number of atoms of each kind in the reactants and in the products:

Reactants	Products
6 carbon atoms	1 carbon atom
12 hydrogen atoms	2 hydrogen atoms
8 oxygen atoms	3 oxygen atoms

Clearly, the equation is not correct as it is written because the law of conservation of mass is violated. There are different numbers of atoms of carbon, hydrogen, and oxygen on both sides of the equation. This equation is not balanced.

Equations are balanced by multiplying reactants and products by numbers called *coefficients*. There is no definite rule that will tell you what coefficients to place in front of each formula. The method you will use is simply "trial and error." But the following guidelines should help:

1. Pick an atom and find the elements or compounds containing that atom on both sides of the equation. Start with those atoms that appear in the fewest substances. Save hydrogen and oxygen for last.

2. Place whole numbers in front of the reactants and products containing that atom to make the total number the same on both sides.

You cannot change the subscripts for any substance in the reaction. Chemical compounds have a definite composition, so you must not rewrite their formulas.

Try to apply these suggestions to the equation for the combustion of glucose. Start with carbon by placing the coefficient 6 in front of CO_2 on the right side:

$$C_6H_{12}O_6 + O_2 \longrightarrow 6CO_2 + H_2O$$

There are now equal numbers of carbon atoms on each side of the equation: 6 in the glucose molecule and 6 in the six molecules of carbon dioxide. The coefficient 6 multiplies the entire formula next to it. Thus, $6CO_2$ means 6×2, or 12, oxygen atoms as well as 6 carbon atoms. This coefficient does not apply to the H_2O, however. The following situation now holds:

Reactants	Products
6 carbon atoms	6 carbon atoms
12 hydrogen atoms	2 hydrogen atoms
8 oxygen atoms	13 oxygen atoms

The carbons are balanced but the other atoms are not.
Next balance the hydrogens by placing a 6 in front of H_2O:

$$C_6H_{12}O_6 + O_2 \longrightarrow 6CO_2 + 6H_2O$$

Now the number of atoms on each side is as follows:

Reactants	Products
6 carbon atoms	6 carbon atoms
12 hydrogen atoms	12 hydrogen atoms
8 oxygen atoms	18 oxygen atoms

Finally, the oxygens are balanced by placing the coefficient 6 in front of O_2:

$$C_6H_{12}O_6 + 6O_2 \longrightarrow 6CO_2 + 6H_2O$$

The equation is now completely balanced:

Reactants	Products
6 carbon atoms	6 carbon atoms
12 hydrogen atoms	12 hydrogen atoms
18 oxygen atoms	18 oxygen atoms

There are the same number of atoms of each kind in the reactants and products, as shown in Figure 4-5.

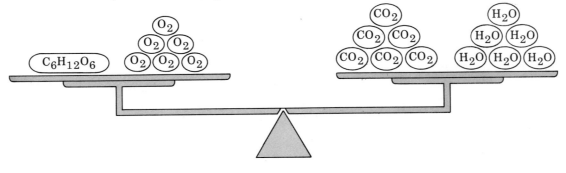

$$C_6H_{12}O_6 + 6O_2 \longrightarrow 6CO_2 + 6H_2O$$

Figure 4-5.   A balanced chemical equation. The number of atoms of each kind is the same on both sides: 6 carbon atoms, 12 hydrogen atoms, and 18 oxygen atoms.

4.7   Interpreting equations

Using the balanced equation, you can interpret this reaction in the following way. *One molecule* of glucose reacts with *six molecules* of oxygen to form *six molecules* of carbon dioxide and *six molecules* of water. You can also read this equation in terms of that huge collection of molecules, the mole. Thus, *1 mole* of glucose molecules reacts with *6 moles* of oxygen molecules to form *6 moles* of carbon dioxide molecules and *6 moles* of water molecules. This interpretation is often most useful because moles of substances are practical amounts that can be easily measured. It is also simple to demonstrate the law of conservation of mass on this basis, as shown in Table 4-6.

Table 4-6   **Demonstration of the Law of Conservation of Mass**

Reactants	Products
$C_6H_{12}O_6 + 6O_2$ $\longrightarrow$	$6CO_2 + 6H_2O$
1 mole $C_6H_{12}O_6$	6 moles CO_2
1 mole × 180 g/mole = 180 g	6 moles × 44 g/mole = 264 g
6 moles O_2	6 moles H_2O
6 moles × 32 g/mole = 192 g	6 moles × 18 g/mole = 108 g
mass of reactants = 372 g =	mass of products = 372 g

More examples of balanced equations interpreted in moles, large numbers of atoms or formula units reacting together, are presented below. You should be able to both read an equation in words and write the equation given the formulas of the reactants and products.

Example: $N_2 + 3H_2 \longrightarrow 2NH_3$
    1 mole of nitrogen molecules reacts with
    3 moles of hydrogen molecules to form
    2 moles of ammonia molecules

Example: $2Na_3PO_4 + 3CaCl_2 \longrightarrow Ca_3(PO_4)_2 + 6NaCl$
    2 moles of sodium phosphate formula units react with
    3 moles of calcium chloride formula units to form
    1 mole of calcium phosphate formula units and
    6 moles of sodium chloride formula units

4.8 Calculations based on equations

Using a balanced chemical equation, you can find out how much product is made from a given amount of reactant or how much reactant is needed to form a certain quantity of product. For example, in the reaction of glucose with oxygen, you may want to know the mass of water produced from 360 g of glucose. To solve this problem, you must think of the equation in terms of moles, using the following steps:

Step 1. Write the balanced chemical equation.

$$C_6H_{12}O_6 + 6O_2 \longrightarrow 6CO_2 + 6H_2O$$

Step 2. Convert the given amount (360 g of glucose) into moles. To convert from grams to moles, divide by the weight of 1 mole.

$$\frac{360 \text{ g}}{180 \text{ g/mole}} = 2 \text{ moles glucose}$$

Step 3. Using the coefficients of the equation, relate the moles of the given substance (2 moles of glucose) to the moles of the substance you must find (water). You can see that 6 moles of water are produced from 1 mole of glucose. Since you have 360 g or 2 moles of glucose, 2×6 moles, or 12 moles, of water must be formed.

Step 4. Convert the found number of moles (12 moles of water) into grams by multiplying by the weight of 1 mole.

$$12 \text{ moles H}_2\text{O} \times 18 \text{ g/mole} = 216 \text{ g H}_2\text{O}$$

An alternate approach is to use the unit-factor method throughout (see Appendix A.4):

$$\text{grams water} = 360 \text{ g glucose} \times \frac{1 \text{ mole glucose}}{180 \text{ g glucose}}$$

$$\times \frac{6 \text{ moles water}}{1 \text{ mole glucose}} \times \frac{18 \text{ g water}}{1 \text{ mole water}} = 216 \text{ g}$$

You now know that 360 g of glucose will form 216 g of water in this reaction. By using this same procedure, you can relate any two substances in a chemical equation.

4.9 Types of reactions

Chemical changes can be classified according to the kind of process taking place. There are four major reaction types.

A **combination** or **synthesis** reaction consists of the formation of one product from two or more reactants. An example is the production of rust (iron oxide) from iron and oxygen; here two elements combine:

$$4Fe + 3O_2 \longrightarrow 2Fe_2O_3$$

Compounds can also combine, as shown in Figure 4-6a, by the reaction of ammonia with hydrogen chloride to make ammonium chloride:

$$NH_3 + HCl \longrightarrow NH_4Cl$$

A second type of reaction is **decomposition**: One substance breaks down into simpler substances. Oxygen can be produced in the laboratory by decomposing potassium chlorate through heating:

$$2KClO_3 \xrightarrow[\text{catalyst}]{\text{heat}} 2KCl + 3O_2$$

Water can be decomposed by applying an electric current to it (a process called electrolysis), as illustrated in Figure 4-6b:

$$2H_2O \xrightarrow[\text{current}]{\text{electric}} 2H_2 + O_2$$

In a **replacement** or **substitution** reaction, one element takes the place of another element in a compound. Thus, iron can replace the hydrogen of hydrogen chloride to form hydrogen gas and iron (ferrous) chloride (Figure 4-6c):

$$Fe + 2HCl \longrightarrow FeCl_2 + H_2$$

Or chlorine can substitute for bromine to make sodium chloride from sodium bromide:

$$2NaBr + Cl_2 \longrightarrow 2NaCl + Br_2$$

The fourth reaction type, **double replacement** or **metathesis**, involves the reaction of two compounds to form two new compounds in which atoms have been exchanged. For example, as shown in Figure 4-6d, sodium sulfate reacts with barium chloride in water to form a white product, barium sulfate, as well as sodium chloride:

$$Na_2SO_4 + BaCl_2 \longrightarrow BaSO_4 + 2NaCl$$

Notice the change: The sulfate ion, which came from sodium sulfate, is now bonded to barium ion, which came from the barium chloride. Another example is the reaction in water of hydrogen phosphate (phosphoric acid) with sodium hydroxide:

$$H_3PO_4 + 3NaOH \longrightarrow Na_3PO_4 + 3H_2O$$

Again, notice that both compounds have switched "partners" (think of H_2O as HOH).

Figure 4-6. Examples of chemical reactions. (a) Synthesis: $NH_3 + HCl \longrightarrow NH_4Cl$; (b) decomposition: $2H_2O \longrightarrow 2H_2 + O_2$ (electrolysis)

(c)   (d)

Figure 4-6.   (c) Replacement or substitution: Fe + 2HCl ⟶ FeCl$_2$ + H$_2$; (d) double replacement or metathesis: BaCl$_2$ + Na$_2$SO$_4$ ⟶ BaSO$_4$ + 2NaCl. (Photos by Al Green.)

4.10 Oxidation–reduction reactions

Many reactions, including the types just mentioned, are based on a *transfer of electrons*. One reactant loses electrons, the process of oxidation, and another gains those electrons, the process of reduction. These reactions are known as **oxidation–reduction** reactions, sometimes called "redox" for short. The formation of an ionic compound, such as sodium chloride from the elements sodium and chlorine, is an example of such an oxidation–reduction reaction:

$$2Na \quad + \quad Cl_2 \quad \longrightarrow \quad 2NaCl$$

$$\text{reducing} \qquad \text{oxidizing}$$
$$\text{agent} \qquad \text{agent}$$

The processes of oxidation and reduction can be written separately as "half-reactions"; in this form, you can see more clearly the loss and gain of electrons.

$$2Na \longrightarrow 2Na^+ + 2e^- \qquad \text{oxidation}$$
$$Cl_2 + 2e^- \longrightarrow 2Cl^- \qquad \text{reduction}$$

The substance losing electrons, sodium in this case, is being oxidized. It is called a **reducing agent** because it provides electrons to another substance, causing reduction. Other reducing agents include hydrogen, H_2, which is used to reduce compounds containing carbon, and sodium thiosulfate, $Na_2S_2O_3$, which is used in photography.

At the same time that one substance is losing electrons, another substance must be gaining them. In all oxidation–reduction reactions, *the number of electrons lost must be the same as the number of electrons gained*. In the sodium chloride reaction, chlorine is accepting the electrons from the sodium, filling up its valence shell. It is thus being reduced. Chlorine is called an **oxidizing agent** because it oxidizes the sodium by taking electrons from it. Oxidizing agents are used medically as antiseptics; they include hydrogen peroxide (H_2O_2), potassium permanganate ($KMnO_4$), iodine (I_2), and sodium hypochlorite ($NaOCl$). One of the best oxidizing agents is oxygen, O_2; in fact, it is the basis for the term "oxidation." You will learn in later chapters how the oxygen you breathe is vital to the breakdown of food and other oxidation processes in the body.

4.11 Energy and chemical reactions

Chemical changes involve energy changes. **Energy**, the ability to do work, can take different forms. Potential energy is the energy a substance has because of its position, such as possessed by a wound-up watch spring or water behind a dam. When the energy is released, it becomes kinetic energy, the energy of a moving object. **Chemical energy** is a form of potential energy that is stored in elements and compounds. It is released or absorbed in a chemical reaction as heat, electrical, or light energy.

For example, some reactions produce light, such as those that take place in a firefly and the burning of a flashbulb. Other reactions, the most important being photosynthesis in plants, absorb light energy. Chemical reactions that create an electric current are used in "dry cells" and car batteries. The breakdown of water molecules by electrolysis is a reaction that requires electrical energy.

Most chemical reactions involve changes in heat energy. Breaking a chemical bond generally requires heat, while forming a bond releases heat energy. If heat is given off during the reaction, it is called **exothermic**. On the other hand, a reaction that absorbs heat is **endothermic**. The change in heat caused by a chemical reaction, the heat of reaction (enthalpy), is measured using a **calorimeter** (Figure 4-7). The reaction takes place in the inner chamber and the heat given off or absorbed changes the temperature of the water surrounding the chamber.

In all cases involving changing energy from one form to another during a chemical reaction, *energy can be neither created or destroyed.* The total amount is the same at the beginning and the end of the reaction. This principle is known as the **law of conservation of energy**.

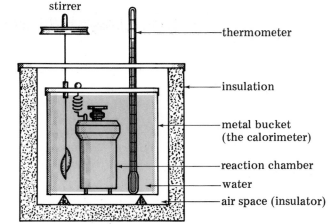

Figure 4-7. A calorimeter. The heat change inside the reaction chamber is measured by a change in temperature of the surrounding water.

4.12 Reversibility of reactions

Most chemical reactions take place by the collision of atoms or molecules. When they hit each other with enough energy, a new chemical bond forms, as shown in Figure 4-8 for the reaction of hydrogen and iodine molecules. But once the amount of hydrogen iodide builds up, the reverse reaction becomes possible. Two hydrogen iodide molecules can collide to form a hydrogen molecule and iodine molecule again. The equation representing this *reversible* reaction is written in the following way:

$$H_2 + I_2 \rightleftharpoons 2HI$$

The double arrow shows that both a forward reaction (forming HI) and a reverse reaction (forming H_2 and I_2) take place.

As the reaction goes on, a point is reached when the rate of formation of hydrogen iodide is the same as the rate of its reconversion to hydrogen and iodine. The *two opposing reactions take place at the same speed*, resulting in a state of balance called **equilibrium** (Figure 4-9). The quantities of H_2, I_2, and HI no longer change even though both the forward and reverse reactions continue. If you started the reaction with 1 mole of H_2 and 1 mole of I_2, you would find 0.2 mole of H_2, 0.2 mole of I_2, and 1.6 moles of HI at equilibrium (at 450°C).

Nearly all the reactions in your body are reversible, like this one. How "far" the reaction goes in the forward direction, the amount of product formed, varies from one reaction to another. Usually, if a compound is formed having stronger chemical bonds, the reverse reaction becomes less likely. In general, *the forward reaction is strongly favored if it is exothermic (heat is given off) and simpler molecules form.*

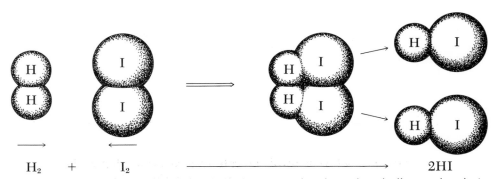

Figure 4-8. The reaction between a hydrogen molecule and an iodine molecule to form two molecules of hydrogen iodide (simplified).

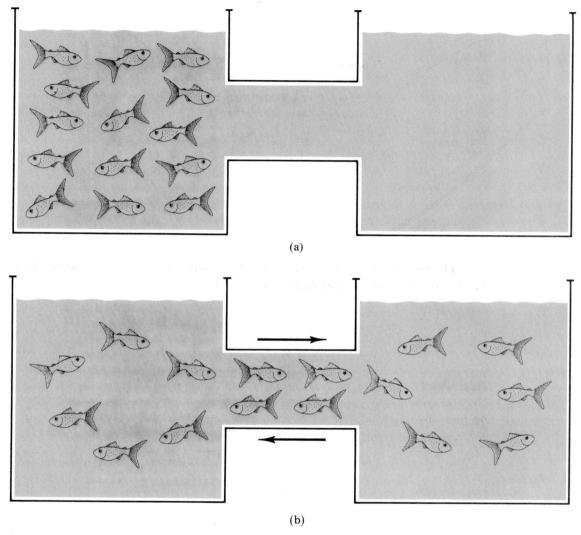

(a)

(b)

Figure 4-9. An example of equilibrium. (a) Before equilibrium; (b) after equilibrium is established. The rate of fish moving to the left is exactly equal to the rate of fish moving to the right at equilibrium. The number of fish on each side need not be the same, however.

4.13   Rate of a reaction

Several factors control the **rate** of a reaction, that is, how fast it takes place. The most important is the *nature of the reactants* themselves. Some substances react very rapidly, such as an antacid tablet and water. Others react

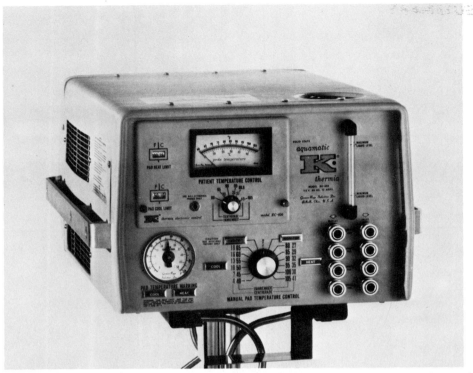

Figure 4-10. Changing the body temperature. The unit shown can raise or lower the body temperature, increasing or decreasing the rate of its chemical reactions. (Photo courtesy of American Hospital Supply.)

very slowly if at all. The second factor is *concentration*, how much of each reactant is present: The greater the concentration, the faster the reaction. Third, *particle size* can influence the rate. Compared to a solid lump, coal dust reacts with oxygen much faster (and sometimes explosively) because of its greater exposed surface area. Finally, the *temperature* has a large effect on reaction rate. A rule of thumb is that the rate approximately doubles for every rise by 10°C. The reactions that take place in the body speed up when a patient has a fever and slow down when the temperature is lowered (hypothermia) (Figure 4-10).

It is possible to speed up a reaction by using a **catalyst.** This substance increases the reaction rate but can be recovered unchanged at the end of the reaction. Life would not be possible without the biological catalysts, or **enzymes,** which are involved in almost every chemical reaction within every living organism. You will learn in Chapter 16 how enzymes speed up reactions.

SUMMARY

The molecular or formula weight of a compound is the sum of the atomic weights of all the atoms (or ions) present in the molecule or formula unit. It is found by multiplying the atomic weight for each element by the number of atoms of that element in the formula and then adding these products.

Every compound has a definite composition. Its percentage composition is the percentage of each element by weight. The percentage composition is determined by dividing the weight of all the atoms of each element by the formula or molecular weight and multiplying by 100.

A mole of a chemical substance consists of a very large number of the particles that make it up, whether atoms, ions, or molecules. This quantity (6.02×10^{23}) is known as Avogadro's number. The weight of 1 mole is equal to the atomic, formula, or molecular weight written in grams.

In contrast to a compound, a mixture has a variable composition. Its parts or components can be physically separated. The properties of each part of a mixture are unchanged after mixing takes place.

A chemical reaction is the formation of new chemical substances, the products, from the starting materials, called the reactants. Bonds are broken and new ones are formed, resulting in new arrangements of atoms. A chemical equation summarizes the changes that occur in a reaction. Chemical reactions follow the law of conservation of mass—the total mass of the reactants must exactly equal the total mass of the products.

The number of atoms of each element is the same before and after the reaction takes place. This principle is the basis for "balancing equations." Equations are balanced by placing numbers called coefficients in front of reactants and products to make the numbers of atoms of each type equal on both sides. Using the balanced equation, you can interpret the reaction in terms of atoms, molecules, and formula units or moles. In addition, you can calculate how much product is made from a given amount of reactant or how much reactant is needed to make a certain amount of product.

Chemical changes can be classified according to the kind of process taking place. A combination or synthesis reaction consists of the formation of one product from two or more reactants. In a decomposition reaction, one substance breaks down into simpler substances. In a replacement or substitution reaction, one element takes the place of another element in a compound. Double replacement or metathesis involves the reaction of two compounds to form the new compounds.

Reactions involving a transfer of electrons are known as oxidation–reduction (redox) reactions. The substance giving up electrons, and therefore being oxidized, is called the reducing agent. The substance that gains the electrons, becoming reduced, is the oxidizing agent.

Chemical changes involve energy changes. Energy, the ability to do work, can take different forms. Chemical energy is a form of potential energy that is stored in elements and compounds. Most chemical reactions involve changes in heat energy—it is either given off (exothermic) or absorbed (endothermic). In all cases, the law of conservation of energy is obeyed—energy can be neither created or destroyed in a chemical reaction.

Most chemical reactions occur through the collision of atoms or molecules. Reactions are often reversible—the products can collide to form the starting materials again. If the forward and reverse reactions take place at the same speed, a state of balance called equilibrium results. In general, the equilibrium favors the forward reaction if it is exothermic and simpler molecules form.

The rate of a reaction is controlled by several factors. These include the nature of the reactants, their concentration, the particle size, and the temperature. A catalyst generally speeds up a reaction and can be recovered unchanged at the end of the reaction.

Exercises

1. (4.1) Find the molecular or formula weight of the following compounds: (a) H_2SO_4; (b) $AgNO_3$; (c) $Mg(OH)_2$; (d) $KHCO_3$; (e) C_6H_6.

2. (4.2) Determine the percentage composition of the compounds in the previous problem.

3. (4.3) What is a mole? Why is it useful?

4. (4.3) Which weighs more, a mole of carbon or a mole of sulfur? Which contains more atoms?

5. (4.3) Find the weight of 1 mole of each of the compounds in problem 1.

6. (4.3) Determine the weight of (a) 2 moles of $NaC_2H_3O_2$; (b) 0.5 mole of H_3PO_4; (c) 10 moles of C_3H_8; (d) 0.1 mole of H_2CO_3.

7. (4.4) Describe at least three ways in which a mixture differs from a compound. Give an example of a mixture and a compound.

8. (4.4) Identify the following as an element, compound, or mixture: (a) ice; (b) milk; (c) silver; (d) carbon monoxide; (e) salt; (f) saline solution (NaCl in water).

9. (4.5) Define chemical reaction, chemical equation.

10. (4.5) Give three examples of chemical reactions you are familiar with.

11. (4.5) Describe the law of conservation of mass.

12. (4.6) State whether each of the following equations is balanced:

    (a) $AgNO_3 + H_2SO_4 \longrightarrow HNO_3 + Ag_2SO_4$
    (b) $H_2O + CO_2 \longrightarrow H_2CO_3$
    (c) $H_3PO_4 + NaOH \longrightarrow Na_3PO_4 + H_2O$
    (d) $C_3H_8 + O_2 \longrightarrow 3CO_2 + H_2O$
    (e) $C_2H_5Cl + NaPb \longrightarrow (C_2H_5)_4Pb + NaCl + Pb$

13. (4.6) Balance each of the above equations if necessary.

14. (4.7) Interpret the following equations first in terms of atoms and molecules (or formula units) and then in terms of moles:

(a) $5C + 2SO_2 \longrightarrow CS_2 + 4CO$

(b) $C_6H_{12}O_6$ (glucose) $\longrightarrow 2C_2H_5OH$ (ethyl alcohol) $+ 2CO_2$

15. (4.8) Using equation 14b above, which represents the fermentation of glucose, how much ethyl alcohol is formed from 90 g of glucose?

16. (4.8) Using equation 14b, what weight of glucose is needed to make 460 g of ethyl alcohol by this process?

17. (4.9) Identify each of the following reactions as combination (synthesis), decomposition, replacement (substitution), or double replacement (methathesis). Explain your answers.

(a) $Cu(NO_3)_2 \longrightarrow 2CuO + 4NO_2 + O_2$

(b) $Al_2(SO_4)_3 + 3Pb(NO_3)_2 \longrightarrow 3PbSO_4 + 2Al(NO_3)_3$

(c) $SO_3 + H_2O \longrightarrow H_2SO_4$

(d) $2Al + 3H_2SO_4 \longrightarrow Al_2(SO_4)_3 + 3H_2$

18. (4.10) What is an oxidation–reduction reaction? How is this process used medically?

19. (4.10) In the following reaction, pick out the oxidizing agent and the reducing agent: $2Al + 3Cl_2 \longrightarrow 2AlCl_3$. Explain.

20. (4.11) How does an exothermic reaction differ from an endothermic reaction?

21. (4.11) What is the law of conservation of energy?

22. (4.12) Carbon dioxide reacts with water in a reversible reaction to form carbonic acid, H_2CO_3, in your blood: $H_2O + CO_2 \rightleftharpoons H_2CO_3$. What are the forward and reverse reactions? What happens at equilibrium?

23. (4.13) What factors determine the rate of a chemical reaction?

24. (4.13) What is the effect of cooling on the reactions inside your body?

25. (4.13) Why are enzymes, the biological catalysts, so important?

Gases and respiration

Gases play a vital role in health. As you read this sentence, gas molecules are passing in and out of your body. In addition to this normal process of breathing, many medical treatments, such as oxygen therapy, depend on gases. As you know, a gas is the state of matter that has neither a definite shape nor a definite volume. It will expand or spread out without limit and can be compressed or squeezed into a small volume. To explain these and other properties, scientists have developed a model of a gas at the molecular level.

5.1 Kinetic molecular theory

The word "kinetic" refers to motion; the **kinetic molecular theory** describes the motion of the molecules of a gas. It is based on several assumptions, statements taken to be correct without proof.

   1. The molecules of a gas are small and far apart from each other; a gas consists mainly of empty space.
   2. The molecules of a gas are constantly moving in an unpredictable or random way.
   3. Gas molecules are moving at very high speeds, colliding with each other and the walls of their container.
   4. Gas molecules do not attract or repel each other. When they collide, no energy is lost. (These are called "elastic" collisions.)

A "picture" of a gas based on this model, an "ideal" gas, is shown in Figure 5-1.

These assumptions should seem reasonable to you. They agree with what you already know about the properties of a gas. For example, a gas quickly expands because the molecules are in rapid motion; they will keep moving if their container is opened. A gas can be compressed because there is much space between the molecules. Since there is no attraction between the mole-

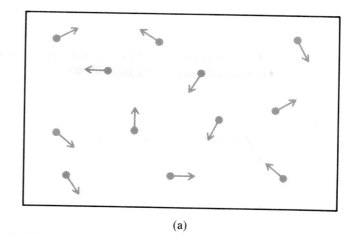

(a)

Figure 5-1. A gas as described by the kinetic molecular theory. (a) Gas molecules in a container. (b) The path taken by a single molecule resulting from random collisions.

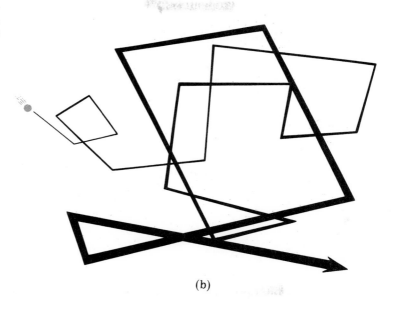

(b)

cules, a gas does not have a definite shape or volume. (Actually, most "real" gas molecules do attract each other, but only very slightly.)

To give you some figures, in oxygen gas at room temperature, over 99.9% of the volume is empty. A molecule must travel over 1000 times its own diameter before hitting another molecule. The average speed of the gas molecules is over 1000 miles/hour.

5.2 Diffusion

An important result of the kinetic theory concerns the "spreading out" or **diffusion** of a gas to fill any volume. This process always occurs when a gas is released because of the motion of its molecules. You are familiar with the effect: Gas molecules escaping from a bottle of perfume or ammonia can soon be smelled throughout the room.

Diffusion takes place from a region of high concentration, where there are many molecules, to a region of low concentration, where there are few. After a while, the molecules will become evenly distributed; the concentration will be the same all over. As you will soon see, diffusion of gases is very important in the body.

How fast a gas diffuses depends on the mass of its molecules. The smaller their mass, the faster diffusion takes place. (The mathematical relationship, known as Graham's law, is that the rate of diffusion is inversely related to the square root of the molecular weight of a gas.) You can compare the rates of diffusion of two gases if you know their molecular weights. For example, oxygen, O_2, with a molecular weight of 32 amu, diffuses faster than heavier carbon dioxide, CO_2, which has a molecular weight of 44 amu (by a factor of $\sqrt{44}/\sqrt{32}$ or 1.2).

An application of this relationship is the administration of an anesthetic gas to a patient whose airway is partially blocked. Normally, the anesthetic is mixed with air, which itself is a mixture of about four-fifths nitrogen and one-fifth oxygen. For these patients, helium is substituted for the nitrogen in the "air" they breathe. This new mixture diffuses faster (by a factor of $\sqrt{28}/\sqrt{4}$ or 2.6) because the mass of helium (4 amu) is much less than the mass of nitrogen gas (28 amu).

5.3 Temperature

The **temperature** of a gas is a measure of the motion of its molecules. Moving objects have a kinetic energy that depends on their mass and the square of their speed (kinetic energy $= \frac{1}{2}mv^2$, where m is mass and v is velocity). Temperature is directly related to this quantity. Thus, the faster gas molecules are moving, the higher the temperature of the gas.

By heating a gas, you give the molecules more energy and therefore increase the temperature. On the other hand, cooling a gas removes heat energy; the molecules move more slowly and the temperature of the gas is lower. The temperature also depends on the mass; if two different gases both have molecules with the same speeds, the heavier gas will have a higher temperature. Its molecules must have more energy to move around at the same speed as the lighter molecules.

When dealing with the temperature of a gas, you must use the Kelvin or absolute scale. The size of a Kelvin degree is the same as a Celsius (centigrade) degree, but the starting point is different. Zero degrees Kelvin (0 K) corresponds to the lowest possible temperature, "absolute zero." The kinetic energy of gas molecules is proportional to the Kelvin temperature. Thus, at absolute zero, the kinetic energy is zero, which essentially means that molecular motion has stopped.

5.4 Pressure

Because gas molecules hit the walls of their container, they "push" against them. This force applied to a certain area is known as **pressure**. Its symbol is P.

$$\text{pressure} = P = \frac{\text{force}}{\text{area}}$$

For example, a force or weight of 200 pounds presses against a square 10 inches long by 10 inches wide; its area is 10 inches × 10 inches or 100 square inches (in.²). The pressure is equal to 200 pounds divided by 100 square inches or 2 pounds/square inch, abbreviated 2 psi. Thus, each square inch "feels" a force of 2 pounds.

Surrounding the earth is the atmosphere. It contains roughly 5 billion billion tons of air pressing down on us, creating atmospheric pressure. You can measure this pressure using a **barometer.** One form of this instrument, the Torricelli barometer, is made by filling up with mercury a long glass tube which is sealed at one end (Figure 5-2). You can close off the open end with your finger and place the tube upside down in a dish of mercury, placing the open end below the surface. After taking away your finger, you might expect all the mercury to flow out. But, in fact, the atmosphere pushing down on the dish of mercury supports a column of mercury in the tube about 760 mm (76 cm) high.

The pressure of the atmosphere changes with the altitude; it is less higher up because there is less air above you. At sea level and 0°C, the atmospheric pressure exactly holds up a 760-mm column of mercury, abbreviated 760 mm Hg. It is also termed 760 torr (after Torricelli), with 1 torr defined as the pressure that supports 1 mm of mercury. These values correspond to what is called 1 atmosphere of pressure, abbreviated atm. *Thus, 760 mm Hg, 760 torr, and 1 atm all mean the same thing.* In terms of pounds per square inch, the atmospheric pressure is 14.7 psi. The SI unit of pressure is the pascal, Pa;

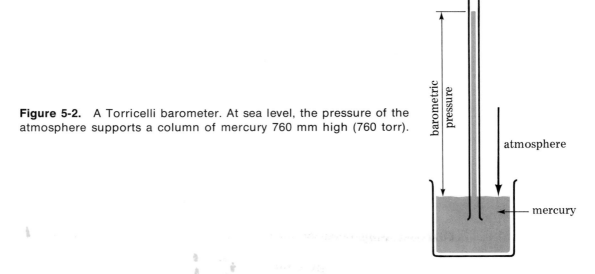

Figure 5-2. A Torricelli barometer. At sea level, the pressure of the atmosphere supports a column of mercury 760 mm high (760 torr).

1 torr = 133.3 Pa = 0.1333 kPa. In terms of this unit, the pressure of the atmosphere is 101.3 kPa.

5.5 The gas laws: pressure and volume

The quantities you use to describe a gas are its pressure (P), volume (V), temperature (T), and amount (in moles). These properties are all connected; if you change any one, you automatically change at least another one. Since you may be working with gases, it is important to understand the relationships, known as the gas laws, between these quantities. It is easiest to look at only two properties at a time, seeing how they affect each other.

To study the effect of pressure on volume, you must keep the temperature and amount of gas fixed. Then, when you increase the pressure on a gas (the external pressure), its volume decreases, as shown in Figure 5-3. By applying a force or weight to the gas, you can squeeze it into a smaller volume, since it has so much empty space. On the other hand, by lowering the pressure on a gas, you allow it to expand and its volume goes up. Thus, *the pressure and volume are inversely related:* If one goes up, the other goes down (at constant temperature).

Similarly, decreasing the volume increases the pressure of the gas molecules because they are hitting a smaller area with the same force. If you increase the volume, the pressure goes down since the effect of the molecules is

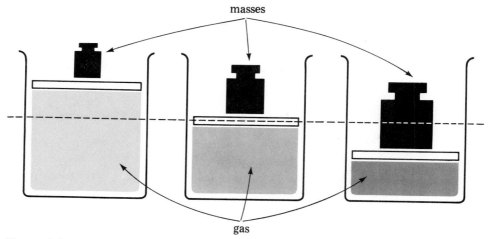

masses

gas

Figure 5-3. The inverse relation between pressure and volume (Boyle's law). As the pressure (resulting from the mass) increases, the volume of the gas decreases at constant temperature.

spread out. Either way you look at the pressure and volume, you see that whatever happens to one, the opposite or inverse happens to the other.

This relationship (known as Boyle's law) can be expressed by saying that, when a change is made in either the pressure or volume at constant temperature, their product, $P \times V$, does not change. Thus, if the pressure is doubled, the volume becomes one-half its original value. Or, if the volume is tripled, the pressure is reduced to one-third of what it was before the change. Mathematically, the relationship is

$$P \text{ (initial)} \times V \text{ (initial)} = P \text{ (final)} \times V \text{ (final)}$$

Thus, if a gas initially has a pressure of 1 atm and a volume of 2 liters, and its volume is decreased to 1 liter, the pressure goes up to 2 atm.

5.6 Breathing

The process of breathing illustrates the pressure–volume relationship. As shown in Figure 5-4, when you breathe in, or inhale (inspiration), your diaphragm moves downward and your rib cage expands, increasing the volume of the thoracic cavity, where the lungs are located. This increase in volume decreases the pressure in your chest to about 3 torr below the normal atmospheric pressure, creating a "negative" pressure. Because the air has a greater pressure outside your body, it is pushed into the lungs. *A gas always moves from a region of higher pressure to one of lower pressure*.

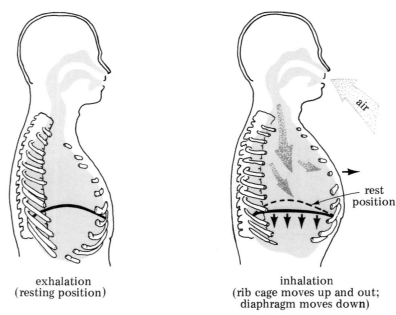

exhalation
(resting position)

inhalation
(rib cage moves up and out;
diaphragm moves down)

Figure 5-4. The process of breathing. Movement of the rib cage and diaphragm changes the volume of the chest cavity and therefore the pressure inside. If the pressure inside decreases, the atmospheric air has a greater pressure and flows into the lungs.

As you breathe out, or exhale (expiration), your thoracic cavity returns to its normal size. This decrease in volume increases the pressure of the air inside the lungs by about 3 torr, creating a "positive" pressure. The air inside thus has a pressure greater than the atmosphere and it is forced out. Normally, you complete this process, inhaling and exhaling about 500 ml (the "tidal volume"), about 12 times each minute.

The tank respirator or "iron lung" is shown in Figure 5-5. It is now used only rarely, but it illustrates the application of the pressure–volume relationship to artificial respiration. The patient's body is placed in a sealed chamber leaving the head exposed. A bellows (the accordionlike object at the bottom of the chamber) moves in and out, changing the volume of air inside. As the bellows moves out, the volume increases, causing the pressure to decrease in the chamber and air to flow into the patient's lungs. Often a slight negative pressure is created to aid this inspiration step. As the bellows is pushed in, the volume of air decreases and its pressure rises, forcing air back out of the patient's lungs.

Figure 5-5. The tank respirator or "iron lung." Notice the bellows at the bottom right of the chamber; it changes the air volume inside and therefore the pressure. (Photo courtesy of National Foundation—March of Dimes.)

5.7 The gas laws: pressure and temperature

The pressure of a gas depends on its temperature as well as its volume. When the volume of a certain amount of gas is kept fixed, the pressure is directly proportional to the Kelvin temperature, as shown in Figure 5-6. If the temperature goes up, the pressure also increases. The molecules have more energy and therefore apply a greater force to the walls of their container. If the temperature is lowered, the molecules must have less energy, so the pressure decreases. At constant volume, *pressure and temperature change in the same direction*—both go up or both go down, unlike pressure and volume. This relationship can be stated mathematically as a proportion (see Appendix A.3):

$$\frac{P \text{ (initial)}}{P \text{ (final)}} = \frac{T \text{ (initial, Kelvin)}}{T \text{ (final, Kelvin)}}$$

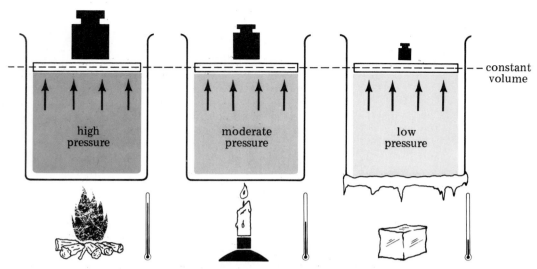

Figure 5-6. The direct relation between pressure and temperature (Kelvin). Decreasing the temperature decreases the pressure of the gas at constant volume.

If a gas with a pressure of 2 atm is lowered in temperature from 400 K (127°C) to 200 K (−73°C), the pressure is reduced to 1 atm.

The direct relationship between temperature and pressure is used in the autoclave (Figure 5-7), which sterilizes medical equipment and supplies. This instrument, having a fixed volume, is filled with steam at pressures above normal atmospheric pressure (up to about 15.5 psi). The temperature of steam is 100°C at 1 atm, but at these greater pressures, it rises to 120°C. Microorganisms are more efficiently destroyed at this higher temperature.

Because of the increase in pressure with temperature, spray cans can explode if overheated. Most, therefore, contain a warning not to incinerate. For the same reason, gas tanks should never be stored in areas where the temperature is high (120°F or higher).

5.8 The gas laws: volume and temperature

Just as the pressure of a gas depends on its Kelvin temperature, so does the volume, as illustrated in Figure 5-8. When a certain amount of gas at a given pressure is heated, the molecules are given more energy and the gas expands. If the gas is cooled, its volume decreases. In other words, *there is a direct relationship between the volume and Kelvin temperature* at constant pressure. These two properties always change together in the same direction when the pressure has a fixed value.

Figure 5-7. The autoclave. The high temperature of steam under pressure kills microorganisms. (WHO photo by G. Pistol.)

You can study this relationship by filling a balloon with air and noting its larger size when heated and smaller size when cooled. Mathematically, it is represented in the following way:

$$\frac{V \text{ (initial)}}{V \text{ (final)}} = \frac{T \text{ (initial, Kelvin)}}{T \text{ (final, Kelvin)}}$$

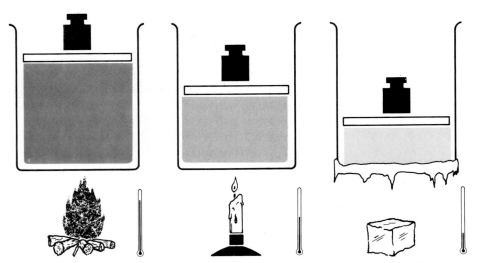

Figure 5-8. The direct relation between volume and temperature (Kelvin). Decreasing the temperature decreases the volume at constant pressure.

A gas occupying 1 liter at 250 K (−23°C) has a volume of 2 liters at 500 K (227°C).

5.9 The gas laws: the quantity of gas

All of the relationships discussed so far hold when the amount of gas is kept constant. But the pressure, volume, and temperature also depend on the number of moles of gas present. For example, increasing the number of molecules, and therefore the number of moles, increases the pressure (at constant volume and temperature). Thus, when blood pressure is measured with a sphygmomanometer, as shown in Figure 5-9, the pressure of the cuff builds up as more molecules are pumped into it. The same principle applies to filling a tire with air.

Increasing the amount of gas increases the volume when pressure and temperature are constant. Thus, a balloon expands when filled with air. Under certain "standard conditions," 1 mole of a gas has a definite volume. It is useful to define such conditions so that the properties of gases can be compared. **Standard temperature and pressure**, abbreviated STP, is 273 K (0°C) and 1 atm (760 torr). At STP, the volume of 1 mole of any gas is the same, 22.4 liters. This fact is made clear when you understand that equal volumes of all gases (at the same temperature and pressure) contain the same number of molecules.

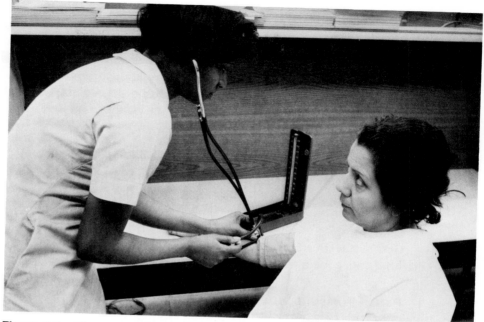

Figure 5-9. The use of a sphygmomanometer to measure blood pressure. Pumping the rubber bulb increases the air pressure in the cuff. (Photo by Al Green.)

5.10 Air and partial pressures

Air is a mixture of gases, as shown in Table 5-1. Most of it is nitrogen, N_2, a gas your body does not use. The important component is oxygen, O_2. Although it makes up only about one-fifth of the air, our form of life could not exist without it. Oxygen is colorless and odorless like nitrogen but is denser; therefore, the density of oxygen (1.43 g/liter) is greater than air (1.30 g/liter).

Table 5-1 Composition of Dry Air

Gas	Percentage (by volume)
nitrogen (N_2)	78.084
oxygen (O_2)	20.946
argon (Ar)	0.934
carbon dioxide (CO_2)	0.033
trace gases[a]	0.003

[a] Neon, helium, methane, krypton, hydrogen, nitrous oxide, xenon.

In a mixture of gases, the properties of each gas, such as its pressure, are not affected by the other gases present. The pressure of each gas is called its **partial pressure** (symbolized p); it contributes only a part of the total pressure of the mixture. *The total pressure is just the combination or sum of the partial pressures of all the gases present.* (This relationship is known as Dalton's law of partial pressures.) Partial pressures for the gases of the atmosphere are presented in Table 5-2. Notice that the sum of the pressures equals the atmospheric pressure.

$$P_{atm} = p_{N_2} + p_{O_2} + p_{Ar} + p_{CO_2} + p_{trace\ gases}$$

The greater the concentration of a gas in a mixture, the greater its partial pressure. In other words, the more molecules of the gas present, the more pressure exerted by that gas. Thus, most of the atmospheric pressure results from nitrogen, as seen in Table 5-2. This principle is used in the administration of anesthesia. To increase the partial pressure of anesthetic gas in the lungs, its concentration in the inhaled air is increased.

Table 5-2   Partial Pressures of the Gases in the Atmosphere

Gas	Partial pressure (atm)	Partial pressure (torr)	Partial pressure (kPa)
nitrogen	0.78084	593.44	79.119
oxygen	0.20946	159.19	21.224
argon	0.00934	7.09	0.945
carbon dioxide	0.00033	0.25	0.033
trace gases	0.00003	0.03	0.004
total pressure	1.00000	760.00	101.325

The intermittent partial pressure breathing apparatus (IPPB unit) is shown in Figure 5-10a. The machine, connected to a face mask or mouthpiece, fills the lungs by increasing the partial pressures of the gases being inhaled by the patient. It is used in cases when a patient cannot breathe normally without assistance. The critical care ventilator, shown in Fig. 5-10b, works on similar principles.

5.11   Respiration

The exchange of gases between your body and the air is part of **respiration.** In this process, oxygen is delivered to the tissues of the body, and carbon dioxide, a waste product, is removed. *Respiration depends on differences in partial pressures.* A gas always diffuses from where there are more molecules,

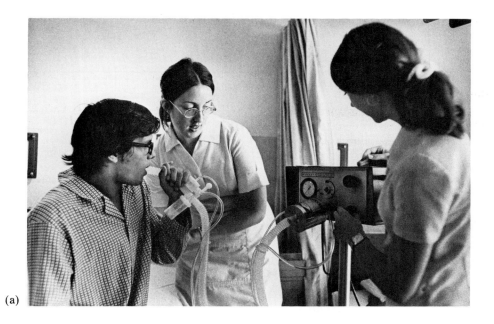

(a)

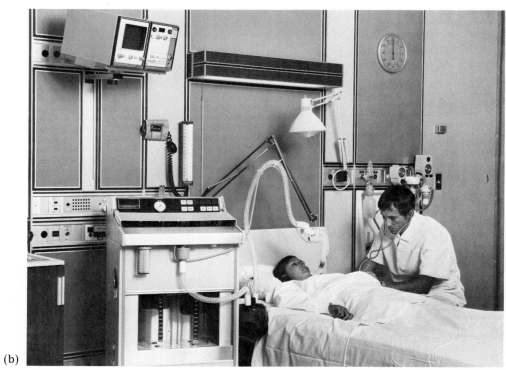

(b)

Figure 5-10. Mechanical ventilators. (a) Intermittent partial pressure breathing (IPPB) unit; (b) critical care ventilator. [Photo (a) courtesy of American Cancer Society; (b) courtesy of Ohio Medical Products (Division of Airco, Inc.), 3030 Airco Drive, P. O. Box 1319, Madison, Wisconsin 53701.]

Table 5-3   Partial Pressures of the Respiratory Gases

Gas	Partial pressure of inspired air (torr)	Partial pressure of alveolar air (torr)	Partial pressure of expired air (torr)
O_2	158.2	101.2	116.2
CO_2	0.3	40.0	28.5
N_2	596.5	571.8	568.3
H_2O	5.0	47.0	47.0

a region of high partial pressure, to where there are fewer, a region of low partial pressure.

The first exchange takes place in the lungs. Table 5-3 compares the partial pressures of inhaled air, exhaled air, and the alveolar air. (The alveoli are the air cells of the lungs.) As you would expect, the expired air has a lower partial pressure of oxygen, p_{O_2}, and a higher partial pressure of carbon dioxide, p_{CO_2}, than the inspired air. Oxygen flows toward the alveoli, since oxygen partial pressure is lower there than in the air breathed in. Carbon dioxide is exhaled because its partial pressure is greater in the alveoli.

Blood circulating through the capillaries of the lungs has a low p_{O_2} (40 torr) and therefore picks up oxygen from the alveolar air. You will learn about the oxygen carrier of the blood, hemoglobin, in later chapters. Carbon dioxide has a greater partial pressure in the blood returning to the lungs (46 torr) and therefore flows into the alveoli to be exhaled.

As shown in Figure 5-11, the arteries carry oxygen from the lungs to the

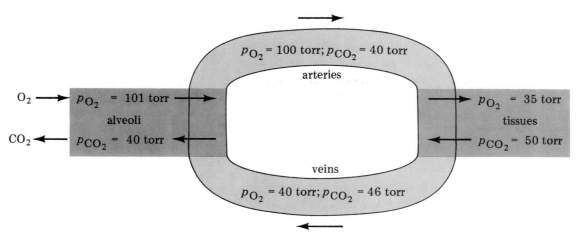

Figure 5-11.   Transfer of oxygen and carbon dioxide in the body. Oxygen is picked up by the blood in the lungs and is released in the tissues; carbon dioxide moves in the opposite direction. Both gases move from regions of higher partial pressure to those of lower partial pressure.

tissues. In the tissues, the oxygen partial pressure is lower (35 torr), resulting in transfer of oxygen from the blood. Carbon dioxide partial pressure is higher in the tissues (50 torr) and flows into the blood. Thus, the venous blood, returning to the lungs, has a lower p_{O_2} and higher p_{CO_2} than the arterial blood. At the lungs, carbon dioxide is again lost from the blood and fresh oxygen is picked up to continue the cycle. At each stage, it is the difference in partial pressures that causes the exchange of gases.

5.12 Oxygen therapy

Oxygen is needed by the tissues of the body. Too little oxygen, hypoxia, even for brief periods, can result in coma and irreversible damage. The brain is especially sensitive because it uses about 20% of the total oxygen of your blood.

As shown in Figure 5-12, oxygen can be administered in various ways. The nasoinhaler or nasal cannula (5-12a) provides an atmosphere of 35 to 50% oxygen, as does the nasal catheter. A larger amount of oxygen, 60 to 90%, is possible with a face mask (5-12b). An oxygen tent supplies 40 to 60% oxygen but is harder to regulate.

If the oxygen is supplied from a compressed gas cylinder, a regulator is needed to lower the high tank pressure (over 2000 psi) to a safe working pressure. When the regulator has two gauges, one shows the cylinder pressure and the other shows the reduced pressure of gas being delivered to the patient. In addition to the pressure, the rate of flow of the gas must be carefully controlled. A valve is used to adjust the amount of gas leaving the cylinder. How fast it escapes is measured by a flowmeter (Figure 5-13). Common flow rates are 6 to 8 liters/minute for a face mask or nasal cannula and 10 to 12 liters/minute for a tent.

In certain cases, **hyperbaric therapy**, which involves exposure to oxygen at pressures above atmospheric pressure, is desired. The patient is placed in a sealed chamber which contains pure oxygen at pressures of 2 to 2.5 atm for periods up to 5 hours. The resulting high partial pressure of oxygen increases the amount of oxygen dissolved in the blood. Hyperbaric oxygen therapy relieves hypoxia such as in cases of carbon monoxide poisoning. It is also used to inhibit bacterial growth as in gas gangrene, increase the effectiveness of radiation treatments, and treat the "bends" (as described in Chapter 7).

You should realize that oxygen therapy can be dangerous. When administered over long periods, oxygen can have toxic (poisonous) effects. Very high oxygen concentrations can result in collapse of the alveoli (atelectasis). Oxygen administration to premature infants has resulted in blindness (retrolental fibrosis). In all cases, oxygen increases the danger of fire, since it sup-

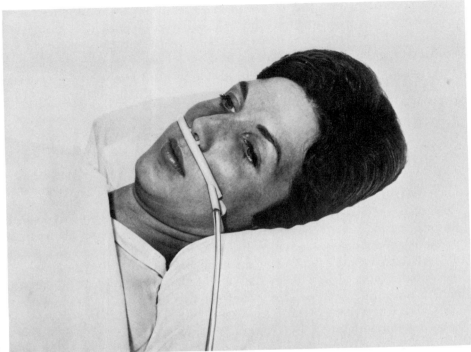

(a)

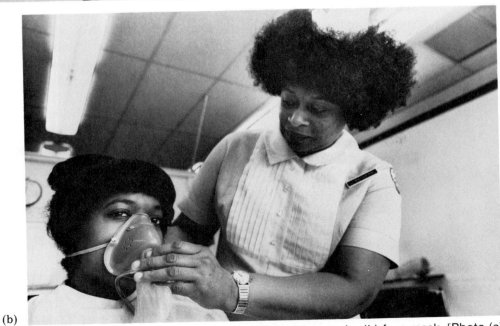

(b)

Figure 5-12. Administration of oxygen. (a) Nasal cannula; (b) face mask. [Photo (a) courtesy of American Hospital Supply; (b) by Al Green.]

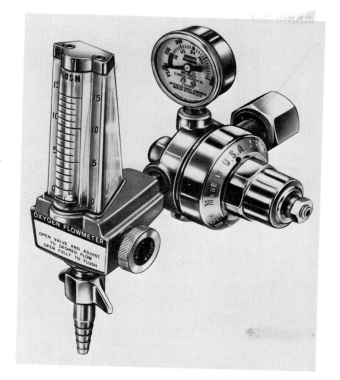

Figure 5-13. A gas regulator and flow-meter. [Photo courtesy of Ohio Medical Products (Division of Airco, Inc.), 3030 Airco Drive, P. O. Box 1319, Madison, Wisconsin 53701.]

ports combustion although it does not itself burn. You must avoid creating sparks or smoking while oxygen is administered.

5.13 Other important gases

In addition to oxygen, several other gases are used in medicine. They are listed in Table 5-4. Carbon dioxide mixed with oxygen serves as a respiratory stimulant but can have serious side effects; it is sometimes used to treat hiccups (hiccoughs or singulation). Given to patients with a blocked airway, a mixture of helium and oxygen diffuses faster than air because of its lower density (0.43 g/liter). Cyclopropane, ethylene, and nitrous oxide are anesthetics; they cause loss of sensation during surgical procedures. The color code in Table 5-4 indicates the bands of color that identify the gas cylinder.

Some gases, present in various amounts in the atmosphere, can be harmful to the body. Known as "air pollutants," they are listed in Table 5-5 along with

Table 5-4   Gases Used Medically

Name	Formula	Color code	Medical use
carbon dioxide	CO_2	gray	respiratory stimulant
carbon dioxide (5%) plus oxygen (95%)	$CO_2 + O_2$ mixture	gray and green	respiratory stimulant
cyclopropane	C_3H_6	orange	anesthetic
ethylene	C_2H_6	red	anesthetic
helium	He	brown	nitrogen substitute (lower density)
helium (80%) plus oxygen (20%)	$He + O_2$ mixture	brown and green	air substitute (lower density)
nitrous oxide	N_2O	light blue	anesthetic ("laughing gas")
oxygen	O_2	green or white	treatment of hypoxia

Table 5-5   Gases Harmful to the Body

Gas	Formula	Properties
carbon monoxide	CO	poisonous, formed by incomplete combustion (car exhausts)
hydrogen sulfide	H_2S	rotten-egg odor, poisonous, formed by decomposing organic matter
nitrogen dioxide	NO_2	brown, irritating, produces smog, formed in car exhausts
ozone	O_3	poisonous, formed by spark from O_2, present in smog, acts as filter in upper atmosphere
sulfur dioxide	SO_2	suffocating odor, formed from burning of sulfur

their properties (described in more detail in Chapter 23). Carbon monoxide, for example, is produced by the incomplete combustion of gasoline in a car engine; this gas can build up to dangerous levels when a car idles in a closed garage. Carbon monoxide poisoning results because the part of your blood that normally carries oxygen (hemoglobin) binds more strongly to CO molecules, preventing enough O_2 molecules from reaching the tissues, causing death by asphyxiation. Figure 5-14 shows an incinerator creating air pollution.

Figure 5-14. Air pollution. Some of the gases listed in Table 5-5 are present in the smoke shown here. (EPA—Documerica photo by Marc St. Gil.)

SUMMARY

The kinetic molecular theory describes the motion of the molecules of a gas. It is based on several assumptions—the molecules are small and far apart, constantly move at very high speeds in a random way, and collide with each other and the walls of their container without losing energy.

Gases diffuse or spread out to fill any volume. This process takes place from a region of high concentration of the gas to a region of low concentration. The less the mass of the molecules, the faster their rate of diffusion (Graham's law).

The temperature of a gas (T) is a measure of the energy and motion of its molecules. It is directly related to the kinetic energy, which depends on the mass of the molecules and the square of their speeds ($\frac{1}{2}mv^2$). The faster the gas molecules are moving, the higher the temperature of the gas. You must use the Kelvin or absolute scale when describing the temperature of a gas.

Because gas molecules hit the walls of their container, they "push" against them. This force applied to a certain area is known as pressure (P); pressure = force/area. The air surrounding the earth creates an atmospheric pressure, which can be measured using a barometer. It supports a column of mercury 760 mm high. Defining 1 mm of mercury as a torr, the pressure exerted at sea level can be expressed as 760 torr or 1 atm (14.7 psi). The SI unit is the pascal, Pa.

The pressure and volume of a fixed amount of gas are inversely related—if one goes up, the other goes down (at constant temperature). This relationship (known as Boyle's law) can also be expressed by saying that, when a change is made in either the pressure or volume of a gas, their product, $P \times V$, does not change. In the process of breathing, you expand the volume of the chest cavity, decreasing the pressure by 3 torr below the atmospheric pressure, creating a flow of air into the lungs. As you exhale, the thoracic cavity decreases in volume, forming a "positive pressure," above atmospheric, which forces air out of the lungs. The tank respirator or "iron lung" also works on this principle.

At constant volume, the pressure of a gas and its Kelvin temperature are directly related. Thus, if the temperature goes up, the volume also increases. In the autoclave, steam at high pressure can reach 120°C and kill microorganisms effectively.

The volume of a gas is directly related to the Kelvin temperature, at constant pressure. Volume also depends on the quantity of gas present. At 273 K and 1 atm, which is standard temperature and pressure (STP), 1 mole of any gas occupies the same volume, 22.4 liters. Pressure also varies directly with the quantity of gas (at constant volume and temperature).

Air is a mixture of gases, approximately four-fifths nitrogen and one-fifth oxygen. The total pressure of such a mixture is equal to the sum of the partial pressures, the individual pressures exerted by each gas present (Dalton's law).

Respiration depends on differences in partial pressures. Oxygen and carbon dioxide diffuse from regions of high partial pressure to regions of low partial pressure. Thus, in the lungs, oxygen moves from the air into the alveoli, while carbon dioxide flows in the opposite direction. Similar transfers take place between alveolar air and the blood circulating through the lung capillaries, and between the blood and the tissues.

Oxygen therapy is required in cases of hypoxia, when the tissues receive too little oxygen. It can be administered by nasoinhaler, face mask, or tent. In hyperbaric therapy, oxygen at pressures above atmospheric pressure is provided in order to increase the amount dissolved in blood.

Other important gases used in medicine include carbon dioxide, as a respiratory stimulant, and cyclopropane, as an anesthetic. Some gases, known as "air pollutants," can have serious medical consequences.

Exercises

1. (5.1) What does the kinetic molecular theory deal with? What are its assumptions?

2. (5.2) What is diffusion? Give an example.

3. (5.2) A mixture contains cyclopropane (C_3H_6) and oxygen. Which gas diffuses faster? Why?

4. (5.3) Explain the meaning of temperature according to the kinetic molecular theory.

5. (5.4) What is pressure? atmospheric pressure?

6. (5.4) Describe the operation of a Torricelli barometer.

7. (5.5) How are the pressure and volume of a gas related (at constant temperature)?

8. (5.5) What happens to the volume of 10 liters of oxygen if the pressure is doubled?

9. (5.6) Explain the process of breathing in terms of Boyle's law.

10. (5.6) Describe the operation of a tank respirator ("iron lung").

11. (5.7) How are the pressure and temperature of a gas related (at constant volume)?

12. (5.7) If the Kelvin temperature is doubled, what happens to the pressure of a quantity of nitrogen initially at 1 atm?

13. (5.7) Describe the operation of an autoclave.

14. (5.8) How are the volume and temperature related (at constant pressure)?

15. (5.8) If the Kelvin temperature of 20 liters of argon is cut by one-half, what is the new volume?

16. (5.9) How is the quantity of a gas related to pressure? to volume?

17. (5.9) What is the volume of 1 mole of any gas at STP?

18. (5.10) Describe the composition of air.

19. (5.10) The following gases are mixed: helium ($p = 400$ torr), oxygen ($p = 200$ torr), nitrogen ($p = 300$ torr). What is the total pressure of the mixture?

20. (5.10) What is an IPPB unit? How does it work?

21. (5.11) Describe the process of respiration in terms of partial pressures of the gases involved.

22. (5.12) When is oxygen therapy called for?

23. (5.12) How is oxygen administered? regulated?

24. (5.12) What is the purpose of hyperbaric therapy?

25. (5.12) What dangers are involved in oxygen therapy? What precautions can you take?

26. (5.13) Describe two gases used medically in addition to oxygen.

27. (5.13) List three air pollutants and their properties.

Water

Water is the most important chemical compound in your body. This colorless, odorless, tasteless liquid makes up from 60 to 70% of your weight. Most activities of the cells take place in environments that consist mainly of water. Many body systems help regulate the amount of water, since it plays such a major role in health.

6.1 The structure of water

The water molecule consists of two atoms of hydrogen covalently bonded to one atom of oxygen (with an angle of 105° between the hydrogen atoms). The Lewis structure appears below (models are shown in Figure 6-1). The two

$$H \overset{\times\times}{:} \overset{..}{O} \overset{\times}{\underset{\overset{\times\times}{H}}{}}$$
$$105° \searrow H$$

the water molecule

single bonds in the molecule are polar covalent. Because the oxygen has a greater attraction for electrons in a bond (electronegativity), the sharing is unequal. The electrons from the hydrogen atoms are pulled toward the oxygen, creating a separation of electrical charge within the molecule. Thus, the entire molecule is polar: The hydrogen "end" of the molecule has a slight positive charge ($\delta+$) and the oxygen "end" has a slight negative charge ($\delta-$). The molecule as a whole, of course, still is neutral.

$$\delta+ \quad \overset{H}{\underset{H}{\diagdown}} O \quad \delta- \quad \text{or} \quad \overset{\frown}{\underset{\smile}{(+ \qquad -)}}$$

Because water molecules are polar, they attract each other through a type of weak force called **hydrogen bonding**. The partially negative oxygen end of one molecule attracts the partially positive hydrogen end of a second water

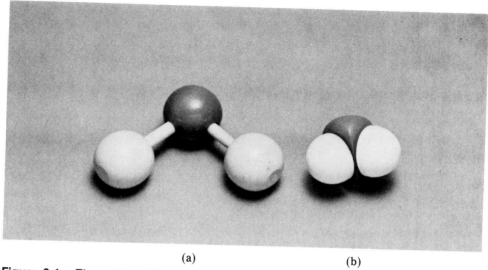

(a)   (b)

Figure 6-1.   The water molecule, H_2O. The ball-and-stick model (a) shows the way the atoms are connected, but the space-filling model (b) presents the shape of the molecule more realistically. (Photo by Al Green.)

molecule. A hydrogen atom is thus shared by two oxygen atoms. The bond in the original molecule is covalent and strong; the new bond, a hydrogen bond, is much weaker. As shown in Figure 6-2, each water molecule forms hydrogen bonds with three or four other molecules. As you will soon learn, water has many unusual properties because of this attractive force between its molecules.

6.2   Kinetic theory of liquids and solids

To understand the properties of water, you must first take a closer look at both liquids and solids. The kinetic theory, which was applied to gases in the previous chapter, can be extended to these two other states of matter. *Atoms,*

Figure 6-2.   Hydrogen bonding between water molecules. The hydrogen atom of one molecule is attracted by the oxygen atom of another molecule.

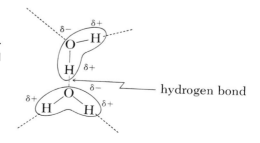

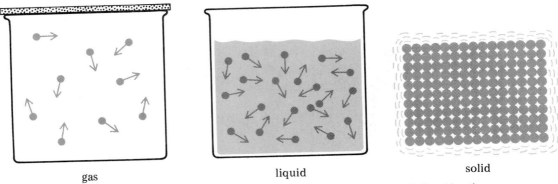

gas   liquid   solid

Figure 6-3. Comparison between a gas, liquid, and solid in terms of the kinetic theory. In a liquid, the molecules are attracted to each other but move freely. In a solid, the atoms, ions, or molecules vibrate in fixed positions close to each other.

ions, or molecules of liquids and solids are moving like gas molecules but their motion is much more restricted.

In a liquid, the molecules are thousands of times closer together than in a gas. Attractive forces between molecules now come into play, giving the liquid definite volume and fixed boundaries. Because of the relative nearness of the molecules, liquids cannot be compressed. But the molecules are still in motion; liquids, therefore, do not have a definite shape and permit diffusion to take place.

The atoms, ions, or molecules of a solid cannot move around; they only vibrate about fixed positions. The attractive forces present give solids a rigid shape. In addition, most solids are highly ordered. They contain repeating fixed arrangements of atoms or ions, as found in crystals of sodium chloride (see Figure 3-2). Solids, liquids, and gases are compared in Figure 6-3.

6.3 Evaporation

Since the molecules in a liquid are moving, some may have enough energy to **evaporate**, to escape from the surface, becoming a gas. The gaseous form of a substance that normally exists as a liquid or solid is known as a **vapor**. Thus, when liquid water evaporates, water vapor forms. Since temperature is a measure of the energy of the molecules of a liquid, the rate of evaporation depends on the temperature. The higher the temperature, the greater the rate of evaporation.

When you exercise heavily, you heat up and perspire. But as you rest, you feel cooler. Water has evaporated; since those molecules having the greatest energy were the ones to form the vapor, those left behind as liquid have less energy. Therefore, the temperature of the remaining water is now lower.

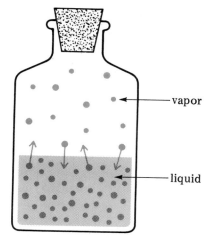

Figure 6-4. Evaporation in a closed container. The rate of molecules leaving the liquid state is equal to the rate of molecules returning. Under these conditions, the pressure exerted by the gaseous form of the liquid is called its vapor pressure.

When evaporation takes place in a closed container as shown in Figure 6-4, molecules leave the surface of the liquid to become vapor. As the amount of vapor increases, some gas molecules return to the liquid. After a while, *the rate of molecules rejoining the liquid becomes equal to the rate of evaporation.* This balance between opposing processes is called equilibrium (like the chemical equilibrium described in Section 4.12). The pressure of the gaseous form of the liquid under these conditions is called its **vapor pressure**.

Liquids have different vapor pressures. Some examples are listed in Table 6-1. The pressure reflects the attraction between molecules in the liquid. The stronger the attraction, the smaller the tendency to evaporate and the lower the vapor pressure. *Water has such a low vapor pressure because hydrogen bonding tends to prevent molecules from escaping from the liquid state.* In cooling a patient, sponging with alcohol is more effective than sponging with water because alcohol evaporates faster; the attractive forces between its molecules are smaller. For the same reason, medications containing alcohol,

Table 6-1 Vapor Pressures of Liquids

Liquid	Vapor pressure (torr) at 20°C
ether	436
acetone	177
chloroform	145
carbon tetrachloride	87
alcohol (ethyl)	43
water	18

or other liquids of relatively high vapor pressure, should be stored in tightly closed containers to prevent evaporation and the resulting change in their strength.

6.4 The calorie and specific heat

Evaporation can be speeded up by adding heat to the liquid. Heat energy is measured using water as the standard. A **calorie**, abbreviated cal, is the amount of energy needed to raise the temperature of 1 g of water by 1°C. The SI unit of energy is the joule, J; 1 joule = 0.239 cal. Since this unit of heat energy is so small, a unit 1000 times larger is commonly used, the kilocalorie, which is abbreviated kcal or Cal; [1 kcal = 1000 cal = 4.184 kJ (kilojoules)].

As you saw in Figure 4-7, the calorimeter is a device that measures exchanges of heat. Any heat released by some change inside the reaction chamber or "bomb" is reflected by an increase in the temperature of the water surrounding it. For example, the energy yield of food is found by oxidation: reacting the food with oxygen in the chamber and measuring the heat given off. One slice of bread produces about 100 kcal (10,000 cal or 418.4 kJ), enough heat to raise the temperature of 1 liter of water by 100°C. The average daily caloric intake is about 2500 kcal (10,460 kJ); when you eat excess food, containing more calories than needed to supply your energy requirements, it is usually stored as fat.

Different substances require various amounts of heat to raise their temperatures. The **heat capacity** of a substance is the quantity of heat needed to increase the temperature of 1 g by 1 Celsius degree. Because of the way calorie is defined, water has a heat capacity of 1 cal/g per degree (4.184 J/g per degree). **Specific heat** is defined as the ratio of the heat capacity of another substance compared to water (at 15°C):

$$\text{specific heat} = \frac{\text{heat capacity of a substance}}{\text{heat capacity of water}}$$

Table 6-2 lists the specific heats of various elements and compounds.

Water has a higher specific heat than almost all other substances. This large value, 1.00, means that much energy must be transferred to change the temperature of water. Thus, water is used in hot water bottles or packs. Since the water contains such a large amount of heat energy, it loses it slowly and the temperature stays high for a relatively long time, commonly from 2 to 4 hours.

The high heat capacity or specific heat of water helps to maintain your normal body temperature. Since so much of you consists of water, relatively large amounts of heat energy are needed to change the body temperature. For

Table 6-2 **Specific Heats
of Various Substances**

Substance	Specific heat
water	1.00
alcohol	0.60
mineral oil	0.50
aluminum	0.21
iron	0.11
glass	0.09
copper	0.09
silver	0.06
lead	0.03
gold	0.03

the same reason, the huge quantity of water on the earth prevents large temperature variations between day and night.

6.5 Heat and the states of water

Although water is a liquid at room temperature, it exists as a solid (ice) below 0°C and as a gas (water vapor or steam), above 100°C. These three forms of water are chemically the same; the composition of the molecule has not changed. *Adding or removing heat merely converts one state of water to another.*

As liquid water is heated, its temperature rises and its vapor pressure increases. At a certain temperature, the vapor pressure becomes equal to the atmospheric pressure. The process that then takes place, formation of bubbles throughout the liquid, is called boiling, and the temperature at which it takes place is the **boiling point**. For pure water at 1 atm, the boiling point is 100°C (212°F). As long as the temperature remains at 100°C, liquid water continues to be converted to water vapor.

The boiling point depends on the pressure above the liquid. If it is below 1 atm, the boiling point is lower. The molecules require less energy, and therefore less heat is needed. At high altitudes, where the atmospheric pressure is lower, water boils at a temperature lower than 100°C. For example, at 18,000 feet (5500 m), the pressure is 380 torr, one-half the value at sea level. Water boils at 82°C at this altitude; cooking times are much longer at this temperature. The boiling point can be raised by using a pressure cooker. Since it is a sealed container of fixed volume, escaping water molecules are trapped, building up the pressure above the liquid. When the vapor pressure is raised to 1000 torr (about 5 psi above normal atmospheric pressure), water boils at 108°C.

The amount of heat required to change 1 g of a substance from the liquid state to a gas at the boiling point is the **heat of vaporization**. For water it is 540 cal/g (2258 J/g). Again, *this value is high compared to other liquids because of the attraction between neighboring water molecules.* Thus, even after water has reached 100°C, an additional 540 cal are required to convert each gram of boiling water to steam. This fact explains why so much heat is needed to boil water.

The high heat of vaporization also explains why you can be badly burned by steam. This same amount of heat, 540 cal/g, is released when the steam condenses back to boiling water. The **heat of condensation**, the energy given off in the reverse process, must be the same as the heat of vaporization because of the law of conservation of energy. Of course, steam is less dense than boiling water so that the amount you might contact in an accident would be smaller. But each gram causes a much more serious burn because of the extra heat energy it contains.

If heat energy is added to a solid, like ice, the solid will melt and become a liquid at a certain temperature, the **melting point**, which is 0°C (32°F) for water. The amount of heat needed to convert 1 g of a substance from a solid to a liquid at the melting point is the **heat of fusion**. The value for water is 80 cal/g, higher than many other solids. An ice pack works by absorbing heat from its surroundings as the ice melts. Every gram of ice absorbs 80 cal (333 J/g), cooling the part of the body where the pack is placed. The opposite process, conversion of a liquid to a solid, must give off the same quantity of heat. Therefore, the **heat of crystallization** of water, which is released when water forms ice, is also 80 cal/g.

Figure 6-5 summarizes the conversion between the three states of water. Heat is needed for the processes of fusion, converting the solid to a liquid, and vaporization, converting the liquid to a gas. Heat is released by the reverse processes, crystallization and condensation.

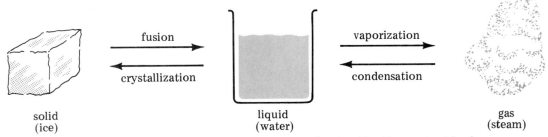

Figure 6-5. The conversion between the three states of water. Heat is required for fusion and vaporization; heat is released by condensation and crystallization.

6.6 Density and specific gravity

The density of water is usually taken as 1 g/ml (or 1 g/cm³). Its value, however, depends on the temperature. At 20°C, the density is 0.998 g/ml. When the temperature is lowered, the density increases slightly (because the volume decreases) until a maximum of 1.000 g/ml is reached at 4°C. As water cools further, the density decreases until it reaches 0.917 g/ml for ice.

Ice floats because it has a lower density, less mass per unit volume, than liquid water. The density of ice is lower because the volume of water increases when freezing takes place. Thus, if a body cell is cooled to 0°C or below, the water expands as it becomes ice and destroys the cell by rupturing it, breaking the cell walls. The same increase in volume of water upon freezing causes water pipes to burst in the winter.

In medical fields, the densities of substances are often related to the density of water by a ratio known as the **specific gravity**:

$$\text{specific gravity} = \frac{\text{density of substance}}{\text{density of water}}$$

Specific gravity simply compares the mass of a substance to the mass of the same volume of water (at a certain temperature). Since the density of water is generally taken as 1.00 g/ml, *the specific gravity has the same numerical value as density;* dividing by 1 does not change a number. But specific gravity has no units (like specific heat) because the grams per milliliter units cancel out. For example, a urine sample may have a density of 1.02 g/ml; its specific gravity is simply 1.02.

$$\text{specific gravity of urine sample} = \frac{1.02 \text{ g/ml}}{1.00 \text{ g/ml}} = 1.02$$

The specific gravities of body fluids are presented in Table 6-3. They are measured with a **hydrometer**, a floating glass bulb with markings on its neck. One example, a **urinometer**, which is used to measure the specific gravity of

Table 6-3 **Specific Gravity of Body Fluids**

Body fluid	Specific gravity
blood, whole	1.052–1.064
blood, plasma (cells removed)	1.024–1.030
cerebrospinal fluid	1.006–1.008
saliva	1.010–1.020
sweat	1.001–1.006
urine	1.008–1.030

Figure 6-6. A urinometer. The depth to which the weighted glass bulb sinks depends on the specific gravity of the urine. (Photo by Al Green.)

urine, is shown in Figure 6-6. It is based on the fact (Archimedes' principle) that a floating object displaces a volume of liquid having a mass equal to the mass of the object. Since the hydrometer has a fixed mass, the volume of liquid displaced, and therefore how far the bulb sinks, depends on the density or specific gravity of the liquid. In pure water, the hydrometer sinks to the 1.00

mark on its neck. In the urine sample described in the previous paragraph, it rises to the 1.02 level.

6.7   Surface tension

Attraction between water molecules through hydrogen bonding results in **cohesion**. The molecules pull toward each other and decrease the amount of exposed surface, the surface area, of the liquid. Therefore, water drops take a spherical shape (like a ball) because a sphere has the least possible surface area, the lowest ratio of surface to volume.

The molecules at the surface of a liquid feel an unbalanced force since there are many molecules attracting them from below the surface but few from above. They are thus pulled inward; this attraction creates **surface tension**. The surface of the water is like a stretched piece of rubber. You can even lay a razor blade on top of the surface; it will not sink, although it is denser than water, because of this tension. *Water has an unusually high surface tension.* This property, as well as many of the others summarized in Table 6-4, results largely from hydrogen bonding.

The attraction of molecules for other substances is **adhesion**. Water in a glass capillary (narrow) tube rises because of the attraction of water molecules to the surface of the tube. You use this effect, known as **capillary action**, when drawing blood samples with a capillary tube (Figure 6-7). Absorbent cotton, wicks, and paper towels also work on this principle, as does the distribution of water in soils.

Because the surface of a glass tube attracts the water surface, the outer edges of the water are higher, forming a curve or **meniscus** (Figure 6-8, right-

Table 6-4   Summary of Properties of Water

colorless, odorless, tasteless
density, 1.00 g/ml (at 4°C)
boiling point, 100°C (at 1 atm)
freezing point, 0°C (at 1 atm)
low vapor pressure (18 torr)
high specific heat (1.00 cal/g°C)
high heat of vaporization (540 cal/g)
high heat of fusion (80 cal/g)
high surface tension
lower viscosity than blood

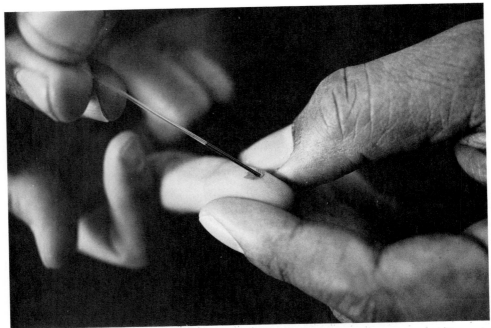

Figure 6-7. A blood sample being taken using a capillary tube. (Photo by Al Green.)

Figure 6-8. A meniscus formed by mercury (left) and water (right). Notice that the mercury meniscus is convex and the water meniscus is concave. (Photo by Al Green.)

hand side). The meniscus becomes more noticeable as the diameter of the tube becomes smaller. When measuring the volume of water in a tube you must always read the value at the lowest point, which is at the bottom of the meniscus. You should also make sure that your eyes are at the same level as the meniscus; otherwise you will have parallax error, an incorrect reading caused by looking from above or below the level of the surface.

As shown in Figure 6-8 (left-hand side), mercury forms a meniscus that is *highest* at the center (convex), compared to the water meniscus, which is *lowest* at the center (concave). Mercury atoms have a greater attraction for each other than for the glass surface; their surface tension is even greater than that of water. When measuring the height of a column of mercury, as in the Torricelli barometer, you must take your reading at the highest point of the meniscus.

Certain substances, known as **surfactants** (surface-active agents), or "wetting agents," can lower the surface tension of water. They reduce the cohesive force between water molecules, increasing their ability to "wet" the surface of their container through adhesion. Such compounds are added to water in the clinical laboratory to permit smoother flow through small-diameter tubing.

6.8 Viscosity

You probably have heard the expression "Blood is thicker than water." It literally means that blood has a greater resistance to flow, or greater **viscosity**, than water. Blood is in fact three to five times as viscous as water. Table 6-5 compares the viscosity of various liquids to that of water, which is given the value of 1.00; these numbers are known as specific viscosities. Viscosity

Table 6-5 Specific Viscosity of Common Liquids

Liquid	Specific viscosity (at 20°C)
glycerol	1490
castor oil	986
ethylene glycol	19.9
mercury	1.55
ethyl alcohol	1.20
water	1.00
chloroform	0.58
ether	0.23

changes with temperature—generally, the higher the temperature, the greater the rate of flow of liquids.

In some conditions, such as shock, it may be necessary to increase the blood volume. If normal blood plasma (the fluid with the cells removed) is not available, a mixture of salt in water containing dextran (a very large molecule consisting of carbon, hydrogen, and oxygen atoms) can be used as a substitute. The presence of this big molecule gives the fluid a viscosity close to that of normal blood.

6.9   Water pressure

The pressure that a liquid like water exerts depends on only two factors—*height* and *density*. The greater the height of a column of liquid and the greater its density, the more pressure exerted. When administering a fluid to a patient, as shown in Figure 6-9, you adjust the pressure by raising or lowering the bottle of fluid. Too high a pressure can be dangerous, while too low a pressure may permit fluid to flow in the reverse direction, from the patient back into the container. Enemas are provided with a fixed length of tubing to prevent you from raising the fluid to a height that would create a harmful pressure. The diameter of the tube does not affect the pressure, but it does influence the rate of flow, which also depends on the length of the tube, the pressure, and the viscosity of the liquid.

When a pressure is applied to a liquid, it is carried unchanged throughout the liquid. This principle, known as Pascal's law, results from the fact that liquids cannot be compressed. When you push the plunger in a syringe, the pressure is transmitted to the liquid inside, which then flows out through the needle. A liquid, like a gas, always moves away from a region of high pressure. If the liquid cannot escape, as in a mattress filled with water, like a "waterbed," the applied pressure is evenly spread throughout, helping to prevent bedsores on a patient.

The pumping action of your heart applies pressure to the blood, which is transmitted throughout the body. But the pressure gradually decreases, as part of the force expands the elastic blood vessels or is converted by friction to heat. Thus, blood pressure is higher in the arteries leaving the heart than in the veins which return the blood supply. The blood pressure is measured by a sphygmomanometer (see Figure 5-9), which balances the pressure in the arteries with the externally applied pressure of a cuff pumped full with air. Normal arterial blood pressure in a young adult has a peak value (systolic pressure) of 120 torr and a minimum value (diastolic pressure) of 70 torr, written 120/70.

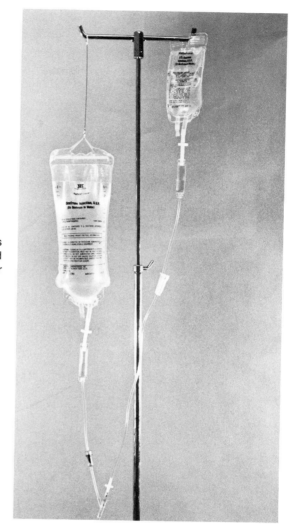

Figure 6-9. Administration of an intravenous fluid. The pressure is controlled by the height and density of the fluid. (Photo courtesy of Baxter Travenol Laboratories.)

6.10 Water of hydration

Certain substances react with water by a process called **hydration,** in which the water molecule acts as a single unit. (In hydrolysis, a reaction you will learn about later, the water molecule is split into two parts.) Compounds called **hydrates** form; they contain water in a definite proportion, known as the "water of hydration." Several different hydrates can sometimes be produced from the same substance. Sodium sulfate, for example, forms hydrates containing one molecule of water ($Na_2SO_4 \cdot H_2O$), seven molecules of water

($Na_2SO_4 \cdot 7H_2O$), and ten molecules of water ($Na_2SO_4 \cdot 10H_2O$). The original substance with no water of hydration (Na_2SO_4) is referred to as anhydrous.

The removal of water, **dehydration**, is generally accomplished by heating. Certain compounds lose their water of hydration on exposure to air, a process known as efflorescence. The hydrate of sodium carbonate, or washing soda ($Na_2CO_3 \cdot 10H_2O$), is an example of an efflorescent compound.

Other compounds, called hygroscopic compounds, pick up water vapor from the air (at a rate that depends on the humidity). If enough water is absorbed so that the compound actually dissolves, the process is called deliquescence. Calcium chloride is a hygroscopic substance used as a "drying agent." As it forms the hydrate $CaCl_2 \cdot 6H_2O$, the anhydrous molecule removes water from the air surrounding it.

One of the most important hydrates is a form of calcium sulfate known as gypsum, $CaSO_4 \cdot 2H_2O$. This hard white substance is used in making casts to hold fractured bones in fixed positions, as shown in Figure 6-10, and in dental work (as dental stone). It is produced by the hydration of a different form of calcium sulfate called **plaster of paris**, $(CaSO_4)_2 \cdot H_2O$:

$$(CaSO_4)_2 \cdot H_2O + 3H_2O \longrightarrow 2CaSO_4 \cdot 2H_2O$$

plaster of paris   gypsum

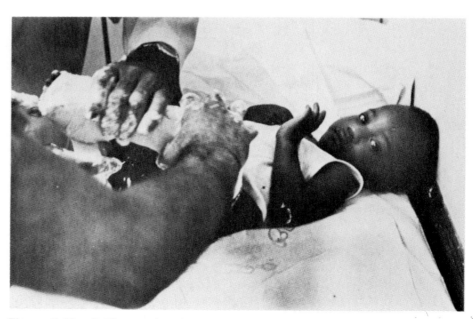

Figure 6-10. Setting a plaster of paris cast. By adding water to plaster of paris, $(CaSO_4)_2 \cdot H_2O$, it is converted to gypsum $CaSO_4 \cdot 2H_2O$. (Photo by A. McGowan, St. Luke's Hospital Center, New York.)

The plaster of paris itself is made from natural gypsum by the reverse process, dehydration.

6.11   Water purification

Pure water does not exist in nature. Natural water contains many types of impurities. Depending on the source, it may contain ionic compounds, oil, dyes, insecticides, foam, sewage, bacteria, viruses, and many other types of water pollutants. To be safe for drinking and medical purposes, water must be treated to remove all possibly harmful substances.

The following procedures are usually part of the steps followed in water conditioning or purifaction. **Sedimentation** is the settling out of suspended matter in tanks or reservoirs over various periods of time. **Coagulation** involves adding aluminum sulfate or iron sulfate to form larger particles from the solids that did not settle, making them easier to eliminate. **Filtration** through sand or gravel removes most suspended matter following coagulation. **Chlorination** employs either chlorine gas, Cl_2, or an ionic compound containing the hypochlorite ion, ClO^-, to kill bacteria by oxidation. After disinfection, the substances causing odors and tastes are removed by passing the water through carbon and by **aeration**, mixing the water with air. Aeration also removes most dissolved gaseous impurities.

Water in certain areas is called "hard" because it contains calcium ions (Ca^{2+}) or magnesium ions (Mg^{2+}). "Hard" water does not form suds with soap and leaves a scale in pipes and boilers, as shown in Figure 6-11. A further treatment step, water "softening," is then necessary to remove these ions and replace them with sodium ions (Na^+). The "soft" water is produced by adding calcium hydroxide (lime), $Ca(OH)_2$, and sodium carbonate (soda ash), Na_2CO_3, which remove Mg^{2+} and Ca^{2+} ions as magnesium hydroxide, $Mg(OH)_2$, and calcium carbonate, $CaCO_3$, compounds that do not dissolve in water.

Another way to remove ions from water is by passing it through an **ion exchanger**. This substance has the ability to replace ions in water with the ions initially present on the exchanger, without itself dissolving. Thus, when "hard" water flows through a sodium cation exchanger, the exchanger picks up the Mg^{2+} and Ca^{2+} ions, putting Na^+ ions in their place. The exchanger can be regenerated to be used over again by passing sodium ions (from NaCl in water) through it.

A technique for water purification that you can carry out in the laboratory is **distillation**. The setup is illustrated in Figure 6-12. Water is boiled, and the vapor is condensed by running cold water in a tube around it. Impurities are left behind in the original flask. In addition, the process of boiling kills most microorganisms.

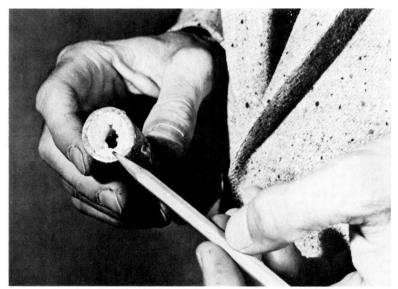

Figure 6-11. ''Hard'' water scale in a pipe. (Photo courtesy of Culligan Water Institute.)

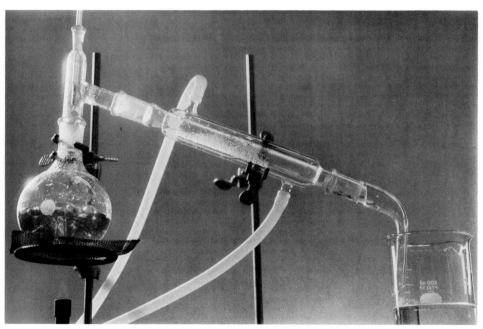

Figure 6-12. Laboratory distillation setup. The liquid vaporizes and then condenses into the beaker on the right. (Photo by Al Green.)

6.12 Water balance

When your body functions normally, *the amount of water taken in is equal to the volume lost.* This balance is illustrated in Table 6-6. Most water enters the body as the pure liquid or in a beverage. A smaller volume by about one-half is contained in the food you eat. In addition, some water is produced as a by-product of the process by which food molecules are broken down in the body (metabolism).

Most excess water is eliminated in the urine. "Insensible water loss" occurs by diffusion through the skin and evaporation from the lungs; both processes take place without your being aware of the change. Normal feces also eliminate a small volume of water each day.

In hot weather and during exercise, the amount of water lost through the skin as sweat increases greatly, to as much as 3500 ml/hour over short periods. This loss cannot continue for long without serious damage. The resulting imbalance, known as **volume depletion**, involves a shortage of both water and sodium ions; it is caused by vomiting, diarrhea, and kidney disease in addition to excessive sweating. The term "dehydration" refers to a loss only of water. In either case, reduction of body water by over 20% can be fatal.

Shock is a special case of fluid loss, which can result from internal or external bleeding. In this condition, blood flow to the tissues is reduced to the point where life can no longer be sustained. Other conditions, such as malnutrition or kidney and liver disease, cause an excess of fluids to build up in the tissue spaces. This abnormal accumulation of fluid, called **edema**, results in swelling of the body tissues (Figure 6-13).

Table 6-6   **Typical Daily Water Balance**[a]

Intake (ml)		Loss (ml)	
liquids	1700	urine	1540
food	900	skin	770
metabolism	300	lungs	490
		feces	100
total intake	2900	equals total loss	2900

[a] Based on a 150-pound male.

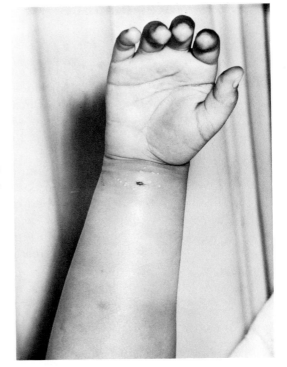

Figure 6-13. The arm of a patient with edema. The swelling results from the accumulation of fluid. (Photo by Martin Rotker.)

6.13 Humidity therapy

Humidity is a measure of the amount of water vapor present in a gas such as air. Absolute humidity is the mass of water contained in a certain volume of gas, generally given as grams per cubic meter; at 20°C, for example, its value is 17.3 g/m³. Relative humidity is the actual quantity of water vapor present compared to the maximum amount that the gas can hold at a given temperature. The relative humidity, stated as a percentage, is the information you hear in weather reports.

Patients with elevated temperatures and poor fluid balance lose abnormally large amounts of water. They are said to have a "humidity deficit" and require replacement of the water. In addition, gases such as oxygen, administered from a tank, are dry and must be humidified to prevent irritation of the mucous membranes and loss of water. In these cases, *water must be added to the gas being breathed by the patient.*

Since the amount of water that air can hold as vapor is limited, several approaches can be used to increase the "dose" of water to the patient. One method is **atomization**, the breaking up of liquid water into very small particles. An atomizer is shown in Figure 6-14a. When air is forced through the tube, it travels faster in the narrow part or constriction (Bernoulli's principle), causing a decrease in pressure at that point. Thus, water is forced up into the airstream in the small connecting tube by the pressure of the atmosphere on the surface of the liquid. This process produces water particles in a wide assortment of sizes.

Nebulization forms a "mist" or "fog" of smaller droplets having more uniform size. It generates an **aerosol**, a suspension of very fine particles sus-

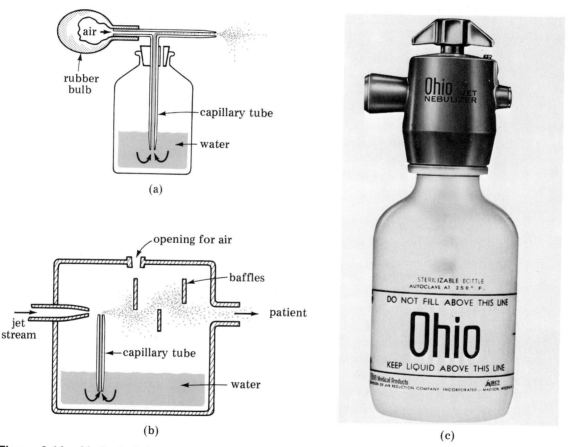

Figure 6-14. Methods for producing small particles of water suspended in a gas. (a) Atomizer; (b,c) jet nebulizer. The baffles prevent the larger droplets from leaving the nebulizer. In both cases, the rush of air above the capillary tube reduces the pressure, causing water to rise into the airstream. [Photo courtesy of Ohio Medical Products (Division of Airco, Inc.), 3030 Airco Drive, P.O. Box 1319, Madison, Wisconsin 53701.]

pended in a gas. Medically, the most useful size is between 0.2 and 0.7 micrometer (μm) in diameter. A jet nebulizer, based on the same principle as the atomizer, is shown in Figure 6-14b,c. The plates or baffles in the airstream knock down the larger droplets but allow the smaller ones through. In an ultrasonic nebulizer, small water particles are produced in a different way; vibrational energy is produced electrically to break up the liquid. The reservoir nebulizer, used for the humidification of oxygen, heats water to 53°C to give the gas maximum humidity at body temperature.

SUMMARY

A water molecule consists of two atoms of hydrogen connected to an oxygen atom by polar covalent bonds. The entire molecule is polar—the oxygen "end" has a slightly negative charge, and the hydrogen "end" has a slightly positive charge. Water molecules engage in hydrogen bonding, a weak force of attraction between opposite ends of the polar molecules.

The kinetic theory can be extended from gases to the two other forms of matter. In a liquid, the molecules have slower speeds than in a gas, and attractive forces exist between them. The atoms in a solid cannot move around but vibrate in fixed positions.

Some molecules of a liquid may have enough energy to evaporate, to escape from the surface and become a gas. The gaseous form of a substance that normally exists as a liquid or solid is known as vapor. Water has a very low vapor pressure because hydrogen bonding tends to prevent molecules from escaping.

A calorie (cal) is the amount of energy needed to raise the temperature of 1 g of water by 1°C. Thus, the heat capacity of water, the quantity of heat needed to raise the temperature of 1 g by 1 Celsius degree, is simply 1 cal (4.184 joule) (per gram per degree). Specific heat is the ratio of the heat capacity of another substance compared to that of water. Because water has a high specific heat (1.00), it helps to maintain your body temperature; relatively large amounts of heat energy must be transferred to cause a change in temperature.

As liquid water is heated to a certain temperature, the boiling point (100°C), its vapor pressure becomes equal to the atmospheric pressure. The amount of heat required to change 1 g from the liquid state to a gas at the boiling point is called the heat of vaporization. Because of the high heat of vaporization of water (540 cal/g), you can be badly burned by steam. If heat energy is added to a solid, like ice, it melts and becomes a liquid at a certain temperature, the melting point (0°C). The amount of heat needed to convert 1 g of solid to liquid at the melting point is the heat of fusion. Water has a high heat of fusion (80 cal/g), making it well suited for ice packs.

The density of water, about 1 g/ml, depends on its temperature. Ice floats because its density is slightly less than that of liquid water. Specific gravity is the ratio of the density of a substance to the density of water. The specific gravity of body fluids such as urine is used in medical diagnosis.

Attraction between water molecules results in cohesion. The inward pull felt by molecules at the surface of the liquid creates "surface tension." The attraction of molecules for other substances is adhesion. Water in a glass capillary tube rises because of adhesion to the walls of the tube—this effect is capillary action. The outer edges are higher than the center, forming a curved surface known as a meniscus.

Viscosity refers to the resistance to flow of a liquid. It is a property that depends on the temperature.

The pressure that a liquid like water exerts depends on its height and density. The diameter of the tube carrying the liquid does not affect the pressure, but it does influence the rate of flow. When a pressure is applied to a liquid, it is carried unchanged throughout (Pascal's law).

Compounds called hydrates contain water in a definite proportion: the "water of hydration." The removal of water, dehydration, generally takes place by heating. A form of calcium sulfate known as gypsum, $CaSO_4 \cdot 2H_2O$, results from the hydration of plaster of paris.

To be safe for drinking and medical purposes, water must be purified. The following steps may be involved in water treatment: sedimentation, coagulation, filtration, chlorination, and aeration. "Hard" water is converted to "soft" water by replacing calcium and magnesium ions with sodium ions.

When your body functions normally, the amount of water taken in is equal to the volume lost. Most enters as pure water or as a beverage; the largest part leaves as urine. Imbalances result from excessive perspiration, vomiting, diarrhea, kidney disease, and shock.

Humidity therapy involves the addition of water to the gas being breathed by the patient. The most effective technique is nebulization, the formation of a "mist" or "fog" by suspending very fine particles of water in a gas as an aerosol.

Exercises

1. (6.1) Draw the Lewis structure of a water molecule. Explain why it is polar.

2. (6.1) Describe the formation of hydrogen bonds.

3. (6.2) How do liquids and solids differ from a gas according to the kinetic theory?

4. (6.3) What is evaporation?

5. (6.3) Why does perspiration cause cooling?

6. (6.3) Why does water have a low vapor pressure?

7. (6.4) Explain how a calorimeter works.

8. (6.4) Describe one important medical application of the high specific heat of water.

9. (6.5) Explain why a steam burn can be dangerous.
10. (6.5) Why is an ice pack useful for cooling a part of the body?
11. (6.6) Why does ice float?
12. (6.6) What is specific gravity? How is it useful medically?
13. (6.7) How does cohesion differ from adhesion?
14. (6.7) What is surface tension?
15. (6.7) Why does blood rise in a capillary tube?
16. (6.8) Which is more viscous, water or molasses?
17. (6.9) When you administer a fluid intravenously, what factors determine its pressure?
18. (6.9) Why is a water-filled mattress useful in preventing bedsores?
19. (6.10) What is water of hydration?
20. (6.10) Describe an application of one important hydrate.
21. (6.11) By what methods can water be purified?
22. (6.11) What is "hard" water? How can it be converted to "soft" water?
23. (6.12) Describe the water balance of the body.
24. (6.12) What conditions can disturb the water balance?
25. (6.13) What is humidity therapy?
26. (6.13) Describe how a nebulizer produces an aerosol.

Solutions

A **solution** is a special kind of mixture. It is uniform or homogeneous, having the same properties throughout. Solutions contain one or more substances, either solid, liquid, or gas, mixed with another substance, usually a liquid. The composition of a solution can vary and its components can be physically separated, but each part is identical to every other part of the mixture.

7.1   Types of solutions

A solution consists of two main parts: the substance being mixed or dissolved, the **solute**, and the substance doing the dissolving, the **solvent**. The solute is also the substance that is present in the smaller amount. Most of the solutions you will use have water as the major component, or solvent. These mixtures are called *aqueous* solutions. For example, a saline solution is an aqueous solution that consists of sodium chloride, the solute, dissolved in water, the solvent.

Solutions can be prepared by mixing elements or compounds, as illustrated in Table 7-1 for the three states of matter. Certain types of solutions have different names. A "tincture," for example, is a solution with alcohol as the solvent, as in tincture of iodine. **Alloys** (Figures 7-1 and 7-2) are solid solutions made by mixing atoms of metals; common examples and their compositions are given in Table 7-2. Many important solutions are used in medicine, such as normal saline solution, replacement solutions for body fluids, dextrose solution for intravenous feeding, and gas solutions for respiratory therapy.

7.2   The process of dissolving

The most common solutions contain a solid mixed in a liquid. When a solid dissolves, its molecules or ions separate from each other. For this process to take place, they must be attracted in some way by the solvent. Generally, like

Table 7-1 **Examples of Solutions**

Solute	Solvent		
	Solid	Liquid	Gas
solid	brass—zinc in copper (alloy)	saline solution—sodium chloride in water	*a*
liquid	amalgam—mercury in silver	liquor—alcohol in water	*a*
gas	hydrogen in platinum metal	soda water—carbon dioxide in water	air—oxygen in nitrogen

a To form a solution in a gas, the solute must generally also exist in the gaseous state.

Figure 7-1. Silver amalgam. The most common solutions consist of a solid dissolved in a liquid. Dental amalgam, on the other hand, consists of a liquid, mercury, dissolved in a solid, silver. (Photo by Al Green.)

Figure 7-2. An alloy used to form a prosthetic device, an artificial knee joint. (Photo courtesy of Pfizer, Inc.)

Table 7-2 **Important Alloys**

Alloy	Composition	Use
amalgam alloy	69% Ag, 26% Sn, 4% Cu, 1% Zn	mixed with mercury for dental restoration
brass	60–85% Cu, 15–40% Zn (or Pb)	plumbing, hardware
bronze	89–92% Cu, 8–11% Sn	artwork, machine parts
gold alloy	70% Au, 15% Ag, 10% Cu, 3% Pd, 1% Pt, 1% Zn	dental casting (inlays, crowns, etc.)
nichrome	80% Ni, 20% Cr	heating coils
pewter	85% Sn, 7% Cu, 6% Bi, 2% Sb	bowls, pots
solder	60% Pb, 40% Sn	joining metal parts
steel	99% Fe, 1% C	building
sterling silver	93% Ag, 7% Cu	cutlery
vinertia alloy	67% Co, 27% Cr, 6% Mo	replacement for body parts (such as hip)

substances dissolve like substances: *Polar solvents tend to attract and dissolve polar (or ionic) solutes, while nonpolar solvents tend to dissolve nonpolar substances.*

Water is a polar molecule having a slightly positive and slightly negative part. Thus, water dissolves many ionic compounds like sodium chloride, as well as many polar covalent molecules such as sugar, ammonia, and hydrogen chloride. Most organic (carbon-containing) substances like oils, greases, and fats are nonpolar; there are no centers of positive and negative charge in their molecules. Thus, they do not dissolve in water but in nonpolar organic substances such as gasoline.

As water attracts a formula unit or molecule, it causes the solid to separate and thus dissolve, as shown in Figure 7-3. Ions, released from ionic com-

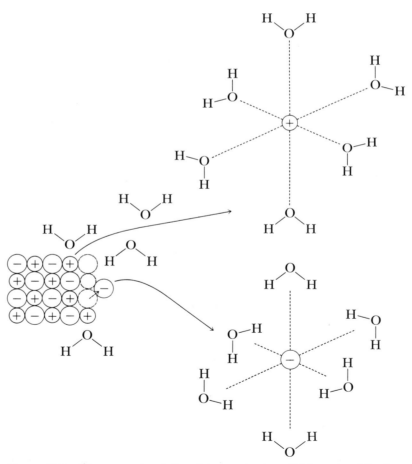

Figure 7-3. The process of dissolving an ionic solid in water. The ions dissociate as they are hydrated, surrounded by water molecules.

pounds or produced by breaking polar covalent bonds, move freely in the water. This process is known as **dissociation** or **ionization**. It takes place because polar water molecules can surround the separated ions in solution. The slightly negative part of the water molecules, the oxygen ends, face the positive cations, while the positive ends of the water molecules, the hydrogen atoms, surround the negative anions. This attraction between the ions and water molecules is another example of **hydration**. (The general term for any solvent is "solvation.") Each ion has about six water molecules around it.

Not all ionic compounds dissolve in water. Some, like silver chloride, AgCl, remain almost entirely as solids. Their ions have a much greater attraction for each other than for the molecules of the solvent, water.

Substances that produced ions when they dissolve in water are **electrolytes**. Their solutions are able to conduct an electric current because of the free charged particles present. Those substances that separate completely into ions are *strong* electrolytes; they conduct electricity well. Sodium chloride

$$NaCl \xrightarrow[\text{water}]{\text{in}} Na^+ + Cl^-$$

exists entirely as sodium ions and chloride ions in solution. Therefore, it is a strong electrolyte. Other compounds, *weak* electrolytes, ionize only partially and therefore conduct electricity poorly. Those substances that do not separate at all into ions when dissolved in water are *non*electrolytes; no electricity at all can be conducted by such a solution, like sugar in water.

7.3   Solubility of solids

The amount of solute that can be dissolved in a certain amount of solvent at a particular temperature is its **solubility**. Solubility is usually given as grams of solute per 100 g (or 100 ml) of solvent. For example, the solubility of sodium chloride in water is 36 g/100 g H_2O at 20°C.

The natures of the solute and solvent are the main factors in determining solubility. Some compounds are very soluble in water, like sucrose or table sugar (204 g/100 g H_2O at 20°C). Others, like sodium chloride, are soluble but to a lesser extent. Certain compounds such as magnesium hydroxide, $Mg(OH)_2$, dissolve to an extremely small degree (0.0009 g/100 g H_2O at 20°C). Substances whose solubility is so low are generally considered very slightly soluble or insoluble; for all practical purposes, they do not dissolve in the solvent.

For many solids, the solubility increases with temperature—more dissolves as you heat the solvent. Several examples are presented in Table 7-3. This table shows you something that you already know. More sugar dissolves in hot water than in cold water. The change here is dramatic; there is a difference in

Table 7-3 **Solubilities of Solids in Water at Different Temperatures**

Temperature (°C)	Solubility of sucrose (g/100 g)	Solubility of sodium chloride (g/100 g)	Solubility of calcium hydroxide (g/100 g)
0	179	35.7	0.185
20	204	36.0	0.165
40	238	36.6	0.141
60	287	37.3	0.116
80	362	38.4	0.094
100	487	39.8	0.077

solubility of more than 300 g as the temperature goes from 0° to 100°C. Sodium chloride increases in solubility by only 4 g over this same range. Some compounds, such as calcium hydroxide, $Ca(OH)_2$, actually decrease in solubility as the temperature is raised, as shown in the table.

The *rate* at which a solid dissolves in a liquid depends on several factors. Grinding up (pulverizing) the solid into smaller pieces increases the surface exposed to the solvent, making it dissolve faster. Stirring (agitating) the solution also increases the rate of dissolving, as you know from mixing sugar with coffee or tea. Heating a solution makes the molecules move faster, thereby speeding up the process of dissolving. Note that stirring and grinding the solid into a powder do not affect the solubility, only the rate of dissolving.

7.4 Saturation

The solubility of any solid in a liquid is limited. You can dissolve only a certain amount of solute at a given temperature; this maximum amount is given by its solubility. When no more solute can be dissolved, the solution is **satu-rated**. At this point, a balance or *equilibrium* exists. The rate of solute dissolving is equal to the rate of solute coming back out of the solution, as shown in Figure 7-4.

If a solution has less solute than the maximum it can hold, it is **unsaturated**. More solute can be dissolved in an unsaturated solution. For example, if a solution is made from 20 g of sodium chloride in 100 g of water, it is unsaturated (see Table 7-3). But when you add more sodium chloride to bring the total mass to 36 g, the solution becomes saturated.

Just because a solution is saturated, do not think that it must contain a large amount of solute. If not much dissolves, in other words, if the solute is only slightly soluble, you can make a saturated solution with a small amount of so-

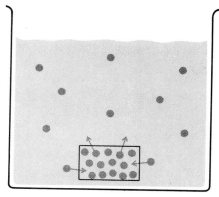

Figure 7-4. A saturated solution. The rate of solid dissolving is equal to the rate of solid crystallizing back out of the solution. The total amount of dissolved solid is the maximum value at a particular temperature.

lute. As shown in Table 7-3, a saturated solution of calcium hydroxide at 20°C has only 0.165 g of solute in each 100 g of water.

Under certain conditions, you can prepare a solution containing more solute than normally possible at that temperature. Such a solution is **supersaturated**. It can be made by taking a saturated solution and slowly lowering the temperature (if the solubility decreases with cooling). The resulting supersaturated solution is unstable; it converts back to a saturated solution at the new temperature by having the "excess" solute crystallize out of the solution.

7.5   The solubility of liquids and gases

Liquids that dissolve in each other, forming a solution like water and alcohol, are said to be **miscible**. These two liquids can be mixed in any proportion. Water and gasoline, however, do not dissolve in each other; they are immiscible. The less dense liquid merely forms a separate layer over the denser liquid, as happens after oil spills (Figure 7-5).

Gases can mix with each other to any extent to form solutions. The most important example is air, consisting mainly of oxygen in nitrogen. The composition of the solution can vary from the normal 20% oxygen all the way up to 100% or pure oxygen.

On the other hand, the solubility of a gas in a liquid is generally quite limited, as shown in Table 7-4. In addition to depending on the nature of the gas and liquid molecules and the temperature, the amount of gas that dissolves is determined by the pressure above the liquid. *A gas becomes more soluble in a liquid as the applied pressure increases;* this relationship is known as Henry's law. Soda water or carbonated water is made by dissolving carbon dioxide, CO_2, in water at pressures greater than 1 atm to force more of the gas to dis-

Figure 7-5. An oil spill. Since oil and water are immiscible, and oil is less dense, it forms a layer on top of the water. (Photo courtesy of U.S. Coast Guard.)

Table 7-4 Solubilities of Gases in Water[a]

Gas	Solubility (g/100 g H_2O at 35°C)
carbon dioxide	0.1105
nitrogen	0.0015
oxygen	0.0033

[a] At body temperature and 1 atm.

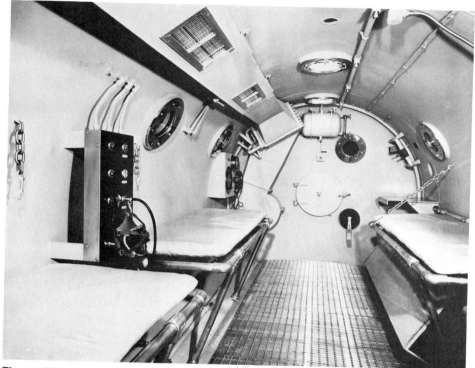

Figure 7-6. A hyperbaric chamber. The pressure inside can be increased to over twice the normal atmospheric pressure. (Photo courtesy of Vacudyne Altair, Inc.)

solve. When you open a bottle, the pressure decreases and some of the gas escapes because less is soluble at atmospheric pressure.

Henry's law is important in the administration of anesthetic gases: The greater the partial pressure of the anesthetic, the greater its solubility in the blood. This principle is also the basis for the hyperbaric chamber (Figure 7-6). Because of the greater external pressure (2 to 2.5 atm), more oxygen dissolves in the patient's blood.

Unlike many solids, gases are *less* soluble at higher temperatures. On heating a liquid solvent, a smaller amount of gas stays dissolved in solution. If you let a cold bottle of soda with its cap removed stand in a room, the carbon dioxide will escape as the temperature rises.

A medical problem caused by the decrease in solubility of a gas in a liquid is the "bends," a condition found in deep sea divers who come up from the bottom too quickly. When they are underwater, they breathe compressed air and are subjected to high pressures from the water above them. The amount of nitrogen that can dissolve in their blood increases greatly under these conditions. If they rise to the surface too fast, the nitrogen, which becomes less soluble as the pressure falls, forms bubbles in the blood, causing pain and possible damage. Artificial breathing mixtures of helium and oxygen are now often used by divers. Because helium is one-fifth as soluble in the blood as nitrogen, less dissolved gas is able to form bubbles as the pressure decreases.

7.6 Concentration of solutions—percentage

The composition of a solution can vary greatly. The term **concentration** refers to the amount of solute present in a given quantity of solvent or solution. If a relatively large amount of solute is in a solution, it is called concentrated; if only a small amount is in a solution, it is dilute. These terms, however, are too vague for describing solutions used in health care. If a nurse gives a patient a medication or intravenous solution, it is important that she or he administer the correct amount. Too little may not be effective and too much may be toxic. Therefore, you must learn more exact methods to express the concentration of a solution.

One of the easiest ways you can express the concentration of a solution is the percentage of solute in the solution. "Percent" means parts per hundred and can be expressed as a ratio or fraction (see Appendix A.1). Five percent, 5%, means 5 parts out of 100 parts and can be written as 5/100. Most commonly, percentage represents the number of grams of solute per 100 ml of solution. Thus, a 5% solution consists of 5 g of solute in every 100 ml of solution. This method of expressing concentration is called **weight–volume percentage** (w/v)% since it compares the weight (mass) of solute to the total volume of the solution.

$$\text{weight–volume percentage (w/v)\%} = \frac{\text{grams of solute}}{100 \text{ ml of solution}}$$

Normal saline solution consists of 0.9% aqueous sodium chloride. Thus, it contains 0.9 g of NaCl per 100 ml of solution. To make 200 ml of 0.9% NaCl, you would need 2×0.9 g, or 1.8 g, of sodium chloride. For 500 ml, you need 5×0.9 g, or 4.5 g, of solute. In general, you can either set up a proportion or use a unit-factor method whenever the volume of solution needed is not exactly 100 ml, as shown below (see also Appendix A.3 and A.4).

Example Prepare 1000 ml of 0.9% NaCl

$$0.9\% = \frac{0.9 \text{ g NaCl}}{100 \text{ ml solution}} \quad \text{(definition)}$$

Proportion method

$$\frac{0.9 \text{ g}}{100 \text{ ml}} = \frac{x}{1000 \text{ ml}}$$

$$(0.9 \text{ g}) \times (1000 \text{ ml}) = (100 \text{ ml}) \times x$$

$$\frac{(0.9 \text{ g}) \times (1000 \text{ ml})}{100 \text{ ml}} = x$$

$$9 \text{ g NaCl} = x$$

Unit-factor method

$$\text{grams} = 1000 \text{ ml solution}$$

$$\times \frac{0.9 \text{ g NaCl}}{100 \text{ ml solution}}$$

$$= 9 \text{ g NaCl}$$

First, you write the definition of 0.9% as a fraction expressed as grams of solute per 100 ml of solution. Then if you write a proportion, the left side is the fraction representing the concentration, 0.9 g/100 ml, and the right side is the unknown weight of solute, x, divided by the new volume, 1000 ml. Notice that grams are on top (in the numerators) and milliliters are on the bottom (in the denominators) on both sides of the equation. You solve the proportion by cross-multiplying. Then divide both sides by the quantity on the same side as the unknown, 100 ml. Finally solve for x. The unit-factor method is simpler because it gives the same answer in one step. You now have the weight of sodium chloride, 9 g, that must be dissolved in water and brought to a final volume of 1000 ml to make a 0.9% solution. This process is illustrated in Figure 7-7.

Occasionally, other forms of percentage concentration are used. **Weight–weight percentage** (w/w)% is the number of grams of solute per 100 g (rather than per 100 ml) of solution. For dilute solutions, such as those you will be using most often, there is practically no difference between this method of expressing concentration and weight–volume percentage because 100 g of aqueous solution has a volume of approximately 100 ml.

Volume–volume percentage (v/v)% is used only for solutions consisting of two liquids. It gives the *milliliters* of solute per 100 ml of solution. For example, 70% alcohol consists of 70 ml of alcohol in a solution whose total volume is 100 ml. You would expect this solution to have 30 ml of water, but it actually contains slightly more. Alcohol and water are similar molecules and form hydrogen bonds with each other, reducing their combined volume. To bring the total volume to 100 ml, more than 30 ml of water are thus needed.

A unit related to percentage is **parts per million**, abbreviated **ppm**. Just as 5% means 5 parts out of 100 parts, 5 ppm means 5 parts out of 1 million parts. For example, if the mercury concentration in a fish is found to be 5 ppm by weight, each kilogram (1000 g) of fish contains 5 mg (0.005 g) of mercury ($0.005/1000 = 5/1,000,000 = 5$ ppm).

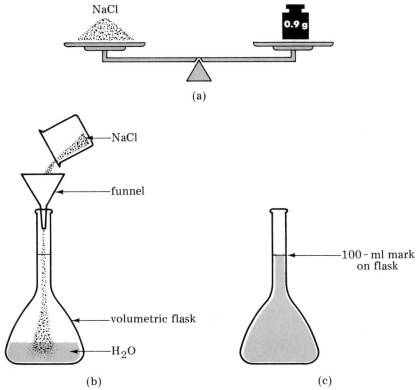

Figure 7-7. Preparing 100 ml of 0.9 (w/v)% saline solution. (a) Weigh 0.9 g NaCl. (b) Add the solid to a 100-ml volumetric flask partially filled with water. (c) Dissolve the solid and fill to the mark with water.

7.7 Molarity

Solutions whose concentrations are given as percentages are easy to prepare but have limitations. A 5% solution of sodium chloride and a 5% solution of dextrose both contain the same weight of solute. But they contain different numbers of moles of solute because sodium chloride and dextrose (glucose) have different formula weights. Since moles are so useful in chemistry, there is a method of expressing concentration, known as molarity, based on the mole. **Molarity**, abbreviated M, is defined as the number of moles of solute per liter (1000 ml) of solution:

$$\text{molarity } (M) = \frac{\text{moles of solute}}{\text{liter (1000 ml) of solution}}$$

Thus, a 1 molar (1 M) solution contains 1 mole of solute per liter of solution; a 2 molar (2 M) solution contains 2 moles of solute per liter of solution; and so on. A $\frac{1}{2}$ molar (0.5 M) solution contains less than 1 mole, 0.5 mole, of solute in each liter of solution.

Example 0.15 M NaCl

$$0.15 \ M \ \text{NaCl} = \frac{0.15 \ \text{mole NaCl}}{\text{liter of solution}}$$

$$0.15 \ \text{mole NaCl} = 0.15 \ \text{mole} \times \frac{58 \ \text{g}}{\text{mole}} = 8.7 \ \text{g}$$

$$0.15 \ M \ \text{NaCl} = \frac{8.7 \ \text{g NaCl}}{\text{liter of solution}}$$

You first write down the meaning of the symbol 0.15 M as shown. Then, since the formula weight of sodium chloride is 58 amu, 1 mole has a mass of 58 g. The mass of 0.15 mole is found by multiplying the number of moles by the mass of 1 mole. A 0.15 M solution of sodium chloride thus contains 8.7 g of sodium chloride per liter of solution.

Look back to the example in the previous section. One liter of 0.9% NaCl contains 9 g of sodium chloride. Thus, 0.15 M NaCl and 0.9% NaCl are different ways of representing approximately the same concentration. One is given in terms of moles per liter and the other in grams per 100 ml, but the amount of sodium chloride in solution is about the same in both cases. The solutions are thus prepared in the same way (Figure 7-7).

If you do not need 1 liter (1000 ml) of solution, you must adjust the weight of solute to match the desired volume. For example, to make 2 liters of a 0.15 M NaCl solution, you need two times the mass required for 1 liter: 2 × 9 g, or 18 g. To prepare less than 1 liter, say 500 ml (0.5 liter), you must take 0.5 × 9 g, or 4.5 g, of sodium chloride. These solutions, containing 9 g in 1 liter, 18 g in 2 liters, and 4.5 g in 0.5 liter, all have the same concentration. The ratio of moles of solute to liters of solution is identical, 0.15 M.

A way of expressing concentrations similar to molarity is known as **molality**, abbreviated m. It is defined as the moles of solute per *kilogram* (1000 g) *of solvent*. Instead of comparing moles to the total volume of the solution as in molarity, this method relates moles of solute to the mass of the solvent. Thus, a 0.15 molal solution of sodium chloride consists of 0.15 mole (9 g) of sodium chloride dissolved in 1000 g of water. For dilute aqueous solutions, *the concentrations in molarity and molality are approximately the same* because 1000 g of water has a volume of 1 liter. Most of the solutions you use fall into this category.

7.8 Dilution of solutions

Very often, you will have to prepare a solution using a more concentrated "stock" solution. It must be diluted by adding more solvent to get the desired concentration. For example, you may have available a 10% solution of sodium chloride but actually need 200 ml of 0.9% sodium chloride. To find out how much of the 10% solution you must dilute to make the desired solution, you can use the following relationship: *stock solution*

initial concentration × initial volume = final concentration × final volume

The initial concentration is the concentration of the solution you have on hand, 10%; the initial volume is the unknown, x, the amount of this solution you need. The new solution will have a final concentration of 0.9% and a final volume of 200 ml. Substituting into the equation and solving for x gives an answer of 18 ml:

$$10\% \times x = 0.9\% \times 200 \text{ ml}$$

$$x = \frac{(0.9\%) \times (200 \text{ ml})}{10\%}$$

$$x = 18 \text{ ml}$$

You must take 18 ml of the 10% solution and add enough water to bring the volume up to 200 ml; the concentration is now 0.9%. You can use the above relationship with any method of expressing concentrations as long as the units for concentration, as well as the units for volume, are the same on both sides of the equation.

When preparing medications for children, you may also have to dilute a

Table 7-5 Formulas for Finding Children's Doses

Name of method	Relationship[a]
Augsburger	$\left(\dfrac{4A + 20}{100}\right) \times \left(\dfrac{\text{adult}}{\text{dose}}\right)$
body surface area (BSA)	$\left(\dfrac{\text{body surface area (m}^2)}{1.7}\right)^{b} \times \left(\dfrac{\text{adult}}{\text{dose}}\right)$
Clark	$\left(\dfrac{\text{weight of child (pounds)}}{150 \text{ pounds}}\right) \times \left(\dfrac{\text{adult}}{\text{dose}}\right)$
Cowling	$\left(\dfrac{A + 1}{24}\right) \times \left(\dfrac{\text{adult}}{\text{dose}}\right)$
Dilling	$\left(\dfrac{A}{20}\right) \times \left(\dfrac{\text{adult}}{\text{dose}}\right)$
Young	$\left(\dfrac{A}{A + 12}\right) \times \left(\dfrac{\text{adult}}{\text{dose}}\right)$

[a] The letter A denotes age of child in years.
[b] Body surface area in square meters is approximately equal to $(4W + 7)/(W + 90)$, where W is the child's weight in kilograms.

more concentrated solution or simply give less of it. Various relationships are used to calculate the smaller dose required for a child. They are based on either the child's age, weight, or body surface area, as shown in Table 7-5. In all cases, you multiply the adult dose by a fraction less than 1 to get a child's dose.

7.9   Osmosis

Osmosis is the diffusion of solvent molecules through a membrane, such as a piece of animal tissue (Figure 7-8). It takes place when two solutions with different concentrations are on opposite sides of such a semipermeable membrane. "Semipermeable" means that only water molecules can pass through, not the solute. They move in the direction of the more concentrated solution, diluting it. If you separate a 1% saline solution from a 2% solution by a semipermeable membrane, water flows from the 1% solution into the 2% solution.

As osmosis takes place, the height of the more concentrated solution increases because of water moving into it. Osmosis ends when the difference in height between the two solutions creates enough pressure to stop any further flow of water. If you now apply pressure to the side containing the more concentrated solution, you can force the water molecules back in the other direction. This applied pressure, needed to keep equal levels in the two solutions, is called the **osmotic pressure**. The osmotic pressure of a solution depends on its concentration as well as the temperature. It can be measured in terms of a height of mercury, just as atmospheric pressure is determined, or as a height of water. The normal osmotic pressure of blood, for example, is 37 cm of water, or 50 torr (6.67 kPa).

The osmotic pressure of a solution depends on the number of particles, ions or molecules of solute. It can be expressed in terms of **osmolarity**, which is simply the molarity times the number of particles produced by ionization of each mole of solute. A 1 M NaCl solution is 86% ionized; out of every 100 formula units of sodium chloride, 86 have dissociated into Na^+ and Cl^- ions. Thus, the osmolarity (abbreviated Osm) is 1.86 Osm—the solution consists of 1.86 osmoles/liter, an osmole (or osmol) being the number of moles of the solute multiplied by the number of particles that are active in osmosis. (The concentrations of Na^+ and Cl^- are each 0.86 M, and the concentration of undissociated NaCl is 0.14 M: 0.86 + 0.86 + 0.14 = 1.86.) Since solutions in the body are dilute, the milliosmole is used; it is 1/1000 of an osmole. In these units, most body fluids contain about 300 milliosmoles/liter. Osmolality, the number of osmoles per kilogram of water, is often used in place of osmolarity; because solutions in the body are dilute, the difference between these terms (as with molarity and molality) can be ignored.

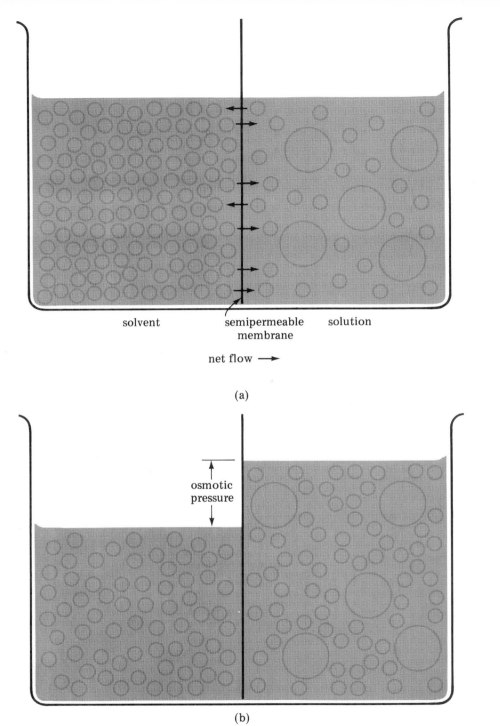

solvent   semipermeable   solution
membrane

net flow ⟶

(a)

osmotic
pressure

(b)

Figure 7-8. The process of osmosis. (a) Solvent flows from pure solvent to the solution. (b) The flow of solvent causes a height difference and pressure difference (osmotic pressure) between the two sides of the membrane. (The small circles represent solvent molecules; the large circles represent solute.)

7.10 Osmosis and the blood

Your blood contains a very large molecule, albumin, which cannot diffuse out of the blood through the membranes. Thus, the blood has a higher protein concentration than the surrounding tissue fluid and a greater osmotic pressure by about 25 torr. This force tends to push water and small dissolved substances into the blood. It counteracts the normal (hydrostatic) pressure of the pumped blood through the capillaries, the tiny vessels that connect arteries with veins. Blood reaching the capillary has a pressure of 35 torr, which pushes water and nutrients out, since the opposing osmotic pressure is only 25 torr.

By the time the blood leaves the capillary, the blood pressure has dropped to about 15 torr, but the osmotic pressure is still roughly the same, resulting in a flow of water back inside along with waste molecules from the tissues. The osmotic pressure caused largely by the albumin thus prevents loss of water from the blood and aids in the flow of food molecules into the tissues and waste molecules into the blood. The relationship between blood pressure and osmotic pressure is illustrated in Figure 7-9.

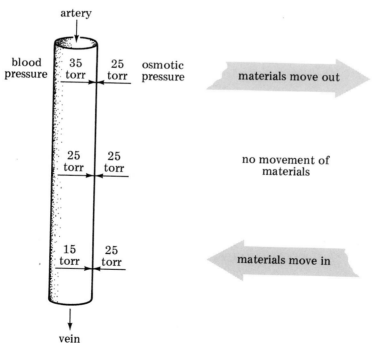

Figure 7-9. The relationship between blood pressure and osmotic pressure in a blood capillary.

The red blood cells themselves are affected by differences in osmotic pressure. If two solutions having equal concentrations and therefore the same osmotic pressure are separated by a semipermeable membrane, there will be no net flow of water from one to the other. These solutions are **isotonic.** A 0.9% (0.15 M) solution of sodium chloride is isotonic with the "solution" inside a red blood cell. When these cells are placed in such a "normal" saline solution, they stay the same size because water does not diffuse across the cell membrane in a favored direction.

When two solutions have different concentrations, water moves through the membrane from the less concentrated, **hypotonic,** solution to the more concentrated, **hypertonic,** solution. If red blood cells are placed in pure water (the most dilute "solution" possible), water molecules diffuse into the cells, as shown in Figure 7-10, causing them to expand and burst. This process is

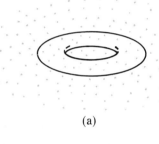

(a)

Figure 7-10. Red blood cells. (a) No change of size in an isotonic solution. (b) Cells expand in hypotonic solution. (c) Cells shrink in hypertonic solution. Arrows indicate direction of water flow.

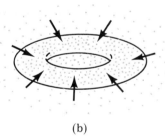

(b)

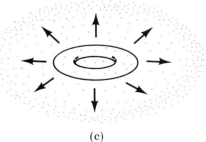

(c)

called **hemolysis**; it would occur if a patient received water injected intravenously instead of normal saline solution. You can compare this change in volume with what happens if you drop a prune in water. The sugars in the prune are solutes, causing water to flow inside and make it swell or burst.

If red blood cells are in a hypertonic solution, such as 5% sodium chloride, osmosis occurs in the opposite direction. Water diffuses out of the cells, resulting in **crenation**, a shrinkage in size. This effect takes place in the pickling process, when cucumbers, for example, are placed in brine, a concentrated salt solution. It is important that body cells are exposed to neither hypertonic or hypotonic solutions unless there are special medical reasons. One such instance is the use of hypertonic nutrient solutions for long-term intravenous feeding in order to prevent buildup of body fluids.

7.11   Colloids

Not all mixtures are solutions, which are formed from small molecules and ions. Mixtures consisting of larger particles may be **colloids**, such as blood and milk. The particle size in a colloid ranges from 1 to 1000 nanometers in diameter, where 1 nanometer (nm) is a billionth (10^{-9}) the size of a meter. One substance in a colloid is said to be dispersed or distributed in another substance; examples are given in Table 7-6.

A **sol** is a dispersion of a solid in a liquid, like milk of magnesia, which contains magnesium hydroxide particles in water. **Emulsions** consist of a liquid dispersed in another liquid, such as milk—butterfat droplets in water. An emulsifying agent is often needed to keep the liquids from separating (it is a protein, casein, in the case of milk). An **aerosol** is a liquid or solid in a gas, like fine water droplets in air. **Foams** consist of a gas dispersed in a liquid or solid; whipped butter, for example, contains air in butter. A **gel** has a rigid open structure with another substance filling the holes; examples are gelatin desserts and jellies.

Table 7-6   Examples of Colloids

Substance being dispersed	Medium in which dispersion takes place		
	Solid	Liquid	Gas
solid	concrete	jelly (gel) india ink (sol)	dust (aerosol)
liquid	cheese	mayonnaise (emulsion)	fog (aerosol)
gas	foam rubber (foam)	whipped cream (foam)	none[a]

[a] Gas mixtures form solutions.

Although colloidal particles are larger than the particles in solutions, they are still too small to be visible under a microscope. They are big enough, however, to reflect and scatter light. Colloids thus show the "Tyndall effect"; you can see a beam of light as it passes through the dispersion (Figure 7-11). When this beam is observed under a microscope, you can see pinpoints of light that move about rapidly. This random, zigzag motion caused by collisions of the colloidal particles with the molecules of the dispersing medium is known as **Brownian motion.** This effect is similar to the movement of molecules in a gas. Unlike a solution, which is transparent, colloids can be either translucent (like a "frosted" light bulb) or cloudy.

Another important property of colloids is **adsorption,** the ability to hold other substances to the surface of the dispersed particles, similar to a dog carrying fleas. (This effect is not the same as absorption, which is the penetration of molecules inside another substance, like a sponge absorbing water.) Some types of colloids (lyophobic) become electrically charged by adsorbing ions onto the surface of the colloidal particles. An example of this type is the gold sol used to test cerebrospinal fluid. The repulsion between like charges normally prevents the colloidal particles from coming together to form larger particles that can settle because of gravity.

Figure 7-11. Comparison between a solution (left) and colloid (right). The beam of light is visible when passing through a colloid—the Tyndall effect. (Photo by Al Green.)

A different type of colloid (lyophilic), such as gelatin, adsorbs a film of molecules, like water, around the dispersed particles. This protective layer also helps prevent the colloidal particles from settling. For this reason, gelatin is used to stabilize the silver bromide colloid on photographic film. Many colloids found in the body also fall into this category. They have the property of forming semisolid gels that take up water and swell (imbibition). Dextran (see Section 6.8) is used as a blood extender because of its ability to hold water by this process.

The adsorption of ions or molecules by colloidal particles results partly from their large surface area. The more surface exposed, the greater the amount of adsorption that can take place. Also, atoms on the surface can often form additional bonds because they are not surrounded by atoms on all sides like those "inside" the particle.

Mixtures with particles even larger than colloids, big enough to see with your naked eye, are **suspensions**. They are heterogeneous (their composition is not uniform), settle upon standing, and can be separated with filter paper. An example is clay in water. Table 7-7 compares the properties of solutions, colloids, and suspensions.

Table 7-7   **Properties of Solutions, Colloids, and Suspensions**

Property	Solution	Colloid	Suspension
particle size	atoms, ions, or small molecules	large molecules or groups of molecules	very large, visible particles
effect of light	transparent	translucent or opaque; shows Tyndall effect	translucent or opaque; shows Tyndall effect
effect of gravity	does not settle	may slowly settle	settles quickly
uniformity	homogeneous	less homogeneous than solution	heterogeneous
separability	cannot separate by filtration	can separate only with special membranes	can be separated with filter paper

7.12   Dialysis

Small molecules and ions can be separated from a colloid by the process of **dialysis**. In contrast to osmosis, which is the movement of solvent, dialysis involves diffusion of *solute* across a dialyzing membrane. The colloidal particles

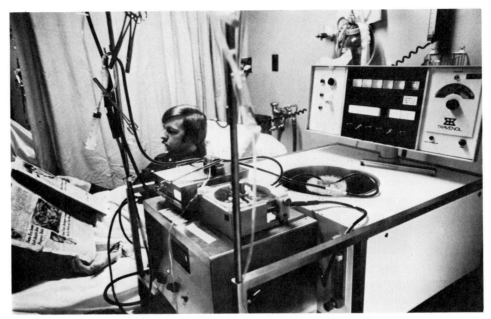

Figure 7-12. Hemodialysis. The patient's blood travels through coils in the artificial kidney machine, allowing wastes to pass through by dialysis. (Photo courtesy of St. Luke's Hospital Center, New York.)

are too large to pass through the small openings in the membrane; they remain on one side. Ions and small molecules, however, can move from the more concentrated side to the more dilute side of the membrane. They flow in the direction that tends to balance the concentrations of solute.

Dialysis is used to purify the blood of patients with kidney failure, a process known as **hemodialysis**. As shown in Figure 7-12, the patient's blood is circulated through tubes in an artificial kidney machine. The tubes act as membranes, permitting the harmful substances that gather in the blood to diffuse out into a surrounding aqueous solution. This bath (consisting of 0.6% NaCl, 0.2% $NaHCO_3$, 0.04% KCl, and 1.5% dextrose) is changed every 2 hours during the total 6-hour dialysis treatment. If the solution were not changed, the waste products that diffused out of the blood would increase in concentration and start flowing back into the blood. The large molecules in the blood are not lost in this process since they cannot pass through the membrane. Unless a patient with kidney failure undergoes hemodialysis several times per week, uremia develops. The body's wastes build up, causing chronic acidosis, anemia, and various systematic and neurological symptoms leading eventually to death.

SUMMARY

A solution is a homogeneous mixture whose composition can vary. It consists of two main parts: the solute, the substance being mixed, and the solvent, the substance doing the dissolving. Many important solutions are aqueous—water is the solvent. Solutions can be prepared from solids, liquids, and gases.

The most common solutions contain a solid mixed in a liquid. Because water is polar, it can attract the ions of an ionic compound, causing the solid to dissolve and dissociate. The separated ions are hydrated, surrounded by water molecules. Water can also cause polar covalent compounds to ionize. Electrolytes are substances that produce ions when dissolved in water.

The solubility of a solute is the amount that can be dissolved in a certain quantity of solvent at a particular temperature. It is usually expressed as grams per 100 g. The nature of the solute and solvent is the main factor in determining solubility. In general, like dissolves like. The rate of dissolving can be increased by pulverizing, agitating, and heating.

When no more solute can be dissolved, the solution is saturated. If a solution has less solute than it can hold, it is unsaturated. A solution is supersaturated if it contains more solute than normally possible at a given temperature.

The solubility of a gas in a liquid is generally limited. A gas becomes more soluble as the applied pressure increases (Henry's law). Gases become less soluble as the temperature increases, unlike many solids. The "bends" results from the formation of nitrogen bubbles in a diver's blood during a too rapid rise to the surface.

The concentration of a solution, the relative amounts of solute and solvent, can be expressed as a percentage. It is given most commonly as weight—volume percentage: grams of solute per 100 ml of solution. Thus, 0.9 (w/v)% sodium chloride (normal saline solution) contains 0.9 g of NaCl in every 100 ml of solution. Other forms of percentage concentration are weight–weight percentage (grams of solute per 100 g of solution) and volume–volume percentage (milliliters of solute per 100 ml of solution).

Molarity is the number of moles of solute per liter of solution. A 0.15 M (molar) solution of sodium chloride contains 0.15 mole of NaCl, or about 9 g, dissolved in a solution whose total volume is 1 liter. The concentration is approximately the same as the 0.9 (w/v)% solution.

Often, a solution must be prepared by diluting a more concentrated solution. In this case, you use the relationship

initial concentration × initial volume = final concentration × final volume

Various formulas are used to dilute medications for children's doses.

Osmosis is the diffusion of solvent molecules through a membrane in the direction of the more concentrated solution. Osmotic pressure is the applied pressure required to prevent movement of solvent; it depends on the concentration of a solution. Two solutions are isotonic if there is no net flow of water from one to the other when they are separated by a semipermeable membrane. Normal saline solution is isotonic with the solution inside red blood cells. These cells hemolyze in a hypotonic (less concentrated) solution and become crenated in a hypertonic (more concentrated) solution.

Mixtures consisting of larger particles (1 to 1000 nm in diameter) are not solutions but colloids. These dispersions exhibit the Tyndall effect, making visible a beam of light as it passes through. Colloids are either translucent or opaque and their particles can adsorb other substances. Mixtures with even larger particles are known as suspensions.

Small molecules and ions can be separated from a colloid by the process of dialysis. Colloidal particles are too large to pass through the openings of the membrane, but ions and small molecules move to the more dilute side. Hemodialysis is the purification of a patient's blood with an artificial kidney machine, which is based on this principle.

Exercises

1. (Intro.) What is a solution?
2. (7.1) Give five examples of solutions with which you are familiar.
3. (7.2) Why is water such a good solvent for ionic and polar covalent compounds?
4. (7.2) What is an electrolyte? How does a strong electrolyte differ from a weak one?
5. (7.3) What factors determine the solubility of a solid in a liquid? Explain.
6. (7.3) How could you dissolve faster a solid medication in a liquid?
7. (7.4) How could you test whether a solution is saturated or unsaturated?
8. (7.4) Is it possible for a dilute solution to be saturated? Explain.
9. (7.5) What are miscible liquids? Give an example.
10. (7.5) What factors determine the solubility of a gas in a liquid? Explain.
11. (7.5) What is the "bends"? How can it be prevented? How do you think this condition could be treated medically?
12. (7.6) What is meant by the "concentration" of a solution?
−13. (7.6) What weight of sodium chloride is needed to prepare the following aqueous solutions: (a) 100 ml of a 2.0 (w/v)% NaCl; (b) 500 ml of 0.45 (w/v)% NaCl; (c) 50 ml of 5.0 (w/v)% NaCl.
14. (7.6) A solution used for skin dressings known as Burrow's solution contains 6.5 g of aluminum acetate in a total volume of 500 ml. Find its concentration as weight–volume percentage.
15. (7.6) What is meant by 60 (v/v)% aqueous alcohol?
16. (7.7) What weight of glucose, $C_6H_{12}O_6$, is needed to prepare the following? (a) 1 liter of 2.0 M solution; (b) 500 ml of a 1.0 M solution; (c) 2 liters of a 0.10 M solution.

17. (7.7) A solution of potassium permanganate, $KMnO_4$, contains 79 g dissolved in 2000 ml of water. Find its molarity.

18. (7.8) How would you prepare 10 ml of a 0.5% solution from a 3% stock solution of hydrogen peroxide?

19. (7.8) If you add enough water to 50 ml of 10% dextrose to make 1 liter of solution, what is the new concentration?

20. (7.8) An adult dose of a medication is 30 ml. Find the dose for a 50-pound child using Clark's formula.

21. (7.9) What is the process of osmosis?

22. (7.9) A 2% glucose solution is separated from pure water by a semipermeable membrane. Describe what takes place.

23. (7.9) What is meant by osmolarity?

24. (7.10) What happens when red blood cells are placed in the following solutions? (a) 0.5% NaCl; (b) 0.9% NaCl; (c) 2.0% NaCl. Explain.

25. (7.11) How does a colloid differ from a solution?

26. (7.11) What is a suspension?

27. (7.12) How does dialysis differ from osmosis?

28. (7.12) Describe the process of hemodialysis in an artificial kidney machine. Why must the solution be changed regularly?

Acids, bases, and salts

Many substances ionize when they dissolve in water. But water itself also dissociates, although only to a very small degree. The *self-ionization of water* produces a hydrogen ion, H^+, and a hydroxide ion, OH^-:

$$H_2O \longrightarrow H^+ + OH^- \qquad \text{(self-ionization of water)}$$

In 10,000,000 liters of water, only 1 g of hydrogen ions and 17 g of hydroxide ions are present. Pure water is considered a nonelectrolyte because so few ions are formed. The concentration of hydrogen ions and the concentration of hydroxide ions are the same in water, 0.0000001 mole/liter ($1 \times 10^{-7} M$), since each water molecule produces one H^+ ion for every OH^- ion.

8.1 Acids

An **acid** is a substance that raises the hydrogen ion concentration of water. When it dissolves in water, *the relative number of H^+ ions becomes larger than the number of OH^- ions.* This effect takes place whether the acid is found in your stomach, in citrus fruits, or in a car battery.

The hydrogen ion is simply a proton; it is the particle formed when a hydrogen atom loses its one electron. In water, the proton is produced by hydration; it is surrounded by water molecules like cations in solution. This hydrated hydrogen ion is often symbolized as H_3O^+ ($H_2O + H^+$), the **hydronium ion.** An acid solution contains excess hydrogen ions, whether you label them H^+ or H_3O^+.

Some acids, such as hydrochloric acid, HCl, contain only one hydrogen atom. They ionize in the following way:

$$HCl \longrightarrow H^+ + Cl^-$$

(This process can also be represented as a reaction of the acid with a water molecule, forming the hydronium ion: $HCl + H_2O \longrightarrow H_3O^+ + Cl^-$.) Acids

Table 8-1   Common Acids

Name	Formula	Ionization[a]	Strength
hydrochloric acid	HCl	over 90%	strong
nitric acid	HNO_3	over 90%	strong
sulfuric acid	H_2SO_4	about 60%	strong
phosphoric acid	H_3PO_4	about 30%	moderate
acetic acid	$HC_2H_3O_2$	about 1%	weak
carbonic acid	H_2CO_3	less than 1%	weak
boric acid	H_3BO_3	less than 1%	weak

[a] In dilute solution.

with more than one proton, polyprotic acids, ionize in several steps, one hydrogen ion being produced each time. With carbonic acid, H_2CO_3, ionization takes place in two steps.

$$H_2CO_3 \longrightarrow H^+ + HCO_3^-$$
$$HCO_3^- \longrightarrow H^+ + CO_3^{2-}$$

First one proton ionizes, producing the bicarbonate ion, HCO_3^-, which then ionizes to liberate a second proton, leaving the carbonate ion, CO_3^{2-}. Phosphoric acid, H_3PO_4, containing three hydrogens, ionizes in three steps.

The *strength* of an acid depends on how completely it is ionized in solution. A strong acid is one that exists almost entirely in the form of ions. Hydrochloric acid is strong because over 90% of the HCl molecules dissociate into H^+ ions and Cl^- ions (in dilute solutions). A weak acid, like acetic acid, is only partially ionized. Most of its molecules do not dissociate into ions at all. Table 8-1 lists the strengths of the most common acids. The stronger an acid, the greater the concentration of hydrogen ions in solution.

8.2   Properties of acids

Acids have a sour taste. The citric acid present in lemons and oranges makes your mouth pucker. In concentrated form, acids are very corrosive, "eating" away most substances they touch. They are therefore dangerous and can cause serious burns if spilled on your skin (Figure 8-1). The affected area must immediately be diluted with large amounts of water to minimize the damage. Because acids react with certain metals, they must not be allowed to come into contact with metallic instruments or containers. The activity series of the metals represents their ability to react with acids. Starting with the most active, the order is

K, Na, Ca, Mg, Al, Zn, Fe, Ni, Sn, Pb, Cu, Hg, Ag, Au

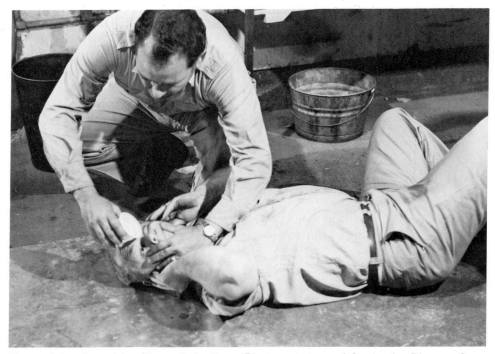

Figure 8-1. An acid spill: eye irrigation. (Photo courtesy of Center for Disease Control, Atlanta, Georgia.)

Properties and uses of specific acids are presented in the paragraphs that follow. Table 8-2 compares their concentrated forms.

Hydrochloric acid, HCl, is found in the gastric juice of your stomach. It is secreted at a concentration of 0.5% or 0.1 M. Too much hydrochloric acid, hyperchlorhydria or hyperacidity, can be treated in mild cases with antacids. Hypochlorhydria or hypoacidity, a deficiency of hydrochloric acid in the stomach, may require the administration of 4-ml doses of 10% HCl. The patient drinks the acid solution through a straw to prevent damage to the teeth. Concentrated hydrochloric acid (12 M) is made by dissolving hydrogen chloride, a gas, in water.

Nitric acid, HNO_3, is a corrosive strong acid that fumes in air. It is sometimes yellow from products of decomposition, such as NO_2. If you spill some on yourself, your skin will become yellow because of a chemical reaction (with protein molecules). Nitric acid has been used for immediate treatment of animal bites and to remove warts.

Sulfuric acid, H_2SO_4, is known as the "workhorse" of the chemical industry. It is used directly or indirectly in nearly every industrial process because of its ability to react with many other chemicals, to dehydrate (remove water),

Table 8-2   Properties of Concentrated Acids

Name	Formula	Molecular weight (amu)	Percentage by weight	Specific gravity	Molarity
acetic acid	$HC_2H_3O_2$	60	100%	1.05	17
hydrochloric acid	HCl	36	37%	1.19	12
nitric acid	HNO_3	63	70%	1.42	16
phosphoric acid	H_3PO_4	98	85%	1.70	5
sulfuric acid	H_2SO_4	98	96%	1.84	18

and to catalyze reactions. The concentrated acid is viscous and very corrosive.

Carbonic acid, H_2CO_3, is an aqueous solution of carbon dioxide gas, CO_2.

$$CO_2 + H_2O \longrightarrow H_2CO_3$$

This weak acid, which exists only in solution, is unstable; boiling completely removes the carbon dioxide. Soda, carbonated water, consists of carbonic acid. This acid and the bicarbonate ion, HCO_3^-, are the forms in which part of the CO_2 produced as waste products from your cells are carried in the blood to the lungs.

Phosphoric acid or orthophosphoric acid, H_3PO_4, is a thick, syrupy liquid in its concentrated form. It is used to acidify soft drinks, such as colas, and to prepare dental cements. In solution, phosphoric acid produces $H_2PO_4^-$, HPO_4^{2-}, and PO_4^{3-} ions through ionization.

Boric acid, H_3BO_3, is a solid that can be dissolved in water to prepare up to a 6% solution at room temperature. Dilute boric acid was formerly used as an antiseptic but is highly poisonous; when ingested, 5 to 20 g can be fatal. Since other antiseptics are more effective, the medical use of boric acid is now limited.

Acetic acid, $HC_2H_3O_2$, is an organic acid, based on carbon. Although it contains four hydrogen atoms, only one can be dissociated. (The other three are firmly bonded to a carbon atom.) The concentrated form of this acid is known as glacial acetic acid because it solidifies to an icelike solid at 17°C. A 36% solution of acetic acid has been used to irritate or redden the skin (a rubefacient). Dilute acetic acid, about 5%, is the main component of vinegar. You will learn about other important organic acids, such as citric acid, lactic acid, and ascorbic acid, in later chapters.

8.3 Bases

A **base** can be defined as a substance that raises the hydroxide ion, OH^-, concentration of water. Most bases are therefore compounds that dissociate in

Table 8-3 Common Bases

Name	Formula	Dissociation[a]	Solubility in H_2O
sodium hydroxide	NaOH	over 90%	high
potassium hydroxide	KOH	over 90%	high
calcium hydroxide	$Ca(OH)_2$	about 100%	low
magnesium hydroxide	$Mg(OH)_2$	about 100%	low
ammonia	NH_3	–	high

[a] In dilute solution.

water to produce hydroxide ions. Examples are sodium hydroxide, NaOH, and magnesium hydroxide, $Mg(OH)_2$:

$$NaOH \longrightarrow Na^+ + OH^-$$
$$Mg(OH)_2 \longrightarrow Mg^{2+} + 2OH^-$$

Ammonia, NH_3, is a base in water even though this molecule contains no hydroxide ion. It acts as a base by reacting with water to form an ammonium ion and a hydroxide ion:

$$NH_3 + H_2O \longrightarrow NH_4^+ + OH^-$$

Essentially, ammonia accepts a hydrogen ion or proton in this reaction. Thus, a basic or alkaline solution has a *larger number of hydroxide ions than hydrogen ions*. This situation is exactly opposite to the one with acid solutions.

Defining the strengths of bases is not as clear-cut as with acids. Table 8-3 lists the most common bases and their percent dissociation in dilute solutions. Sodium hydroxide and potassium hydroxide are both strong bases; they are very soluble in water and dissociate into ions almost completely. Calcium hydroxide and magnesium hydroxide are also highly dissociated but only slightly soluble in water. Thus, a small amount of hydroxide ions is present in solution, even though they are "strong" in the sense of dissociation. Ammonia is a weak base because it has a relatively poor ability to accept hydrogen ions, forming a small amount of ammonium ions and hydroxide ions in water.

8.4 Properties of bases

Solutions of bases have a bitter taste and a slippery feeling. Like strong acids, strong bases are corrosive. Their ability to react with fats and oils makes bases useful as cleaning agents but dangerous if spilled on your skin or eyes (Figure 8-2). Bases also react with certain metals, such as zinc and aluminum, so you should make sure they do not come into contact with materials made from

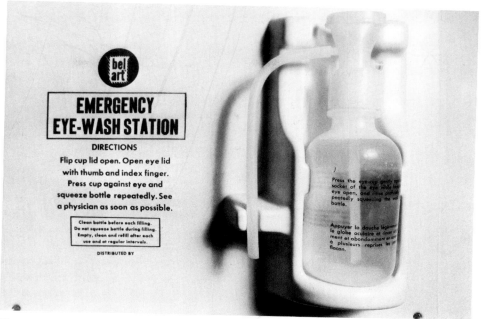

Figure 8-2.   An emergency eye-wash station. (Photo by Al Green.)

these metals. The properties of the most important bases are presented in the following paragraphs.

Sodium hydroxide (lye), NaOH, and **potassium hydroxide**, KOH, are both solids. Each is known as a caustic alkali because it forms a strongly basic, corrosive solution. Because they slowly react with glass, solutions of these bases should not be stored in bottles with glass stoppers. They are used in industry in such processes as making soaps and textiles.

Magnesium hydroxide, $Mg(OH)_2$, is a solid that dissolves only slightly in water. At a concentration of 7 to 8.5%, magnesium hydroxide dispersed in water is known as milk of magnesia. It is used as a laxative and antacid.

Calcium hydroxide, $Ca(OH)_2$, which is also only slightly soluble, is used as an astringent; it causes contraction of the skin and stops discharges by shrinking tissue. Its solution is known as limewater.

Ammonia, NH_3, is a gas that is very soluble in water. Solutions of ammonia are often mistakenly called "ammonium hydroxide." Although a small amount of ammonium and hydroxide ions are present, no undissociated NH_4OH actually exists. Next to sulfuric acid, ammonia is the most important chemical compound used in industry. Two percent ammonia is present in a mixture called aromatic spirits of ammonia, which is used as an inhalant to revive a person who has fainted.

8.5 The pH of acids and bases

The acidity or basicity (alkalinity) of a substance depends on two factors—its *solubility* in water and its *ability to ionize* in solution. Neither of these taken alone, the concentration or the degree of dissociation, gives you enough information. For example, acetic acid is extremely soluble in water and solutions having a high concentration can be prepared. These solutions are weakly acidic, however, because only a small fraction of the molecules dissociate. On the other hand, magnesium hydroxide is a base that dissociates almost completely in water. Yet only a small amount of hydroxide ions is present because relatively little of the solid base can be dissolved.

An acid is strong if the solution contains a large excess of hydrogen ions and a base is strong if the solution contains a large excess of hydroxide ions. The concentration of hydrogen ion and of hydroxide ion, and therefore the strength of an acid or base, is expressed by a quantity known as **pH**. The pH of a solution is a direct measure of the *concentration of hydrogen (or hydronium) ions in solution;* it is a scale running from 0, strong acid, to 14, strong base, in water. The midpoint, pH 7, represents a neutral solution, neither acidic nor basic, because the concentrations of H^+ and OH^- are equal. Table 8-4 illustrates the meaning of the pH scale; concentrations are given in terms of moles per liter.

Table 8-4   The pH Scale

pH	H^+ ion concentration[a]		OH^- ion concentration[a]	
0	1.		0.000 000 000 000 01	
1	0.1		0.000 000 000 000 1	
2	0.01		0.000 000 000 001	
3	0.001	acid	0.000 000 000 01	
4	0.000 1	(excess	0.000 000 000 1	
5	0.000 01	H^+ ions)	0.000 000 001	
6	0.000 001		0.000 000 01	
7	0.000 000 1		0.000 000 1	
8	0.000 000 01		0.000 001	
9	0.000 000 001		0.000 01	
10	0.000 000 000 1		0.000 1	base
11	0.000 000 000 01		0.001	(excess
12	0.000 000 000 001		0.01	OH^- ions)
13	0.000 000 000 000 1		0.1	
14	0.000 000 000 000 01		1.	

[a] Concentration in moles per liter (molarity).

Pure water is neutral; both the hydrogen ion concentration and the hydroxide ion concentration are 0.0000001 (10^{-7}) mole/liter. An acid solution has a greater hydrogen ion concentration and smaller hydroxide ion concentration than pure water; its pH is *less* than 7. The more hydrogen ions present, the more acidic the solution and the lower its pH. In a base, the reverse is true. It contains a larger number of hydroxide ions and a smaller number of hydrogen ions than a neutral solution. The pH of a basic solution is *greater* than 7; the more hydroxide ions present, the more basic the solution and the larger the value of pH above 7.

Tables 8-5 and 8-6 list the pH values of common solutions and body fluids. Most of the values given are not whole numbers. A pH of 1.5, for example, means that the hydrogen ion concentration is between 0.1 mole/liter (corresponding to pH 1) and 0.01 mole/liter (corresponding to pH 2). The actual relationship between pH and H^+ ion concentration is slightly complicated because it depends on a relationship based on logarithms: $pH = -\log[H^+]$; a logarithm (log) is the power to which 10 must be raised to equal the given number. Because of this relationship, a difference of 1 unit in pH reflects a difference of 10 in the hydrogen ion concentration. A solution with pH 1 has 10 times more hydrogen ions per liter than a solution with pH 2.

Table 8-5   The pH of Common Solutions

Solution	pH
hydrochloric acid[a]	1.1
sulfuric acid[a]	1.2
phosphoric acid[a]	1.5
citric acid (lemon juice)[a]	2.2
acetic acid (vinegar)[a]	2.9
carbonic acid (soda water, saturated)	3.8
tomato juice	4.2
coffee, black	5.0
boric acid[a]	5.2
rainwater	6.2
milk	6.5
water, pure	7.0
sodium bicarbonate[a]	8.4
magnesium hydroxide (milk of magnesia, saturated)	10.5
ammonia[a]	11.1
calcium hydroxide (saturated)	12.4
sodium hydroxide[a]	13.0

[a] A 0.1 N solution.

Table 8-6   **Typical pH Values of Body Fluids**

Fluid	pH
gastric juice	0.9
vaginal secretion	3.8
urine	6.0
milk	6.8
saliva	7.2
aqueous humor	7.2
blood	7.4
cerebrospinal fluid	7.4
intestinal juice	7.7
bile	7.8
pancreatic juice	8.0

8.6   Measurement of pH

Certain molecules called dyes, whose solutions are colored, can be used to measure the pH of a solution. Many have more than one color, depending on the pH. Examples of these **indicators** are listed in Table 8-7. Each has a certain region of about 2 pH units over which it changes from one color to another. Litmus paper, which contains a dye, is a commonly used indicator that changes from red to blue over a relatively wide region, from pH 4.4 to 8.3. Its color identifies a solution as a relatively strong acid (red color) or base (blue color). Litmus paper and other indicators can be used to test the pH of urine or body fluids.

Table 8-7   **Common Indicators**

Indicator	pH range	Acid color	Base color
thymol blue	1.2–2.8	red	yellow
bromphenol blue	3.0–4.6	yellow	blue
methyl orange	3.1–4.4	red	yellow
bromcresol green	3.8–5.4	yellow	blue
methyl red	4.8–6.0	red	yellow
bromthymol blue	6.0–7.6	yellow	blue
phenol red	6.4–8.0	yellow	red
thymol blue	8.0–9.6	yellow	blue
phenolphthalein	8.2–10.0	colorless	red
alizarin yellow	10.2–12.0	yellow	red
nitramine	10.8–13.0	colorless	brown

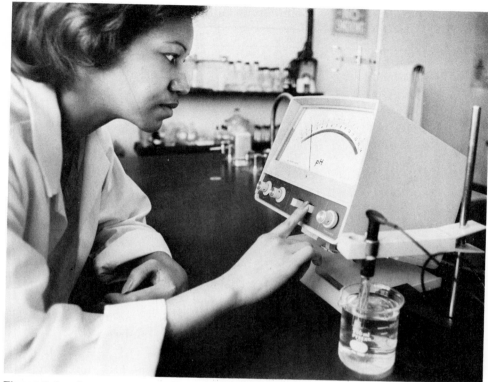

Figure 8-3. A technician using a laboratory pH meter. The scale is calibrated to measure the hydrogen ion concentration in pH units. (Photo by Al Green.)

The pH can be measured more exactly with an electronic instrument called a **pH meter**. Special glass tubes called electrodes are placed in the solution; the instrument detects the concentration of hydrogen ions in the solution and a pointer shows the pH value. The meter must be adjusted first, or standardized, with a solution of known pH. A pH meter is shown in Figure 8-3.

The pH of the stomach can be measured approximately by having the patient swallow a dye, azure A. It is administered bound to granules of a cation-exchange resin (such as used in converting "hard" water to "soft" water). If the stomach pH is below 3, the hydrogen ion concentration is great enough to replace much of the dye on the resin. The dye is thus released, absorbed, and excreted into the urine, giving it a blue or green color.

8.7 Neutralization and titration

The reaction of an acid with a base is called **neutralization.** In water, neutralization can be written as the reaction between a hydrogen ion and a hydroxide

Table 8-8 **Composition of Commercial Antacids**

Commercial name	Main antacid
Alka-seltzer	$NaHCO_3$
Bromo-seltzer	$NaHCO_3$
Di-Gel	$Al(OH)_3$, $Mg(OH)_2$
Eno	$NaHCO_3$
Gelusil	$Al(OH)_3$, $Mg_2Si_3O_8$
Maalox	$Al(OH)_3$, $Mg(OH)_2$
Mylanta	$Al(OH)_3$, $Mg(OH)_2$
Pepto-Bismol (tablet)	$CaCO_3$
Phillips' Milk of Magnesia	$Mg(OH)_2$
Rolaids	$NaAl(OH)_2CO_3$
Tums	$CaCO_3$

ionization

ion; it is the reverse of the dissociation process:

$$H^+ + OH^- \longrightarrow H_2O \qquad \text{(neutralization)}$$

(In terms of the hydronium ion, $H_3O^+ + OH^- \longrightarrow 2H_2O$.) A neutral water molecule forms, thus removing a H^+ ion and OH^- ion from the solution. One hydroxide ion exactly "cancels out" the effect of one hydrogen ion. If the solution is initially acidic, adding a base will decrease the concentration of hydrogen ions and raise the pH of the solution.

This reaction is the basis for commercial antacids, listed in Table 8-8. Antacids are used in cases of hyperacidity, excess acidity, as in patients with ulcers. These products often have effects other than their ability to neutralize acids. Aluminum hydroxide is constipating and magnesium hydroxide acts as a laxative; they are often combined to minimize these side effects. Prolonged use of antacids (especially sodium bicarbonate and calcium carbonate) can be dangerous without medical supervision because of other possible side effects.

The neutralization reaction can be used in a procedure called **titration** to find the concentration of either the acid or base, if the concentration of the other is known. For example, you could do a titration to find the concentration of acetic acid in different brands of vinegar. Using a buret, you slowly add a solution of known base concentration, such as sodium hydroxide, until all of the hydrogen ions of the acid have been neutralized by the added hydroxide ions. An indicator is used to signal the completion (end point) of the titration (Figure 8-4).

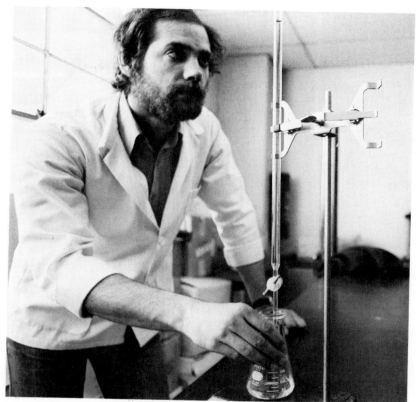

Figure 8-4. A technician performing a titration. A base is being slowly added from a buret to neutralize a known amount of acid of unknown concentration. (Photo by Al Green.)

8.8 Normality

At the point of neutrality, the concentration of hydrogen ions is the same as the concentration of hydroxide ions. The number of moles of H^+ ions is equal to the number of moles of OH^- ions in solution. But the moles of acid and base present are not necessarily equal. An acid can have more than one hydrogen atom in its molecule, and a base can have more than one hydroxide ion in its formula unit. A 1 M solution of sulfuric acid, H_2SO_4, can provide 2 moles of hydrogen ions when reacted with a base. Similarly, 1 M calcium hydroxide, $Ca(OH)_2$, contains 2 moles of hydroxide ion in solution.

To take this factor into account, a different method of expressing concentration is sometimes used. Instead of dealing with moles of a compound, you concentrate on moles of hydrogen ion or hydroxide ion. For acids, an **equivalent** or **equivalent weight** is the weight in grams that will provide 1 mole of hydrogen ions. It is found by dividing the weight of 1 mole by the number of hydrogens in the molecule that can react with a base. One equivalent of H_2SO_4 corresponds to the weight of 1 mole, 98 g, divided by 2, the number of replaceable hydrogens, or 49 g. Thus 49 g of sulfuric acid provide 1 mole of H^+ ions; 1 mole, 98 g, yields 2 moles of H^+ ions, or 2 equivalents.

For bases, the procedure is similar. The equivalent weight of a base is the weight in grams that reacts with 1 mole of hydrogen ions. You can find the weight of 1 equivalent by dividing the number of grams in 1 mole by the number of hydroxide ions in the formula unit. One equivalent of $Ca(OH)_2$ weighs 74 g divided by 2, or 37 g.

The number of equivalents in 1 liter of solution is called its **normality** (N):

$$\text{normality } (N) = \frac{\text{equivalents of solute}}{\text{liter of solution}}$$

You work with normality in exactly the same way as molarity except that mole is replaced by equivalent. A 1 normal (1 N) solution contains 1 equivalent/liter; a 2 normal (2 N) solution has 2 equivalents/liter, and so on. To prepare a liter of 0.01 N $Ca(OH)_2$, you would dissolve 0.01 equivalent (0.37 g) in enough water to make 1 liter of solution. If you need a different volume, the weight must be adjusted accordingly as with molarity.

The normality of a solution is always greater than or equal to its molarity. A 1 molar (1 M) solution of HCl is also 1 normal (1 N) because the weights of 1 mole and 1 equivalent are the same in this case. But a 1 molar (1 M) solution of H_2SO_4 is 2 normal (2 N) because each mole of acid provides 2 moles of hydrogen ions; the weight of a mole is two times the weight of an equivalent.

In a titration, *the number of equivalents of acid is equal to the number of equivalents of base* at neutrality because the number of moles of H^+ is the same as the number of moles of OH^-. You can use the following relationship to find the concentration of the acid or base:

$$\left(\begin{array}{c}\text{normality}\\\text{of acid}\end{array}\right) \times \left(\begin{array}{c}\text{volume}\\\text{of acid}\end{array}\right) = \left(\begin{array}{c}\text{normality}\\\text{of base}\end{array}\right) \times \left(\begin{array}{c}\text{volume}\\\text{of base}\end{array}\right)$$

For example, for the vinegar titration described in the previous section, you might need 20 ml of 0.5 N NaOH to titrate 10 ml of the vinegar. The solution is as follows:

$$\left(\begin{array}{c}\text{normality}\\\text{of acid}\end{array}\right) \times (10 \text{ ml}) = (0.5 \text{ } N) \times (20 \text{ ml})$$

$$\text{normality of acid} = \frac{(0.5 \text{ } N) \times (20 \text{ ml})}{10 \text{ ml}} = 1 \text{ } N$$

8.9 Salts and hydrolysis

When an acid reacts with a base, it forms an ionic compound, a **salt**, as well as water molecules. For example, the reaction of hydrochloric acid and sodium hydroxide produces sodium chloride and water:

$$HCl + NaOH \longrightarrow NaCl + H_2O$$

acid   base   salt   water

The salt consists of a *cation from the base* (Na^+) and an *anion from the acid* (Cl^-). If you carry out this reaction and boil off the water, you will see white crystals of sodium chloride. More examples of the formation and naming of salts are presented in Table 8-9. (See Table 3-10 for common salts and their medical uses.)

When dissolved in water, some salts react with the water in a process called **hydrolysis**, changing the pH from the neutral value 7. For example, when sodium acetate, $NaC_2H_3O_2$, dissolves, the resulting acetate ion can react with water to form slightly ionized acetic acid:

$$C_2H_3O_2^- + H_2O \longrightarrow HC_2H_3O_2 + OH^-$$

The net result is the breakup of water molecules, with the H^+ ions being tied up as weakly dissociated acetic acid and the OH^- ions left free in solution. The solution is thus basic and the pH is above 7.

On the other hand, when ammonium chloride, NH_4Cl, dissolves, the ammonium ion reacts with water to form ammonia:

$$NH_4^+ + H_2O \longrightarrow NH_3 + H_3O^+$$

A hydronium ion, the hydrated hydrogen ion, is produced, making the solution acidic with a pH below 7. Hydrolysis occurs whenever one of the ions of the salt can react with water to form a weak acid or a weak base.

Table 8-10 lists those common ions that make an aqueous solution acidic

Table 8-9 Formation of Salts

Base	Acid	Formula of salt	Name of salt
$NaOH$	H_2SO_4	Na_2SO_4	sodium sulfate
KOH	H_3PO_4	K_3PO_4	potassium phosphate
$Mg(OH)_2$	$HC_2H_3O_2$	$Mg(C_2H_3O_2)_2$	magnesium acetate
$Ca(OH)_2$	HCl	$CaCl_2$	calcium chloride
NH_3	HNO_3	NH_4NO_3	ammonium nitrate

Table 8-10 **Effect of Ions on pH**

Form acidic solutions	Form neutral solutions	Form basic solutions
HSO_4^-	Cl^-	CO_3^{2-}
$H_2PO_4^-$	Br^-	HCO_3^-
NH_4^+	I^-	PO_4^{3-}
metal cations	NO_3^-	HPO_4^{2-}
(Zn^{2+}, Cu^{2+}, etc.)	SO_4^{2-}	$C_2H_3O_2^-$

and those that make it basic by hydrolysis, as well as those that do not hydrolyze at all.

8.10 Body electrolytes

The salts present in the body exist in solution as ions and are therefore called **electrolytes** (they are able to conduct an electric current). The major electrolytes and their roles are summarized in Table 8-11. They are essential for maintaining a fluid balance and acid–base balance in the body as well as for the normal functioning of the cells.

The concentration of electrolytes in body fluids is expressed in terms of equivalents. The equivalent weight of an ion is equal to the weight in grams of 1 mole of the ion divided by its charge (ignoring the sign). For ions with a

Table 8-11 **Functions of Major Electrolytes**

Ion	Function
Na^+	primary extracellular (outside cell) cation; maintains osmotic pressure and water balance in blood and tissue spaces; needed for nerve and muscle activity
K^+	primary intracellular (inside cell) cation; maintains osmotic pressure in cells; needed for nerve and muscle activity
Ca^{2+}	provides framework for bones and teeth; needed for blood clotting and muscle activity
Mg^{2+}	needed for enzyme activity and neuromuscular system
Cl^-	primary extracellular anion; needed for gastric HCl secretion; involved in blood transport of O_2 and CO_2
HPO_4^{2-}	primary intracellular anion; present in bones with Ca^{2+}; acid–base buffer
HCO_3^-	required as buffer to maintain acid–base balance of blood
SO_4^{2-}	present in cells with proteins

Table 8-12   **Electrolyte Composition of Body Fluids**

	Intracellular[a] (mEq/liter)	Interstitial[b] (mEq/liter)	Plasma[c] (mEq/liter)
Cation			
Na$^+$	15	147	142
K$^+$	150	4	5
Ca^{2+}	2	2.5	5
Mg^{2+}	27	1	2
total cations	194	155.5	154
Anion			
Cl$^-$	1	114	105
HCO$_3^-$	10	30	24
HPO$_4^{2-}$	100	2	2
SO$_4^{2-}$	20	1	1
organic acids	0	7.5	6
proteins	63	0	16
total anions	194	155.5	154

[a] Fluid inside the cells.
[b] Fluid in the spaces between the cells.
[c] The liquid portion of blood.

single charge, such as Na$^+$ or Cl$^-$, the equivalent weight is the same as the weight of 1 mole. For ions with two charges, like Ca^{2+} or SO$_4^{2-}$, the equivalent weight is equal to one-half the weight of a mole of the ion. Because relatively small amounts of ions are present in the body, the milliequivalent (mEq), 1/1000 the size of an equivalent, is used. The use of the milliequivalents is gradually being replaced by the preferred SI unit, the millimole (mmol).

Table 8-12 lists the composition of the electrolytes of the major body fluids in milliequivalents per liter (mEq/liter). The ions found mainly within the body cells are potassium ion (K$^+$), magnesium ion (Mg^{2+}), monohydrogen phosphate ion (HPO$_4^{2-}$), sulfate ion (SO$_4^{2-}$), and proteins (which contain anionic groups). The ions present in greater concentration outside the cells are sodium ion (Na$^+$), calcium ion (Ca^{2+}), chloride ion (Cl$^-$), bicarbonate ion (HCO$_3^-$), and organic acids (in the form of their carboxylate anions).

Just as 1 equivalent of acid neutralizes 1 equivalent of base, 1 milliequivalent of any cation chemically balances 1 milliequivalent of any anion. In the body fluids, an **electrolyte balance** exists. The total number of milliequivalents of positive ions, cations, equals the total number of milliequivalents of negative ions, anions.

Table 8-13 presents the causes and symptoms of various **electrolyte imbalances**. These conditions result from either a deficiency or an excess of one of

Table 8-13 Electrolyte Imbalances

Ion	Normal range[a] (blood)	Deficiency		Excess	
		Causes	Symptoms	Causes	Symptoms
Na^+	135–143 mEq/liter (135–143 mmol/liter)	*hyponatremia* high H$_2$O intake, adrenal or hypophysis insufficiency	low blood pressure, weakness, lethargy, vomiting, cramps	*hypernatremia* water deficit, diabetes insipidus	central nervous system affected, edema
K^+	3.7–4.7 mEq/liter (3.7–4.7 mmol/liter)	*hypokalemia* excessive losses through kidney or bowel (diarrhea, diuretics, vomiting)	muscular weakness, paralysis, mental changes	*hyperkalemia* acidosis, poor excretion	heart failure, weakness
Ca^{2+}	9–10.5 mg/100 ml (2.2–2.7 mmol/liter)	*hypocalcemia* lack of vitamin D, increased body need, hormonal imbalance	tetany: spasms and convulsions	*hypercalcemia* excessive milk or vitamin D	gastrointestinal and neuromuscular systems disturbed, renal calculi
Mg^{2+}	1.4–2.3 mEq/liter (0.7–1.2 mmol/liter)	*hypomagnesemia* poor absorption, alcoholism, diuretics	variable: confusion, convulsions, tetany	*hypermagnesemia* magnesium drugs	hypotension, stupor, respiratory failure

[a] SI units given in parentheses.

Table 8-14 Common Electrolyte Solutions

Name	Cations (mEq/liter)	Anions (mEq/liter)	Comparison to blood cells	Use
normal saline (0.9% NaCl) solution	Na^+ (154)	Cl^- (154)	isotonic	expand extracellular fluid
Ringer's solution	Na^+ (147), K^+ (4), Ca^{2+} (5)	Cl^- (156)	isotonic	replace K^+ and Ca^{2+}, in addition to Na^+ and Cl^-
Hartmann's (lactated Ringer's) solution	Na^+ (130), K^+ (4), Ca^{2+} (3)	Cl^- (109), lactate$^-$ (28)	isotonic	replace fluid lost as bile, diarrhea, or in burns
Butler's solution (electrolyte #2) balanced	Na^+ (57), K^+ (25), Mg^{2+} (5–6)	Cl^- (49–50), lactate$^-$ (25), HPO_4^{2-} (13)	hypotonic	supply H_2O, maintain Na^+, Mg^{2+}, K^+, Cl^-
isotonic solution	Na^+ (140), K^+ (5), Mg^{2+} (3)	Cl^- (98), acetate$^-$ (27), gluconate$^-$ (23)	isotonic	replace extracellular fluid, substitute for normal saline

the ions required by the body. Because the electrolyte concentration affects the osmotic pressure of the fluid, an electrolyte disturbance may also cause a water imbalance. Table 8-14 lists common electrolyte solutions administered intravenously to replace body fluids and supply needed ions.

8.11 Buffers

Your body works properly only when the pH values of the body fluids are within their normal range. To keep these fluids from becoming too acidic or too basic, your body uses a number of **buffer** systems. They are substances that protect you, reducing the shock of a sudden change in pH. A buffer maintains an almost constant pH by neutralizing small amounts of acid or base that may be added to your system.

Blood is kept at pH 7.4 by several buffers. The most important consists of a weak acid, carbonic acid, H_2CO_3, and an anion of that acid, bicarbonate ion, HCO_3^-, in the ratio 1 to 20. The two parts of the buffer enable it to react with either added acid or added base. If acid enters the blood, the bicarbonate ion picks up H^+ ions to form more carbonic acid:

$$H^+ + HCO_3^- \longrightarrow H_2CO_3$$

Since carbonic acid is only slightly ionized, the added H^+ ions are effectively removed from solution, keeping the hydrogen ion concentration from changing. If a base enters the blood, OH^- ions react with the carbonic acid part of the buffer to form more bicarbonate ion:

$$OH^- + H_2CO_3 \longrightarrow HCO_3^- + H_2O$$

The added base is thus neutralized and cannot change the pH of the blood. Adding small amounts of acid or base changes only very slightly the relative amounts of carbonic acid and bicarbonate ion in the buffer. Thus, the pH stays nearly constant.

An additional buffer in the blood consists of the monohydrogen phosphate ion, HPO_4^{2-}, and the dihydrogen phosphate ion, $H_2PO_4^-$. These ions work in a similar way to the carbonic acid/carbonate ion buffer, "neutralizing" any added acid or base:

$$H^+ + HPO_4^{2-} \longrightarrow H_2PO_4^-$$
$$OH^- + H_2PO_4^- \longrightarrow HPO_4^{2-} + H_2O$$

As you will learn later, large molecules called proteins also help to keep the blood pH from changing. Figure 8-5 illustrates a blood pH meter.

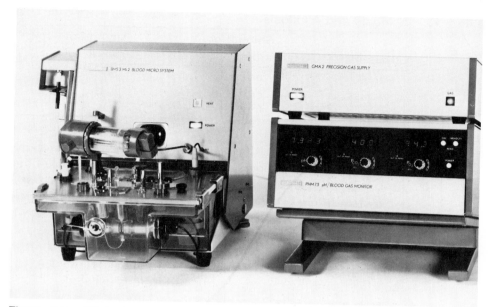

Figure 8-5. Clinical blood pH meter. This instrument determines the pH as well as partial pressure of carbon dioxide and oxygen of a 130-μl blood sample. (Photo courtesy of The London Company.)

8.12   Acidosis and alkalosis

Blood has a pH of 7.35–7.45 [corresponding to a hydrogen ion concentration of 36–45 nanomoles (10^{-9} mole) per liter] when the bicarbonate ion concentration is 20 times the concentration of carbonic acid. Changing the amount of one of the components of the buffer relative to the other changes the pH of the blood. A pH lower than 7.35 is known as **acidosis**; this condition results from a decrease in the bicarbonate ion concentration or an increase in the concentration of carbonic acid. If untreated, acidosis causes disorientation, coma, and death.

Alkalosis is an increase in the blood pH above 7.45. The normal ratio of the buffers is changed by an increase in the bicarbonate ion and a decrease in carbonic acid concentration. Symptoms include weak, irregular breathing, muscle cramps, and convulsions. Table 8-15 compares the conditions of acidosis and alkalosis with the body's normal state.

Respiration is the most important means of controlling the blood buffer system. The concentration of carbonic acid depends on the partial pressure of carbon dioxide in your lungs, since the acid can form by a reaction between this gas with water:

$$CO_2 + H_2O \longrightarrow H_2CO_3$$

Table 8-15   Acidosis and Alkalosis

Condition	Amount of bicarbonate ion	Amount of carbonic acid	Blood pH
normal	normal	normal	7.35–7.45
acidosis	low	high	below 7.35
alkalosis	high	low	above 7.45

If you hold your breath, the increased partial pressure of carbon dioxide in the lungs causes more carbonic acid to form, lowering the pH of the blood. If you did not breathe for 5 minutes, the blood pH would fall to 6.3.

This condition, known as **respiratory acidosis**, also results from pneumonia, emphysema, anesthesia, poliomyelitis, or heart failure, all of which cause hypoventilation (reduced respiration). The nervous system responds by increasing the depth and rate of breathing to decrease the partial pressure of carbon dioxide in the lungs. Carbonic acid breaks down, reducing its concentration in the blood, and the resulting carbon dioxide is exhaled.

Respiratory alkalosis results in an increase of the blood pH because of hyperventilation. The excessive loss of carbon dioxide reduces the carbonic acid concentration, raising the pH to 7.6 or 7.7 in several minutes. To compensate for this condition, found in hysterics, the body lowers the respiratory rate. This drop increases the partial pressure of carbon dioxide in the lungs, causing a rise in the carbonic acid concentration and therefore lowering the pH.

Metabolic acidosis is a condition in which the blood plasma pH drops below 7.35 through a decrease in the bicarbonate ion concentration. This change is more common and more dangerous than respiratory acidosis. Plasma HCO_3^- ion concentration falls whenever any acid stronger than carbonic acid is added to the extracellular fluid. One such acid is acetoacetic acid, formed in cases of untreated diabetes mellitus. This acid converts bicarbonate ions into molecules of carbonic acid, lowering the pH. Metabolic acidosis is also caused by severe dehydration because of accompanying failure in hydrogen ion excretion and by diarrhea, from loss of bicarbonate ion.

To counteract this decrease in pH, the kidney increases the amount of bicarbonate ion by excreting acidic urine and ammonium ion, NH_4^+. Acid is excreted as $H_2PO_4^-$, the dihydrogen phosphate ion, made from the monohydrogen phosphate ion, HPO_4^{2-}, of the blood. Ammonium ions, made by protonating ammonia, replace sodium ions in the blood.

Metabolic alkalosis is an increase in the blood pH from an abnormally high concentration of bicarbonate ion. It may result from vomiting or even an overdose of antacids used to relieve "heartburn." In this case, the kidney excretes alkaline urine to make the extracellular fluid more acidic. Thus, sodium ions along with either HPO_4^{2-} or HCO_3^- are excreted to decrease the bicarbonate ion concentration of the blood.

SUMMARY

Pure water dissociates very slightly into an equal number of hydrogen ions, H^+, and hydroxide ions, OH^-. An acid is a substance that raises the hydrogen ion concentration of water. The relative number of H^+ ions becomes larger than the number of OH^- ions. A hydrogen ion in water can be represented by the hydronium ion, H_3O^+. The strength of an acid depends on how completely it is ionized in solution. Important acids include hydrochloric acid (HCl), nitric acid (HNO_3), sulfuric acid (H_2SO_4), carbonic acid (H_2CO_3), phosphoric acid (H_3PO_4), boric acid (H_3BO_3), and acetic acid ($HC_2H_3O_2$).

A base is a substance that raises the hydroxide ion concentration of water. A basic or alkaline solution thus has a larger number of hydroxide ions than hydrogen ions. Some bases are "strong" in terms of dissociation but provide few hydroxide ions because of their limited solubilities. Important bases include sodium hydroxide (NaOH), potassium hydroxide (KOH), magnesium hydroxide [$Mg(OH)_2$], calcium hydroxide [$Ca(OH)_2$], and ammonia (NH_3).

The pH of a solution is a direct measure of its hydrogen ion concentration. It is a scale running from 0, strongly acidic, to 14, strongly basic, in water. The midpoint, pH 7, represents a neutral solution, neither acidic nor basic. An acid has a pH below 7, while a base has a pH above 7. The pH of a solution can be determined using certain dyes, called indicators, or by a pH meter.

Neutralization is the reaction of an acid with a base. One hydroxide ion exactly "cancels out" the effect of one hydrogen ion, forming a neutral water molecule. Antacids work by neutralizing acids in this way. A titration is a neutralization reaction performed to find the concentration of an acid or a base.

An equivalent or equivalent weight of an acid is the weight in grams that provides 1 mole of hydrogen ions. For a base, it is the weight that reacts with 1 mole of hydrogen ions. Normality is defined as the number of equivalents of acid or base per liter of solution. The normality of a solution is always greater than or equal to its molarity.

A salt, an ionic compound, forms when an acid and base react. It consists of a cation from the base and an anion from the acid. Some salts react with water in a process called hydrolysis, changing the pH from the neutral value of 7.

The salts present in the body exist in solution as ions and are therefore called electrolytes. The most important electrolytes are Na^+, K^+, Ca^{2+}, Mg^{2+}, Cl^-, HCO_3^-, HPO_4^{2-}, and SO_4^{2-}. They are essential for maintaining fluid balance and acid–base balance, as well as for the normal functioning of the cells. Their concentrations are expressed as milliequivalents per liter.

Buffers keep the pH values of the body fluids within their normal range. They keep the pH from changing by neutralizing small amounts of acid or base that may be added to the system. The most important buffer in the blood consists of carbonic acid, H_2CO_3, and bicarbonate ion, HCO_3^-.

Blood has a normal pH of 7.4 when the bicarbonate ion concentration is 20 times the concentration of carbonic acid. Respiration is the major method of maintaining this ratio. Acidosis is a condition in which the blood pH drops below 7.35; respiratory acidosis can be caused by hypoventilation. Respiratory alkalosis, which may result from hyperventilation, is an increase in the blood pH above 7.45. Metabolic acidosis is a condition in which the blood pH drops below 7.35 because of a decrease in the bicarbonate ion concentration. Metabolic alkalosis is the increase of the pH above 7.45 because of an abnormally high concentration of bicarbonate ion. The kidney reacts to these conditions by adjusting the excretion of ions in the urine.

Exercises

1. (Intro.) What is meant by the "self-ionization" of water?
2. (8.1) Define an acid.
3. (8.1) Write the equations showing the ionization of phosphoric acid, H_3PO_4.
4. (8.1) "Carbonic acid is a weak acid." Explain this statement.
5. (8.2) Describe three properties of acids.
6. (8.2) For each of the following, write its formula and state its use: hydrochloric acid, nitric acid, sulfuric acid, carbonic acid, phosphoric acid, boric acid, acetic acid.
7. (8.3) What is a base?
8. (8.3) Why is ammonia a base even though it contains no hydroxide ions?
9. (8.4) Describe three properties of bases.
10. (8.4) For each of the following, write its formula and state its use: sodium hydroxide, potassium hydroxide, magnesium hydroxide, calcium hydroxide, ammonia.
11. (8.5) What factors determine the acidity or basicity of a substance?
12. (8.5) What does pH mean?
13. (8.5) Refer to Tables 8-5 and 8-6 and identify the following as acidic or basic: tomato juice, coffee, blood, urine, milk of magnesia.
14. (8.5) What is the hydrogen ion concentration and hydroxide ion concentration in a solution of pH 5 (refer to Table 8-4).
15. (8.6) What is an indicator? How does it work?
16. (8.7) Describe how an antacid functions.
17. (8.7) What is a titration?
18. (8.8) Determine the equivalent weight of (a) H_2CO_3; (b) $Mg(OH)_2$; (c) H_3PO_4.
19. (8.8) How many grams of H_2SO_4 are needed to prepare 500 ml of a 4 N solution?

20. (8.8) What is the normality of the following solutions: (a) 2 M H_3PO_4; (b) 0.5 M HCl; (c) 1 M $Ca(OH)_2$; (d) 0.1 M HNO_3.

21. (8.8) Find the concentration of vinegar if 20 ml of 0.5 N NaOH are needed to neutralize 40 ml.

22. (8.9) What is a salt? How is it formed?

23. (8.9) Write the formula of the salt formed from (a) hydrochloric acid and magnesium hydroxide; (b) nitric acid and potassium hydroxide; (c) ammonia and acetic acid.

24. (8.9) Predict whether the solutions containing the following salts will be acidic, basic, or neutral: (a) Na_3PO_4; (b) $KHCO_3$; (c) $LiNO_3$; (d) NH_4Br; (e) $CsHSO_4$.

25. (8.9) Describe five salts and their medical applications.

26. (8.10) What are electrolytes of the body fluids and what are their functions?

27. (8.10) What is meant by electrolyte balance?

28. (8.10) Compare the electrolyte composition of the intracellular fluid, the interstitial fluid, and the blood plasma.

29. (8.10) What is hypercalcemia? hypokalemia? hypernatremia? hypomagnesemia?

30. (8.11) How is the blood kept at pH 7.4?

31. (8.12) What is acidosis? alkalosis?

32. (8.12) How is respiration related to acidosis and alkalosis?

33. (8.12) How do metabolic acidosis and alkalosis differ from respiratory acidosis and alkalosis?

34. (8.12) How does the kidney compensate for (a) metabolic acidosis? (b) metabolic alkalosis?

Nuclear chemistry and radiation

So far, you have learned about the chemistry of elements and compounds. Their properties and reactions depend mainly on the number and arrangement of valence electrons of the atoms involved. In contrast, nuclear chemistry deals with changes in the *nuclei* of atoms, their number of protons and neutrons. These nuclear changes can create a large amount of energy as demonstrated by the explosion of an atomic bomb. They also provide valuable methods for the diagnosis and treatment of disease, such as cancer.

9.1 Radioactivity

Certain atoms are unstable: Their nuclei break down, giving off particles or energy known as **radiation**. This process of nuclear decomposition or disintegration is called **radioactivity**. Many elements are naturally radioactive, most commonly when the atomic number is larger than 83. For example, uranium (atomic number 92) is a radioactive element.

When a radioactive atom decomposes, the radiation it gives off can have three possible forms, called alpha radiation, beta radiation, and gamma radiation. **Alpha radiation** is symbolized by the Greek letter α, for which it is named. It consists of particles, called alpha particles, that have two protons and two neutrons; they comprise the nuclei of helium atoms. Thus, alpha particles have a mass of 4 amu and a charge of 2+. They are easily stopped because of their large size and cannot even penetrate the layer of dead cells on the surface of the skin.

Beta radiation, symbolized by the Greek letter β, simply consists of electrons given off with high speed and energy. These beta particles thus have a charge of 1− and a very small mass (1/1827 amu), which is usually ignored and called zero. Because they are smaller than alpha particles, beta particles can pene-

Table 9-1   Properties of Radiation

Name	Symbol	Composition	Mass (amu)	Charge	Penetrating ability
alpha	α	helium nucleus (2 protons and 2 neutrons)	4	2+	low
beta	β	electron	1/1827	1−	moderate
gamma	γ	ray of energy	0	0	very high

trate the skin (to a depth of 4 mm), causing burns. They are stopped before reaching any internal organs.

Gamma radiation, given the symbol γ, differs from the other two kinds of radiation. It does not consist of particles, but of rays having extremely high energy. Gamma rays therefore have no mass and no charge. They penetrate deeply into the body and can cause serious damage. This type of radiation can even pass through lead or concrete. The properties of the different types of radiation given off by a radioactive atom are compared in Table 9-1.

9.2   Nuclear reactions

The process of nuclear change can be represented in the form of a **nuclear equation**, just as chemical changes are expressed by chemical equations. Each nucleus is written as the symbol for that element with a superscript, or raised number, on its left side giving the mass number and a subscript below it giving the atomic number and therefore the charge. For example, radium, atomic number 88, has an isotope of mass 226 which can be symbolized in the following way:

$$(\text{mass number} = 226) \quad \longrightarrow \quad {}^{226}_{88}\text{Ra}$$
$$(\text{atomic number} = 88) \quad \longrightarrow$$

Sometimes the atomic number is left out because the symbol for the element already gives you this information. You would call the nucleus described by the above symbol radium-226.

Radium is radioactive, breaking down to form another element, radon (Rn), and giving off an alpha particle:

$$\begin{array}{ccccc} {}^{226}_{88}\text{Ra} & \longrightarrow & {}^{222}_{86}\text{Rn} & + & {}^{4}_{2}\text{He} \\ \text{radium} & & \text{radon} & & \text{alpha} \\ & & & & \text{particle} \end{array}$$

This process, a nuclear reaction, involves one element changing or transforming into a different element. By emitting an alpha particle from its nucleus, radium loses two protons, becoming radon. Its mass decreases by 4 because the alpha particle has two neutrons in addition to the two protons. Therefore, the isotope of radon formed has a mass of 226 − 4, or 222 amu.

In a nuclear reaction, the protons and neutrons are rearranged, just as in a chemical reaction the atoms are rearranged. But *their total number remains the same.* The sum of the mass numbers, the superscripts, on one side of the equation (226 amu) must be the same as the sum on the other side (222 + 4 = 226 amu). Also, the sum of the atomic numbers, the subscripts, on the left side (88) has to be equal to the sum on the right side of the equation (86 + 2 = 88) because electrical charge cannot be destroyed.

9.3 Natural radioactivity

The reaction in which $^{226}_{88}\text{Ra}$, radium-226, breaks down into $^{222}_{86}\text{Rn}$, radon-222, is an example of radioactive decay or disintegration. It is part of the chain of reactions called a **radioactive disintegration series.** All the naturally occurring radioactive elements fall into one of four such series. This one begins with $^{238}_{92}\text{U}$, uranium-238, as shown in Table 9-2.

Each nucleus in the series is radioactive and decays by giving off either an alpha particle or a beta particle. It is easy to understand how an unstable nucleus can give off an alpha particle consisting of two protons and two neutrons. Since the nucleus contains no electrons, it is more difficult to explain how a beta particle, an electron, can be emitted by the nucleus. According to the present theory of the nucleus, protons and neutrons are held together by small particles (having about 300 times the mass of an electron) called pi-mesons or pions. Protons and neutrons share these pions in a way similar to

Table 9-2 Uranium Disintegration Series[a]

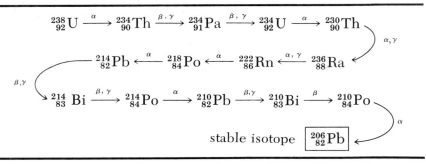

[a] Symbols above arrow indicate types of radiation emitted by transformation.

atoms sharing electrons in a molecule. Separation of the particles in the nucleus breaks these nuclear bonds and may result in a pion being given off. This pion then converts into an electron (beta particle) as well as other particles (neutrinos). At the same time, a neutron is converted to a proton. Thus, although electrons are not present in the nucleus, they can be formed by this process and emitted as beta particles. As shown in Table 9-2, gamma rays are also given off in many of these transformations.

All isotopes of the elements having atomic weights greater than bismuth (atomic number 83) are naturally radioactive. In addition, some stable elements have particular isotopes that are radioactive (such as potassium-40 and rubidium-87). Each of these spontaneously decays to form another nucleus and therefore emits one or more types of radiation.

9.4 Artificial radioactivity

Most of the radioactive isotopes—called **radioisotopes,** for short—used in medicine are produced artificially. One of the most common methods is bombarding a stable nucleus with neutrons, forming a radioactive isotope and causing the emission of gamma rays (Figure 9-1). This process is known as **neutron activation** or **neutron capture.** For example, radioactive cobalt-60 is made by hitting stable cobalt-59 nuclei with neutrons:

$$\ce{^{59}_{27}Co} + \ce{^{1}_{0}n} \longrightarrow \ce{^{60}_{27}Co}$$

The radioisotope gold-198 is prepared from gold-197 in a similar way:

$$\ce{^{197}_{79}Au} + \ce{^{1}_{0}n} \longrightarrow \ce{^{198}_{79}Au}$$

In these cases, the radioactive isotope cannot be readily separated from the stable form of the element, called the carrier.

By hitting a stable nucleus with smaller particles, like protons, neutrons, or alpha particles, it can be changed into a nucleus of a different element, which is radioactive. This process is **transmutation;** it became possible only after the development of nuclear reactors, which produce neutrons and other particles, and huge machines called accelerators, which can give charged particles great energy. An example of a transmutation is the production of radioactive phosphorus-32 from stable sulfur-32:

$$\ce{^{32}_{16}S} + \ce{^{1}_{0}n} \longrightarrow \ce{^{32}_{15}P} + \ce{^{1}_{1}H}$$

In this case, because different elements are involved, they can be separated. The resulting radioisotope is said to be "carrier free."

A convenient method for generating radioisotopes in a health center or laboratory is with a generator, as shown in Figure 9-2. It consists of a shielded column filled with alumina (aluminum oxide), which holds a commercially

"I Thought Thorium Gave Off A <u>Gamma Ray</u> When It Absorbed A Neutron"

Figure 9-1. Reprinted with permission from *Industrial Research,* October 1973.

prepared radioactive isotope. This "parent" nucleus continually decays, forming its "daughter" nucleus, the radioisotope of interest. The "daughter" can be easily separated from the column when needed by elution, that is, "washing" or "milking" it from the resin. Examples are the generation of technetium-99m from molybdenum-99 and indium-113m from tin-113. (The "m" stands for metastable, meaning that the isotope is in an "excited" state that decays quickly.)

Over 1500 radioisotopes have been prepared artificially. The transuranium elements, those having atomic numbers greater than uranium (92), are all produced by nuclear reactions in particle accelerators or nuclear reactors.

9.5 Half-life

One of the most important properties of a radioactive nucleus, whether natural or artificial, is its *rate of decay.* The time required for the decay of one-half

(a)

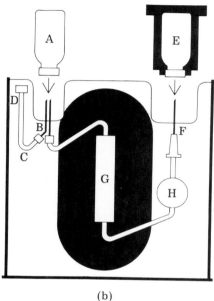

(b)

Figure 9-2. (a) Photograph of a generator used to produce samples of technetium-99m. (b) A charge vial A containing 20 ml of isotonic saline solution is placed onto the double needle B and vented through tube C fitted with a cotton pledget filter D. Elution begins automatically when a shielded evacuated collection vial E is placed onto needle F. As the saline is removed from charge vial A, it passes through shielded column G loaded with the parent, molybdenum-99. Technetium-99m is selectively eluted and the eluate then passes through filter H and needle F into shielded collection vial E. (Photo and line drawing courtesy of New England Nuclear.)

of a given number of nuclei is called a **half-life**, abbreviated $t_{1/2}$. The half-life can vary from a fraction of a second for a very unstable radioisotope to millions of years for a stable isotope. For example, the half-life of iodine-131 is 8 days. If you initially had 100 g of this radioisotope, after 8 days you would have 50 g of iodine-131 left. After another 8 days, only 25 g of iodine-131 would remain. In the following 8 days, decay would continue, leaving 12.5 g of iodine-131, and so on. With the passing of each 8 days, the half-life of this nucleus, one-

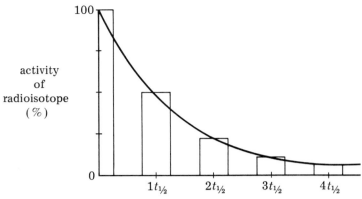

Figure 9-3. The decay curve of a radioisotope. After each half-life, the radioactivity decreases by one-half.

half of the remaining radioisotope disintegrates, producing another nucleus and radiation. This relationship is shown by the graph in Figure 9-3.

Table 9-3 contains the values of the half-life for important radioisotopes. For medical applications, a *short* half-life is necessary. The less time the radioactive isotope is present in the body, the less possible damage it can do. A short half-life permits larger amounts of an isotope to be administered.

Do not confuse this physical half-life of a radioisotope with what is known as the **biological half-life**. This term represents the rate of elimination of the isotope from the body. The biological half-life is determined by where an isotope ends up and how quickly it gets replaced by nonradioactive atoms. Strontium-90, which is released in nuclear explosions, is a dangerous radioisotope because it can replace calcium and become concentrated in the bones. Here its biological half-life is nearly 50 years. Since the half-life, $t_{1/2}$, of this isotope is also long, 21 years, a child absorbing this isotope carries some inside the body for a long time, increasing the possibility of damage from the radiation it gives off.

9.6 Nuclear energy

The energy given off by a nuclear reaction is based on the *difference in mass between the products and reactants*. Unlike a chemical reaction, a nuclear change does not obey the law of conservation of mass. Instead, a *small* amount of mass is converted to a *large* quantity of energy. (The relationship between matter and energy is the Einstein equation $E = mc^2$, where E is the energy

Table 9-3
Radioactive Isotopes Used in Medicine

Name	Radiation emitted	Decay product	Physical half-life	Biological half-life	Uses
chromium-51	gamma	^{51}V	28 days	616 days	red blood cell studies
cobalt-60	beta, gamma	^{60}Ni	5.3 years	9.5 days	cancer therapy
gold-198	beta, gamma	^{198}Hg	2.7 days	120 days	cancer therapy
iodine-131	beta, gamma	^{131}Xe	8 days	138 days	thyroid studies, organ scans
indium-113m	gamma	^{113}In	100 minutes	48 days	liver and spleen scans
phosphorus-32	beta	^{32}S	14 days	257 days	leukemia therapy
radium-226	alpha, gamma	decay chain	1600 years	22 years	cancer therapy
selenium-75	beta, gamma	^{75}As	120 days	11 days	pancreas scans
strontium-85m	gamma	^{85}Rb	70 minutes	35.6 years	bone scans
technetium-99m	gamma	^{99}Ru	6 hours	1 day	organ scans
ytterbium-169	gamma	^{169}Tm	32 days	685 days	brain scans
xenon-133	beta, gamma	^{133}Cs	5 days	short	lung studies

given off, m is the mass, and c is a constant, the speed of light.) For example, in the decay of radium-226 (mass of reactant = 226.0254 amu) to radon-222 and an alpha particle (mass of products = 222.0176 + 4.0026 = 226.0202 amu), mass is "lost" (0.0052 amu)—it is transformed into an equivalent amount of energy. In a nuclear reaction, mass and energy are conserved only when considered together.

A special reaction that releases vast energy is **fission**, the "splitting" of a heavy atomic nucleus into two or more pieces when hit by a neutron. The uranium-235 nucleus is "fissionable"; it can absorb a neutron to form an unstable form of uranium-236, which then breaks into two fragments, such as barium-142 and krypton-91:

$$\ce{^{235}_{92}U} + \ce{^{1}_{0}n} \longrightarrow [\ce{^{236}_{92}U}] \longrightarrow \ce{^{142}_{56}Ba} + \ce{^{91}_{36}Kr} + 3\ce{^{1}_{0}n}$$

Only one neutron starts the reaction but two to three are formed. These additional neutrons can react with three other uranium-235 atoms, which in turn each form three more neutrons that hit still other uranium atoms, and so on, creating a **chain reaction**. If uncontrolled, this process can result in a nuclear explosion, the release of a tremendous amount of energy in a short period of time, as in the atomic bomb.

In a nuclear reactor (Figure 9-4) fission is controlled by substances that absorb neutrons. Here the energy is released more slowly and can be used to

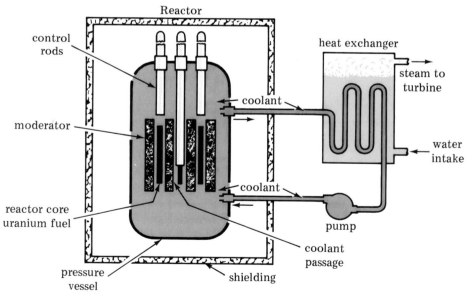

Figure 9-4. Diagram of a nuclear reactor used to generate electricity. The coolant, heated to a high temperature by the nuclear fuel in the reactor, flows through a heat exchanger, where it turns water into steam. The steam drives a turbine, which converts mechanical energy to electrical energy. (Courtesy of ERDA.)

generate electricity. About one-half of a gram of uranium-235 provides the same amount of energy through fission as the burning of 1 ton of coal. Fission products from a nuclear reactor are often useful radioisotopes. In addition, the neutrons released by fission can be channeled into reactions that form other radioisotopes.

Nuclear **fusion** is a reaction that releases even more energy than does fission, although very high temperatures, over 1 million degrees Celsius, are needed to start it. This process consists of the reaction of two light elements to form a heavier one, such as the following reaction:

$$\underset{1}{^2}\text{H} + \underset{1}{^3}\text{H} \longrightarrow \underset{2}{^4}\text{He} + \underset{0}{^1}\text{n}$$

Fusion takes place on the sun, where hydrogen atoms are converted to helium atoms.

9.7   X-ray radiation and photography

Although not produced by radioactivity, **x rays** are a form of radiation like gamma rays (Figure 9-5). They are generated when a beam of electrons strikes a metal plate in an evacuated (low-pressure) tube. X rays are similar to light rays except they are invisible and have much greater energy. They easily pass through soft tissue like your flesh but are partially absorbed by denser parts of

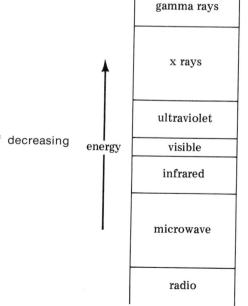

Figure 9-5.   The types of radiation, listed in order of decreasing energy.

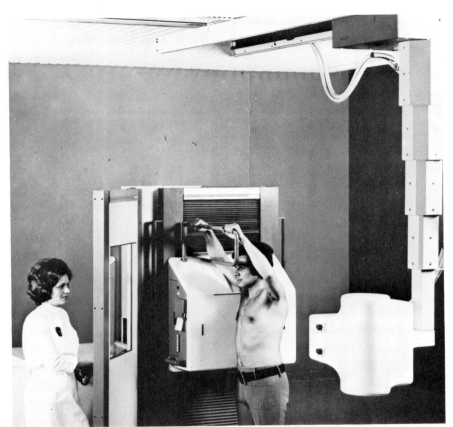

Figure 9-6. Administration of diagnostic x rays. (Photo courtesy of Siemens Corporation.)

your body like bones or teeth. Thus, if a piece of photographic film is placed behind the part of the body receiving x-ray radiation, a "picture" is created (Figure 9-6).

A photographic film consists of a plastic backing that contains an emulsion of silver bromide in gelatin. The silver bromide is sensitive to light or radiation. X-rays strike certain of the grains of AgBr, changing them and forming a latent (hidden) image. This process, *exposure,* occurs only in those parts of the film that the radiation can reach. Visible light causes the same type of change.

Development of the film involves reduction, conversion of the changed form of the exposed AgBr grains to a black form of silver metal. The reducing agent is generally an organic molecule (hydroquinone). This process is carried out in a dark room so that the entire film is not exposed by ordinary light. The film now contains silver, and is therefore dark, in those regions where radiation passed through the tissue and onto the emulsion. It is light where radiation was absorbed by bone or tooth. The film is therefore a negative: The image is a reversal of the true picture.

After being developed, the film is placed in a "stop bath" to prevent all the silver from later being reduced and the whole picture turning black. This bath consists of dilute acetic acid; it stops development by lowering the pH.

The next step is *fixing* the image. This process removes all the unchanged silver bromide to make sure that light will not expose the film any further. The fixing agent is the thiosulfate ion, $S_2O_3^{2-}$. (Solutions of sodium thiosulfate are known as "hypo.") The insoluble silver bromide forms a soluble silver salt in this solution, removing it from the film. It has no effect on the silver image.

$$AgBr + 2S_2O_3^{2-} \longrightarrow Ag(S_2O_3)_2^{3-} + Br^-$$

The film is ready to be examined after it is rinsed with water and hung up to dry.

X-ray photographs are used to locate bone fractures as well as dental caries, tooth decay. The decayed part of the tooth is softer and absorbs more x rays, showing up as a darker region on the film, as seen in Figure 9-7a. To observe

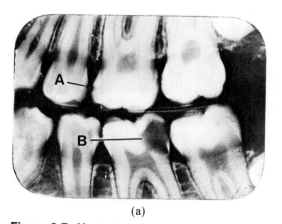

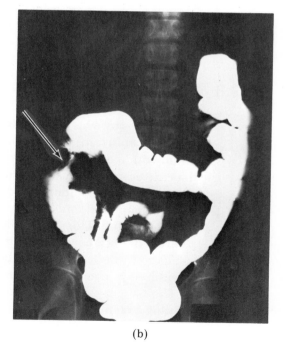

(a)   (b)

Figure 9-7. X-ray photographs. (a) A tooth with decay: small cavity (A) and large cavity (B). (b) A colon filled by barium sulfate enema showing the presence of a tumor (arrow). [Photo (a) copyright by the American Dental Association, reprinted by permission; (b) courtesy of American Cancer Society.]

Table 9-4 **X-Ray Exposures**[a]

Area examined	Exposure (mR)
head and neck	279
chest, radiographic	45
chest, photofluorographic	504
abdomen	790
extremities	117
teeth	1138

[a] From J. N. Gitlin and P. S. Lawrence, "Population Exposure to X-rays," U.S. Public Health Service Publication No. 1519, Government Printing Office, Washington D.C., 1964.

the internal organs such as the stomach and intestine, you must first administer a suspension of barium sulfate, $BaSO_4$, in water. It coats the walls of the gastrointestinal tract, making them appear lighter on the film because x rays cannot penetrate the layer of barium sulfate. Figure 9-7b reveals a tumor (growth) blocking the colon. Values for the x-ray exposure in different types of examinations are listed in Table 9-4. (The units, milliroentgens, are defined in Section 9.9.)

9.8 Detection of radiation

One way you can measure radiation is with a **Geiger–Müller counter,** called Geiger counter for short. Shown in Figure 9-8a, it consists of a metal tube filled with molecules of gas under low pressure and connected to a source of electric current such as a battery. When radiation enters the "window" in the front of the tube, it ionizes the gas, allowing it to conduct electricity. This current is amplified and results in the movement of a needle on a meter or a clicking sound. The more radiation entering the tube, the greater the swing of the needle or the number of clicks. Gamma rays can pass right through a Geiger–Müller counter, but beta rays are easily detected, the number being measured in counts per minute (cpm).

The Geiger–Müller counter is used to survey an area for radiation, determining its intensity at different locations. To simply measure a single quantity or dose of radiation, a dose meter, or **dosimeter**, is employed. The ionization type of dosimeter (Figure 9-8b) is based on the same principle as the Geiger–Müller counter. It is first charged up and then discharges, releasing its stored electricity when gas molecules in the tube are ionized. The amount of discharge is related to the quantity of radiation entering the tube.

(a)

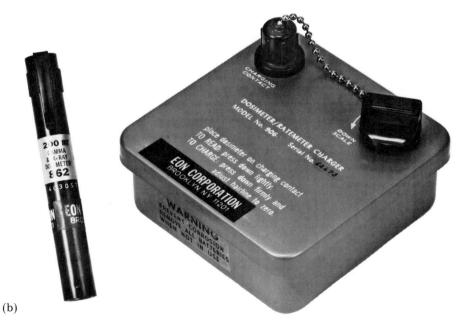

(b)

Figure 9-8. Examples of radiation detectors. (a) Geiger-Müller counter; (b) dosimeter. (Photos courtesy of Eon Corporation.)

Figure 9-8. (c) Film badge. (Photo courtesy of the American Cancer Society.)

(c)

A **film badge** (Figure 9-8c) consists of a small piece of photographic film in a plastic holder that you wear on your body. As radiation hits the film, it becomes exposed. The film is removed at least monthly and developed; the darker the film, the more radiation the badge and therefore its wearer received during this period of time.

The technique of **thermoluminescent dosimetry** (TLD) is gradually replacing the film badge method. A solid crystal like lithium fluoride (LiF) or calcium fluoride (CaF_2) changes in structure slightly when it absorbs radiation at room temperature, the amount of change depending on the energy received. The crystal can be removed periodically and heated rapidly; it returns to the original state, giving off the stored energy as visible light. This process is called thermoluminescence. The amount of light produced is a measure of the radiation received by the person wearing the crystal detector.

The most important detector for medical purposes is the **scintillation counter.** It contains a crystal, such as sodium iodide with a small amount of thallium iodide, which gives off flashes of light, or scintillations (do not pronounce the "c") when hit by radiation. A counting device (photomultiplier tube) then electrically records and amplifies these scintillations. Scintillation counters are part of a machine called the rectilinear scanner (Figure 9-9a), which moves the detector back and forth in parallel lines, measuring radioac-

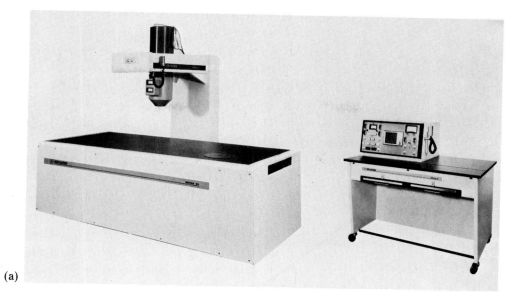

(a)

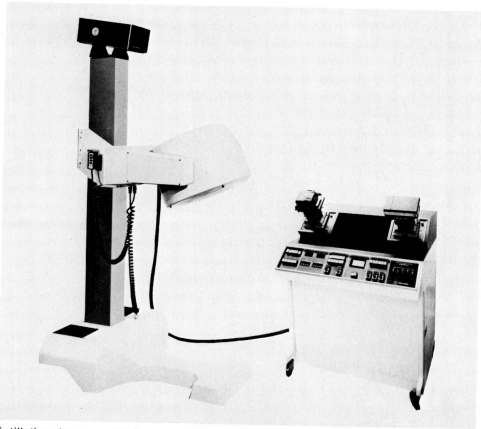

(b)

Figure 9-9. Scintillation detectors. (a) Scanner; (b) gamma-ray camera. (Photos courtesy of Ohio-Nuclear, Inc.)

tivity in the patient as a series of points. The gamma ray or scintillation "camera" (Figure 9-9b), on the other hand, is stationary; it views all parts of the organ being studied continuously, taking a "scinphoto."

9.9 Units of radiation

Like any medication, too much radiation is harmful and too little may not be effective. Several different units are available to measure the radiation being administered to a patient.

The **curie**, abbreviated Ci, describes the activity of a radioisotope. One curie is equal to 37 billion disintegrations per second, the rate of decay of 1 g of radium. Commonly used units are the millicurie (mCi), one-thousandth of a curie, and the microcurie (μCi), one-millionth of a curie.

The exposure of gamma rays or x rays is measured by the **roentgen**, abbreviated R. One roentgen of radiation produces a known amount of ionization, 2.1 billion pairs of ions, in a given volume of dry air, 1 cm^3, at standard temperature and pressure (equivalent to a charge of 0.000258 coulomb/kilogram, where the coulomb is a unit of charge). Geiger–Müller counters are often calibrated to read in milliroentgens per hour (mR/hour).

The dose of radiation actually absorbed is expressed using the **rad**, which stands for radiation absorbed dose. One rad is defined as the absorption of a certain small amount of energy (100 ergs, an amount of energy equivalent to 0.0000024 cal) per gram of tissue receiving the radiation. The rad is one of the most useful dose units because it applies to all types of radiation and takes into account the energy received by the exposed tissue. As a rough guide, the value of an exposure in roentgens is the same numerically as the absorbed dose in rads. A dose of 600 rads is lethal, or deadly, for most people.

The **dose equivalent** (DE) measures the quantity of absorbed radiation on a common scale for different types of radiation. Each kind of radiation is assigned a number called the quality factor (QF) that depends on its biological effect in tissue. The quality factors for beta particles and x rays, for example, are 1, while alpha particles have a value of 10 because their larger size and higher charge give them a greater effect. The dose equivalent is equal to the absorbed dose in rads multiplied by the quality factor, as well as by any other modifying factors. The unit of dose equivalent is the rem, which stands for roentgen equivalent for man, the quantity of absorbed radiation that has the same biological effect as the absorption of 1 R (or 1 rad). A recommended dose limit for the general public is no more than 0.5 rem/year.

The toxicity of radiation can be expressed as the **lethal dose**, LD$_{50}^{30}$ (or LD 50/30). It is the dose that kills 50% of those exposed within 30 days. Typical values for various organisms are given in Table 9-5.

Table 9-5   **Lethal Doses for Various Organisms**

Organism	LD_{50}^{30} (R)
protozoans	3,000,000
viruses	100,000–200,000
bacteria	5000–13,000
algae	4000–8500
goldfish	750
rats	600
monkeys	450
mice	400
humans	400
rabbits	300

9.10   Radioisotopes in diagnosis

A radioisotope behaves chemically like the same element in its stable form. But since the atoms are radioactive, *they can be located in the body using a radiation detector* such as a scintillation counter or Geiger–Müller counter. Depending on the way they are administered, certain radioisotopes tend to concentrate in particular types of cells or organs, allowing them to be examined by the gamma rays given off.

Iodine normally gathers in an organ located in the neck called the thyroid gland. The rate at which this gland removes iodine (in the form of iodide ion, I^-) from the blood can be measured if radioactive iodine-131 (about 5 μCi) is administered to the patient. If the gland is working normally, about 15 to 35% of this iodine is picked up by the thyroid gland within 24 hours, as measured by a detector placed at the neck. If this thyroid uptake test shows less than that amount, the patient has an underactive gland and hypothyroidism; if more is removed, the patient has an overactive gland and hyperthyroidism. The effects of these conditions are described in detail in Chapter 21. Iodine-131 (100 μCi) as a drink or intravenous technetium-99m (1 to 5 mCi) can be used to obtain a thyroid scan. The resulting "picture" shows the distribution of the radioisotope in that organ, as illustrated in Figure 9-10a. If a part of the thyroid does not show up, it did not absorb the isotope, indicating that it may be diseased.

In a thyroid scan, a "cold spot," the lack of radioactivity in a region of the gland, may reveal a pathological, or diseased, condition. In a brain scan, an abnormality is shown by the presence of a "hot spot," radiation being given off where none should be present as shown in Figure 9-10b. Technetium-99m (10 mCi) is administered in the form of a salt, sodium pertechnecate,

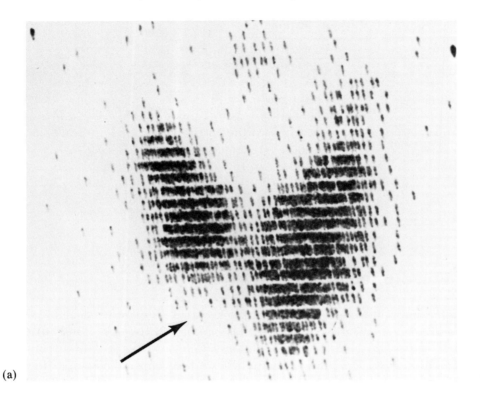

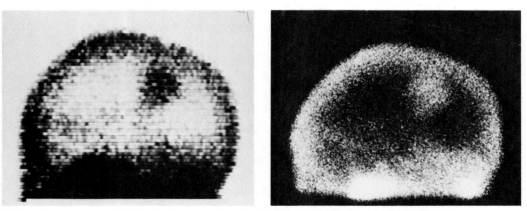

Figure 9-10. Organ scans. (a) Thyroid scan: gland with cancer. (b) Brain scans: tumor shown by rectilinear scan (left) and scintillation camera (right). In the thyroid scan, the abnormality, shown by the arrow, is a "cold spot," while in the brain scan it is a "hot spot." [Scans (a) courtesy of Oak Ridge National Laboratories; (b) from "Clinical Scintillation Imaging" (L. M. Freeman and P. M. Johnson, eds.). Grune & Stratton, New York, 1975; courtesy of Dr. G. V. Taplin.]

$Na^{99m}TcO_4$. In a normal individual, the "blood–brain barrier" keeps ionic compounds in the blood from entering the brain. In certain conditions, however, such as in the case of a tumor or the blocked blood vessel of a stroke, this barrier "leaks." The radioisotope enters the brain and locates the diseased or damaged part of the brain by causing it to show up in a scan of that region.

The liver and spleen can also be scanned using technetium-99m (1 to 3 mCi), administered intravenously in the form of a technetium–sulfur colloid ($^{99m}Tc_2S_7$ and sulfur). This colloid is removed from the blood by cells in the liver and spleen (as well as bone marrow). In diseases such as cirrhosis or hepatitis, the damaged parts of these organs do not pick up radioactivity and show up as "cold spots" in a scan. The pancreas is scanned using radioactive selenium-75, which acts as a substitute for sulfur. Technetium-99m, iodine-131, and chromium-51 can be used to study kidney function.

In addition to organs, body fluids can be studied using radioisotopes. Technetium-99m (10 to 20 mCi), administered intravenously, allows blood flow to be monitored, revealing defects in the heart. The rate of disappearance of red blood cells can be determined using chromium-51. Cells are removed from the blood, "tagged" with this radioisotope, and then reinjected into the patient. The normal lifetime of these labeled cells is 30 days before being destroyed; a shortened life span indicates a pathological condition. The cerebrospinal fluid, which surrounds the brain and spinal cord, can be labeled using iodine-131 (100 μCi) with a spinal tap. A blockage indicates hydrocephalus, the buildup of fluid in the cranium, also called "water on the brain."

The functioning of the lungs can be studied by a perfusion examination; small particles are "tagged" with iodine-131 (250 μCi) or technetium-99m (5 mCi) and injected into a vein. The radioactive particles go to the lungs, where they are trapped. If there is a clot, a semisolid mass, or another pathological condition, no blood flows to that region and therefore it cannot become radioactive and show up in a lung scan. In a ventilation, or "wash-in wash-out" study, xenon-133, a radioactive gas, is used to study the flow of air into and out of the lungs.

Bone scans are obtained with either technetium-99m, fluorine-18, barium-131, strontium-87m, or strontium-85 (if cancer is definitely known to be present). A pathological condition shows up as a "hot spot" because greater activity takes place in that region compared to normal bone. These and other applications of radioisotopes are summarized in Table 9-3.

Certain tests are performed with radioisotopes outside the body using a sample obtained from the patient. For example, the volume of blood in the body can be determined by injecting red blood cells labeled with chromium-51 (as in the life span study) or particles "tagged" with iodine-131 or technetium-99m. After 5 to 10 minutes, a blood sample of known volume is withdrawn and measured for radioactivity. The blood volume, reported in

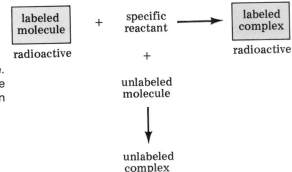

Figure 9-11. The radioimmunoassay technique. The ratio of the amount of free labeled molecule to labeled complex measures the concentration of the reactant of interest.

milliliters per kilogram of body weight, is related to the number of counts: The greater the volume, the smaller the radioactivity of the sample. (Normal values range from 70 to 90 ml/kg.)

Another important test is known as **radioimmunoassay** (RIA); the procedure is illustrated in Figure 9-11. It is used to find the concentration in the blood of a substance like insulin, which is present in low concentrations. The specific reactant being tested for can form a "complex" either with the radioactive molecule that is added to the blood sample or with an unlabeled molecule; the ratio of the amount of free labeled molecule to labeled complex is a measure of the concentration of the substance of interest in the blood sample.

9.11 Radiation therapy and cancer

The most important application of radioactivity in the treatment of disease is related to cancer therapy. **Cancer,** the common name for malignant tumors or neoplasms, refers to the development of new tissue composed of cells that grow without normal controls. These unspecialized cells take over the nutrients needed by normal cells, thereby killing them. In addition, cancer cells can metastasize, or spread, to other parts of the body. Radiation is used to destroy cancer cells, but normal cells are usually also destroyed. The success of radiation therapy depends to a large degree on the relative rates of tissue repair in normal cells compared to cancer cells.

A common approach is to irradiate the site of the cancer either with x rays or with gamma rays from an external source containing cobalt-60, as shown in Figure 9-12. This technique is known as **teletherapy.** Another method is to place a tube, needle, or surface application containing a source of gamma rays, such as radium-226 or cobalt-60, next to the tumor.

Certain radioisotopes are used internally in cancer therapy. Iodine-131, since it gathers in the thyroid gland, is used to treat cancer of this organ, as

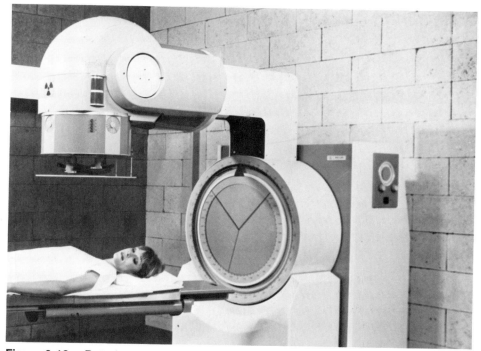

Figure 9-12. Rotational teletherapy. A patient is receiving radiation treatment from a source of cobalt-60. (Photo courtesy of Philips Medical Systems, Inc., Shelton, Conn.)

well as hyperthyroidism. Phosphorus-32 has been used in the treatment of chronic leukemia, cancer of the white blood cells, the leukocytes. Phosphorus builds up in those areas that require phosphate, such as the bone marrow, where leukocytes are formed.

9.12   Effects of radiation

Alpha particles, beta particles, gamma rays, and x rays are **ionizing radiation**. When passing through matter, they interact with the molecules, creating ions and high-energy fragments called free radicals, which cause chemical changes harmful to the cells. The cells most sensitive to radiation are those in the lymphatic tissue, bone marrow, intestinal mucous membranes, gonads, and lens of the eye. Far more resistant to damage are muscle and nervous tissue; their cells are harder to destroy but are also more difficult to replace.

Overexposure to ionizing radiation, either in one large dose or many small doses, may result in **radiation sickness**. A whole-body exposure of less than 25 rems is not detectable. A slight drop in the white cell count follows exposure from 25 to 100 rems. A dose of 100 to 200 rems causes nausea and fatigue; 200 to 400 rems result in nausea and vomiting on the first day. Exposure to 450 rems is fatal within 30 days for about 50% of those receiving this amount of radiation, while exposure to 600 rems is fatal for all individuals. If the dose is localized to the gonads, 600 rems cause sterility; exposure to the eyes produces cataracts, clouding of the lens. Figure 9-13 shows a burn caused by radiation.

The first sign of radiation damage appears in those cells that divide most rapidly, such as the blood-forming tissue of the bone marrow. Overexposure is therefore first seen as a *decrease in the white blood cells*. This effect increases the danger of infection to the exposed individual since resistance is thus reduced. In addition, the blood platelets and red cells also decrease in number, resulting in an increased tendency to bleed (hemorrhage) and anemia. Bone marrow transplants sometimes make it possible for persons to survive otherwise lethal doses of radiation.

Exposure to very large doses has the immediate effect of death in a few minutes to a few weeks from failure of the blood-forming system, gastrointestinal system, and central nervous system. Smaller doses have effects that may be observed only later, such as impaired fertility, life shortening, or cancer. For example, those Japanese that survived the explosion of the atomic bomb at Hiroshima in World War II later had a higher rate of leukemia; the increase was directly related to their distance from the site of the blast.

The most sensitive "target" of radiation is the genetic material. *Radiation damage causes changes in the molecules that carry information from one generation to the next*. The effects are thus long range and are generally lethal or undesirable, producing what are known as mutations. (This process is described in more detail in Chapter 20.) In addition, since the developing fetus is very sensitive to radiation, exposure may result in abortion, malformation of the body, or leukemia.

9.13 Radiation safety

You cannot completely avoid exposure to radiation. For example, the earth is constantly being hit by cosmic rays, radiation of high energy from outer space, which contribute to the **background radiation** always present. Another source of radiation from the environment is fallout, the radioactive particles produced by nuclear explosion that fall to the earth from the atmosphere. As shown in Table 9-6, the estimated dose rates from these sources are larger than those from medical applications for the average individual.

According to recommendations from the National Council on Radiation Protection and Measurements, the general public should not be exposed to more

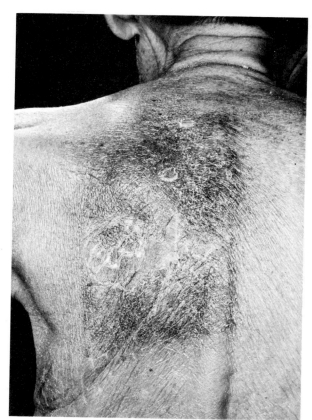

Figure 9-13. A radiation burn. (Photo by Martin Rotker.)

Table 9-6   Estimated Dose Rates in the United States (1970)[a]

Source	Average dose rate (mrem/year)
environmental	
natural	102
fallout	4
nuclear power	0.003
medical	
diagnostic	72
radiopharmaceuticals	1
occupational	0.8
miscellaneous	2
total	182

[a] From "The Effects on Populations of Exposure to Low Levels of Ionizing Radiation." National Academy of Sciences, National Research Council, Washington, D.C., 1972.

than 0.5 rem (500 mrems) in one year. For those persons who use radiation in their occupation, the dose limit is 5 rems in one year for whole-body exposure. Since radiation is invisible, you usually cannot tell when you are being exposed. Therefore, it is important to take certain precautions to protect yourself and the patient.

The exposure received depends on the *distance* from the source of radiation. Intensity of radiation follows an inverse-square law; it decreases as the square of the distance from the source. If you stand 3 feet away, the exposure is cut down to $1/(3)^2$, or 1/9, of the intensity 1 foot away. Every time you double your distance from the source, the intensity is reduced to one-quarter of its original value. You should always try to stay at least 3 feet away from radioactive materials whenever possible.

The *length of exposure* also determines the possible danger from radiation. The shorter the amount of time you or the patient is exposed, the smaller the dose absorbed. Radioisotopes used internally are chosen because they have short half-lives and therefore become nonradioactive in a short time. The families of radioactive patients have limited visiting hours to prevent excessive exposure. (Their yearly dose limits are 0.5 rem if under 45 years old or 5 rems if above 45, which is over the normal childbearing age.)

Shielding provides protection from radiation. Its effect depends on the density and thickness of the shielding material and the energy of the radiation. Alpha particles are absorbed by a sheet of paper, and beta particles are absorbed by 1/4 inch of plastic or 1 inch of wood. Gamma rays are much more penetrating and require lead or concrete shielding. The **half-thickness** of lead, the thickness in millimeters needed to absorb one-half the radiation emitted, is listed in Table 9-7 for commonly used radioisotopes.

The greatest danger results from radioisotopes taken internally by accident. You should not eat, drink, or smoke in areas where radioactive materials are handled. These areas should be marked by the standard three-bladed design of purple or magenta on a yellow background (Figure 9-14). The sign says "Caution" followed by one of three descriptions depending on the exposure

Table 9-7   Half-Thickness Value of Lead for Common Radioisotopes

Radioisotope	Half-thickness of lead (mm)
technetium-99m	0.25
chromium-51	1.8
indium-113m	2.7
gold-198	2.8
selenium-73	3.0
iodine-131	6.0

Figure 9-14. The caution symbol for radiation. (Courtesy of Nuclear Associates, Inc.)

in the area in 1 hour: "Radioactive Materials"(0.6 to 5 mR), "Radiation Area" (5 to 100 mR), or "High Radiation Area" (more than 100 mR). You should wear protective clothing and rubber gloves if you handle radioactive material. Radioactive wastes must be disposed of in special containers. Finally, if you are regularly in contact with radioactivity, you should have your film badge or other detector checked at least monthly and your blood count taken every 3 months to monitor your exposure to radiation.

SUMMARY

Nuclear chemistry deals with changes in the nuclei of atoms. Certain atoms have unstable nuclei; they break down, giving off particles or energy known as radiation. This process of nuclear decomposition or decay is called radioactivity. The radiation emitted can consist of alpha particles (helium nuclei), beta particles (electrons), or gamma rays.

Nuclear changes can be represented by nuclear equations. The sum of the mass numbers of the nuclei must be the same on both sides of the equation, as must the atomic numbers (charges).

All the naturally occurring radioactive elements fit into a chain of reactions called a radioactive disintegration series. They spontaneously decay to form another nucleus and thereby emit one or more types of radiation. All isotopes of elements having atomic weights greater than bismuth (atomic number 83) are naturally radioactive.

Most of the radioactive isotopes, or radioisotopes, used in medicine are produced artificially. By hitting a stable nucleus with smaller particles, like protons, neutrons, or alpha particles, it can be changed into a different element which is radioactive. This process is called transmutation. Over 1500 radioisotopes have been prepared artificially.

The half-life ($t_{1/2}$) of a radioactive nucleus is the time required for the decay of one-half of a given number of nuclei. It can vary from a fraction of a second for a very unstable nucleus to millions of years for a stable isotope. For internal medical applications, a short half-life is necessary to reduce possible damage.

The energy given off by a nuclear reaction is based on the difference in mass between the products and reactants. Fission is the "splitting" of a heavy nucleus by a neutron into two or more pieces. Fusion is the reaction of light nuclei to form a heavier one.

Although not produced by radioactivity, x rays are a form of radiation like gamma rays. X rays are used to take "pictures" of the body because they are partially absorbed by the denser parts, like bones or teeth. A photographic film is exposed when radiation strikes grains of silver bromide. The process of development produces a negative image, which is darkest in those regions receiving most exposure.

Radiation can be detected by a Geiger–Müller counter for surveying areas and by a dosimeter, film badge, or thermoluminescent detector for measuring doses. The most important detector for medical purposes is the scintillation counter, used in scanners and gamma ray "cameras."

Units for measuring radiation include the curie (activity), roentgen (exposure), rad (radiation absorbed dose), and dose equivalent (absorbed dose on common scale). The toxicity of radiation is specified by its lethal dose.

A radioisotope behaves chemically like the same element in its stable form, but it can be located in the body using a radiation detector. Depending on the method of administration, the radioisotope concentrates in certain types of cells or organs, allowing them to be examined. Examples include thyroid scans, brain scans, liver and spleen scans, blood circulation tests, bone scans, and lung function tests. The most commonly used radioisotopes are technetium-99m and iodine-131; others are employed for specific types of examinations.

The most important application of radioactivity in the treatment of disease is related to cancer therapy. The tumor is irradiated either externally using a source of gamma rays such as cobalt-60 or internally with a radioisotope like iodine-131.

Alpha particles, beta particles, gamma rays, and x rays are ionizing radiation. When passing through matter, they interact with the molecules, creating high-energy fragments called free radicals, which cause chemical changes harmful to the cells. Overexposure may result in radiation sickness; the first sign of exposure is a drop in the white blood cell count. The most sensitive target is the genetic material.

To minimize exposure from radiation, you should stay as far from the source as possible (at least 3 feet), reduce the time of exposure, and use shielding. To avoid taking in radioisotopes internally, you should not eat, drink, or smoke in areas where radioactive materials are handled.

Exercises

1. (9.1) What is radioactivity?

2. (9.1) For each of the three kinds of radiation, give its symbol, composition, charge, mass, and penetrating ability.

3. (9.2) What element does X represent in the symbol $^{242}_{94}X$?

4. (9.2) Complete the following equations:

   (a) $^{14}_{7}N + ^{4}_{2}He \longrightarrow ^{17}_{8}O + ?$
   (b) $^{27}_{13}Al + ^{1}_{0}n \longrightarrow ? + ^{1}_{1}H$

5. (9.3) Which elements are naturally radioactive?

6. (9.4) How are radioisotopes produced?

7. (9.4) What is transmutation?

8. (9.5) The half-life of technetium-99m is 6 hours. If 50 μg is initially present, how much of the radioisotope is left after 24 hours?

9. (9.5) Why is a short half-life necessary for medical applications?

10. (9.6) What is the basis for the energy released by a nuclear reaction?

11. (9.6) How does fission differ from fusion?

12. (9.7) What are x rays?

13. (9.7) Explain why bone appears lighter than the surrounding tissue on an x-ray picture.

14. (9.7) Describe the chemical steps involved in photography.

15. (9.8) How do each of these detectors work and how are they used? Geiger–Müller counter, scintillation detector, film badge, dosimeter, thermoluminescent detector.

16. (9.9) Which unit of radiation would you use to describe the following? (a) toxicity; (b) exposure; (c) activity; (d) absorbed dose.

17. (9.10) Why is iodine-131 used to measure thyroid function?

18. (9.10) Explain the difference between a "cold spot" and a "hot spot."

19. (9.10) Describe three applications of radioisotopes to medical diagnosis.

20. (9.11) How is radiation used in cancer therapy?

21. (9.12) Explain the term "ionizing radiation."

22. (9.12) What is radiation sickness?

23. (9.12) What are the short-range and long-range effects of exposure to radiation?

24. (9.13) What can you do to avoid excessive exposure to radiation?

25. (9.13) Why is it especially harmful if radioisotopes enter your body by accident?

Organic chemistry— hydrocarbons

complete combustion of alkane: $CO_2 + H_2O$

Chemistry is divided into two main branches, **organic chemistry** and **inorganic chemistry**, depending on the type of substance studied. Organic chemistry was originally defined as the study of compounds of biological origin, in contrast to inorganic compounds, which are largely mineral in origin. The idea that organic compounds contain a mysterious "vital force" was abandoned in the nineteenth century when an organic molecule, urea, was produced artificially in the laboratory. Organic chemistry is now simply defined as the *study of compounds of carbon.*

Carbon forms a greater number of different compounds (over 2 million) than any other element; there are about 100 times more organic compounds than inorganic compounds. The properties of typical members of each of these classes of compounds are compared in Table 10-1. Most of the compounds in your body consist of organic molecules, ranging in size from several atoms to many thousand. They contain carbon combined by covalent bonds to other elements, generally hydrogen and oxygen or nitrogen.

Table 10-1 Comparison between Typical Organic and Inorganic Compounds

Organic compounds	Inorganic compounds
based on carbon	based on elements other than carbon
largely covalent bonding	largely ionic bonding
relatively large number of atoms in molecule	relatively few atoms in formula unit
low boiling and melting points	high boiling and melting points
low solubility in water	high solubility in water
nonelectrolytes	electrolytes
flammable	nonflammable

10.1   The carbon atom

Carbon has the following atomic structure:

$$
\begin{array}{ll}
\text{atomic number} = 6 \\
\text{atomic mass} = 12 \text{ amu}
\end{array}
\qquad
\boxed{\begin{array}{c} 6p \\ 6n \end{array}} \quad 2e \quad 4e
$$

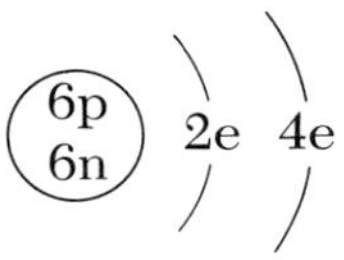

carbon atom

It contains four valence electrons and forms bonds by sharing electrons either with atoms of different elements or with other carbon atoms. This ability of carbon to form bonds with itself (called catenation) is very important because it permits carbon atoms to join together in long chains or in rings. In all of these compounds, carbon forms *four covalent bonds* in any of the combinations shown in Table 10-2.

   The long dash (—) or bond means a pair of electrons being shared, one from the carbon and one from the other atom. With four valence electrons of its own plus the electron from each of the four covalent bonds, carbon has a complete outer shell of eight electrons like the noble gas neon. If you remember that carbon always has four covalent bonds (long dashes) around it, you will find it easy to draw the structures of organic compounds. (A more detailed discussion of bonds formed by carbon is presented in Appendix C.2.)

Table 10-2   The Possible Bonding Arrangements of a Carbon Atom

Arrangement	Description	Total number of bonds
—C—	4 single bonds	4
C=	2 single bonds, 1 double bond	4
=C=	2 double bonds	4
—C≡	1 single bond, 1 triple bond	4

10.2 Alkanes

The simplest carbon-containing compounds are those composed only of atoms of carbon and hydrogen, the **hydrocarbons**. The simplest one is methane, or marsh gas. It consists of one carbon atom and four hydrogen atoms; the formula is therefore CH_4. This way of describing the compound is, of course, the molecular formula. It tells you the number of each kind of atom present in one unit, the molecule. Note that carbon is always written first in the molecular formula. In organic chemistry, you must know more than the molecular formula, the number of each type of atom present. You must be able to write the **structural formula**, which is really a simple picture of the molecule since it shows how the atoms are arranged. The structural formula of methane is as follows:

$$\begin{array}{ccc}
\overset{\textstyle H}{\underset{\textstyle H}{H\!:\!C\!:\!H}} & \text{or} & \overset{\textstyle H}{\underset{\textstyle H}{H\!-\!C\!-\!H}}
\end{array}$$

methane

Each of the four hydrogen atoms is attached by a single covalent bond to the central carbon atom.

Actually, methane is not flat as the structural formula indicates. More realistic models are presented in Figure 10-1. The first is a "ball and stick" model; it shows how the atoms are connected in three dimensions. The second is a "space-filling" model, which represents more accurately the sizes and distances between the atoms. The four bonds formed by carbon in methane and other organic molecules are **tetrahedral** because, when the atoms around the carbon are connected, they form a tetrahedron, the regular geometric figure with four faces and six edges, as shown in Figure 10-1c. The angles between the bonds in a tetrahedral molecule are 109.5°. Because of their simplicity, "flat" structural formulas are used even though they do not represent the way organic molecules "really" look.

The next compound after methane has two carbons attached together, each sharing one electron with the other. Thus, the carbons now have five outer electrons, and each gets three more by combining with three hydrogen atoms:

$$\begin{array}{cc}
\overset{\textstyle H\ \ H}{\underset{\textstyle H\ \ H}{H\!:\!C\!:\!C\!:\!H}} & \overset{\textstyle H\ \ \ H}{\underset{\textstyle H\ \ \ H}{H\!-\!C\!-\!C\!-\!H}}
\end{array}$$

ethane

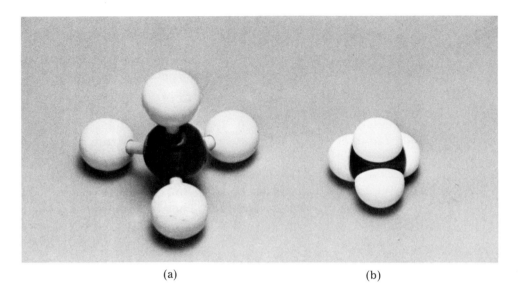

(a)   (b)

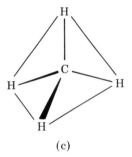

(c)

Figure 10-1. Models of methane, CH_4. (a) Ball and stick; (b) space-filling; (c) formation of tetrahedron by connecting hydrogen atoms. (Photo by Al Green.)

This molecule, C_2H_6, is ethane; it is shown in Figure 10-2. Instead of writing out the complete structure, you can use a shortcut, known as the **condensed structural formula**. For ethane, it is written in the following way:

$$CH_3—CH_3 \quad \text{or} \quad CH_3CH_3$$

ethane

In this way of representing formulas, the atoms bonded to each carbon are written on its right side. Thus, CH_3CH_3 means that the first carbon has three hydrogen atoms and is connected by a single bond to another carbon atom that also has three hydrogen atoms attached. If no dash is written between atoms, a single bond is understood to be present. Because condensed structural for-

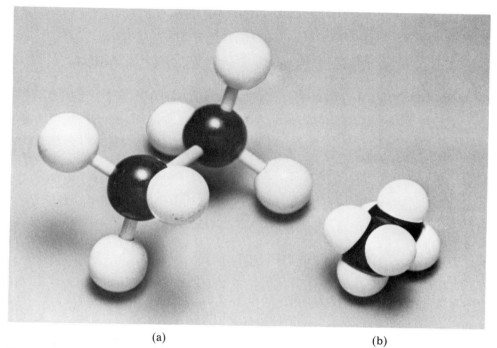

Figure 10-2. Models of ethane, CH_3CH_3. (a) Ball and stick; (b) space-filling. (Photo by Al Green.)

mulas are easy to draw, they are the most common way of representing organic molecules.

The hydrocarbon formed by joining three carbon atoms with single bonds has the following structure (see also Figure 10-3):

$$
\begin{array}{ccc}
& H\ H\ H & \\
H\!:\!\overset{\times\bullet}{\underset{\bullet\times}{C}}\!:\!\overset{\bullet\times}{\underset{\times\bullet}{C}}\!:\!\overset{\times\bullet}{\underset{\bullet\times}{C}}\!:\!H & \quad \text{or} \quad \\
& H\ H\ H &
\end{array}
\qquad
\begin{array}{ccc}
H & H & H \\
| & | & | \\
H-C-C-C-H \\
| & | & | \\
H & H & H
\end{array}
$$

propane

The middle carbon has only two hydrogen atoms attached because it already has two covalent bonds—one to each of the other carbons. This molecule, C_3H_8, is propane. Its condensed structural formula is

$$CH_3{-}CH_2{-}CH_3 \qquad \text{or} \qquad CH_3CH_2CH_3$$

propane

(a)   (b)

Figure 10-3. Models of propane, $CH_3CH_2CH_3$. (a) Ball and stick; (b) space-filling. (Photo by Al Green.)

As more and more carbon atoms join together in a straight chain, forming single covalent bonds with other carbon atoms and with hydrogen atoms to complete their valence shells, a whole series of compounds known as the **alkanes** (or paraffins) is formed. The name of each depends on the number of carbons but all have the ending "-ane." Table 10-3 lists the first ten members of this class of hydrocarbons.

Notice that each end carbon forms bonds to three hydrogen atoms but that the middle carbons form only two. The next member in the series is made by inserting "—CH_2—" into the one before it. The alkanes form a **homologous series**: One member differs from the next by a constant number of atoms. Each alkane has the general formula C_nH_{2n+2}, which means that the number of hydrogen atoms is always two more than twice the number of carbon atoms.

Table 10-3 The Alkanes

Name	Molecular formula	Condensed structural formula
methane	CH_4	CH_4
ethane	C_2H_6	CH_3CH_3
propane	C_3H_8	$CH_3CH_2CH_3$
butane	C_4H_{10}	$CH_3CH_2CH_2CH_3$
pentane	C_5H_{12}	$CH_3CH_2CH_2CH_2CH_3$
hexane	C_6H_{14}	$CH_3CH_2CH_2CH_2CH_2CH_3$
heptane	C_7H_{16}	$CH_3CH_2CH_2CH_2CH_2CH_2CH_3$
octane	C_8H_{18}	$CH_3CH_2CH_2CH_2CH_2CH_2CH_2CH_3$
nonane	C_9H_{20}	$CH_3CH_2CH_2CH_2CH_2CH_2CH_2CH_2CH_3$
decane	$C_{10}H_{22}$	$CH_3CH_2CH_2CH_2CH_2CH_2CH_2CH_2CH_2CH_3$

Thus, an alkane with 50 carbons ($n = 50$) would have $(2 \times 50) + 2 = 100 + 2 = 102$ hydrogen atoms, giving the formula $C_{50}H_{102}$.

You are already familiar with a number of the alkanes, for example, methane in natural gas, propane as bottled gas for cooking, butane as fuel for cigarette lighters, and octane in gasoline. As the number of carbons in these alkanes increases, the molecular weight gets larger and the properties change: Propane is a gas but octane is a liquid; the longest alkanes are solids.

Alkanes are found as a complex mixture in petroleum, the natural deposits of oily liquid under the surface of the earth. The components of this mixture are separated by distillation—heating and then condensing the various "fractions" at different temperatures. A typical fractionating tower is shown in Figure 10-4. Each fraction consists of alkanes having several different numbers of carbon atoms. Examples are listed in Table 10-4. Better known as vaseline or petroleum jelly, the lubricants (26 to 38 carbons) are used as an ointment base and protective dressing.

Table 10-4 Typical Petroleum Fractions

Number of carbon atoms	Name	Boiling point (°C)
1–5	natural gases	less than 40
6–10	gasoline	40–180
11–12	kerosene	180–230
13–17	light gas oil	230–305
18–25	heavy gas oil	305–405
26–38	lubricants	405–515
over 39	asphalts	over 515

Figure 10-4. A fractionating tower. Petroleum is separated into various components here on the basis of their boiling points (see Table 10-4). (Photo courtesy of Mobil Oil Corporation.)

10.3 Alkyl groups

By removing one hydrogen atom (along with its electron) from an alkane, you form a fragment called an alkyl group. It is an "incomplete" molecule since the outer shell of the carbon atom is no longer filled. The first six simple alkyl groups are shown in Table 10-5.

Table 10-5 **Alkyl Groups**

Name	Formula
methyl	CH_3—
ethyl	CH_3CH_2—
propyl	$CH_3CH_2CH_2$—
butyl	$CH_3CH_2CH_2CH_2$—
pentyl (amyl)	$CH_3CH_2CH_2CH_2CH_2$—
hexyl	$CH_3CH_2CH_2CH_2CH_2CH_2$—

The methyl group can be drawn in the following way:

$$
\begin{array}{ccc}
\text{H} & & \text{H} \\
\overset{\cdot\times}{\text{H}:\text{C}\cdot} & \text{or} & \text{H}-\text{C}- \\
\underset{\times\cdot}{\text{H}} & & \text{H}
\end{array}
$$

methyl group

Notice that the carbon atom has only seven electrons. It must form another covalent bond to fill its valence shell. A long dash is placed next to the carbon atom to show that another bond must form to make a complete molecule. *Alkyl groups cannot exist by themselves;* they are always attached to some other atom. Thus, the methyl group can join with a chlorine atom, for example, to form the methyl chloride molecule:

$$
\begin{array}{ccccc}
\text{H} & & & & \text{H} \\
\text{H}:\text{C}\cdot & + & \times\text{Cl}\times & \longrightarrow & \text{H}:\text{C}:\text{Cl}\times \\
\text{H} & & & & \text{H}
\end{array}
$$

| methyl | chlorine | methyl chloride |
| group | atom | molecule |

Look at the names of the alkyl groups. They all have the ending "-yl." The first part of the name comes from the alkane and shows you the number of

carbon atoms in the group. Thus, the group derived from the four-carbon alkane, butane, is the butyl group:

$$\frac{\text{but} \ \big| \ \text{yl}}{\begin{array}{c|c}\text{based} & \text{ending} \\ \text{on butane} & \text{for alkyl} \\ \text{(four carbons)} & \text{group}\end{array}}$$

Many important organic compounds contain alkyl groups. You should be able
to recognize them by counting the number of carbon atoms connected
together.

10.4 Isomers

Not all organic compounds have carbon atoms joined in a straight chain; some
have branches. The following are examples of branched alkanes:

(A) 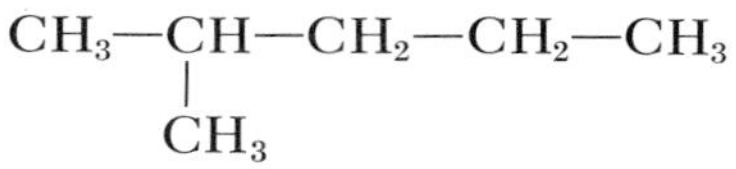 $CH_3-CH-CH_2-CH_2-CH_3$ (B) 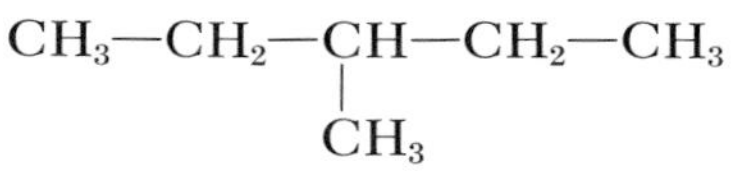 $CH_3-CH_2-CH-CH_2-CH_3$
         $\quad\quad\ \ |$ $\quad\quad\quad\quad\ \ |$
         $\quad\quad\ \ CH_3$ $\quad\quad\quad\quad\ \ CH_3$

         $CH_3\ \ CH_3$ $\quad\quad\quad CH_3$
         $\ \ |\quad\ \ |$ $\quad\quad\quad\ |$
(C) 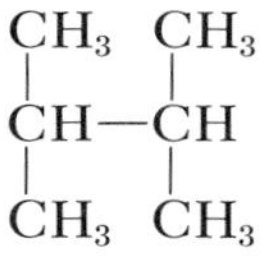 $CH-CH$ (D) 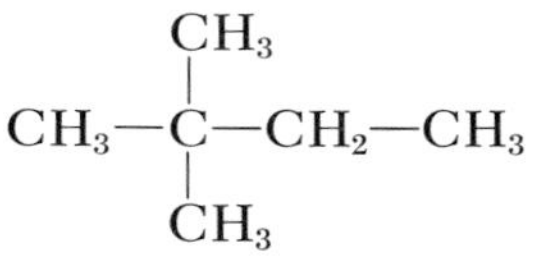 $CH_3-C-CH_2-CH_3$
         $\ \ |\quad\ \ |$ $\quad\quad\quad\ |$
         $CH_3\ \ CH_3$ $\quad\quad\quad CH_3$

You cannot draw a line through all the carbons without lifting your pencil off
the page or retracing part of the line. Each of these molecules is different. And
because the atoms are not connected to each other in the same way, these molecules all have different properties. But if you count up the number of carbon
atoms and hydrogen atoms, you will find that each has the same molecular formula, C_6H_{14}. They are **isomers**, molecules with *identical* numbers of atoms
having *different* structures.

These isomers are related to the straight-chain alkane hexane (E).

(E) $CH_3CH_2CH_2CH_2CH_2CH_3$

The five isomers all cannot be called hexane, even though their molecular
formulas are C_6H_{14}. You must have a way to name each one separately. For
example, E, the simplest isomer, can be named normal hexane, abbreviated
n-hexane, because all the carbon atoms are joined in a straight line. The
branched isomers, A to D, are named on the basis of the *longest* possible chain

of carbon atoms. Carbon atoms coming off the chain are named as alkyl groups and are numbered to show from which carbon they form a branch. Thus, isomer A is called 2-methylpentane. Note that it is identical to the structure A′,

$$\text{(A)} \quad \overset{1}{C}H_3 - \overset{2}{C}H - \overset{3}{C}H_2 - \overset{4}{C}H_2 - \overset{5}{C}H_3 \longleftarrow \text{longest chain has five carbons}$$
$$\overset{|}{C}H_3 \longleftarrow \text{methyl group on second carbon}$$

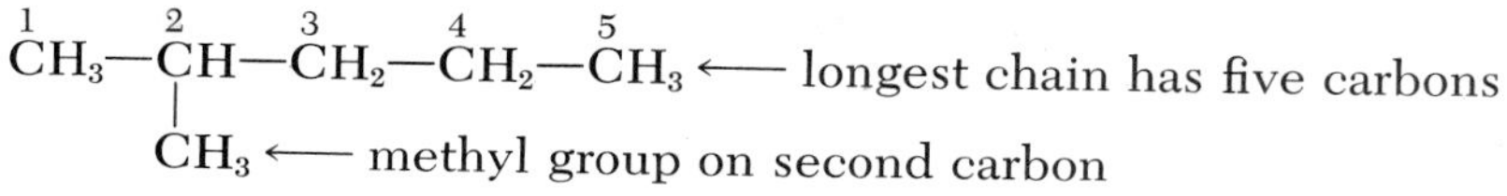

$$\text{(A′)} \quad \overset{5}{C}H_3 - \overset{4}{C}H_2 - \overset{3}{C}H_2 - \overset{2}{C}H - \overset{1}{C}H_3 \longleftarrow \text{longest chain has five carbons}$$
$$\overset{|}{C}H_3 \longleftarrow \text{methyl group on second carbon}$$

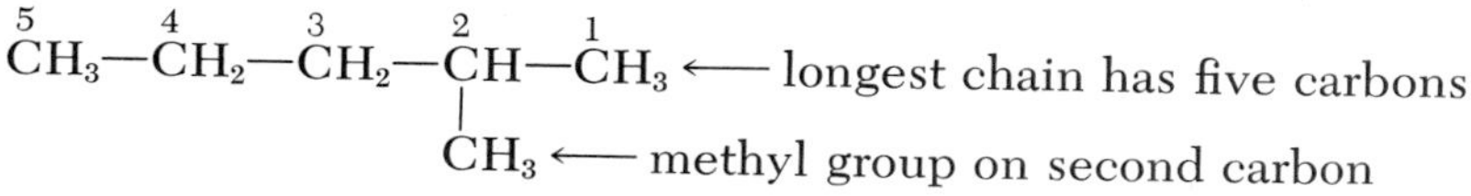

which is obtained by simply turning the molecule around, writing it "backward." Since the atoms are connected the same way, this molecule is also 2-methylpentane; it is not called 4-methylpentane because you always give the alkyl group the lowest possible number. Isomer B is called 3-methylpentane.

$$\text{(B)} \quad \overset{1}{C}H_3 - \overset{2}{C}H_2 - \overset{3}{C}H - \overset{4}{C}H_2 - \overset{5}{C}H_3$$
$$\overset{|}{C}H_3$$

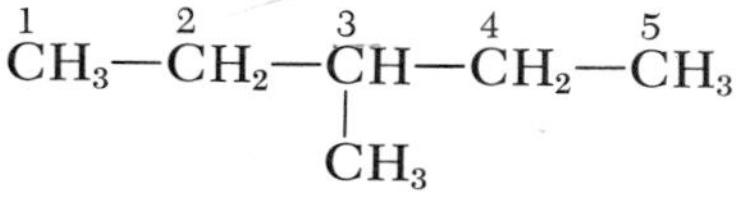

Isomer C is more complicated. Here it is easier to see the longest chain if you redraw the structure:

$$\text{(C)} \quad \overset{1}{C}H_3 - \overset{2}{C}H - \overset{3}{C}H - \overset{4}{C}H_3 \longleftarrow \text{longest chain has four carbons}$$
$$\overset{|}{C}H_3 \ \overset{|}{C}H_3 \longleftarrow \text{methyl groups on second and third carbons}$$

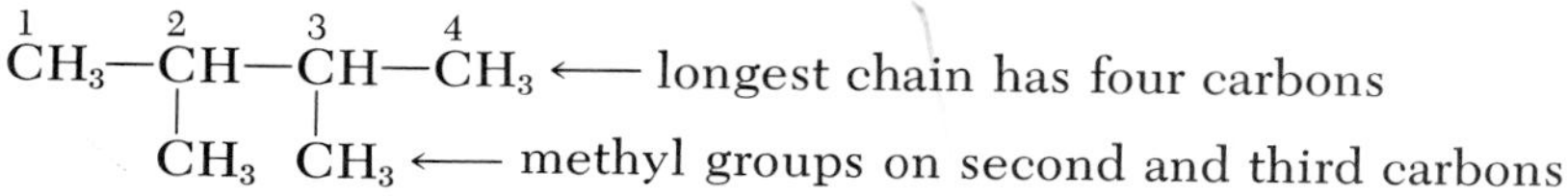

The connections have not been changed, only the way the molecule is pictured. It is now clear that the longest chain has four carbon atoms and that two alkyl groups form branches. The name is 2,3-dimethylbutane. Since the alkyl groups are the same and there are two of them, the prefix "di" is used. Note that the numbers are separated by commas and followed by a hyphen. Finally, isomer D is also named as a butane: 2,2-dimethylbutane.

$$\text{(D)} \quad \overset{1}{C}H_3 - \overset{2}{\underset{|}{\overset{|}{C}}} - \overset{3}{C}H_2 - \overset{4}{C}H_3$$

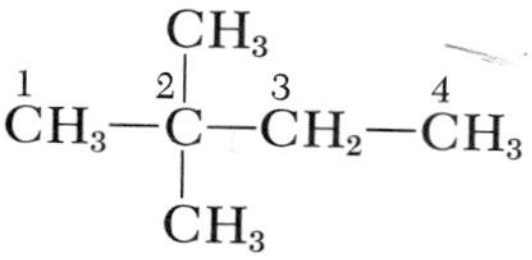

with CH_3 above and CH_3 below carbon 2.

Table 10-6 compares the melting point and boiling point of these isomers. As you can see, even though they all have exactly the same composition, C_6H_{14}, their properties are quite different.

As the size of the molecule increases, the number of possible isomers also rises. For example, the 15-carbon alkane $C_{15}H_{32}$ has nearly 5000 isomers. The

Table 10-6 **Isomers of Hexane**

Name	Formula	Melting point (°C)	Boiling point (°C)
n-hexane	$CH_3CH_2CH_2CH_2CH_2CH_3$	-95	69
2-methylpentane	$CH_3CH(CH_3)CH_2CH_2CH_3$	-153	60
3-methylpentane	$CH_3CH_2CH(CH_3)CH_2CH_3$	-118	63
2,3-dimethylbutane	$(CH_3)_2CHCH(CH_3)_2$	-129	58
2,2-dimethylbutane	$CH_3C(CH_3)_2CH_2CH_3$	-100	50

Table 10-7 **Isomers of Common Alkyl Groups**

Name	Structure
isopropyl	$\begin{matrix} H_3C \\ \diagdown \\ CH- \\ \diagup \\ H_3C \end{matrix}$
isobutyl	$\begin{matrix} H_3C \\ \diagdown \\ CHCH_2- \\ \diagup \\ H_3C \end{matrix}$
secondary butyl (sec-butyl)	$\begin{matrix} CH_3CH_2 \\ \diagdown \\ CH- \\ \diagup \\ H_3C \end{matrix}$
tertiary butyl (tert-butyl)	$\begin{matrix} CH_3 \\ | \\ H_3C-C- \\ | \\ CH_3 \end{matrix}$

possibility of isomers contributes to the tremendous variety of organic compounds.

Just as there are structural isomers of molecules, the alkanes, there are also isomers of the alkyl groups. They begin with the three-carbon propyl group. Important ones are listed in Table 10-7.

10.5 Reactions of alkanes

An alkane is "saturated" with hydrogen atoms. All the bonds not connecting carbon atoms are used in bonding hydrogen atoms. Alkanes are relatively inert; they do not generally react with other substances under normal conditions. One important exception, however, is their ability to react with oxygen—combustion. The alkanes burn to form carbon dioxide and water by oxidation. It is because of this property that propane is used for bottled gas, butane for cigarette lighters, octane for gasoline, and higher alkanes in kerosene for lanterns. A typical reaction, representing complete combustion, is as follows:

$$2C_4H_{10} + 13O_2 \longrightarrow 8CO_2 + 10H_2O$$

Incomplete combustion, burning in a limited supply of oxygen, may produce carbon monoxide and carbon as well.

When combustion takes place slowly, the alkane burns, producing heat and light. But when hydrocarbons such as gasoline, the hydrocarbon mixture based on five to ten carbon atoms, are mixed with oxygen in certain proportions, they react much faster, creating an explosion. The heat generated by this exothermic reaction is released suddenly, rather than slowly over a longer period of time.

In a car, air is mixed with gasoline vapor in the carburetor and ignited by a spark plug. Sometimes, when the mixture ignites too soon, the explosive force of rapid combustion is heard as "knocking" in the engine. Highly branched alkanes such as 2,2,4-trimethylpentane, or isooctane, burn more smoothly than the straight-chain alkanes and prevent this condition. The octane number of a gasoline states its tendency to "knock" compared to isooctane, which is given the value 100.

10.6 Alkenes

A series of hydrocarbons exists in which the molecules contain a *double bond* between two carbon atoms. These are the **alkenes**, or olefins. The simplest member is ethene or, as it is more commonly called, ethylene, a gas that has been used as an anesthetic. Models of this molecule are shown in Figure 10-5.

$$
\begin{array}{cc}
H & H \\
\overset{\times}{\underset{\times}{C}} \!::\! \overset{\times}{\underset{\times}{C}} & \quad \text{or} \quad CH_2 {=\!=} CH_2 \\
H & H
\end{array}
$$

ethene
(ethylene)

To get the four electrons needed to complete its valence shell, each carbon atom receives two electrons from the double bond and two electrons from

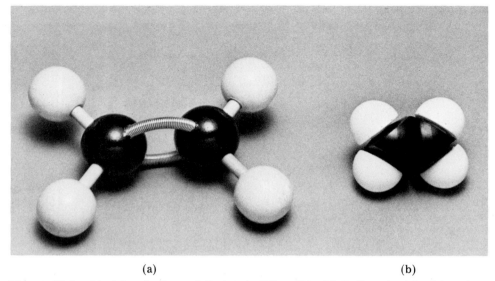

Figure 10-5. Models of ethene (ethylene), $CH_2\!=\!CH_2$. (a) Ball and stick; (b) space-filling. (Photo by Al Green.)

single bonds to two hydrogen atoms. The second member of the group has three carbons, C_3H_6.

$$
\begin{array}{ccc}
\mathrm{H} & \mathrm{H} & \mathrm{H} \\
\mathrm{C}\!\times\!\!\times\!\mathrm{C}\!:\!\mathrm{C}\!:\!\mathrm{H} & \text{or} & CH_2\!=\!CH\!-\!CH_3 \\
\mathrm{H} & \mathrm{H} &
\end{array}
$$

propene
(propylene)

The names and formulas of simple alkenes are summarized in Table 10-8. Notice that all have the ending "-ene," meaning one double bond in the molecule. The general formula is C_nH_{2n}: The number of hydrogen atoms is exactly twice the number of carbon atoms. The "common" name is the one given in parentheses after the "formal" name, which is based on a system approved by the International Union of Pure and Applied Chemistry (IUPAC).

Starting with butene (butylene), isomers are possible since the double bond can appear in different places. The two possibilities for butene are the following:

$$
\overset{1}{C}H_2\!=\!\overset{2}{C}H\!-\!\overset{3}{C}H_2\!-\!\overset{4}{C}H_3 \qquad \overset{1}{C}H_3\!-\!\overset{2}{C}H\!=\!\overset{3}{C}H\!-\!\overset{4}{C}H_3
$$

1-butene   2-butene

Table 10-8 **Alkenes**

Molecular formula	Name[a]	Structural formula
C_2H_4	ethene (ethylene)	$CH_2{=}CH_2$
C_3H_6	propene (propylene)	$CH_2{=}CHCH_3$
C_4H_8	butene (butylene)	$CH_2{=}CHCH_2CH_3$[b]
C_5H_{10}	pentene	$CH_2{=}CHCH_2CH_2CH_3$[b]
C_6H_{12}	hexene	$CH_2{=}CHCH_2CH_2CH_2CH_3$[b]

[a] Common name in parentheses.
[b] One of several possible isomers.

The number of the first carbon in the double bond is placed in front of the name. If a molecule has two double bonds, it is called a diene; the structure of 1,3-butadiene (four carbons, two double bonds) is

$$\overset{1}{C}H_2{=}\overset{2}{C}H{-}\overset{3}{C}H{=}\overset{4}{C}H_2$$

1,3-butadiene

In addition to the location of the double bond, another possible type of isomerism exists. Consider the following two molecules of 2-butene:

cis-2-butene   trans-2-butene

Notice how the end carbons are located with respect to each other. In the first case, they are on the same side of the double bond; in the second molecule, they are on the opposite side. This situation, called **cis–trans isomerism**, occurs because the double bond makes a molecule rigid; atoms cannot twist or rotate around a multiple bond as they can around a single bond. The molecule with identical groups on the same side is the cis isomer, while the trans isomer has them on opposite sides, diagonally across the double bond.

10.7 Reactions of alkenes

Alkenes are much more reactive than alkanes. They are considered **unsaturated** because the presence of a double bond allows the molecule to pick up two more atoms in an addition reaction. For example, an alkene can react with hydrogen gas, H_2, to form an alkane; this process is known as **hydrogenation**:

$$CH_2{=}CH_2 + \quad H_2 \quad \longrightarrow CH_3{-}CH_3 \quad \text{(hydrogenation)}$$

an alkene   hydrogen   an alkane
(ethene)   (ethane)

The electrons from one of the carbon–carbon bonds are now shared with the hydrogen atoms that have been added. Thus, only a single bond remains; the unsaturated alkene was converted to a saturated alkane. (When you learn about fats, you will see that those with single bonds in the hydrocarbon part of the molecule are also called saturated, while those with double bonds are unsaturated.) Besides H_2, many other molecules can add to alkenes. The reaction with water is known as **hydration**:

$$CH_2{=}CH_2 + H_2O \longrightarrow CH_3{-}CH_2{-}OH \quad \text{(hydration)}$$

(Compounds containing oxygen, like the alcohol formed in this reaction, will be studied in the next chapter.) The hydration of double bonds is an important biological reaction.

10.8 Alkynes

There is a third series of hydrocarbons, the **alkynes**, whose molecules contain a *triple* bond. Two carbons thus share six electrons (three pairs) between them and each has only one left to form a bond with hydrogen. The structure of the simplest, C_2H_2, is as follows (see also Figure 10-6):

$$H{:}C{:}{:}{:}C{:}H \qquad \text{or} \qquad CH{\equiv}CH$$

ethyne (acetylene)

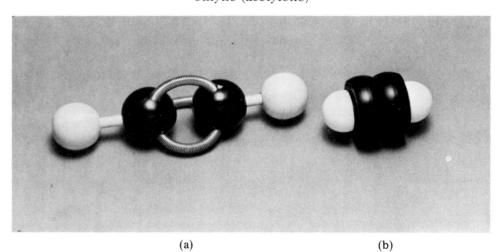

(a)   (b)

Figure 10-6. Models of ethyne (acetylene), $CH{\equiv}CH$. (a) Ball and stick; (b) space-filling. (Photo by Al Green.)

Table 10-9 Alkynes

Molecular formula	Name[a]	Structural formula
C_2H_2	ethyne (acetylene)	$CH{\equiv}CH$
C_3H_4	propyne	$CH{\equiv}CCH_3$
C_4H_6	butyne	$CH{\equiv}CCH_2CH_3$[b]
C_5H_8	pentyne	$CH{\equiv}CCH_2CH_2CH_3$[b]
C_6H_{10}	hexyne	$CH{\equiv}CCH_2CH_2CH_2CH_3$[b]

[a] Common name in parentheses.
[b] One of several possible isomers.

The first members of this series are listed in Table 10-9.

The general formula is C_nH_{2n-2}; the number of hydrogen atoms is twice the number of carbon atoms minus 2. The ending that tells you the molecule has a triple bond is "-yne." The first part of the name, as before, depends on the number of carbon atoms. Alkynes also have isomers; the number in front describes the location of the triple bond.

$$\overset{1}{C}H_3-\overset{2}{C}H_2-\overset{3}{C}{\equiv}\overset{4}{C}-\overset{5}{C}H_2-\overset{6}{C}H_3$$
3-hexyne

This class of hydrocarbons is very reactive. Acetylene burns readily and serves as a fuel for welding. Alkynes undergo addition reactions, like alkenes, but since they contain a triple bond, four atoms can be added to form an alkane:

$$CH{\equiv}CH + 2H_2 \longrightarrow CH_3-CH_3 \quad \text{(addition)}$$
an alkyne an alkane
(ethyne) (ethane)

Few reactions of the alkynes are of any biological importance in the body. Certain drugs, however, including oral contraceptives and antibiotics, contain the carbon–carbon triple bond.

10.9 Cycloalkanes

So far, the hydrocarbons you have learned about all consist of chains of carbon atoms, either connected in a straight line or having branches. It is also possible to form a ring by joining the first carbon to the last; this type of compound is a **cyclic hydrocarbon**. The simplest cyclic alkane, or **cycloalkane**, has three carbon atoms. Since it is an alkane like propane except that it forms a

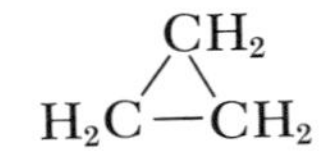

$$CH_2$$
$$H_2C-CH_2$$
cyclopropane

ring, the name is cyclopropane. A whole series of cyclic compounds exists; they are all named by placing the prefix "cyclo-" in front of the name of the corresponding straight-chain alkane. Each consists of a ring of carbon atoms with two hydrogen atoms connected to every carbon, as shown in Table 10-10. Cyclohexane, C_6H_{12}, whose ring is "puckered," is shown in Figure 10-7.

Cycloalkanes, especially the smaller ones, are reactive because the bonds are "strained"; the internal angles are smaller than the normal carbon bond angle of 109.5°. Therefore, the ring breaks readily as another atom becomes attached to each of the two end carbons. For example, cyclopropane converts to propane by the addition of a hydrogen molecule:

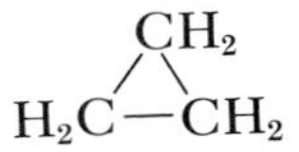

$$CH_2$$
$$H_2C-CH_2 \quad + H_2 \longrightarrow CH_3CH_2CH_3$$
cyclopropane propane

Cyclopropane is a sweet-smelling colorless gas used widely as an anesthetic. It acts quickly as an anesthetic but can be explosive when mixed with oxygen.

Table 10-10 Cycloalkanes

Molecular formula	Name	Structural formula
C_3H_6	cyclopropane	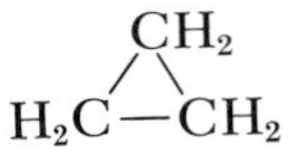 CH_2 / H_2C-CH_2
C_4H_8	cyclobutane	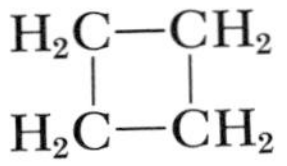 H_2C-CH_2 / H_2C-CH_2
C_5H_{10}	cyclopentane	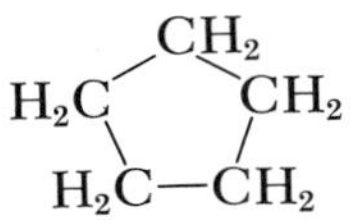 CH_2 / $H_2C \quad CH_2$ / H_2C-CH_2
C_6H_{12}	cyclohexane	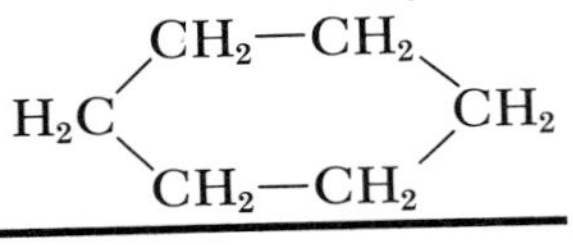 CH_2-CH_2 / $H_2C \quad CH_2$ / CH_2-CH_2

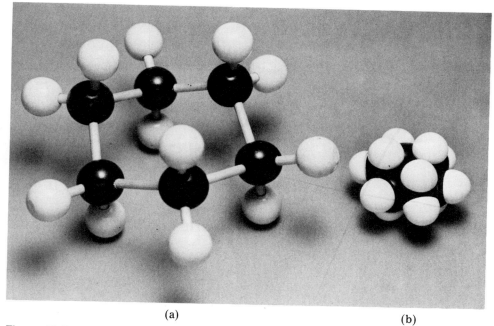

(a)   (b)

Figure 10-7. Models of cyclohexane, C_6H_{12}. (a) Ball and stick; (b) space-filling. (Photo by Al Green.)

10.10 Aromatic hydrocarbons—benzene

A special class of cyclic hydrocarbons contains what looks at first sight like alternating double bonds and single bonds in the same ring. These are known as **aromatic hydrocarbons**. The best example is benzene, C_6H_6. This molecule

$$\begin{array}{c} \text{H} \\ \text{C} \\ \text{HC} \quad \text{CH} \\ | \quad\quad || \\ \text{HC} \quad \text{CH} \\ \text{C} \\ \text{H} \end{array}$$

benzene (I)

can be drawn as shown (I), with three double bonds and three single bonds located in alternating positions in this flat ring. Each of the six carbon atoms is also covalently bonded to a hydrogen atom. Compare the model of benzene in Figure 10-8 with the models of cyclohexane (Figure 10-7).

You might expect this compound to be called a cycloalkene. But it has very unusual properties and does not react in the same way as the alkenes. For ex-

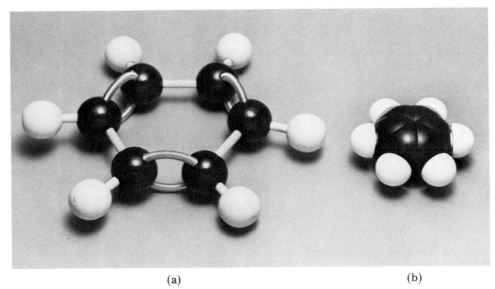

Figure 10-8. Models of benzene, C_6H_6. (a) Ball and stick; (b) space-filling. Compare with Figure 10-7. (Photo by Al Green.)

ample, benzene does not undergo addition reactions at the double bonds. Also, all the bonds in the molecule are chemically the same, even though the structure shown above has distinct single and double bonds. Compare formulas I and II. They both represent benzene, but the double bonds and single

benzene (II)

bonds have been shifted. The actual structure of benzene is a combination of these two equally likely forms. In a sense, the electrons of the three double bonds can be thought of as being *spread out* (delocalized) over all six carbon atoms. Thus, each bond is now more than a single bond but less than a double

III   or   IV

bond; it is said to have partial double-bond character. This situation, which is characteristic of all aromatic compounds, can be represented by III or IV. Notice that in IV neither the carbon atoms nor hydrogen atoms are labeled. A carbon atom is located at each vertex (the point at which two lines meet), and one hydrogen atom is attached to it. Sometimes the benzene ring is drawn with the double bonds in fixed positions, as in V and VI, even though they are spread out over the entire hexagon of carbon atoms.

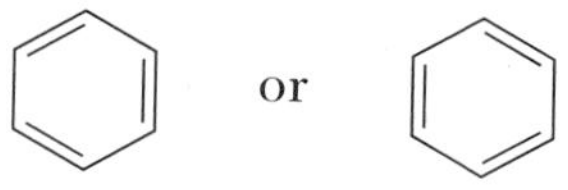

V   or   VI

10.11 Properties of benzene

Benzene is a nonpolar molecule, insoluble in water. Like other hydrocarbons, it is extremely flammable but burns with the smoky flame characteristic of aromatic molecules. The vapors are toxic; inhalation can cause respiratory failure and death.

As mentioned before, benzene does not undergo the typical addition reactions of alkenes because of the stability of its ring system. Instead, the hydrogens can be replaced by other groups in *substitution* reactions. For example, the following molecule, toluene, is obtained by substituting a methyl group for one hydrogen atom on the ring:

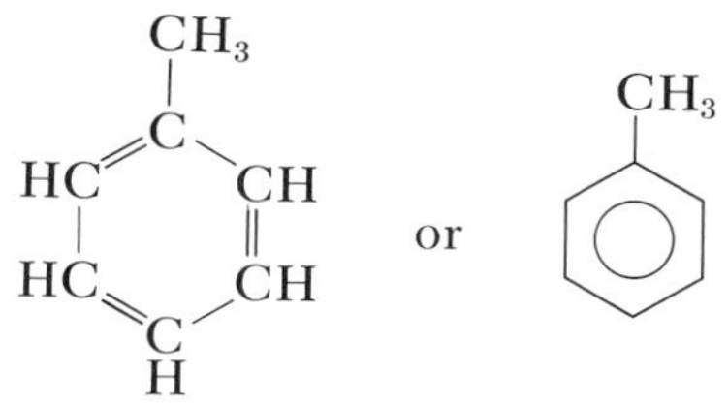

or

toluene

Toluene is the ingredient in airplane glue that is "sniffed" for an alcohol-like "high." It also produces double vision, slurred speech, and lack of coordination and can result in stupor, coma, and death.

Groups can be added to the benzene ring including those with a double bond, as in styrene. Since all carbon atoms in the ring are the same, it makes

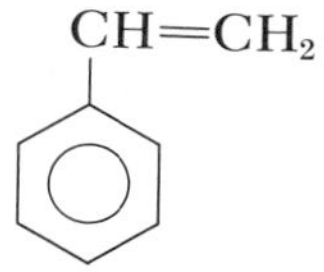

styrene

no difference at which position you draw the substituting group, called the **substituent**.

When you replace two hydrogen atoms around the ring by other atoms or groups, three different isomers are possible, as shown below for xylene (the "x" in the name is pronounced like a "z"), or dimethylbenzene:

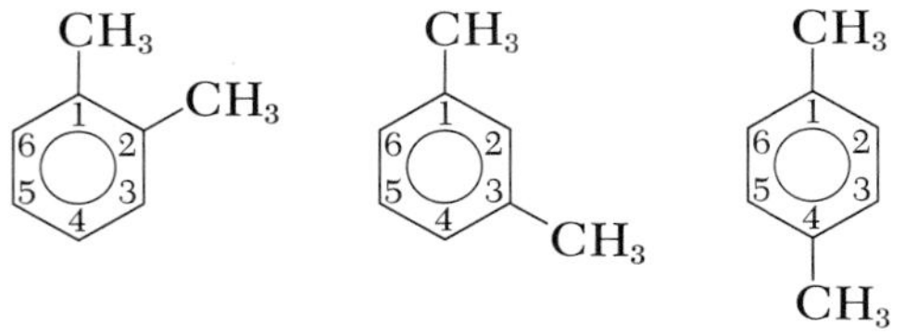

ortho	meta	para
(1,2)	(1,3)	(1,4)
isomer	isomer	isomer

The first isomer, called *ortho*-xylene, abbreviated *o*-xylene, is 1,2-dimethylbenzene. The two methyl groups are on carbons directly next to each other. The second one is *meta*-xylene, *m*-xylene, or 1,3-dimethylbenzene; the methyl groups are on carbons separated by a single carbon atom. The third possibility, *para*-xylene, *p*-xylene, or 1,4-dimethylbenzene, has the two substituents directly opposite each other across the ring. It does not matter on which carbon atoms you place the two groups as long as they have the correct relative orientation. Thus, each of the following is the same molecule, *p*-xylene:

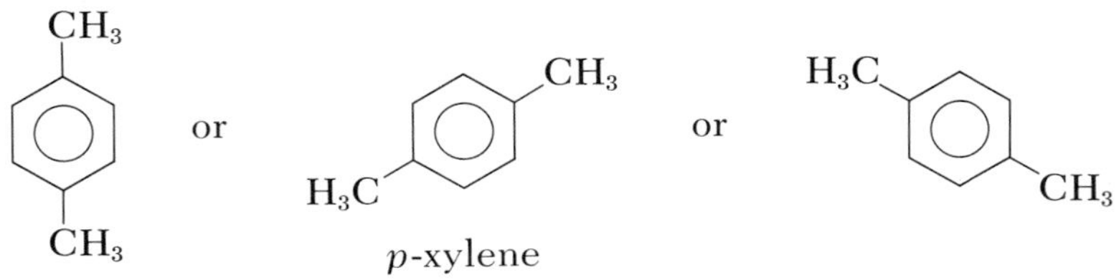

p-xylene

Groups can be formed by aromatic hydrocarbons in a manner similar to the nonaromatic (aliphatic) hydrocarbons, such as the alkanes. By removing a hydrogen from benzene, the phenyl group is produced:

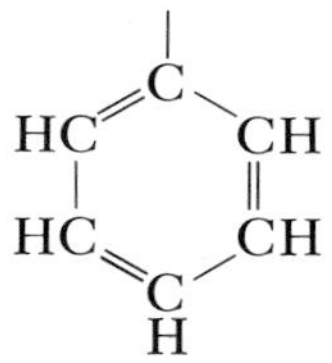

phenyl group

Similarly, the benzyl group is formed from toluene:

benzyl group

These fragments are known as **aryl groups**; like the alkyl groups, they cannot exist by themselves. In the next chapters, you will learn about the different types of groups to which they can be attached to form a large number of compounds.

10.12 Polynuclear aromatic hydrocarbons

Aromatic hydrocarbons other than substituted benzenes may consist of benzene rings "fused" together, that is, having carbon atoms in common. An example you are familiar with is naphthalene, $C_{10}H_8$, the substance from which mothballs are made:

naphthalene

It is called a **polynuclear aromatic hydrocarbon**. Others are made by fusing three or more benzene rings:

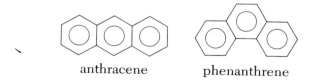

anthracene   phenanthrene

In these molecules, the electrons from the double bonds are spread out over the entire aromatic ring system. These types of molecules are known to be **carcinogenic,** cancer producing. A particularly dangerous carcinogen is benz-

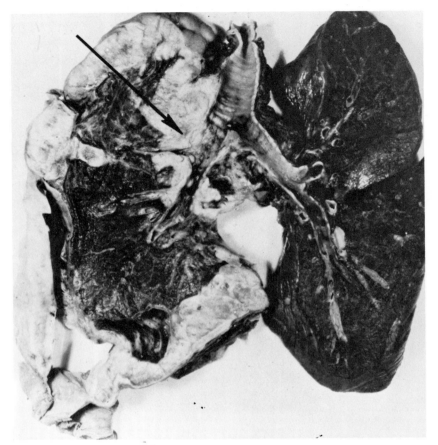

Figure 10-9. The lungs of a heavy smoker. An arrow indicates the area of lung cancer. (Photo courtesy of American Cancer Society.)

pyrene, formed by the incomplete combustion of tobacco, coal, and oil. It is found in the "tar" of cigarette smoke and may be a factor in the relationship between cigarette smoking and lung cancer (Figures 10-9 and 10-10), cancer of the larynx and oral cavity, and possibly cancer of the bladder and pancreas. Benzpyrene is also present in heavily charcoal-broiled meats and smoked fish as well as in the air of large cities.

benzpyrene

Figure 10-10. Copyright 1975, American Heart Association, Inc., reprinted with permission.

Figure 10-11 (p. 230) contains a chart listing this type of hydrocarbon, along with the others described in the chapter.

SUMMARY

Organic compounds contain carbon, generally combined with hydrogen and oxygen or nitrogen by covalent bonds. Inorganic compounds are substances based on elements other than carbon, largely of mineral origin. Organic chemistry is the study of molecules based on carbon.

The carbon atom forms four covalent bonds to attain the noble gas arrangement. It can bond to other carbon atoms (catenation) or to different types of atoms.

Hydrocarbons are composed only of atoms of carbon and hydrogen. The simplest one is methane, CH_4. Its structural formula describes how the atoms in this molecule are connected. The four bonds around carbon are called tetrahedral. Methane is a member of the alkane class, which has the general formula C_nH_{2n+2}. Other members are ethane, propane, butane, pentane, and hexane; all end in "-ane."

By removing one hydrogen atom (along with its electron) from an alkane, you form a fragment called an alkyl group, like the methyl group from methane. Alkyl groups cannot exist by themselves but must be attached to another atom.

Isomers are molecules with identical numbers of atoms having different structures. Branched isomers are named on the basis of the longest chain of

Saturated with H atoms

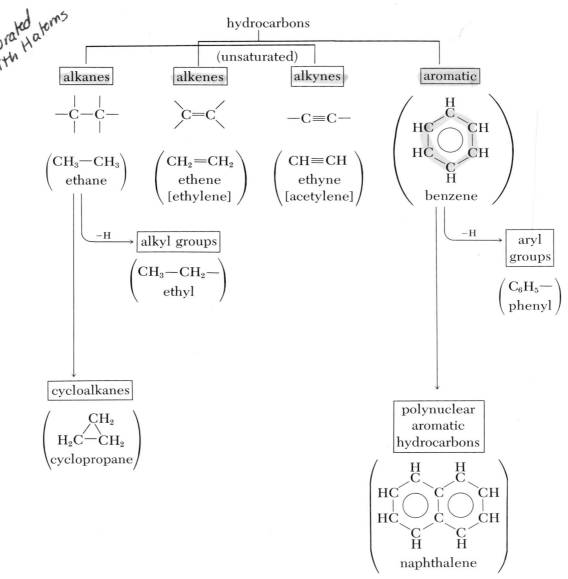

Figure 10-11. Summary chart of hydrocarbons. Compounds in parentheses are typical examples of classes listed.

carbon atoms. As the size of a molecule increases, the number of possible isomers also increases.

Alkanes are "saturated" with hydrogen atoms and are therefore relatively unreactive. The most important exception is their ability to react with oxygen—combustion; they produce carbon dioxide and water, liberating heat. Many alkanes are thus used as fuels.

Alkenes are hydrocarbons that contain a double bond between two carbon atoms. The simplest member is ethene or ethylene, C_2H_4. All end in "-ene" and have the general formula C_nH_{2n}. Alkenes form cis and trans isomers.

The alkenes are much more reactive than alkanes. They are "unsaturated" because the double bond allows the molecule to pick up two more atoms in an addition reaction. Examples are hydrogenation, addition of H_2, and hydration, addition of water.

Alkynes are hydrocarbons that contain a triple bond between two carbon atoms. They have the general formula C_nH_{2n-2} and end in "-yne." The simplest alkyne is ethyne, or acetylene, C_2H_2. This class of compounds is very reactive but relatively unimportant biologically.

Cycloalkanes are cyclic hydrocarbons formed by joining the first and last carbon atoms of an alkane to form a ring. The simplest one is cyclopropane, C_3H_6. These molecules are reactive, particularly the smaller ones, because of ring strain.

Aromatic hydrocarbons, a special class of cyclic hydrocarbons, contain alternating double bonds and single bonds in a ring. The best example is benzene, C_6H_6. The electrons of the three double bonds of benzene can be thought of as being spread out (delocalized) over all six carbon atoms of the ring.

The hydrogen atoms of benzene can be replaced by substitution reactions. When two hydrogens are replaced, isomers known as ortho, meta, and para can form. The group produced by removing a hydrogen from benzene is the phenyl group.

Polynuclear aromatic hydrocarbons may consist of benzene rings "fused" together. In these molecules, such as naphthalene, the electrons from the double bonds are spread out over the entire ring system. Benzpyrene and other polynuclear aromatic hydrocarbons are carcinogenic, cancer producing.

Exercises

1. (Intro.) Identify each as organic or inorganic: (a) HNO_3; (b) C_5H_{10}; (c) CH_3OH; (d) $CaCl_2$; (e) CO_2.

2. (10.1) What possible combinations give carbon four covalent bonds? Why does carbon form four bonds?

3. (10.2) Write the names, molecular formulas, and structural formulas of the first five alkanes.

4. (10.2) Name the following: (a) C_6H_{14}; (b) C_2H_6; (c) $CH_3CH_2CH_2CH_3$; (d) C_5H_{12}; (e) $CH_3CH_2CH_3$.

5. (10.2) If an alkane has 60 carbon atoms, how many hydrogen atoms are in the molecule?

6. (10.2) What is petroleum?

7. (10.3) Write the name and structural formula of an alkyl group formed from (a) propane; (b) ethane; (c) hexane; (d) butane.

8. (10.4) What are isomers?

9. (10.4) Draw as many isomers as possible having the formula C_7H_{16}.

10. (10.4) Name each of the isomers drawn in problem 9.

11. (10.4) Draw structural formulas for (a) 2-methylpentane; (b) 3,3-dimethylhexane; (c) 2,3-dimethylbutane.

12. (10.5) What is combustion? How is it used?

13. (10.6) How do alkenes differ from alkanes?

14. (10.6) Write the names and structural formulas of the first five alkenes.

15. (10.6) Identify the following: (a) C_4H_8; (b) $CH_2{=}CHCH_3$; (c) C_2H_4; (d) $CH_3CH_2CH_2CH{=}CH_2$.

16. (10.6) Which is the cis isomer and which is the trans isomer?

(a) Cl, Br on left C; Br, Cl on right C; $C{=}C$

(b) Cl, Cl on left C; Br, Br on right C; $C{=}C$

17. (10.6) Draw structural formulas for (a) 2-pentene; (b) 3-hexene; (c) 2,3-butadiene.

18. (10.7) Why are alkenes more reactive than alkanes?

19. (10.7) Define unsaturated, addition reaction, hydration, hydrogenation.

20. (10.7) Draw a product of (a) hydrogenation and (b) hydration of propene.

21. (10.8) Which of the following are alkynes? (a) C_3H_4; (b) ethylene; (c) $CH_3CH_2C{\equiv}CH$; (d) C_5H_8.

22. (10.8) Write the names and structural formulas of the first five alkynes.

23. (10.9) Write the names and structural formulas of three cycloalkanes.

24. (10.9) How is cyclopropane used medically?

25. (10.10) Why doesn't benzene act like an alkene even though it has double bonds?

26. (10.10) Draw a shorthand notation for benzene.

27. (10.11) How does a substitution reaction differ from an addition reaction?

28. (10.11) Identify as ortho, meta, or para:

(a) benzene ring with Cl, H, H, Cl substituents

(b) benzene ring with Cl, Cl, H, H substituents

(c) benzene ring with H, Cl, Cl, H substituents

29. (10.12) Give an example of a polynuclear aromatic hydrocarbon.

Oxygen derivatives of the hydrocarbons

All the compounds described in the previous chapter were hydrocarbons, consisting only of carbon and hydrogen atoms. You saw that, with just these two elements, many different kinds of molecules exist: alkanes, alkenes, alkynes, cycloalkanes, and aromatic hydrocarbons. When oxygen atoms are substituted into a hydrocarbon, an even greater variety of molecules results.

11.1 Functional groups

The oxygen atom may appear in a number of different arrangements of atoms, each of which is known as a **functional group**. These groups cannot stand by themselves; they exist in molecules in combination with one or more hydrocarbon groups. Whether located at the beginning, middle, or end of the hydrocarbon, *the functional group controls the properties of the molecule*. But as the size of the molecule increases, the hydrocarbon part begins to play a greater role.

Table 11-1 lists the major functional groups containing oxygen. Remember, the long dash (—) means that a bond must be formed at that atom to complete the molecule. Notice that some functional groups accept one hydrocarbon group (alcohol, aldehyde, acid), while others must form bonds with two (ether, ketone, ester).

Because oxygen can be substituted into a hydrocarbon in so many ways, a large number of isomers are possible. For example, consider the two mole-

Table 11-1 **Functional Groups**

Name	Structure
alcohol	—OH
ether	—O—
aldehyde	$-CH$ with $=O$
ketone	$-C-$ with $=O$
acid	$-C-OH$ with $=O$
ester	$-C-O-$ with $=O$

cules with the same molecular formula, C_2H_6O, shown in the following listing:

Molecule	Structure	Boiling point (°C)
ethyl alcohol	CH_3CH_2—OH	78 (liquid at room temperature)
dimethyl ether	CH_3—O—CH_3	−25 (gas at room temperature)

The first compound is the alcohol that people drink; the functional group, —OH, is attached to an ethyl group. In the second case, two methyl groups are attached to the functional group, —O—, to form an ether similar to the one used as an anesthetic. Even though both molecules have the same number of atoms (and therefore identical molecular formulas), they are very different molecules having different properties, as shown by their boiling points. Thus, they are *isomers*, just as you saw for the hydrocarbons.

11.2 Alcohols

Alcohols are molecules that contain the **hydroxyl group**, —OH, connected to one hydrocarbon group. They are *not* bases, however, like the hydroxides you learned about before. Look at the following two Lewis diagrams:

$$^- \ddot{\text{O}} \colon \text{H} \qquad\qquad \cdot \ddot{\text{O}} \colon \text{H}$$

hydroxide ion   hydroxyl group

The hydroxide ion is a polyatomic anion in which the oxygen, as well as the hydrogen, has a complete valence shell. The hydroxyl group is incomplete.

Table 11-2 Alcohols

Formula	Name[a]
CH_3OH	methanol (methyl alcohol)
CH_3CH_2OH	ethanol (ethyl alcohol)
$CH_3CH_2CH_2OH$[b]	propanol (propyl alcohol)
$CH_3CH_2CH_2CH_2OH$[b]	butanol (butyl alcohol)
$CH_3CH_2CH_2CH_2CH_2OH$[b]	pentanol (amyl alcohol)

[a] Common name in parentheses.
[b] One of several possible isomers.

The oxygen has only seven electrons in its valence shell and must form a covalent bond to get the inert gas arrangement.

Now examine the following two structures, a base and an alcohol:

$$Na^+ \ ^- :\ddot{O}:H \qquad H:\overset{\displaystyle H}{\underset{\displaystyle H}{\overset{\times}{C}}}:\ddot{O}:H$$

a base   an alcohol

Sodium hydroxide is a base because it contains an ionic bond between a sodium ion and hydroxide ion that dissociates in water. The alcohol has a *covalent bond* between a carbon and an oxygen atom. It does not ionize in water to form any hydroxide ions. This bond is polar, however, because oxygen has a greater attraction for electrons (electronegativity) than does carbon. Therefore, alcohols are generally soluble in water and form hydrogen bonds.

The simplest alcohols are presented in Table 11-2. Each of the alcohols can have two names, a systematic name and a common name (written in parentheses) just as you saw for some of the hydrocarbons. The common name is made from the name of the hydrocarbon part followed by the name of the functional group. For example, the molecule formed from a methyl group and an alcohol group is called methyl alcohol. The systematic name is based on adding an ending that identifies the group to a root that tells you the number of carbons. Under this international (IUPAC) system, the ending for alcohols is "—ol." The roots are derived from the corresponding alkane by dropping the final "e," as shown in Table 11-3. Thus, the one-carbon alcohol CH_3OH is called methanol using this way of naming organic molecules.

In addition to the straight-chain alcohols with a hydroxyl group on the last carbon atom, other possible arrangements exist. The —OH group may be located on a middle carbon, as in the following molecule:

$$\overset{1}{C}H_3 - \overset{2}{C}H - \overset{3}{C}H_3$$
$$| $$
$$OH$$

2-propanol (isopropyl alcohol)

Table 11-3 **Roots for Formal Names**

Number of carbons	Root
1	methan-
2	ethan-
3	propan-
4	butan-
5	pentan-
6	hexan-

Its systematic name is 2-propanol; the number in front tells you to which carbon the hydroxyl group is attached. The common name is isopropyl alcohol or *sec*-propyl alcohol, where *sec* stands for secondary, meaning that the alcohol group is attached to a carbon that has two other carbon atoms attached to it. In contrast, the straight-chain alcohols are called primary alcohols because the —OH group is on a carbon that is connected to only one other carbon atom. If the hydroxyl group is attached to a carbon connected to three other carbon atoms, a tertiary alcohol results.

$$CH_3CH_2CH_2CH_2OH \qquad CH_3CHCH_2CH_3 \qquad H_3C-\underset{\underset{CH_3}{|}}{\overset{\overset{CH_3}{|}}{C}}-OH$$
$$\underset{\text{OH}}{|}$$

a primary	a secondary	a tertiary alcohol
alcohol	alcohol	(*tert*-butyl
(*n*-butyl	(*sec*-butyl	alcohol)
alcohol)	alcohol)	

11.3 Examples of alcohols

Commonly known as wood alcohol, methyl alcohol or methanol is poisonous. Drinking small amounts can cause blindness, paralysis, and death. Its vapors are also harmful.

When people say "alcohol," they are referring to ethyl alcohol or ethanol. It is formed from the fermentation of grains and fruits, a process in which sugars are chemically converted to ethyl alcohol and carbon dioxide (with the help of an enzyme from yeast). Ethanol is the only alcohol that is "safe" to drink. All

$$C_6H_{12}O_6 \xrightarrow{\text{enzyme}} 2CH_3CH_2OH + 2CO_2 \qquad \text{(fermentation)}$$

whiskey, wines, and beer contain ethyl alcohol. Their concentration is given by the "proof." Since 100% ethyl alcohol (absolute alcohol) is defined as 200 proof, a bottle marked 80 proof is therefore 40% alcohol in water.

When taken internally, ethyl alcohol is a readily available source of energy since each gram releases 7000 cal. It is used in some hospitals to overcome shock or collapse. In alcoholic beverages, it gives the drinker a "lift," but in large amounts it acts as a depressant on the central nervous system, causing confusion, lack of coordination, drowsiness, and eventually stupor (partial unconsciousness). Continual drinking may lead to the destruction of the liver, a condition known as cirrhosis (Figure 11-1). Alcoholism is a serious illness that affects millions of people in this country (see Chapter 23 for further discussion).

Since ethyl alcohol is a good solvent, it is used to prepare medicines in the form of "tinctures." Mixed with water to form a 50 to 70% solution, it acts as a disinfectant or antiseptic. When it is to be used externally only, alcohol is usually "denatured": Methyl alcohol or other poisonous substances are added. (In this way, the government forces those who wish to drink alcohol to buy the highly taxed form.)

Isopropyl alcohol (2-propanol) is the "rubbing alcohol compound" for sponge baths. It acts as an astringent, causing the tissues to contract, hardening the skin, and limiting secretions. Its evaporation causes lowering of the skin temperature (see Section 6.3). Isopropyl alcohol is toxic and cannot be taken internally (Figure 11-2).

Another important alcohol, menthol, has the hydroxyl group attached to a substituted cyclohexane ring. This alcohol, with a mintlike odor, acts as a

menthol

counterirritant and is found in medications as well as household products like shaving cream.

In addition to the alcohols formed with alkyl groups, aromatic alcohols are also important. Substituting one alcohol group on a benzene ring produces phenol, C_6H_5OH:

phenol

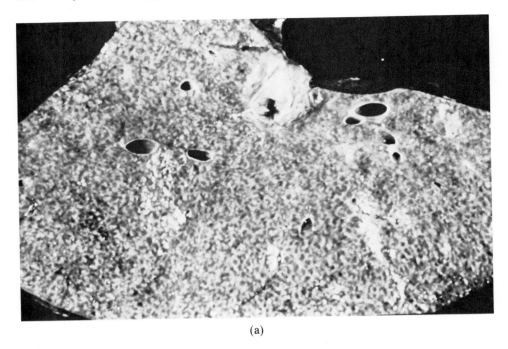

(a)

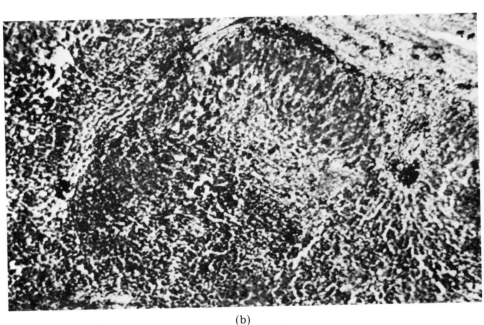

(b)

Figure 11-1. Cirrhosis of the liver. (a) Gross view; (b) microscopic view. Small nodules (collections of cells), fibrosis (scar formation), and destruction of liver architecture result. (Photos courtesy of Dr. Frank A. Seixas, National Council on Alcoholism.)

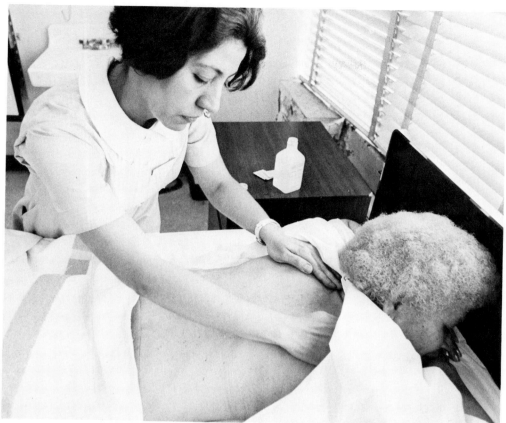

Figure 11-2. Application of rubbing alcohol. (Photo by Al Green.)

The hydrogen of this alcohol group is slightly acidic and can ionize:

$$C_6H_5OH \longrightarrow C_6H_5O^- + H^+$$

Thus, phenol is also known as carbolic acid (not to be confused with carbonic acid, H_2CO_3). Phenol is very corrosive, causing severe burns to skin and tissue. A 2% solution is used as a germicide for decontaminating instruments. Other germicides are compared to this alcohol by their "phenol coefficients." For example, a 1% solution of a germicide that has the same germ-killing effect as a 2% solution of phenol is given a phenol coefficient of 2. Most modern antiseptics have phenol coefficients between 500 and 1000.

Cresols are alcohols formed from toluene. Three isomers are possible—ortho, meta, and para. They are less corrosive than phenol and are

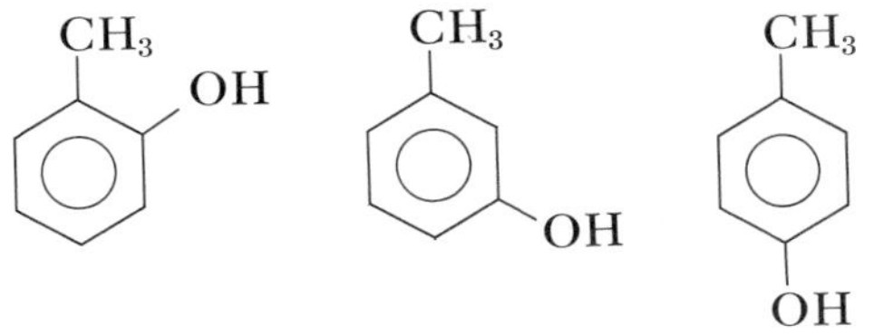

o-cresol   m-cresol   p-cresol

better antiseptics. Lysol is a mixture of the cresols with soap in a dilute aqueous solution. Thymol is a substituted cresol that is also an antiseptic used in dental formulations to sterilize a tooth before filling and in the treatment of hookworm. Its structure is related to that of menthol:

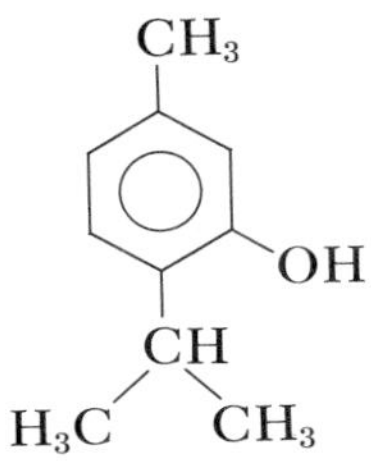

thymol

Another related alcohol is butylated hydroxytoluene (BHT), an antioxidant used to preserve foods.

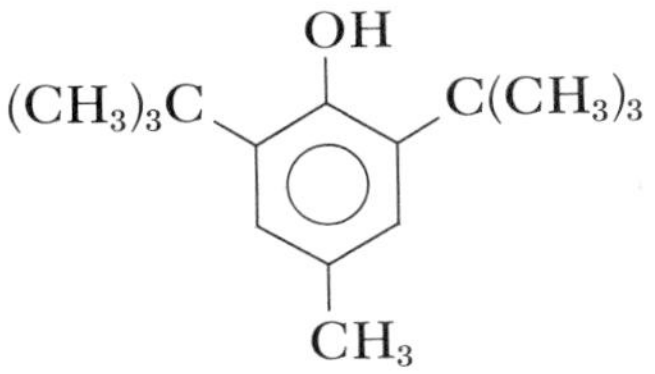

BHT

Two alcohol groups can appear in the same molecule as in ethylene glycol (1,2-dihydroxyethane). It is used in preparations to moisten the skin and as

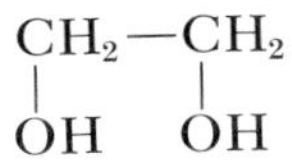

$$CH_2-CH_2$$
$$\;\;|\qquad\;\;|$$
$$OH\quad OH$$

ethylene glycol

an antifreeze in car radiators because it lowers the freezing point of water. Terpin hydrate, found in cough medications, is a cyclic hydrocarbon with two hydroxyl groups. It serves as an expectorant, stimulating the output of fluids

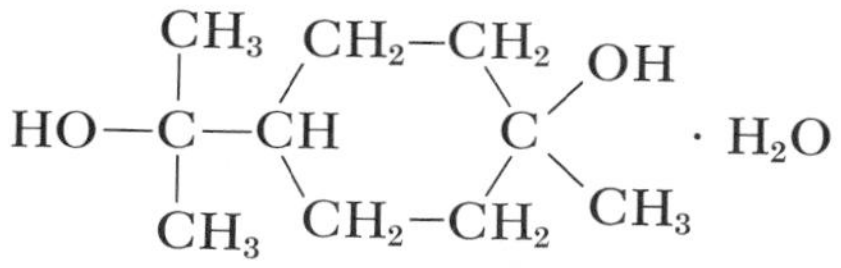

$$CH_3 \quad CH_2-CH_2 \quad OH$$
$$HO-\overset{|}{\underset{|}{C}}-\overset{}{CH} \qquad \overset{}{C} \qquad \cdot H_2O$$
$$CH_3 \quad CH_2-CH_2 \quad CH_3$$

terpin hydrate

in the respiratory tract. Resorcinol, *m*-dihydroxybenzene, is an antiseptic used in skin diseases. Its hexyl-substituted derivative, hexylresorcinol, is an excel-

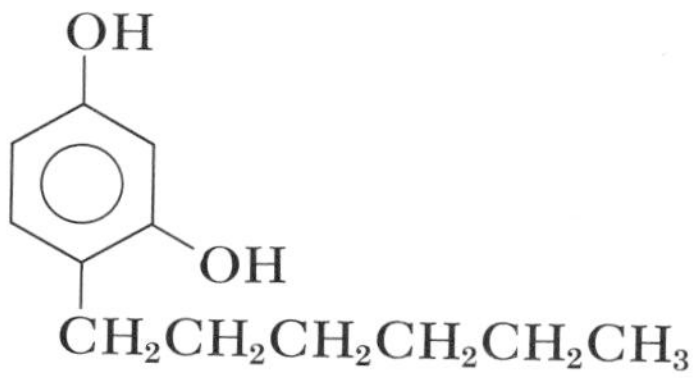

OH

OH

$CH_2CH_2CH_2CH_2CH_2CH_3$

hexylresorcinol

lent antiseptic and germicide. Diethylstilbestrol (DES), an alcohol with two hydroxyl groups on different ends of the molecule, has effects similar to those of female sex hormones. Cancer of the uterus has appeared in young

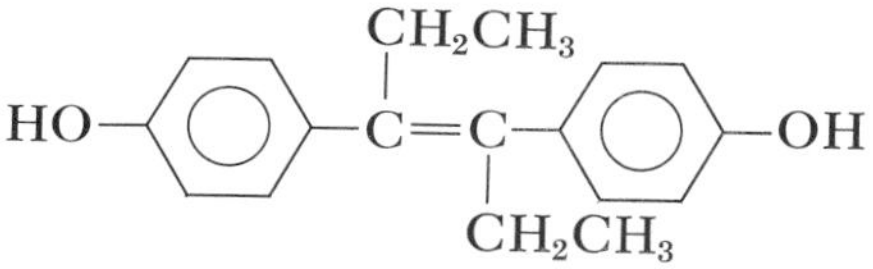

$$CH_2CH_3$$
$$HO-\bigcirc-\overset{|}{C}=\overset{}{C}-\bigcirc-OH$$
$$CH_2CH_3$$

diethylstilbestrol

girls whose mothers received this compound as a drug during pregnancy. Glycerol or glycerin (glycerine), trihydroxypropane, is an important molecule with three alcohol groups. Glycerol is a colorless, syrupy (viscous) liquid

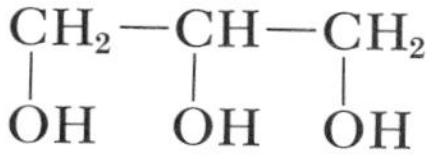

$$CH_2-CH-CH_2$$
$$\overset{|}{O}H \quad \overset{|}{O}H \quad \overset{|}{O}H$$

glycerol

which is nontoxic and hygroscopic (takes up water). It is used for softening the skin and for introducing drugs into the body as in the form of suppositories; it forms part of the molecular structure of the fats and oils of your body.

11.4   Reactions of alcohols

The alcohol group can be converted to other oxygen-containing functional groups by *oxidation*. In organic chemistry, oxidation generally involves *the addition of oxygen to or the removal of hydrogen from a molecule.* Depending on the type of alcohol, different products are possible. Primary alcohols first form an aldehyde and then an organic acid, as shown for the oxidation of ethanol:

$$CH_3CH_2OH \xrightarrow{\text{oxidation}} CH_3\overset{\displaystyle O}{\overset{\|}{C}}H \xrightarrow{\text{oxidation}} CH_3\overset{\displaystyle O}{\overset{\|}{C}}-OH$$

a primary alcohol   an aldehyde   an acid

Secondary alcohols are generally oxidized only to a ketone, described later:

$$\begin{array}{c} H_3C \\ {}^{\diagdown} \\ {}^{\diagup}CH-OH \\ H_3C \end{array} \xrightarrow{\text{oxidation}} \begin{array}{c} H_3C \\ {}^{\diagdown} \\ {}^{\diagup}C=O \\ H_3C \end{array}$$

a secondary alcohol   a ketone

Tertiary alcohols cannot be oxidized at all under normal conditions.

Alcohols undergo dehydration, the removal of water; they lose the hydroxyl group and a hydrogen atom to form an alkene:

$$CH_3CH_2OH \longrightarrow CH_2{=}CH_2 + H_2O \qquad \text{(dehydration)}$$

an alcohol   an alkene

They can also react together in the presence of sulfuric acid (acting as a dehydrating agent) to form ethers, compounds described in the next section:

$$CH_3OH + CH_3OH \longrightarrow CH_3-O-CH_3 + H_2O$$

an alcohol   an alcohol   an ether

11.5 Ethers

Ethers are organic molecules that contain an oxygen atom covalently bound to two hydrocarbon groups. The simplest is dimethyl ether:

$$\begin{array}{c} \quad H \quad\ H \\ H\!:\!\overset{\times}{\underset{\times}{C}}\!:\!\overset{..}{\underset{..}{O}}\!:\!\overset{\times}{\underset{\times}{C}}\!:\!H \quad \text{or} \quad CH_3-O-CH_3 \\ \quad H \quad\ H \end{array}$$

dimethyl ether

The groups on either side of the oxygen can be the same, as in this molecule, or they can be different, as in ethyl methyl ether:

$$CH_3CH_2-O-CH_3 \qquad \text{ethyl methyl ether}$$

As in this case, the common name consists of simply listing the names of the two hydrocarbon groups (in alphabetical order) followed by the word "ether." The systematic names are quite complicated and are rarely used for the simple ethers.

The most important ether medically is diethyl ether, also known as ethyl ether, or simply "ether."

$$CH_3CH_2-O-CH_2CH_3 \quad \text{diethyl ether ("ether")}$$

It is an anesthetic that is easy to administer and that relaxes muscles effectively. Diethyl ether only slightly affects blood pressure, pulse rate, and respiration. Its disadvantages, however, are vapors that are irritating to the respiratory passages, possible aftereffects of nausea, and occasionally pneumonia after surgery (Figure 11-3).

"Ether" is a volatile liquid and its fumes are very combustible. The vapors, which are denser than air, accumulate at the floor and can form an explosive mixture with air. Nylon clothing is forbidden in the operating room to prevent the buildup of static electricity which could create a spark that would ignite the fumes. Another ether used as an anesthetic is divinyl ether or Vinethene. (The vinyl group is formed from ethylene, or ethene.)

$$CH_2{=}CH-O-CH{=}CH_2 \quad \text{divinyl ether (Vinethene)}$$

Ethers can also be formed from aromatic groups, such as in methyl phenyl ether (anisole) or diphenyl ether. Eugenol is a substituted anisole found in oil

anisole   diphenyl ether

of clove that acts as an anesthetic and antiseptic. It is used to relieve the pain from toothache. Eugenol and zinc oxide form a plasterlike material used in

eugenol

dentistry for making impressions and for temporary fillings. Another aromatic ether, mephenesin, is used as a skeletal muscle relaxant.

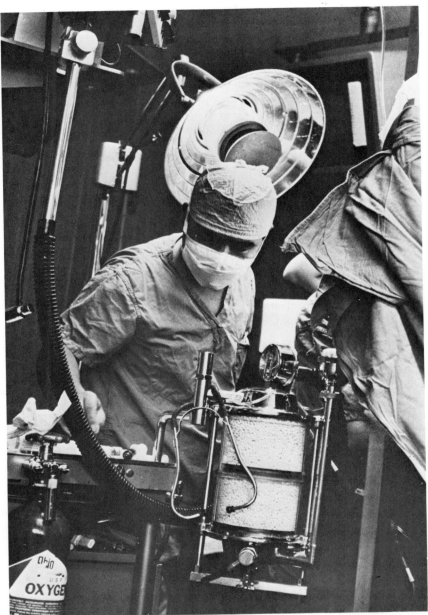

Figure 11-3. Administration of anesthesia. (Photo by A. McGowan, St. Luke's Hospital Center, New York.)

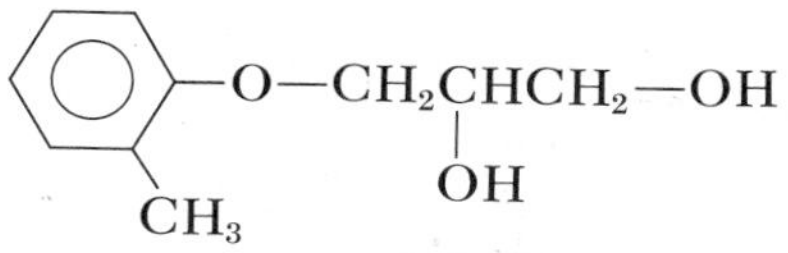

$$\text{(benzene ring)}-O-CH_2CHCH_2-OH$$
$$\underset{CH_3}{|} \qquad \underset{OH}{|}$$

mephenesin

Ethers can also exist as cyclic molecules. Three-membered rings containing two carbons and one oxygen are **epoxides**, such as ethylene oxide. This ether

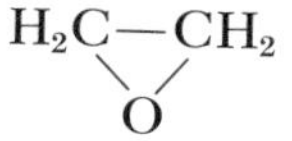

$$\underset{\diagdown O \diagup}{H_2C - CH_2}$$

ethylene oxide
(an epoxide)

is used for sterilization because it can readily penetrate many materials, and at a concentration of 1 g/liter it is an effective germicide. Tetrahydrofuran and dioxane are cyclic ethers with larger rings than epoxides.

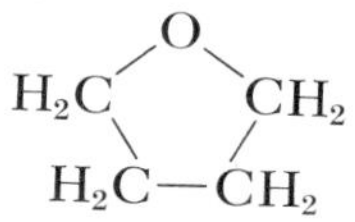

$$\underset{H_2C-CH_2}{\overset{O}{H_2C \qquad CH_2}}$$

tetrahydrofuran 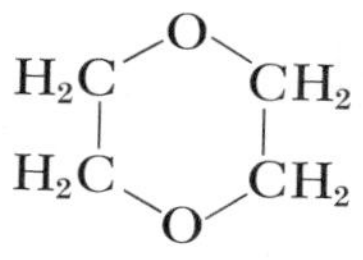   dioxane

When you study sugars, you will see molecules that have two functional groups connected to the same carbon atom. When two ethers share a common carbon, the molecule is an **acetal** (or a ketal if the ethers are identical). When a carbon with an ether group also has a hydroxyl group attached, the molecule is a **hemiacetal**:

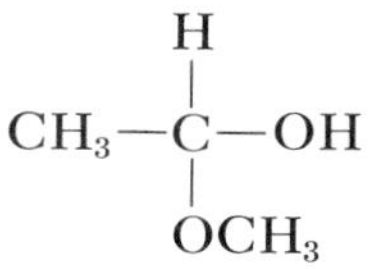

$$\overset{H}{\underset{OCH_3}{\underset{|}{CH_3-\overset{|}{C}-OH}}} \qquad \overset{H}{\underset{OCH_3}{\underset{|}{CH_3-\overset{|}{C}-OCH_2CH_3}}}$$

a hemiacetal 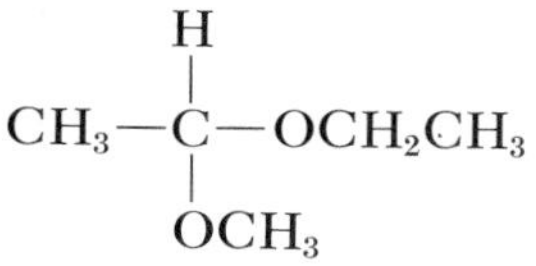   an acetal

11.6 Reactions of ethers

Ethers are relatively stable chemically and do not generally react with acids, bases, or active metals. They do have a tendency, however, to become slowly oxidized to **peroxides**, compounds containing the functional group —O—O—. Since these molecules are very explosive, ethers should be treated carefully.

An exception to the general inertness of ethers is the class of epoxides. These molecules are extremely reactive because the three-membered ring is strained (like cyclopropane) and can readily open up, as in the following hydration reaction:

$$H_2C\!-\!CH_2 \quad + H_2O \longrightarrow \quad CH_2\!-\!CH_2$$

with O bridging in ethylene oxide, and OH, OH on ethylene glycol.

ethylene oxide   ethylene glycol

11.7 Aldehydes

Both aldehydes and ketones have a functional group consisting of a carbon atom double bonded to an oxygen atom, the **carbonyl group.** In the case of al-

$$
\cdot \overset{\times \overset{\times}{\underset{\times}{O}} \times}{C} \cdot \quad \text{or} \quad -\overset{\overset{O}{\|}}{C}-
$$

the carbonyl group

dehydes, one side of the carbon atom is attached to a hydrogen atom, leaving only one bond available for a hydrocarbon group. **Ketones,** on the other hand,

$$-\overset{\overset{O}{\|}}{C}H \qquad\qquad -\overset{|}{\underset{|}{C}}-\overset{\overset{O}{\|}}{C}-\overset{|}{\underset{|}{C}}-$$

the aldehyde group   the ketone (keto) group

have two hydrocarbon groups attached on either side of the carbonyl group. The simplest aldehydes are listed in Table 11-4. The formal names are based

Table 11-4   **Aldehydes**

Formula	Name[a]
$\overset{\overset{O}{\|}}{\text{HCH}}$	methanal (formaldehyde)
$\overset{\overset{O}{\|}}{\text{CH}_3\text{CH}}$	ethanal (acetaldehyde)
$\overset{\overset{O}{\|}}{\text{CH}_3\text{CH}_2\text{CH}}$	propanal (propionaldehyde)
$\overset{\overset{O}{\|}}{\text{CH}_3\text{CH}_2\text{CH}_2\text{CH}}$	butanal (butyraldehyde)

[a] Common name in parentheses.

Table 11-5   Roots for Common Names

Number of carbons	Root
1	form-
2	acet-
3	propion-
4	butyr-

on the roots described in Section 11.2 on alcohols; the ending for aldehydes is "-al." The common names (in parentheses) all end in "-aldehyde," and the first part comes from the roots listed in Table 11-5.

Formaldehyde or methanal is the simplest aldehyde. It is a colorless gas that is made into a 40% aqueous solution, called formalin. In this form it is a germicide used as a disinfectant and is a preservative that hardens tissues (Figure 11-4). The fumes are irritating to the mucous membranes.

By joining three molecules of acetaldehyde together, a cyclic compound, paraldehyde, is formed. As you can see, it is no longer an aldehyde but really a cyclic ether. Paraldehyde is used to depress the central nervous system and to desensitize the gums of the mouth against heat or cold.

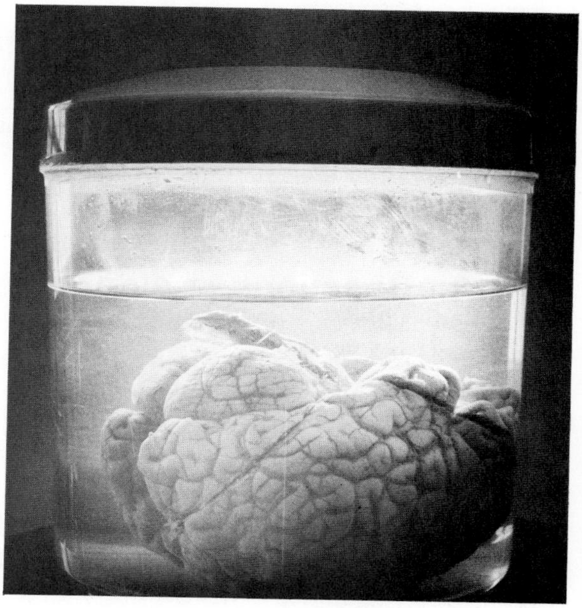

Figure 11-4.   Brain preserved in formaldehyde solution (formalin). (Photo by Al Green.)

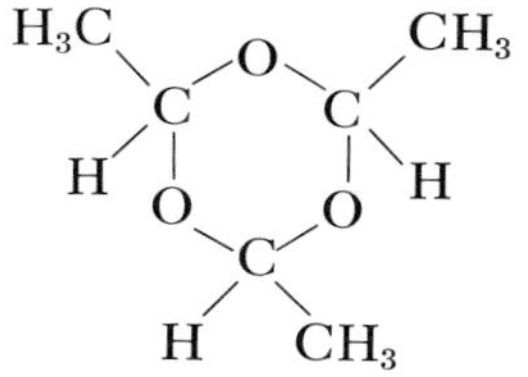

paraldehyde

Aldehydes can exist in molecules with other functional groups as in acrolein, which also has a double bond. It is formed when fats are burned. Glyoxal

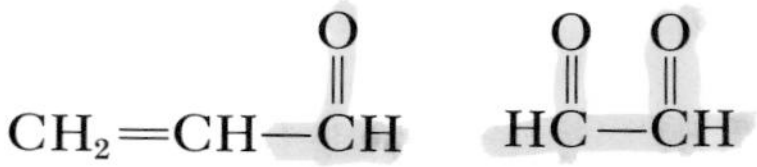

$$CH_2\text{=}CH\text{--}CH$$
acrolein

glyoxal

has another aldehyde group (a dialdehyde). Aldehydes also exist as aromatic compounds, such as benzaldehyde, used in making drugs, dyes, perfumes, and flavorings. A derivative of benzaldehyde, vanillan (3-methoxy-4-

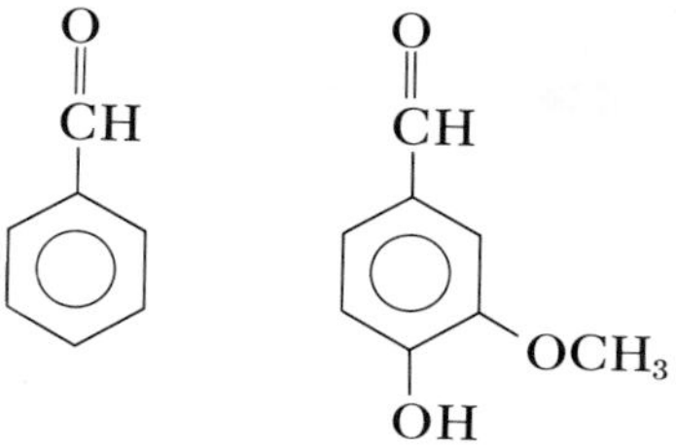

benzaldehyde   vanillan

hydroxybenzaldehyde), is used as vanilla flavoring.

11.8 Reactions of aldehydes

As you have seen, primary alcohols can be oxidized to form aldehydes, and the aldehyde can undergo further oxidation to an acid:

$$\text{primary alcohol} \xrightarrow{\text{oxidation}} \text{aldehyde} \xrightarrow{\text{oxidation}} \text{acid}$$

Since aldehydes can be oxidized, they are good *reducing agents*: Electrons are lost from the aldehyde when oxygen is added (or hydrogen is lost). The ability of an aldehyde to reduce other substances, becoming oxidized itself in the process, is the basis for several laboratory tests for this class of molecules.

Benedict's solution and Fehling's solution both contain the cupric ion, Cu^{2+}. Adding certain types of compounds with aldehyde groups to either of these clear solutions causes a precipitate of cuprous oxide, Cu_2O. The copper has been reduced from Cu^{2+} to Cu^+ by gaining an electron as the aldehyde has been oxidized to an acid:

$$\text{aldehyde + cupric ion} \xrightarrow[\text{solution}]{\text{basic}} \text{acid + cuprous oxide}$$

Another type of test based on the reducing ability of aldehydes uses Tollen's solution, which contains silver ions (combined with ammonia). When an aldehyde is added, the silver is reduced from Ag^+ to its metallic form as the aldehyde undergoes oxidation. The result is the formation of a "silver mirror" on the inside of the tube where the test was performed.

$$\text{aldehyde + silver ions} \longrightarrow \text{acid + silver metal}$$

These tests can be used to find whether a large amount of glucose is present in the urine. Because the sugar glucose contains an aldehyde group, it gives a "positive" test with any of these solutions. For example, with a 1 to 10 ratio of urine to Benedict's solution, a green color means about 0.25% glucose, a yellow-orange color means about 1% glucose, and a brick-red color indicates over 2% glucose (see Chapter 13 for further discussion).

11.9 Ketones

Ketones differ from aldehydes in that the carbonyl group has two hydrocarbon groups attached to the carbon atom. Examples are listed in Table 11-6. The formal name comes from the hydrocarbon root, followed by the ending "-one"; a number in front indicates which carbon has the oxygen. The common name (in parentheses) is based on first listing the alkyl groups and then the word "ketone."

Acetone (propanone or dimethyl ketone), the simplest and most important ketone, has two methyl groups attached to the carbonyl group. It is normally found in the blood in very small quantities (as one of the three "ketone bodies"). Large amounts indicate faulty breakdown of fats as occurs in diabetes mellitus. Under these conditions, its sweetish odor may be found when the patient exhales ("acetone breath"). Acetone is also an excellent solvent and is used in nail polish removers.

Table 11-6 **Ketones**

Formula	Name[a]
$$CH_3\overset{\displaystyle O}{\overset{\|}{C}}CH_3$$	propanone (dimethyl ketone, acetone)
$$CH_3\overset{\displaystyle O}{\overset{\|}{C}}CH_2CH_3$$	butanone (ethyl methyl ketone)
$$CH_3CH_2\overset{\displaystyle O}{\overset{\|}{C}}CH_2CH_3$$	3-pentanone (diethyl ketone)
$$CH_3\overset{\displaystyle O}{\overset{\|}{C}}CH_2CH_2CH_3$$	2-pentanone (methyl propyl ketone)

[a] Common name in parentheses.

Camphor is a ketone in which the carbonyl group is part of a substituted cyclic hydrocarbon. It is a mild antiseptic used in liniments. Aromatic ketones

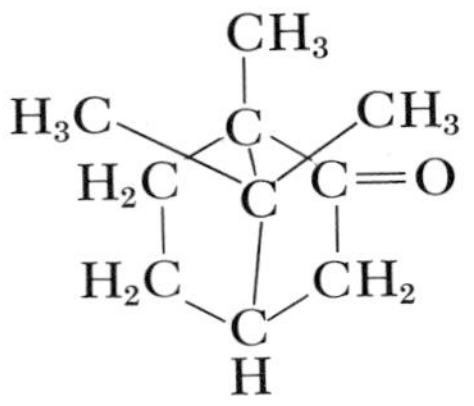

camphor 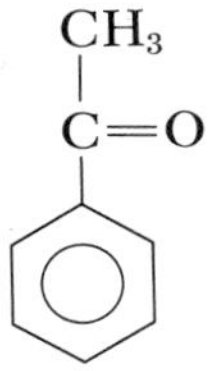   acetophenone

include acetophenone (methyl phenyl ketone), which is a hypnotic, a substance that induces sleep.

Ketones are generally unreactive. They do not react with Benedict's, Fehling's, or Tollen's solutions as do the aldehydes.

11.10 Acids

The combination of a carbonyl group and a hydroxyl group results in another functional group—the **carboxyl group**. The carboxyl group is *acidic* because

$$-\overset{\displaystyle O}{\overset{\|}{C}}-OH \quad \text{or} \quad -COOH$$

the carboxyl group

hydrogen ions are produced by ionization in water. Molecules containing a carboxyl group are referred to as carboxylic acids, in contrast to the mineral acids such as HCl or H_2SO_4, which do not contain carbon. The bond between the oxygen atom and hydrogen atom of the hydroxyl group is weakened by the double-bonded oxygen atom of the carbonyl group. Therefore, it is possible for a water molecule to break off the hydrogen, forming a hydronium ion (hydrated H^+):

$$\underset{\substack{\| \\ O}}{CH_3C}-OH + H_2O \longrightarrow \underset{\substack{\| \\ O}}{CH_3C}-O^- + H_3O^+$$

This process occurs to a slight degree; generally only a few percent of the molecules ionize. Thus, carboxylic acids are all *weak* acids. Note that the hydrogen atoms attached to carbon do not ionize.

The simplest examples are listed in Table 11-7. The common name (in parentheses) comes from the common root plus the ending "-ic acid," while the formal name comes from the formal root plus the ending "-oic acid." In either case, the carbon atom of the carboxyl group is counted as part of the number of atoms in the hydrocarbon.

The simplest acid, formic acid or methanoic acid, is the irritating substance found in insect bites. The two-carbon acid, acetic acid or ethanoic acid, is the acid you have been writing as $HC_2H_3O_2$. Now you should understand why one hydrogen atom is written separately from the others in the formula. It is this hydrogen atom that can dissociate because it is attached to the oxygen, while the other hydrogens are firmly bound to a carbon atom. The name of its con-

Table 11-7   Carboxylic Acids

Formula	Name[a]
$\underset{\substack{\| \\ O}}{HC}-OH$	methanoic acid (formic acid)
$\underset{\substack{\| \\ O}}{CH_3C}-OH$	ethanoic acid (acetic acid)
$\underset{\substack{\| \\ O}}{CH_3CH_2C}-OH$	propanoic acid (propionic acid)
$\underset{\substack{\| \\ O}}{CH_3CH_2CH_2C}-OH$	butanoic acid (butyric acid)

[a] Common name in parentheses.

Table 11-8 **Carboxylic Acids Found in Blood**[a]

Name	Formula	Concentration in blood (mg/100 ml)
lactic acid	$\underset{\underset{OH}{\vert}}{CH_3CH}-\overset{\overset{O}{\vert\vert}}{C}-OH$	9.0
citric acid	$\begin{array}{c} CH_2-\overset{\overset{O}{\vert\vert}}{C}-OH \\ \vert \\ HO-C-\overset{\overset{O}{\vert\vert}}{C}-OH \\ \vert \\ CH_2-\overset{\overset{O}{\vert\vert}}{C}-OH \end{array}$	2.2
β-hydroxybutyric acid	$\underset{\underset{OH}{\vert}}{CH_3CH}CH_2\overset{\overset{O}{\vert\vert}}{C}-OH$	1.5
pyruvic acid	$CH_3\overset{\overset{O}{\vert\vert}}{C}-\overset{\overset{O}{\vert\vert}}{C}-OH$	1.1
acetoacetic acid	$CH_3\overset{\overset{O}{\vert\vert}}{C}CH_2\overset{\overset{O}{\vert\vert}}{C}-OH$	0.9
α-ketoglutaric acid	$HO-\overset{\overset{O}{\vert\vert}}{C}CH_2CH_2\overset{\overset{O}{\vert\vert}}{C}-\overset{\overset{O}{\vert\vert}}{C}-OH$	0.6
succinic acid	$HO-\overset{\overset{O}{\vert\vert}}{C}CH_2CH_2\overset{\overset{O}{\vert\vert}}{C}-OH$	0.5
malic acid	$HO-\overset{\overset{O}{\vert\vert}}{C}CH_2\underset{\underset{OH}{\vert}}{CH}\overset{\overset{O}{\vert\vert}}{C}-OH$	0.5
		16.3

[a] Normal fasting adult.

centrated form is glacial acetic acid; it freezes at 17°C and resembles ice—hence its name. Vinegar is the dilute form (about 5%) of acetic acid.

Carboxylic acids found in the blood are listed in Table 11-8. Lactic acid, a substituted propanoic acid, is produced in your muscles when you exercise;

this acid is present in sour milk. Citric acid, found in citrus fruits, is an important part of the breakdown of sugars in your body (the "citric acid cycle"). It is a tricarboxylic acid, consisting of three carboxylic acid groups, as well as a hydroxyl group. β-Hydroxybutyric acid is the common name for 2-hydroxybutanoic acid; beta, the second letter of the Greek alphabet, is sometimes used in naming instead of the number 2. Both this acid and acetoacetic acid may be found in abnormally high amounts, along with acetone, in the blood of persons who are eating or breaking down too much fat. Pyruvic acid and those remaining in Table 11-8 are dicarboxylic acids. They all have two carboxyl groups in the molecule.

A special category of acids will be described more fully in the chapter on lipids, namely, the **fatty acids**. These are straight-chain acids, usually with 12 to 20 carbon atoms, containing up to four double bonds. They include stearic acid ($C_{17}H_{35}$—COOH), palmitic acid ($C_{15}H_{32}$—COOH), and oleic acid ($C_{17}H_{33}$—COOH).

The simplest aromatic acid is benzoic acid, in which one hydrogen atom of the benzene ring is replaced by a carboxyl group. It is a stimulant, antiseptic,

benzoic acid

and diuretic (a substance that increases the flow of urine). Salicylic acid (o-hydroxybenzoic acid) can be considered a phenol with a carboxyl group or benzoic acid with a hydroxyl group.

salicylic acid

11.11 Reactions of acids

Since these molecules are acidic, they react with bases in neutralization reactions to form a salt and water. The salt that is produced from a carboxylic acid and a base consists of the cation of the base plus the anion of the acid. For example, if acetic acid reacts with sodium hydroxide, water is formed plus the sodium salt of the anion of acetic acid:

$$
\underset{\text{carboxylic acid}}{CH_3\overset{\displaystyle O}{\overset{\|}{C}}-OH} + \underset{\text{base}}{NaOH} \longrightarrow \underset{\text{salt}}{CH_3\overset{\displaystyle O}{\overset{\|}{C}}-O^-\ {}^+Na} + \underset{\text{water}}{H_2O}
$$

The anions of the acids have the carboxylate group, which is simply the

$$
\overset{\displaystyle O}{\overset{\|}{-C}}-O^-
$$

the carboxylate group

carboxyl group minus the hydrogen ion, H^+. It has a negative charge because the electron from the hydrogen atom remains on the oxygen. Carboxylate

Table 11-9   Carboxylate Anions

Formula	Name[a]	Formed from[a]
$HC\overset{O}{\overset{\|}{-}}O^-$	methanoate (formate) ion	methanoic (formic) acid
$CH_3C\overset{O}{\overset{\|}{-}}O^-$	ethanoate (acetate) ion	ethanoic (acetic) acid
$CH_3CH_2C\overset{O}{\overset{\|}{-}}O^-$	propanoate (propionate) ion	propanoic (propionic) acid
$CH_3CH_2CH_2C\overset{O}{\overset{\|}{-}}O^-$	butanoate (butyrate) ion	butanoic (butyric) acid

[a] Common name in parentheses.

Table 11-10   Common Names of Sodium Salts of Carboxylic Acids

Acid	Sodium salt of acid	Use
benzoic acid	sodium benzoate	bread preservative (inhibits yeast and bacteria)
citric acid	sodium citrate	anticoagulant (prevents clotting)
propionic acid	sodium propionate	food preservative (antimicrobial)
salicylic acid	sodium salicylate	antipyretic (reduces fever), analgesic (reduces pain)

anions are named by dropping the "-ic acid" ending from the acid and replacing it by "-ate ion." Look at the examples in Table 11-9. The salt formed from acetic acid and sodium hydroxide is therefore sodium acetate or sodium ethanoate. The common names of other salts from carboxylic acids are listed in Table 11-10.

11.12 Esters

The reaction of a carboxylic acid with an alcohol produces an **ester**. This type of molecule has the following functional group:

$$\overset{\displaystyle O}{\overset{\displaystyle \|}{-C-O-}}$$

the ester group

The hydrocarbon attached to the carbon end comes from the acid, while the group attached to the oxygen atom comes from the alcohol. For example, look at this **esterification**, the synthesis of an ester, from acetic acid and methyl alcohol:

$$\underset{\text{acid}}{CH_3C-OH} + \underset{\text{alcohol}}{HOCH_3} \longrightarrow \underset{\text{ester}}{CH_3C-OCH_3} + \underset{\text{water}}{H_2O}$$

Notice that a molecule of water is removed in order to form the ester.

The name of an ester consists of two separate words. The first is the name of the hydrocarbon group of the alcohol, methyl in this case. Next comes the name of the carboxylate anion of the acid (replacing "-ic acid" by "-ate"), acetate. The full name of the ester formed from methyl alcohol and acetic acid is methyl acetate (or methyl ethanoate in the formal system). Other examples are presented in Table 11-11. Remember that *the carbon atom of the ester group is part of the acid, and the alcohol part is attached to the oxygen end.*

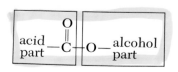

Esters are important as fragrances and flavorings. Table 11-12 lists some of the most common ones.

Table 11-11 **Common Names of Esters**

Alcohol	Acid	Name	Formula
methyl alcohol	formic acid	methyl formate	$HC{-}OCH_3$ with $\overset{O}{\overset{\|}{}}$
methyl alcohol	propionic acid	methyl propionate	$CH_3CH_2C{-}OCH_3$ with $\overset{O}{\overset{\|}{}}$
ethyl alcohol	formic acid	ethyl formate	$HC{-}OCH_2CH_3$ with $\overset{O}{\overset{\|}{}}$
ethyl alcohol	acetic acid	ethyl acetate	$CH_3C{-}OCH_2CH_3$ with $\overset{O}{\overset{\|}{}}$
propyl alcohol	acetic acid	propyl acetate	$CH_3C{-}OCH_2CH_2CH_3$ with $\overset{O}{\overset{\|}{}}$

Table 11-12 **Flavors of Esters**

Flavor	Ester
apricot	amyl butyrate
banana	amyl acetate
orange	octyl acetate
pear	isoamyl acetate
pineapple	ethyl butyrate
raspberry	isobutyl formate
rum	ethyl formate

Aromatic rings can also be a part of an ester, in either the alcohol part or the acid part. For example, benzoic acid reacts with methyl alcohol to form methyl benzoate, which has the odor of freshly mown hay. When salicylic

methyl
benzoate

acid is esterified with methyl alcohol, methyl salicylate is produced. This

$$\underset{\substack{\text{methyl}\\\text{salicylate}}}{\text{[structure of methyl salicylate: benzene ring with } -\!\!\overset{\overset{\text{O}}{\|}}{C}\!-\!OCH_3 \text{ and } OH \text{ substituents]}}$$

ester is known as "oil of wintergreen" and is found in many liniments for muscles and joints. Acetylsalicylic acid is the ester formed from salicylic acid and acetic acid. In this case, it is the alcohol part of the salicylic molecule

$$\underset{\substack{\text{acetyl-}\\\text{salicylic acid}\\\text{(aspirin)}}}{\text{[structure of acetylsalicylic acid: benzene ring with } -\!\!\overset{\overset{\text{O}}{\|}}{C}\!-\!OH \text{ and } -\!O\!-\!\!\overset{\overset{\text{O}}{\|}}{C}\!-\!CH_3 \text{ substituents]}}$$

that is esterified. This ester is *aspirin.* It is useful for reducing fever (antipyretic effect) and for reducing pain (analgesic effect). Because it is acidic, aspirin irritates the stomach; it is often combined with buffers to try to minimize this side effect ("buffered" aspirin).

11.13   Reactions of esters

Esters can be broken back into an acid and alcohol through **hydrolysis**, the addition of water (in the presence of a catalyst). This reaction is just the reverse of esterification. For example, amyl acetate can be hydrolyzed into amyl alcohol and acetic acid:

$$\underset{\text{amyl acetate}}{CH_3\overset{\overset{\text{O}}{\|}}{C}\!-\!OCH_2CH_2CH_2CH_2CH_3} + \underset{\text{water}}{H_2O} \longrightarrow \underset{\substack{\text{acetic}\\\text{acid}}}{CH_3\overset{\overset{\text{O}}{\|}}{C}\!-\!OH} + \underset{\text{amyl alcohol}}{CH_3CH_2CH_2CH_2CH_2OH}$$

Hydrolysis of an ester is the reaction that takes place when fats and oils are digested by the body.

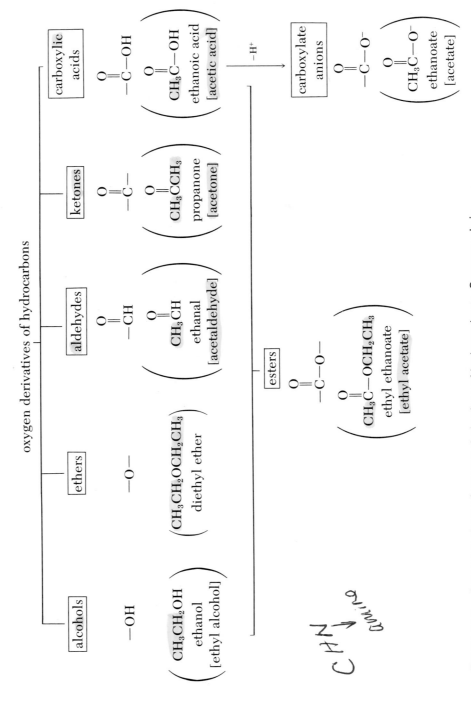

Figure 11-5. Summary chart of oxygen derivatives of hydrocarbons. Compounds in parentheses are typical examples of classes listed.

A related reaction occurs when a strong base like sodium hydroxide is added to an ester. Instead of the ester breaking down to form an acid and alcohol as in hydrolysis, the salt of the acid is formed. This reaction is known as **saponification**. Thus, when amyl acetate is saponified with sodium hydroxide, sodium acetate and amyl alcohol are produced:

$$\underset{\text{amyl acetate}}{CH_3\overset{\overset{O}{\|}}{C}-OCH_2CH_2CH_2CH_2CH_3} + \underset{\text{base}}{NaOH} \longrightarrow \underset{\substack{\text{sodium} \\ \text{acetate}}}{CH_3\overset{\overset{O}{\|}}{C}-O^-\ ^+Na} + \underset{\text{amyl alcohol}}{CH_3CH_2CH_2CH_2CH_2OH}$$

Saponification produces soaps when the carboxylate ion comes from a long-chain fatty acid (as described in Section 14.2).

Figure 11-5 is a chart containing esters and other oxygen derivatives described in this chapter.

SUMMARY

The oxygen atom can be combined with a hydrocarbon in the form of several functional groups. These groups determine to a large degree the properties of the resulting molecule.

Alcohols contain the hydroxyl group, —OH, connected to a hydrocarbon group. The simplest alcohol is methanol, or methyl alcohol, CH_3OH. The compound known simply as "alcohol" is generally ethanol, or ethyl alcohol, CH_3CH_2OH. "Rubbing alcohol compound" is 2-propanol, or isopropyl alcohol. Other important alcohols include menthol, phenol, ethylene glycol, hexylresorcinol, and glycerol.

The alcohol can be converted to other oxygen-containing groups by oxidation. In organic chemistry, oxidation generally involves the addition of oxygen or removal of hydrogen from a molecule. Alcohols also undergo dehydration, the removal of water.

Ethers are organic molecules that contain an oxygen atom covalently bound to two hydrocarbon groups. Diethyl ether, or simply "ether," $CH_3CH_2OCH_2CH_3$, is an important anesthetic. Other ethers include divinyl ether, eugenol, and ethylene oxide (an epoxide). Ethers are relatively stable chemically but can oxidize to form explosive peroxide compounds.

Aldehydes contain a carbonyl group

$$-\overset{\overset{O}{\|}}{C}-$$

attached to a hydrogen atom at one end and a hydrocarbon group at the other.

Methanal, or formaldehyde

$$\underset{\text{HCH}}{\overset{\displaystyle\overset{\text{O}}{\|}}{}}$$

is the simplest aldehyde. Tests for aldehydes, using Benedict's, Fehling's, or Tollen's solution, are based on the ability of aldehydes to act as reducing agents, becoming oxidized (to acids) in the process.

Ketones have a carbonyl group attached at both ends to hydrocarbon groups, as in propanone (dimethyl ketone) or, more commonly, acetone. Ketones are generally unreactive.

Acids contain the carboxyl group

$$\underset{-\text{C}-\text{OH}}{\overset{\displaystyle\overset{\text{O}}{\|}}{}}$$

a combination of the carbonyl and hydroxyl groups. Compounds containing this group, carboxylic acids, are weak acids because hydrogen ions are produced to a small extent by ionization. The simplest members of this class are methanoic (formic) acid, and ethanoic (acetic) acid.

The salt produced by a neutralization reaction of a carboxylic acid and a base contains the carboxylate group

$$\underset{-\text{C}-\text{O}^-}{\overset{\displaystyle\overset{\text{O}}{\|}}{}}$$

The reaction of acetic acid and sodium hydroxide, for example, forms sodium acetate and water.

The reaction of a carboxylic acid and an alcohol produces an ester, which contains the functional group

$$\underset{-\text{C}-\text{O}-}{\overset{\displaystyle\overset{\text{O}}{\|}}{}}$$

The hydrocarbon attached to the carbon end comes from the acid, while the group at the oxygen end comes from the alcohol. The ester resulting from methyl alcohol (methanol) and acetic acid (ethanoic acid) is methyl acetate (methyl ethanoate). Esters are important as flavors and fragrances. Aspirin is the ester acetylsalicylic acid. Esters can be hydrolyzed, broken back into the acid and alcohol, by a reaction with water.

Exercises

1. (11.1) What is a functional group?
2. (11.2) How does the hydroxyl group differ from the hydroxide ion?

3. (11.2) Write the names and structural formulas for five alcohols.

4. (11.3) Describe the medical applications of three alcohols.

5. (11.3) Why are alcohols used for sponge baths?

6. (11.4) Draw the product of the dehydration of propanol (propyl alcohol).

7. (11.5) Write the names and structural formulas of five ethers.

8. (11.5) What are the advantages and disadvantages of diethyl ether as an anesthetic?

9. (11.6) Why should ethers be carefully handled and not be stored for very long periods of time?

10. (11.7) How does an aldehyde differ from a ketone?

11. (11.7) Write the names and structural formulas of five aldehydes.

12. (11.8) Describe how Benedict's test works.

13. (11.9) Write the names and structural formulas of five ketones.

14. (11.9) Which of the following are ketones and which are aldehydes?

(a) $CH_3CH_2CCH_3$ (b) CH_3CH (c) $HCCH_2CH_2CH_3$ (d) $CH_3CH_2CH_2CCH_2CH_3$

15. (11.9) Name the compounds in question 14.

16. (11.10) Write an equation for the ionization of propanoic (propionic) acid in water.

17. (11.10) Why are carboxylic acids considered weak? How do they differ from mineral acids?

18. (11.10) Write the names and structural formulas of five carboxylic acids.

19. (11.10) Which carboxylic acids are present in blood?

20. (11.11) Write an equation for the reaction of butanoic acid and potassium hydroxide.

21. (11.11) Name the salt formed by the reaction of sodium hydroxide with (a) formic acid; (b) acetic acid; (c) pentanoic acid.

22. (11.12) Write the names and structural formulas of five esters.

23. (11.12) Write an equation for the reaction between butanol (butyl alcohol) with acetic (ethanoic) acid.

24. (11.12) Name the ester formed from (a) methanol and propanoic acid; (b) butanol and pentanoic acid; (c) propyl alcohol and formic acid.

25. (11.12) Write the chemical name and structure of aspirin.

26. (11.13) Write an equation for the (a) hydrolysis and (b) saponification of ethyl formate.

27. Identify the class of the following compounds (alcohol, ether, etc.):

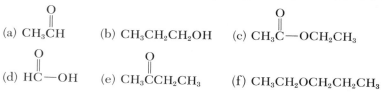

(a) CH_3CH (b) $CH_3CH_2CH_2OH$ (c) $CH_3C-OCH_2CH_3$

(d) $HC-OH$ (e) $CH_3CCH_2CH_3$ (f) $CH_3CH_2OCH_2CH_2CH_3$

28. Name the compounds in question 27.

Other organic derivatives and polymers

In addition to carbon, hydrogen, and oxygen, other elements are often found in organic compounds. The most important are the halogens (fluorine, chlorine, bromine, and iodine), sulfur, and nitrogen. As in the case of oxygen derivatives, each of these additional elements can be substituted into a hydrocarbon in various ways, resulting in classes of compounds containing functional groups with characteristic properties.

12.1   Halogen derivatives

Atoms of fluorine, chlorine, bromine, or iodine can replace the hydrogen atoms of a hydrocarbon, since each needs one more electron to fill its outer shell. For example, the four chlorine derivatives of methane, CH_4, are shown in Table 12-1. When a halogen is substituted into an alkane, the resulting molecule is an **alkyl halide**; the chlorine derivatives are then called **alkyl chlorides**. The bond between the carbon and the halogen is *covalent;* there is no dissociation as with ionic chlorides like NaCl.

Methyl chloride (chloromethane), as well as ethyl chloride, is used as a local anesthetic. It is a gas at atmospheric pressure but is stored under high pressure as a liquid. When sprayed onto the skin, it readily evaporates, "freezing" the surface by removing heat energy, thus reducing sensation. It is used for minor operations only.

Chloroform (trichloromethane) was once widely used as a general anesthetic administered by inhalation, but it can cause respiratory and circulatory failure. It is slowly oxidized by atmospheric oxygen to the poisonous gas phosgene, $COCl_2$. Commercial chloroform contains about 1% ethyl alcohol to destroy any phosgene that may form.

Table 12-1 **Chlorine Derivatives of Methane**

Formula	Lewis structure	Name[a]
CH_3Cl	H H:C:Cl: H	chloromethane (methyl chloride)
CH_2Cl_2	H H:C:Cl: :Cl:	dichloromethane (methylene chloride)
$CHCl_3$	H :Cl:C:Cl: :Cl:	trichloromethane (chloroform)
CCl_4	:Cl: :Cl:C:Cl: :Cl:	tetrachloromethane (carbon tetrachloride)

[a] Common name in parentheses.

Carbon tetrachloride (tetrachloromethane) is a volatile, dense liquid. Although its use has been limited, it is an excellent solvent for cleaning since it is nonpolar and dissolves fats, oils, and greases. Because it does not burn, "carbon tet" has been used in extinguishing some types of fires, such as petroleum fires. It cannot be used in a closed area, however, since this molecule also forms toxic phosgene. You should avoid breathing vapors of carbon tetrachloride because they can be lethal.

Other alkyl halides of importance involve different halogens attached to a carbon atom. Iodoform (triodomethane), CHI_3, is the iodine equivalent of chloroform. It is a pale-yellow solid sometimes used as an antiseptic in ointments or as a dusting powder. Freon, CCl_2F_2, is made by substituting two chlorine atoms of carbon tetrachloride by fluorine atoms. It is used as a refrigerant and aerosol propellant in spray cans. Some scientists think this gas may undergo chemical reactions in the upper atmosphere that result in partial destruction of the ozone (O_3) layer, which protects us by filtering harmful ultraviolet radiation from the sun.

Halogenated hydrocarbons can have more than one carbon, as in halothane (2-bromo-2-chloro-1,1,1-trifluoroethane), a popular inhalation anesthetic. A

$$\begin{array}{cc} F & Cl \\ | & | \\ F-C-C-H \\ | & | \\ F & Br \end{array} \qquad \begin{array}{cc} Cl & OH \\ | & | \\ Cl-C-C-H \\ | & | \\ Cl & OH \end{array}$$

halothane   chloral hydrate

molecule that contains hydroxyl groups in addition to chlorine atoms is chloral hydrate, a drug that induces sleep and prevents convulsions. (It is the basis for the "Mickey Finn" or "knockout drops.") A fluorinated ether, hexafluorodiethyl ether (Flurothyl, Indoklon), produces convulsions and is used as a substitute for electroshock therapy in the treatment of certain mental disorders.

$$CF_3CH_2OCH_2CF_3$$
hexafluorodiethyl ether

Some halogenated hydrocarbons are used to make plastics and other big molecules, as described later in this chapter. One of these is tetrafluoroethylene, used to make Teflon. Another is vinyl chloride, chloroethylene. In addi-

$$CF_2{=}CF_2 \qquad CH_2{=}CHCl$$
tetrafluoroethylene   vinyl chloride

tion to being used to make a plastic (polyvinyl chloride, PVC), it has been used until recently as an aerosol propellant. Vinyl chloride is now believed to cause certain types of cancer.

The simplest aromatic halogen derivative is chlorobenzene or phenyl chloride. A more complicated example is hexachlorophene, a disinfectant used in

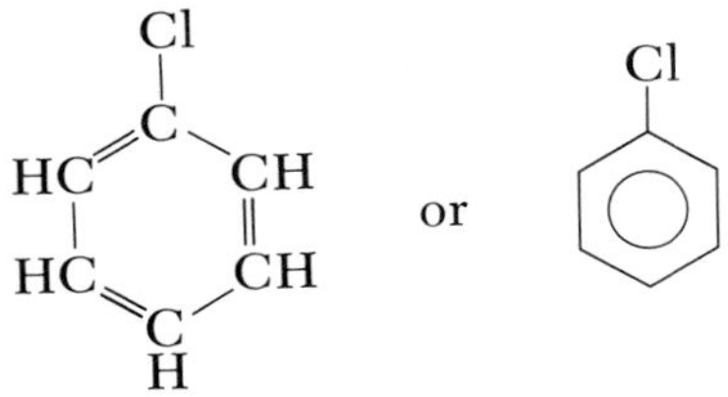

chlorobenzene

germicidal cleansers, such as pHisoHex (as a 3% solution). It is now available

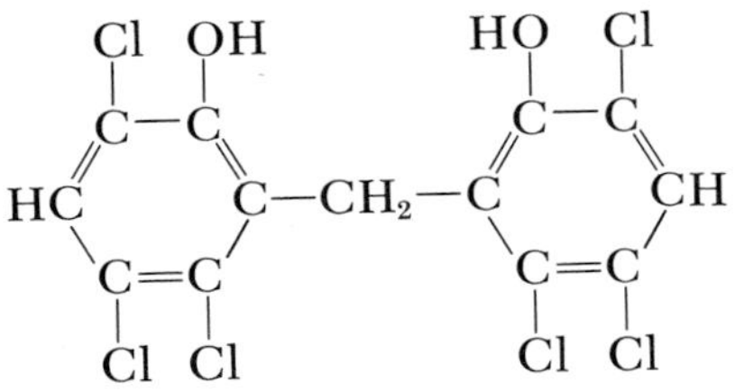

hexachlorophene

only on a prescription basis because of possible harmful effects on the nervous system. The insecticide DDT (dichlorodiphenyltrichloroethane) is a chlorinated aromatic hydrocarbon that effectively kills the mosquito that carries malaria. Its use is now banned in many parts of the world because it

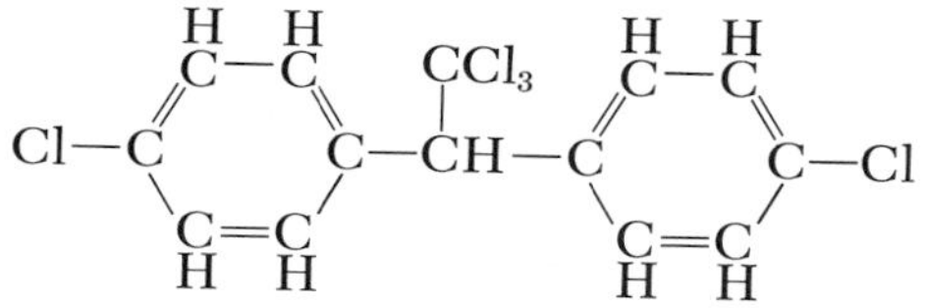

DDT

poisons fish and can disrupt the natural balance of the environment.

Various organoiodine compounds are **radiopaque**. They cause soft tissue to become visible in x-ray studies by absorbing the radiation instead of allowing it to pass through. They are used in studying blood flow (angiography) and urine flow (urography). Two examples are sodium diatrizoate (Hypaque) and iopanoic acid.

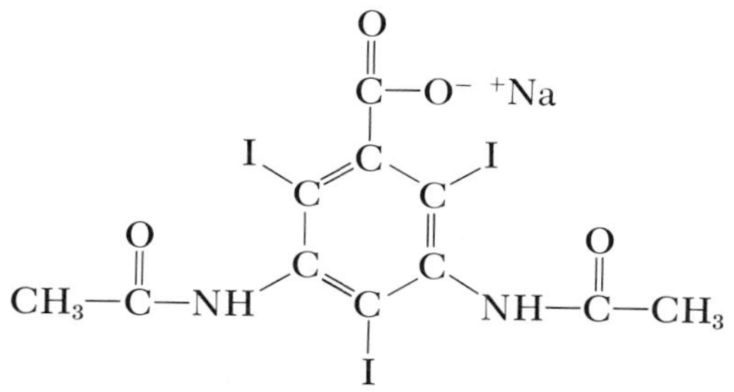

sodium diatrizoate

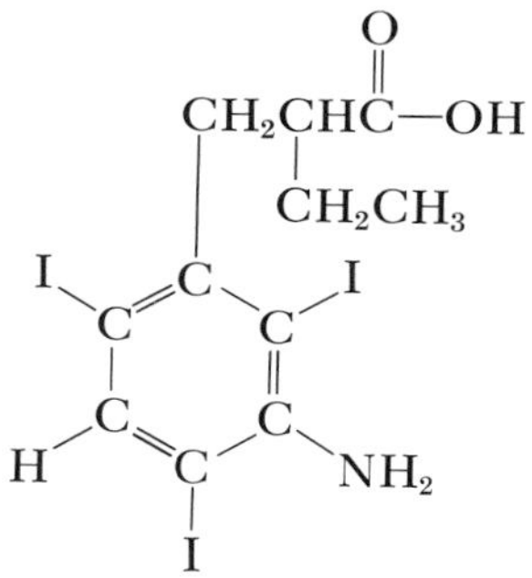

iopanoic acid

12.2 Sulfur derivatives

Sulfur is found in the same group in the periodic table as oxygen; it also has six valence electrons. Many *organic sulfur compounds therefore resemble the oxygen derivatives* you have already studied.

The equivalent of an alcohol is the **thiol**, thioalcohol (or mercaptan). The simplest example is methanethiol (or methyl mercaptan), CH_3SH. Notice that the formula is the same as methanol or methyl alcohol, CH_3OH, except that a sulfur atom takes the place of the oxygen. These compounds, particularly those with low molecular weights, have disagreeable odors. Methanethiol is partly responsible for the odor of feces, while butanethiols form part of the dreaded weapon of the skunk.

An important reaction of the —SH or **sulfhydryl group** is the formation of a **disulfide** from the oxidation of two thiols. This reaction is important because it helps to maintain the structure of proteins, as you will learn later.

$$CH_3-SH + HS-CH_3 \xrightarrow{\text{oxidation}} CH_3-S-S-CH_3 + H_2O$$

a disulfide

Table 12-2 **Sulfur Derivatives**

Type	Example	Name
thiol	CH_3SH	methanethiol (methyl mercaptan)
thioether	CH_3SCH_3	methylthiomethane (dimethyl sulfide)
thioketone		thiobenzophenone
thioacid	$CH_3-\overset{\displaystyle O}{\overset{\|}{C}}-SH$	thioacetic acid
sulfoxide	$CH_3-\overset{\displaystyle O}{\overset{\|}{S}}-CH_3$	dimethyl sulfoxide (DMSO)
sulfonic acid	$CH_3-\overset{\displaystyle O}{\underset{\underset{\displaystyle O}{\|}}{\overset{\|}{S}}}-OH$	methanesulfonic acid
sulfone		diphenyl sulfone

The thioether has a sulfur atom bridging two hydrocarbon groups. For example, the analog of diethyl ether is called diethyl sulfide, diethyl thioether, or ethylthioethane.

$$CH_3CH_2-S-CH_2CH_3$$
ethylthioethane

These and other types of sulfur derivatives are shown in Table 12-2. Whenever you see "thio-" or "sulf-" in a name, you should recognize that the compound contains a functional group based on sulfur.

Certain organic sulfur compounds have characteristic colors and are used as dyes. Phenol red, phenolsulfonphthalein, is the basis for a test of kidney function. Amaranth, a sulfonic acid salt, is known as Red Number Two. It was the

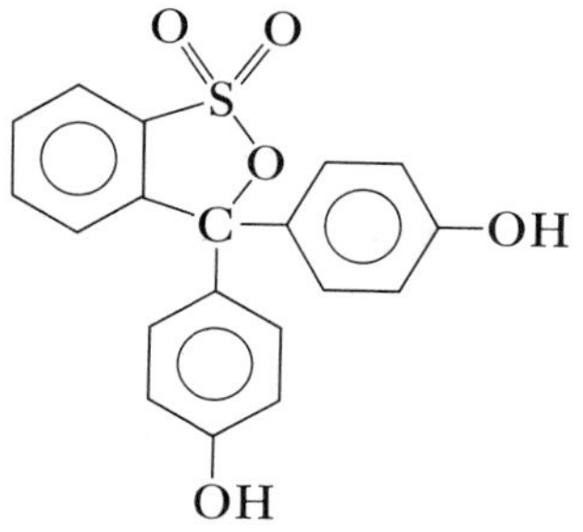

phenol red

most commonly used dye in foods as well as cosmetics. Recent studies link this compound with cancer in animals, and its use in food is now banned in several countries, including the United States.

12.3 Nitrogen derivatives—amines

Amines are substituted ammonia molecules. They are produced by replacing one, two, or all three hydrogens of NH_3 by hydrocarbon groups. When one hydrogen is replaced, a *primary* amine results. Two are replaced in a *secondary* amine, and all three are replaced in a *tertiary* amine. For example, replacing hydrogen atoms by methyl groups makes the following amines:

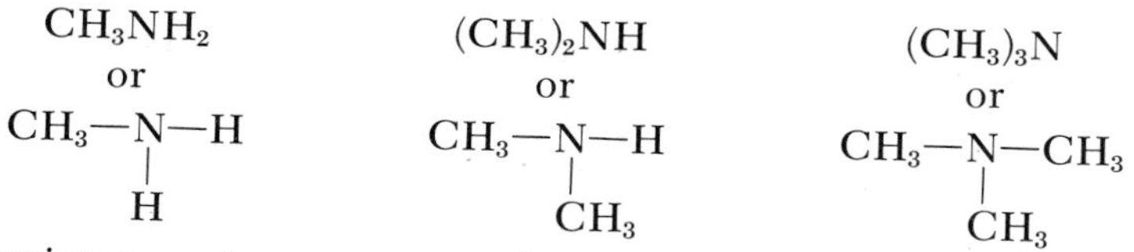

$$CH_3NH_2$$
or
$$CH_3{-}N{-}H$$
$$|$$
$$H$$
a primary amine
(methylamine)

$$(CH_3)_2NH$$
or
$$CH_3{-}N{-}H$$
$$|$$
$$CH_3$$
a secondary amine
(dimethylamine)

$$(CH_3)_3N$$
or
$$CH_3{-}N{-}CH_3$$
$$|$$
$$CH_3$$
a tertiary amine
(trimethylamine)

Their names are written as one word, ending in "-amine." When the three hydrocarbon groups are the same, the prefix "di-" or "tri-" shows you that two or three of the hydrogen atoms of ammonia have been replaced, respectively. Lack of any prefix, as in methylamine, means that only one hydrocarbon group is present. Other examples are listed in Table 12-3.

Table 12-3 **Amines**

Formula	Name
$CH_3CH_2NH_2$	ethylamine
$(CH_3CH_2)_2NH$	diethylamine
$(CH_3CH_2)NHCH_3$	methylethylamine
$(CH_3CH_2)_3N$	triethylamine

The functional group of amines is the **amino group**, often appearing as —NH_2. Several may appear in the same molecule, as in ethylenediamine.

$$NH_2CH_2CH_2NH_2$$

ethylenediamine

A derivative of this amine, ethylenediaminetetraacetic acid (or EDTA), is used to trap metal poisons, such as lead, in the body.

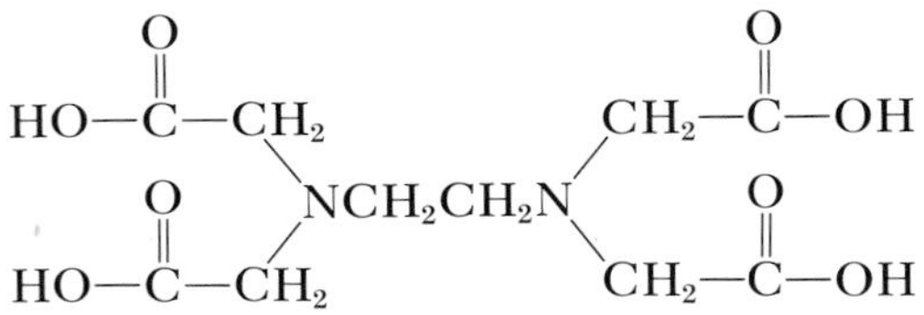

ethylenediaminetetraacetic acid
(EDTA)

Many important drugs are amines. The primary amine amphetamine, or benzedrine ("speed"), is a central nervous system stimulant that can be addictive. Ephedrine is a secondary amine related to amphetamine; it is used to

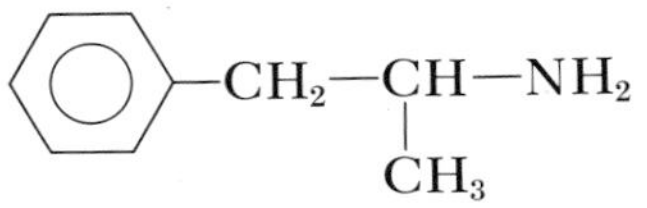

amphetamine 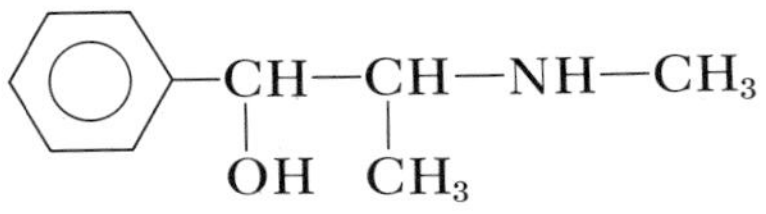   ephedrine

open bronchial passages and also to dilate the pupils in eye examinations. An important tertiary amine is methadone, used as a substitute for heroin.

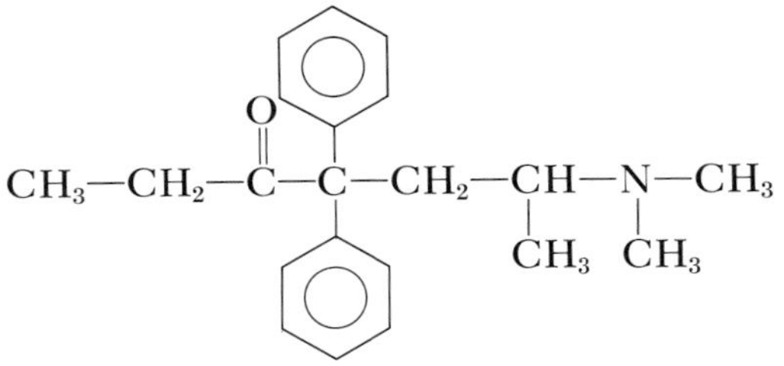

methadone

Several halogen derivatives of amines are cytotoxic agents (cell poisons) administered in cancer therapy. An example is nitrogen mustard, or mechloreth-

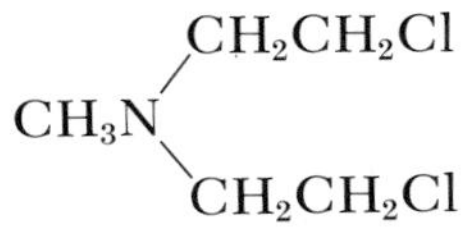

nitrogen mustard
(mechlorethamine)

amine. An amine formed from a thiol, cysteamine or 2-aminoethanethiol, is a radioprotective drug; it helps reduce the effect of radiation if given before exposure.

$$NH_2CH_2CH_2SH$$

cysteamine

One example of a cyclic amine is piperazine, a vermifuge, which kills pinworms and round worms. A more complex example is urotropine (hexamethyl-

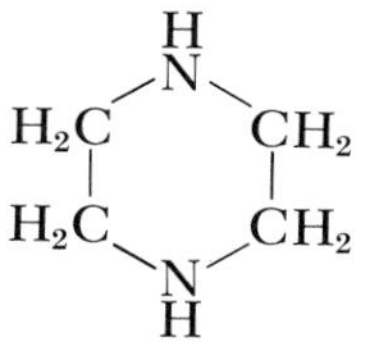

piperazine

enetetramine), a urinary antiseptic formed by reacting ammonia with formaldehyde. Aromatic amines are also common, such as aniline (phenylamine) or

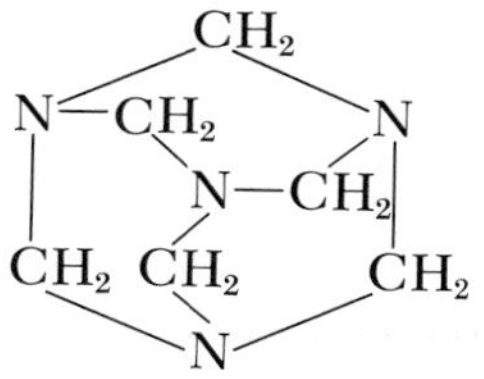

urotropine

p-aminobenzoic acid (PABA), used in suntan preparations to prevent ultraviolet radiation from reaching the skin. Several aromatic amines (such as those

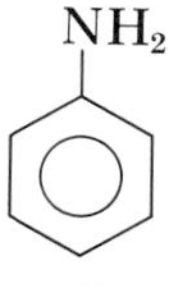

aniline 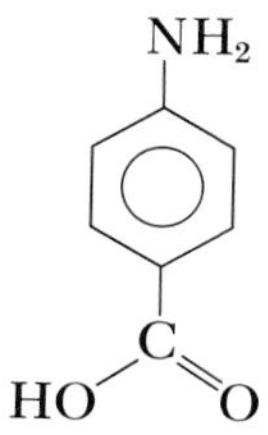   p-aminobenzoic
acid (PABA)

based on naphthalene, naphthylamines) are known to be carcinogenic, cancer producing.

An extremely important class of amines consists of the **amino acids**. These molecules contain both the amine functional group and the carboxylic acid group. The simplest example is glycine. Amino acids are the basic units of all proteins; you will learn about them in Chapter 15.

$$NH_2-CH_2-\overset{\overset{\textstyle O}{\|}}{C}-OH$$

glycine, an amino acid

12.4 Reactions of amines

The most important property of amines, many of which have an unpleasant fishy odor, is that they are *basic*, like ammonia. The nitrogen forms three covalent bonds, to hydrogens or carbons, but has two electrons left over. This unshared pair of electrons is called a **lone pair**. Amines are basic because the

$$\text{lone pair} \longrightarrow \overset{..}{H} \overset{..}{\underset{\underset{H}{\times}}{N}} \overset{\times}{H}$$

ammonia

lone pair enables them to accept a hydrogen ion, which has no valence electrons. The hydrogen ion shares these two electrons on the nitrogen to form a (coordinate) covalent bond:

$$-\overset{..}{\underset{|}{N}}- + H^+ \longrightarrow \left[-\overset{H}{\underset{|}{\overset{..}{N}}}-\right]^+$$

The amine is now positively charged since it gained a proton. The cation that forms is named by replacing the "-amine" ending by "-ammonium ion." Thus, when methylamine is protonated, it forms the methylammonium ion. More examples are listed in Table 12-4.

Table 12-4   Ions Formed by Protonation of an Amine

Amine	Cation formed	Formula
ethylamine	ethylammonium ion	$CH_3CH_2NH_3^+$
diethylamine	diethylammonium ion	$(CH_3CH_2)_2NH_2^+$
ethylmethylamine	ethylmethylammonium ion	$(CH_3CH_2)\overset{+}{N}H_2CH_3$
triethylamine	triethylammonium ion	$(CH_3CH_2)_3NH^+$

Amines react with any substance that can donate a proton, including water:

$$CH_3NH_2 + H_2O \longrightarrow CH_3NH_3^+ + OH^-$$

This process occurs only to a very slight degree, just as ammonium ions and hydroxide ions are formed in small amounts by the reaction of ammonia with water. Amines react with mineral acids to form ammonium salts as follows:

$$CH_3NH_2 \quad + HCl \longrightarrow \quad CH_3NH_3^+Cl^-$$

methylamine   methylammonium
                                      chloride

Salts formed from amines are named like any other ionic compound; the name of the cation, in this case a substituted ammonium ion, is placed before the anion. Because these salts are much more soluble in water than the original amine, many drugs containing nitrogen are administered in their ammonium form. An example is the local anesthetic Novocain, which is procaine hydrochloride (Figure 12-1).

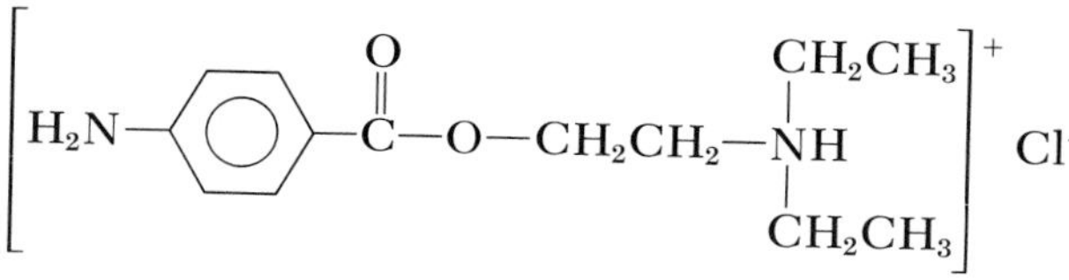

procaine
hydrochloride
(Novocain)

Another important reaction of amines occurs with alkyl halides. In this case, the hydrocarbon group occupies the fourth position on the nitrogen atom, also forming a **quaternary ammonium salt**:

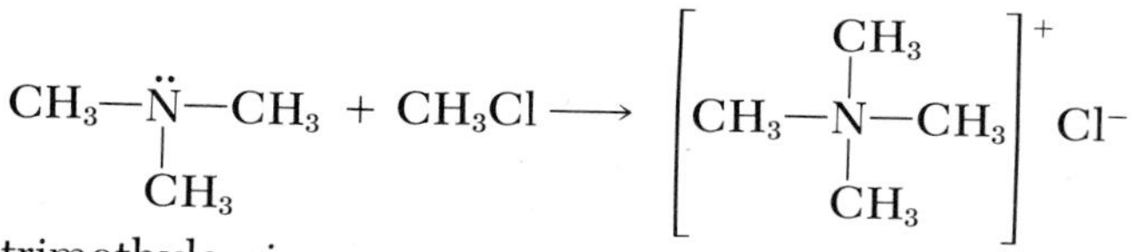

trimethylamine   tetramethylammonium
                                         chloride

It is called quaternary because there are four carbon atoms surrounding the nitrogen atom. Members of this class of compounds are sometimes called

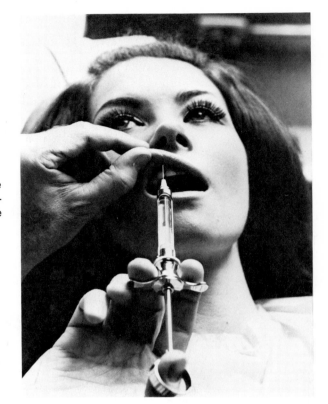

Figure 12-1. Administration of procaine hydrochloride (Novocain), a local anesthetic. (Photo courtesy of Cook-Waite Laboratories.)

"quats." One example, benzalkonium chloride (Zephiran Chloride) is used in

$$\left[\bigcirc\!\!\!\!-CH_2-\overset{\overset{\displaystyle CH_3}{|}}{\underset{\underset{\displaystyle CH_3}{|}}{N}}-R \right]^+ Cl^-$$

benzalkonium chloride
(R = mixture of alkyl groups
with 8 to 18 carbons,
C_8H_{17} to $C_{18}H_{37}$)

solution for skin preparation (1:500 to 1:1000), for cleaning mucous membranes (1:2000 to 1:10,000), and for sterilizing instruments (1:1000) (Figure 12-2). Choline is an important naturally occurring quaternary ammonium salt.

$$\left[\begin{array}{c} CH_2-CH_2-\overset{\displaystyle CH_3}{\underset{\displaystyle CH_3}{N}}-CH_3 \\ OH \end{array} \right]^+ OH^-$$

choline

It is used to prevent excess fat accumulation and acts as a stimulant. Acetyl-choline, an ester of choline, is a related molecule. This molecule is vital in transmitting nerve impulses in the body.

$$\left[CH_3-\overset{\displaystyle O}{\overset{\|}{C}}-O-CH_2CH_2-\overset{\displaystyle CH_3}{\underset{\displaystyle CH_3}{N}}-CH_3 \right]^+ OH^-$$

acetylcholine

Figure 12-2. Dental instruments being sterilized in a "quat" solution. (Photo by Al Green.)

12.5 Nitrogen derivatives—amides

Another class of nitrogen compounds forms by the reaction of an amine with a carboxylic acid. The resulting **amide** contains the following functional group:

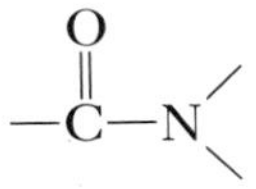

$$-\overset{\overset{\displaystyle O}{\|}}{C}-N\diagup^{\diagdown}$$

the amide group

For example, if acetic acid reacts with ammonia, an amide is formed by the nitrogen losing one hydrogen and the acid losing the hydroxyl group, thus forming water.

$$CH_3-\overset{\overset{\displaystyle O}{\|}}{C}-OH \ + \ NH_3 \ \longrightarrow \ CH_3-\overset{\overset{\displaystyle O}{\|}}{C}-NH_2 \ + \ H_2O$$

acetic acid   ammonia   an amide

The simple amides are named by replacing the ending "-ic acid" (or "-oic acid") by "-amide." Thus, the amide formed in the reaction above is acetamide (or ethanamide). Other simple amides appear in Table 12-5.

Table 12-5 Amides

Formula	Name[a]
$HC\overset{\overset{\displaystyle O}{\|}}{-}NH_2$	methanamide (formamide)
$CH_3\overset{\overset{\displaystyle O}{\|}}{C}-NH_2$	ethanamide (acetamide)
$CH_3CH_2\overset{\overset{\displaystyle O}{\|}}{C}-NH_2$	propanamide (propionamide)

[a] Common name in parentheses.

More complicated amides are formed from amines that contain hydrocarbon groups at the nitrogen:

$$CH_3-\overset{\overset{\displaystyle O}{\|}}{C}-OH \ + \ CH_3NH_2 \ \longrightarrow \ CH_3-\overset{\overset{\displaystyle O}{\|}}{C}-\underset{\underset{\displaystyle CH_3}{|}}{N}H \ + \ H_2O$$

acetic acid   methylamine   N-methylacetamide

The amide from methylamine and acetic acid is called *N*-methylacetamide; the *N* in front shows that the nitrogen of acetamide has a methyl group. If di-methylamine reacts with acetic acid, *N,N*-dimethylacetamide results.

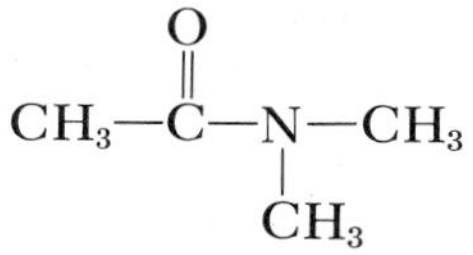

$$CH_3-\overset{\overset{\textstyle O}{\|}}{C}-\underset{\underset{\textstyle CH_3}{|}}{N}-CH_3$$

N,N-dimethylacetamide

One of the most important simple amides is urea (carbamide), an end product of the breakdown of nitrogen-containing molecules in the body. As you can see, it is really two amides (a diamide), since two molecules of ammonia are attached to a central carbonyl group.

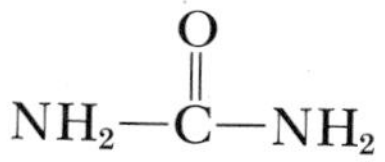

$$NH_2-\overset{\overset{\textstyle O}{\|}}{C}-NH_2$$

urea

An aromatic amide, acetanilide, reduces pain and fever. It is formed from acetic acid and aniline. Phenacetin, a derivative of acetanilide that is less toxic, is found in commercial headache preparations. Another amide, derived from phenacetin, acetaminophen (*p*-hydroxyacetanilide), is sold as an aspirin substitute under such names as Tylenol.

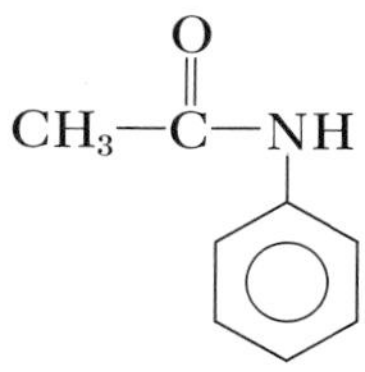

$$CH_3-\overset{\overset{\textstyle O}{\|}}{C}-NH$$

acetanilide

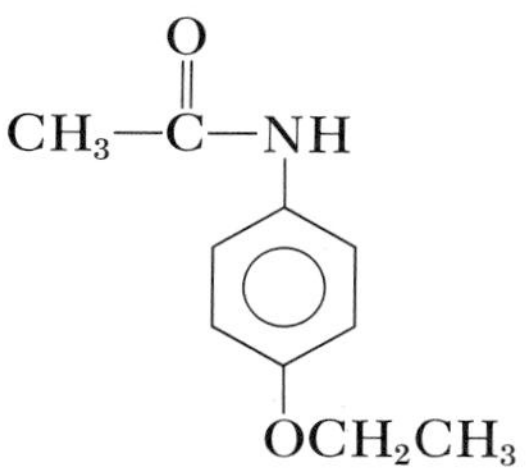

$$CH_3-\overset{\overset{\textstyle O}{\|}}{C}-NH$$

OCH$_2$CH$_3$

phenacetin

$$CH_3-\overset{\overset{\textstyle O}{\|}}{C}-NH$$

OH

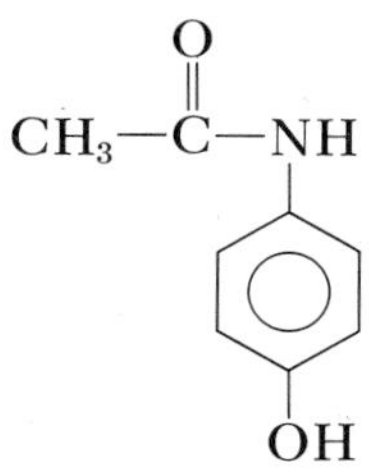

acetaminophen

A special class of amides is formed by the combination of an amine, not with a carbonyl group, but with a group containing sulfur and oxygen. This class, the **sulfonamides**, has the following functional group:

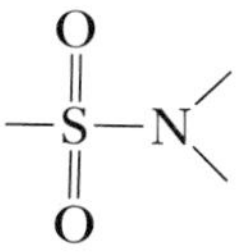

$$-\overset{\overset{\textstyle O}{\|}}{\underset{\underset{\textstyle O}{\|}}{S}}-N\big\langle$$

the sulfonamide group

Sulfa drugs, which all have the general formula given below, are important examples of sulfonamides.

$$H_2N-\underset{}{\bigcirc}-\overset{\displaystyle O}{\underset{\displaystyle O}{\overset{\|}{\underset{\|}{S}}}}-\underset{R}{N}H$$

sulfa drugs
(R = hydrocarbon group)

The most important reaction undergone by amides is *hydrolysis*, the breakdown of the amide by water back into an acid and amine (in the presence of a catalyst). This type of process takes place when proteins are digested.

$$\underset{\underset{CH_3}{|}}{\overset{\overset{\displaystyle O}{\|}}{CH_3C}}-NH + H_2O \longrightarrow \overset{\overset{\displaystyle O}{\|}}{CH_3C}-OH + CH_3NH_2 \qquad \text{(hydrolysis)}$$

an amide   an acid   an amine

12.6 Heterocyclic nitrogen compounds

When another atom replaces a carbon in a cyclic molecule, a **heterocycle**, a ring made from two or more different kinds of atoms, results. You have already seen this type of molecule in the previous chapter. Epoxides and molecules like dioxane, the cyclic ethers, are oxygen heterocycles; as you will learn later, sugars also fall into this category. The heterocyclic compounds of nitrogen are extremely important since they are found in proteins and nucleic acids (the genetic material).

Pyrrole is a five-membered unsaturated nitrogen heterocycle. Both hemo-

$$\begin{array}{c} HC-CH \\ \| \quad\quad \| \\ HC \quad\quad CH \\ \diagdown N \diagup \\ | \\ H \end{array} \quad \text{or} \quad \begin{array}{c} \bigcirc \\ N \\ | \\ H \end{array}$$

pyrrole

globin in the blood and chlorophyll in plants contain a ring system based on four pyrroles connected together. When a benzene ring is fused to pyrrole, indole results. This substance, along with skatole (indole with a methyl group),

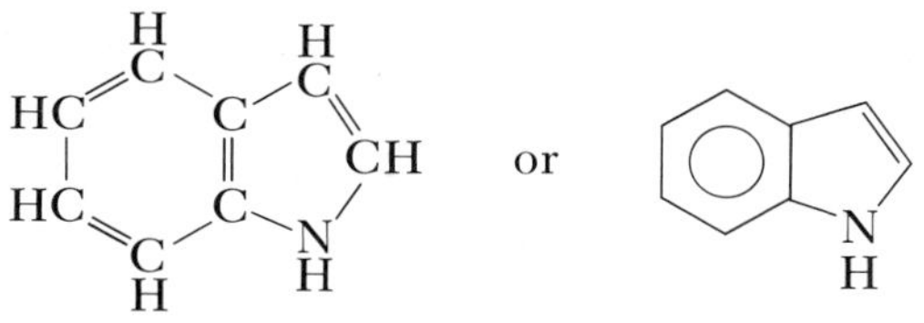

indole

contributes to the odor of feces. Pyridine is a six-membered aromatic nitrogen

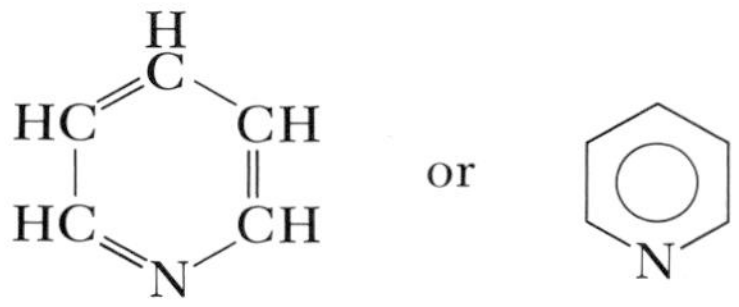

pyridine

heterocycle. When fused to a benzene ring, the resulting heterocycle is quinoline.

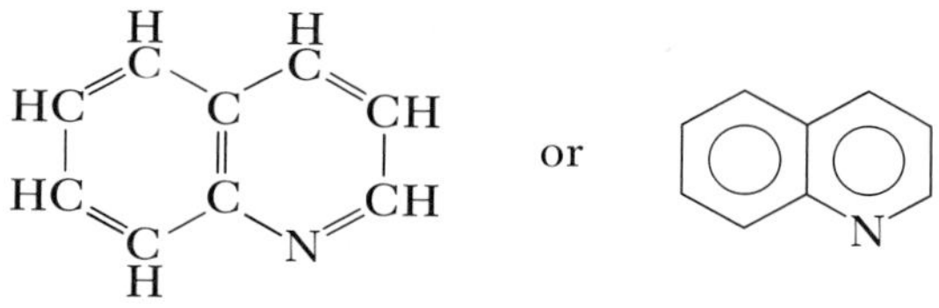

quinoline

Other important heterocycles have two nitrogen atoms in the same ring. For
example, pyrimidine has two nitrogens in a six-membered aromatic ring.

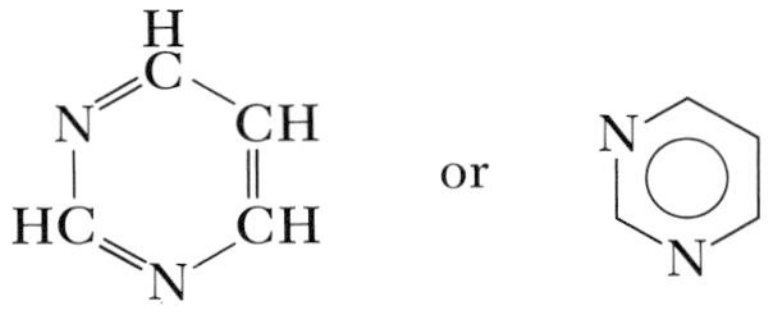

pyrimidine

Purine is formed when a five-membered ring with two nitrogen atoms is fused
to pyrimidine. Pyrimidine and purine are the nitrogen parts of the nucleic

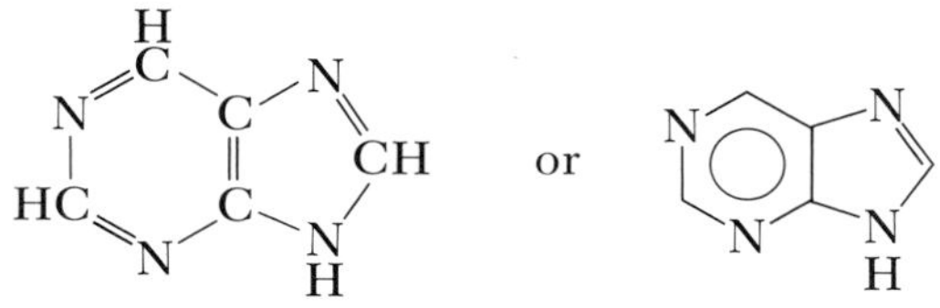

purine

acids (DNA and RNA), as explained in Chapter 20.

12.7 Nitrogen derivatives—alkaloids

Alkaloids are nitrogen-containing molecules of vegetable origin (often having a bitter taste); they generally have a heterocyclic ring system and are physiologically active. In fact, many are used as drugs because of their effects as tranquilizers, stimulants, muscle relaxants, pain killers, sedatives, or anesthetics.

Lysergic acid is an alkaloid produced from a fungus known as ergot that grows on rye. Its amide, lysergic acid diethylamide, known as LSD, is well known as a hallucinogen.

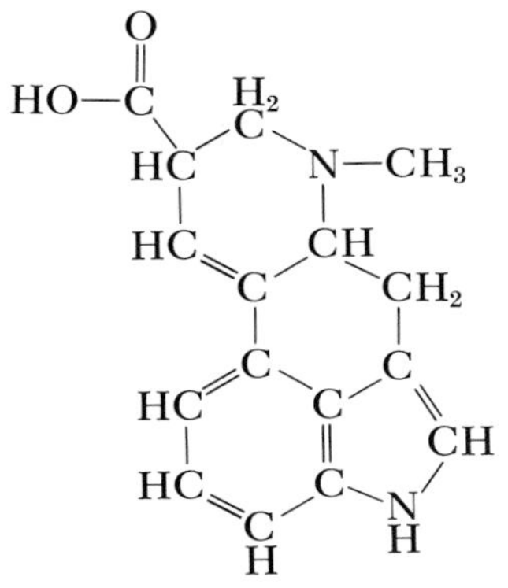

lysergic
acid

Nicotine, a pyridine-based alkaloid, is found in tobacco leaves and is therefore present in cigarettes. Nicotine is a poison and is used as an insecticide.

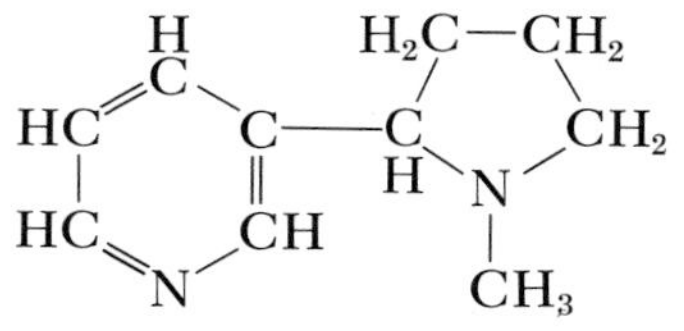

nicotine

Papavarine, an alkaloid related to quinoline, is found in opium poppy. Its effect is to relax the muscles.

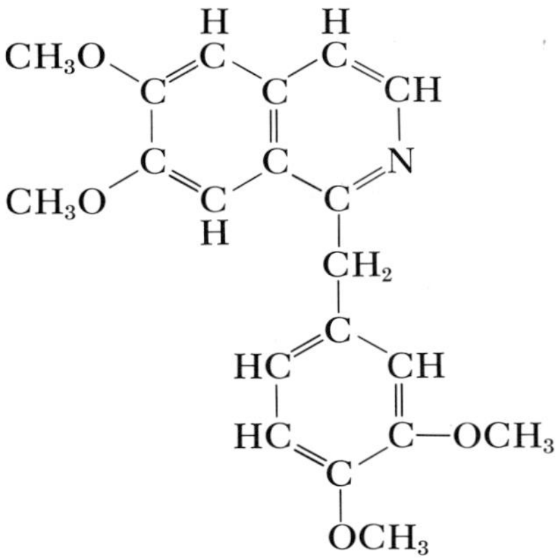

papavarine

Caffeine contains the purine ring. It is found in coffee beans and tea leaves and acts as a mild stimulant.

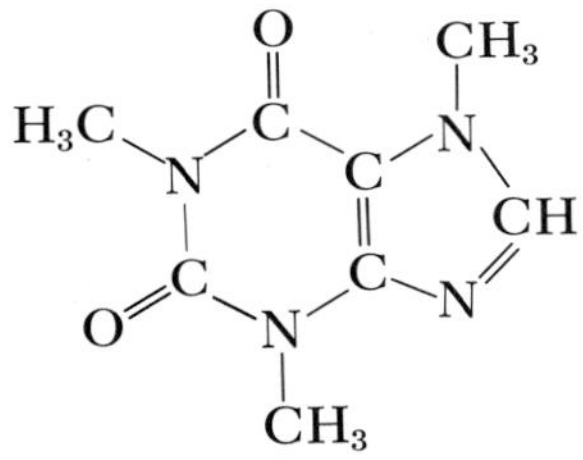

caffeine

These and other alkaloids are listed in Table 12-6 (see also Figure 12-3).

A number of the alkaloids and synthetic compounds derived from them, such as heroin, which is diacetylmorphine, are addictive.

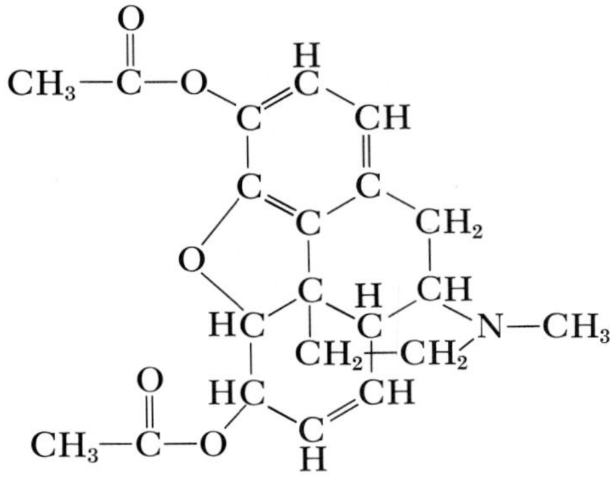

heroin

Table 12-6   **Alkaloids**

Name	Effect
atropine	relaxes muscles, reduces secretions, dilates pupils
caffeine	mild stimulant
cocaine	anesthetic
codeine	pain killer, relieves coughing
colchicine	relieves gout
emetine	kills amoebas (amoebicide)
morphine	narcotic, sedative, pain killer
nicotine	toxic
papavarine	muscle relaxant
quinine	antimalarial, reduces pain and fever
reserpine	tranquilizer, reduces high blood pressure
strychnine	stimulant, toxic

Figure 12-3.   A source of an alkaloid. The tobacco leaf contains nicotine. (Photo courtesy of The Tobacco Institute.)

Other synthetic drugs related to alkaloids are based on barbituric acid, such as veronal and phenobarbital. These drugs act as hypnotics (sleep inducing) and sedatives and can also be habit-forming. They are discussed in detail in Chapter 23.

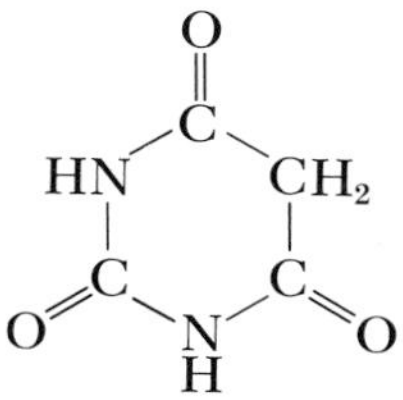

barbituric acid 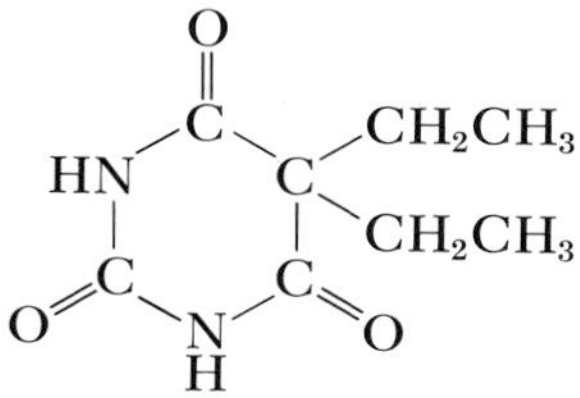   veronal 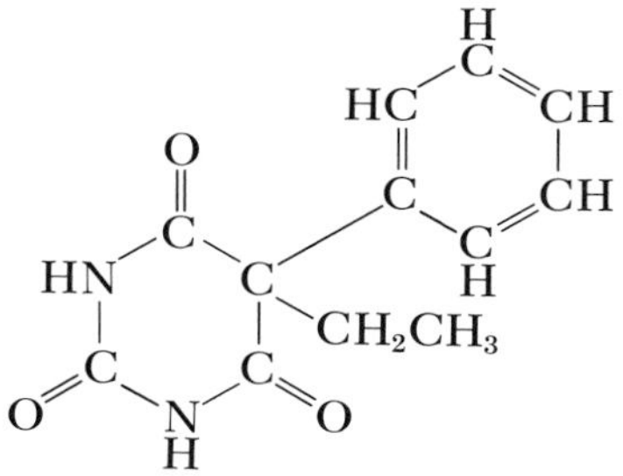   phenobarbital

12.8 Other nitrogen derivatives

A number of organic molecules contain nitrogen combined with oxygen in the form of the **nitro group**, $-NO_2$. Examples are trinitrotoluene (TNT) and the related trinitrophenol (or picric acid), both of which are explosive. Picric acid is used for the treatment of burns.

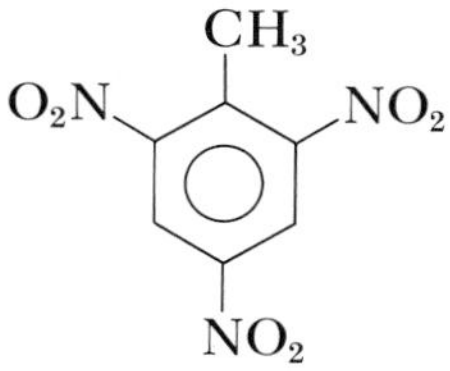

TNT 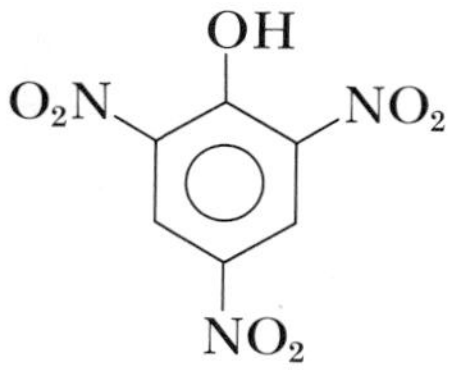   picric acid

The **nitrite group**, $-ONO$, is found in amyl nitrite (actually isoamyl nitrite), which is used to relax arterial spasms in angina pectoris.

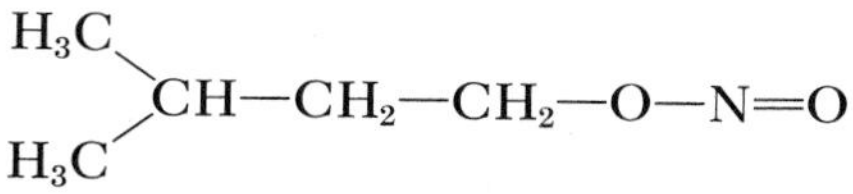

$$H_3C{\diagdown} CH-CH_2-CH_2-O-N{=}O$$
$$H_3C{\diagup}$$

isoamyl nitrite

Nitrates contain the functional group $-ONO_2$, as in glycerol trinitrate or nitroglycerin. This compound is also used in angina to dilate or widen the

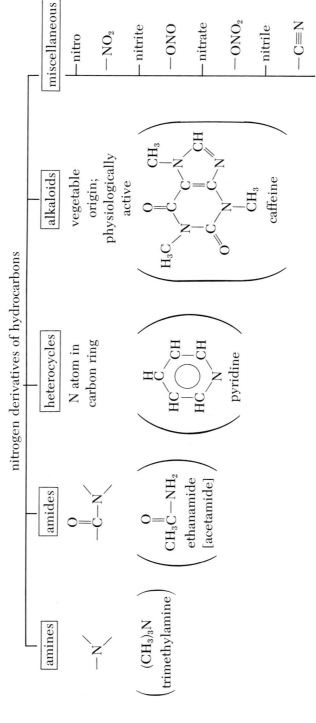

Figure 12-4. Summary chart of nitrogen derivatives of hydrocarbons. Compounds in parentheses are typical examples of classes listed.

coronary arteries but does not have the unpleasant odor of amyl nitrite, although side effects such as headache may occur.

$$CH_2—ONO_2$$
$$CH—ONO_2$$
$$CH_2—ONO_2$$

nitroglycerin

Nitriles have a functional group containing a triple bond between a carbon and nitrogen, $—C\equiv N$. Acrylonitrile is one example.

$$CH_2=CH—C\equiv N$$

acrylonitrile

Figure 12-4 summarizes the nitrogen derivatives described in this chapter.

12.9 Organic polymers

The organic molecules you have studied so far are all relatively small in size and molecular weight. Very large molecules can be produced by linking together many such small molecules. Each individual unit is called a **monomer**; the resulting molecule is a **polymer**, as shown in Figure 12-5. You can think of a polymer as an extremely long railroad train, with each car being a monomer. As you will see, the most important molecules of the body, proteins and nucleic acids, are polymers. In addition, commercially prepared or synthetic polymers are of great value in health care; they are classified as plastics, rubbers, or fibers.

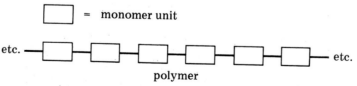

Figure 12-5. Representation of a polymer; the repeating unit is the monomer.

In a health center, much equipment is of the "throwaway" type—it is used once and discarded. Examples are syringes, tubing, and drug containers, and the equipment shown in Figure 12-6. They offer the advantages of convenience and sterility. Most of the disposable supplies are made of plastic because of its relatively low cost. A **plastic** is any material based on an organic substance of high molecular weight that is solid in its final form but has been shaped by flow at some stage in its manufacture.

Plastics can be divided into different types, depending on their properties and how they are made. **Thermoplastic** polymers (as well as elastomers—rubbers or rubberlike elastic substances) generally have long polymer chains that are not connected to each other. Their shape can be changed by heating, as when a record buckles if put in a hot place. The other type of polymer is **thermosetting**. It cannot be reshaped or remolded in this way because of **cross-links**, chemical bonds between the chains, giving the plastic a more rigid structure, as shown in Figure 12-7.

Plastics can be made by two different types of polymerization reactions: addition and condensation. In *addition* reactions, all of the atoms of the monomer molecules become part of the polymer. They simply get "hooked" together by chemical bonds. In *condensation* reactions, some of the atoms of the monomer are split off when they join together. You have already seen this on a small scale, when a water molecule is removed in the formation of an ester or an amide (dehydration). Examples of synthetic body replacement parts, made largely from plastics, are shown in Figure 12-8.

Figure 12-6. Examples of disposable medical supplies. (Photo by Al Green.)

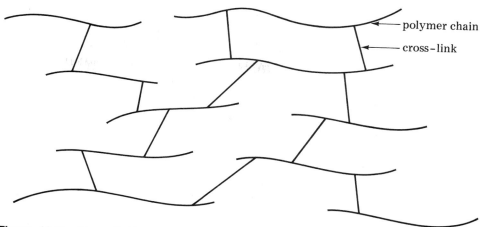

Figure 12-7. Cross-linking—chemical bonds linking polymer chains.

Figure 12-8.   A surgeon holds a synthetic aortic valve. This and the other heart valves and blood vessels are made from plastics. (WHO photo; USIS, NIH/OMS.)

12.10 Addition polymers

The simplest type of polymer is formed by the addition of molecules of ethylene (ethene).

$$CH_2\!=\!CH_2$$
ethylene

One of the bonds of the double bond is "opened up," allowing molecules of ethylene to join together by forming single covalent bonds:

$$\text{forms bond} \longrightarrow \cdot \overset{\text{H}}{\underset{\text{H}}{C}} \overset{\times}{\underset{\times}{:}} \overset{\text{H}}{\underset{\text{H}}{C}} \cdot \longleftarrow \text{forms bond}$$

The product, made by joining together many such units, is polyethylene. The following is a fragment:

$$+CH_2\!-\!CH_2\!\!+\!\!CH_2\!-\!CH_2\!\!+\!\!CH_2\!-\!CH_2\!\!+\!\!CH_2\!-\!CH_2\!+$$
polyethylene

(Note: the brackets in the long-chain polymer are used only to show the original ethylene monomers.) This plastic is made in two forms, one of which is flexible and the other of which is rigid. It is the plastic produced in the largest volume and used in squeeze bottles, film, tubing, and many other applications.

The halogen derivatives of ethylene can also be used to make plastics. Vinyl chloride polymerizes to form clear, flexible polyvinyl chloride (PVC):

$$n\,CH_2\!=\!CHCl \longrightarrow +CH_2\!-\!CHCl+_n$$
PVC

(Note: the n stands for a large number of such monomer units.) It is used in film, insulation, tubing, and fibers. One form is marketed under the trade name Saran. Tetrafluoroethylene polymerizes to form the well-known polytetrafluoroethylene, or Teflon, used to reduce friction or "sticking" as in frying pans.

$$n\,CF_2\!=\!CF_2 \longrightarrow +CF_2\!-\!CF_2+_n$$
Teflon

Polystyrene, the foamlike plastic, is a polymer of styrene. Ethylene with a

$$n\,CH_2\!=\!CH \longrightarrow +CH_2\!-\!CH+$$

polystyrene

nitrile group, acrylonitrile, polymerizes to form, not a plastic, but a fiber, poly-acrylonitrile or Orlon, used in clothing. Acrylic plastics, such as Lucite or

$$n\text{CH}_2\!\!=\!\!\underset{\underset{\text{C}\equiv\text{N}}{|}}{\text{CH}} \quad\longrightarrow\quad \left[\!\!\begin{array}{c}\text{CH}_2\!-\!\underset{\underset{\text{C}\equiv\text{N}}{|}}{\text{CH}}\end{array}\!\!\right]_n$$

Orlon

Plexiglas, are made from methyl methacrylate, also a derivative of ethylene.

$$n\text{CH}_2\!\!=\!\!\underset{\underset{\underset{\text{OCH}_3}{|}}{\underset{\text{C}=\text{O}}{|}}}{\overset{\overset{\text{CH}_3}{|}}{\text{C}}} \quad\longrightarrow\quad \left[\!\!\begin{array}{c}\text{CH}_2\!-\!\underset{\underset{\underset{\text{OCH}_3}{|}}{\underset{\text{C}=\text{O}}{|}}}{\overset{\overset{\text{CH}_3}{|}}{\text{C}}}\end{array}\!\!\right]_n$$

poly(methyl
methacrylate)

Poly(methyl methacrylate) is strong and transparent. It is used in dentistry for restorations and denture bases, as well as cements.

Natural rubber is an addition polymer of isoprene (2-methyl-1,3-butadiene):

$$n\text{CH}_2\!\!=\!\!\underset{\underset{\text{CH}_3}{|}}{\text{C}}\!-\!\text{CH}\!\!=\!\!\text{CH}_2 \quad\longrightarrow\quad \left[\;\;\right]\left[\;\;\right]\text{etc.}$$

The polymer, *cis*-1,4-polyisoprene, is the natural form. (Guttapercha has the same composition but is the trans isomer.) Synthetic rubbers can be made from related monomers. Neoprene, for example, is polychloropropene. Rubbers are hardened by vulcanization, the addition of sulfur, which forms cross-links between the chains. The most important commercial rubber is made from styrene and butadiene. A polymer known as rubber impression material (Figure 12-9) is used in dentistry for making metal or porcelain restorations. It is formed by cross linking the sulfhydryl groups and chain lengthening of:

$$\text{HS}\!-\!(\text{R}\!-\!\text{S}\!-\!\text{S})_{23}\!-\!\text{R}\!-\!\text{SH} \quad \text{R} = -\text{C}_2\text{H}_4\!-\!\text{O}\!-\!\text{CH}_2\!-\!\text{O}\!-\!\text{C}_2\text{H}_4-$$

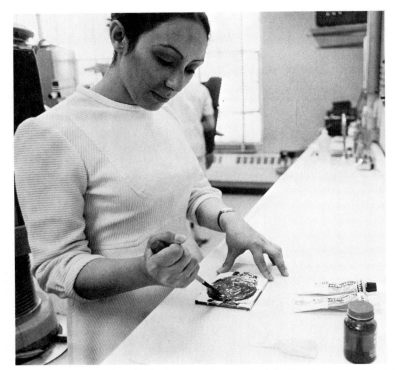

Figure 12-9. Polysulfide impression material. The base and catalyst are mixed to form the polymer, used for dental impressions. (Photo by Al Green.)

12.11 Condensation polymers

Polyesters consist of two different kinds of monomers joined together as esters. An example is Dacron (polyethylene glycol terephthalate), formed from dimethyl terephthalate and ethylene glycol; a molecule of methyl alcohol is lost for each ester linkage formed. Dacron is used to make fibers and films.

Polyamides are made by reacting diamines with dicarboxylic acids. Nylon results from the reaction of adipic acid and hexamethylenediamine. Nylon is made into fibers as well as molded articles.

$$\underset{\text{adipic acid}}{HO-\overset{O}{\overset{\|}{C}}-CH_2CH_2CH_2CH_2-\overset{O}{\overset{\|}{C}}-OH} + \underset{\text{hexamethylenediamine}}{NH_2CH_2CH_2CH_2CH_2CH_2CH_2NH_2} \longrightarrow$$

$$\underset{\text{nylon}}{\left[\overset{O}{\overset{\|}{C}}(CH_2)_4-\overset{O}{\overset{\|}{C}}-NH(CH_2)_6-NH\right]_n} + H_2O$$

One of the oldest thermosetting plastics is made by condensing phenol with formaldehyde, resulting in Bakelite, a polymer cross-linked in three dimensions. This hard plastic is used in electronic components. Another type of cross-linked condensation polymer is formed from formaldehyde and a molecule with amino groups, such as urea or melamine. In the latter case, the resulting plastic is Melmac, used for dishes.

$$H_2N-\overset{N}{\underset{N}{\overset{C}{\underset{\|}{\overset{\|}{C}}}}}\overset{NH_2}{\underset{NH_2}{\overset{C}{\underset{N}{\overset{\|}{C}}}}}$$

melamine

The properties of polymers, plastics included, vary greatly, depending on the type of monomer, on the presence or absence of cross-links, and even on how the chains are arranged. Polymers range from the disorganized rubbers to highly oriented fibers, with the plastics somewhere in between. The proper care of polymers, especially plastics, depends on recognizing the characteristics of the type you are working with. For example, certain plastics like Teflon can be sterilized by heating because they are stable at the temperature of the autoclave. Others, such as polyvinyl chloride, polyethylene, or polystyrene, must be sterilized by disinfectant solutions, such as quaternary ammonium halides, because they deform readily with heat.

Table 12-7 summarizes the most important organic functional groups and

Table 12-7
Summary of Major Types of Organic Compounds

Class	Functional group (identifying characteristic)	Example Formula	Example Name[a]
alkane	saturated hydrocarbon, $-\overset{|}{\underset{|}{C}}-\overset{|}{\underset{|}{C}}-$	CH_3CH_3	ethane
alkene	hydrocarbon (double bond), $\overset{|}{C}=\overset{|}{C}$	$CH_2=CH_2$	ethene (ethylene)
alkyne	hydrocarbon (triple bond), $-C\equiv C-$	$CH\equiv CH$	ethyne (acetylene)
cycloalkane	cyclic saturated hydrocarbon	$H_2C\overset{CH_2}{\underset{CH_2}{\diagdown\diagup}}CH_2$	cyclohexane
aromatic hydrocarbon	based on a benzene type of ring system	benzene ring structure	benzene
alcohol	hydroxyl group, —OH	CH_3CH_2OH	ethanol (ethyl alcohol)
ether	oxygen atom, —O—	$CH_3CH_2OCH_2CH_3$	diethyl ether
aldehyde	carbonyl group with one alkyl group, $-\overset{O}{\overset{\|}{CH}}$	$CH_3\overset{O}{\overset{\|}{CH}}$	ethanal (acetaldehyde)

ketone	carbonyl group with two alkyl groups, $-\overset{\displaystyle O}{\overset{\|}{C}}-$	$CH_3\overset{\displaystyle O}{\overset{\|}{C}}CH_3$	propanone (acetone, dimethyl ketone)
carboxylic acid	carboxyl group, $-\overset{\displaystyle O}{\overset{\|}{C}}-OH$	$CH_3\overset{\displaystyle O}{\overset{\|}{C}}-OH$	ethanoic acid (acetic acid)
ester	$-\overset{\displaystyle O}{\overset{\|}{C}}-O-$ group	$CH_3\overset{\displaystyle O}{\overset{\|}{C}}-OCH_3$	methyl ethanoate (methyl acetate)
alkyl halide	halogen atom	CH_3CH_2Cl	chloroethane (ethylene chloride)
thiol	sulfhydryl group, $-SH$	CH_3CH_2SH	ethanethiol (ethyl mercaptan)
amine	nitrogen with one, two, or three alkyl groups	$CH_3CH_2NH_2$	ethylamine
amide	$-\overset{\displaystyle O}{\overset{\|}{C}}-N\big\langle$ group	$CH_3\overset{\displaystyle O}{\overset{\|}{C}}-NH_2$	ethanamide (acetamide)
nitrogen heterocycle	N atom replacing carbon in ring	pyridine ring structure	pyridine

ᵃ Common name in parentheses.

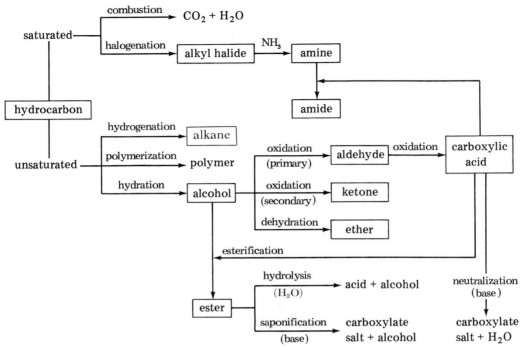

Figure 12-10. Summary chart of organic reactions.

the classes of compounds that you have studied in the last three chapters. Figure 12-10 presents a summary of the most important types of organic reactions.

In the next chapters of the book, you will learn about the important biological molecules, including carbohydrates, proteins, and nucleic acids. They contain many of the functional groups described here and can exist as polymers.

SUMMARY

When a halogen (fluorine, chlorine, bromine, or iodine) is substituted into an alkane, the resulting molecule is an alkyl halide. The chlorine derivatives of methane are methyl chloride (chloromethane), methylene chloride (dichloromethane), chloroform (trichloromethane), and carbon tetrachloride (tetrachloromethane). Other halogen derivatives include iodoform, Freon, halothane, chloral hydrate, and DDT.

Since sulfur is found in the same group of the periodic table (VIa) as oxygen, many of its derivatives are similar to those described in Chapter 11. The thiol (mercaptan) is the sulfur equivalent of an alcohol. Thioethers, thioketones, and thioacids are some other sulfur derivatives of hydrocarbons.

Amines are substituted ammonia molecules. They are produced by replacing one, two, or all three hydrogens of NH_3 by hydrocarbon groups. Examples are methylamine, CH_3NH_2; dimethylamine, $(CH_3)_2NH$; and trimethylamine, $(CH_3)_3N$. Other amines include amphetamine, ephedrine, and methadone.

Amines are basic because they can accept hydrogen ions to form ammonium ion derivatives. Quaternary ammonium salts or "quats" result from the reaction of amines with alkyl halides. In these compounds, four carbon atoms surround the nitrogen atom, instead of the usual three.

Amides form by the reaction of amines (or ammonia) with carboxylic acids. The simplest example is formamide (methanamide) produced from formic acid and ammonia. Urea is one of the most important amides. Sulfa drugs are

$$NH_2-\overset{\overset{\textstyle O}{\|}}{C}-NH_2$$

examples of a class of compounds known as sulfonamides.

When another atom replaces a carbon in a cyclic molecule, a heterocycle, a ring containing two or more different kinds of atoms, results. Nitrogen heterocycles include pyrrole, indole, pyridine, quinoline, pyrimidine, and purine.

Alkaloids are nitrogen-containing molecules of vegetable origin; they generally have a heterocyclic ring system and are physiologically active. Some alkaloids are nicotine, papavarine, caffeine, atropine, cocaine, and morphine.

A number of organic molecules contain nitrogen combined with oxygen in the form of the nitro group, $-NO_2$, as in picric acid. The nitrite group, $-ONO$, is found in amyl nitrite. Nitrates, like glycerol trinitrate or nitroglycerin, contain the functional group $-ONO_2$.

Very large molecules can be produced by linking together many small units. The most important substances in the body are such molecules, called polymers. Plastics are examples of synthetic polymers. They are materials that are solid in their final form but are shaped by flow at some stage in their processing. Plastics can be formed by addition reactions, in which all the atoms of the monomer units become part of the polymer. Examples are polyethylene, polyvinyl chloride, polytetrafluoroethylene (Teflon), polystyrene, poly(methyl methacrylate), and rubber. Condensation polymers form with the elimination of atoms when the monomers join. These molecules include polyesters, polyamides (like Nylon), and Bakelite.

Exercises

1. (12.1) Give the names, structural formulas and medical uses of three organic halogen derivatives.
2. (12.2) Why are the organic sulfur derivatives similar to the oxygen derivatives?
3. (12.2) What is a disulfide and how is it formed?
4. (12.3) How are amines related to ammonia?
5. (12.3) Identify each as a primary, secondary, or tertiary amine:
   (a) $(CH_3CH_2CH_2)_2NH$; (b) $(CH_3CH_2)N(CH_3)_2$; (c) $CH_3CH_2NH_2$.
6. (12.3) Name the compounds in question 5.
7. (12.3) Describe three medical applications of amines.
8. (12.4) Why are amines basic?
9. (12.4) Write an equation for the reaction of triethylamine with (a) hydrochloric acid; (b) ethyl chloride (monochloroethane).
10. (12.4) What are "quats" and how are they used medically?
11. (12.5) Write an equation for the formation of an amide from ammonia and propanoic acid.
12. (12.5) How does an amide differ from an amine?
13. (12.5) Write the structural formula of urea.
14. (12.5) What are sulfonamides? What is their medical role?
15. (12.6) Draw the structural formulas of three nitrogen heterocycles.
16. (12.7) What is an alkaloid? Give three examples and their effects on the body.
17. (12.8) Give an example of a compound with a (a) nitrite group; (b) nitrate group; (c) nitro group.
18. (12.9) What is a polymer? a plastic?
19. (12.9) What is the effect of cross-linking on the properties of a polymer?
20. (12.9) How does an addition polymer differ from a condensation polymer?
21. (12.10) Give three examples of addition polymers and draw two repeating units of each.
22. (12.11) Draw two repeating units each for two condensation polymers.
23. (Table 12-7) Identify the class of the following compounds:

(a) $CH_3CH_2\overset{\displaystyle O}{\overset{\displaystyle \|}{C}}-NH_2$

(b)
$$
\begin{array}{c}
\text{H} \quad \text{H} \\
\text{C}-\text{C} \\
\text{HC} \qquad \text{N} \\
\text{C}=\text{C} \\
\text{H} \quad \text{H}
\end{array}
$$

(c) CH_3CH_2SH

(d) CH_3Br

(e) CH_3OCH_3

(f) $CH_3CH_2\overset{\displaystyle O}{\overset{\displaystyle \|}{C}}CH_3$

(g) $CH_3CH{=}CH_2$

(h) $(CH_3)_3N$

(i) CH_3OH (j) $CH_3C\equiv CCH_3$ (k) ⬡   (l) $\overset{\overset{\displaystyle O}{\|}}{H C}CH_3$ (m) $\overset{\overset{\displaystyle O}{\|}}{H C}{-}OCH_3$

(n) $CH_3CH_2CH_2CH_2CH_3$ (o) $CH_3\overset{\overset{\displaystyle O}{\|}}{C}{-}OH$ (p) $H_2C\overset{\overset{\textstyle CH_2}{\diagup\diagdown}}{-}CH_2$

24. (Table 12-7) Name the compounds in question 23.

Carbohydrates

Carbohydrates are one of your major sources of energy. They are synthesized by plants in a process called photosynthesis. Plants use the energy of sunlight to convert carbon dioxide, a waste product of your body, along with water into carbohydrate molecules. Carbohydrates serve as an important "fuel" in humans because energy from the sun is stored in their chemical bonds. It is released when carbohydrates are "burned" in your body. To follow this process (which is described in detail in Chapter 17), you must understand the structures and properties of the most important carbohydrate molecules.

13.1 Classification of carbohydrates

Carbohydrates are organic molecules that contain carbon, hydrogen, and oxygen. At one time, it was thought that they were hydrates of carbon—therefore, the name carbohydrate. Although many carbohydrate molecules do have the ratio of two hydrogen atoms to one oxygen atom (as in H_2O) for every carbon atom, this definition is not correct. Carbohydrates are related to either aldehydes or ketones and also contain hydroxyl groups. Thus, they can be defined as *polyhydroxyl aldehydes* or *polyhydroxyl ketones* ("poly" means many) or as substances that produce these compounds upon hydrolysis (when they react with water) and their derivatives.

Carbohydrates are classified according to their size. The largest carbohydrates are polymers called **polysaccharides**; they contain many monomers linked together. These basic units, the simplest carbohydrates, are called **monosaccharides.** Polysaccharides can be hydrolyzed, broken down in a reaction with water (in the presence of acid) to produce monosaccharides. The monosaccharides, however, cannot be changed to simpler molecules. In between the large polysaccharides and the single monosaccharides are the **oligosaccharides.** They contain a small number of monosaccharide units, generally from two to ten, bonded together. The most important oligosaccharides

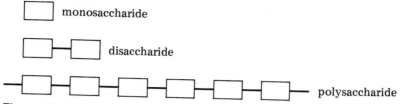

Figure 13-1. Representations of a monosaccharide, disaccharide, and polysaccharide.

are the **disaccharides**, which consist of two monosaccharides joined by a chemical bond. These definitions are illustrated in Figure 13-1. The term **sugar** generally applies only to those monosaccharides and oligosaccharides that are soluble in water and taste sweet.

13.2 Monosaccharides

Monosaccharides are further classified according to their functional group. Those based on an aldehyde group are called **aldoses**, while those containing a ketone group are **ketoses**. The ending "-ose" is characteristic of the simple carbohydrates.

The monosaccharides are also divided into groups on the basis of the number of carbon atoms in the molecule. A **triose**, a three-carbon monosaccharide, is the smallest possible carbohydrate. A **tetrose** contains four carbon atoms, and a **pentose** has five. The **hexoses**, molecules with six carbon atoms, are the most important group of monosaccharides.

These two systems of classification can be combined. For example, a five-carbon monosaccharide with a ketone group is called a ketopentose. Similarly, an aldose with six carbon atoms is an aldohexose.

Glyceraldehyde, the simplest of all monosaccharides, is an aldotriose,

$$
\begin{array}{c}
\overset{\textstyle O}{\overset{\|}{C}H} \\
| \\
H-C-OH \\
| \\
CH_2OH
\end{array}
$$

glyceraldehyde

formed from the breakdown of hexoses in muscle tissue. It contains three carbon atoms, one of which forms part of an aldehyde group. Each of the other two carbon atoms is bonded to a hydroxyl group. You can think of larger monosaccharides (that are aldoses) as derived from glyceraldehyde by lengthening the chain of carbon atoms.

Tetroses are not of major importance biologically, but two pentoses, five-carbon monosaccharides, make up part of the complex molecules in your genes, the nucleic acids. The structures of these aldopentoses, **ribose** and **deoxyribose**, are as follows:

$$
\begin{array}{cc}
\underset{\text{CH}}{\overset{\text{O}}{\parallel}} & \underset{\text{CH}}{\overset{\text{O}}{\parallel}} \\
\text{H}-\text{C}-\text{OH} & \text{H}-\text{C}-\text{H} \\
\text{H}-\text{C}-\text{OH} & \text{H}-\text{C}-\text{OH} \\
\text{H}-\text{C}-\text{OH} & \text{H}-\text{C}-\text{OH} \\
\text{CH}_2\text{OH} & \text{CH}_2\text{OH} \\
\text{ribose} & \text{deoxyribose}
\end{array}
$$

The prefix "deoxy-" in deoxyribose means that the molecule contains one less oxygen atom (at carbon 2), as you can see from its structural formula.

13.3 Open and closed forms of monosaccharides

The structures drawn for ribose and deoxyribose, open-chain formulas, are simple to draw but are not strictly correct for most monosaccharide molecules. The aldehyde group and one of the hydroxyl groups on another carbon atom can react to form a closed-chain or cyclic molecule, as in ribose:

$$
\overset{5}{\text{CH}_2\text{OH}} \qquad \overset{5}{\text{CH}_2\text{OH}} \qquad \overset{5}{\text{CH}_2\text{OH}}
$$

alpha   beta   hemiacetal forms

Carbon 1 no longer forms an aldehyde group; instead it contains an ether linkage to carbon 4 and a hydroxyl group, in addition to a bond to another carbon and a hydrogen atom. "Closing" of the carbon chain in this way creates a **cyclic hemiacetal**. It results from the addition of an *alcohol* to an *aldehyde* group. The ring can form in two ways, called alpha (α) and beta (β), as shown. One oxygen atom becomes part of the heterocyclic ring. Since both ribose and deoxyribose form five-membered rings similar to the cyclic ether furan, they are known as **furanoses**.

$$\underset{\text{furan}}{\overset{\displaystyle O}{\underset{\displaystyle HC-CH}{HC \quad\quad CH}}}$$

Figure 13-2 presents two ways of drawing the closed forms of ribose. In the Fischer formula, in which the carbon chain is written vertically, the alpha form has the hydroxyl group on the right side of carbon 1. In the Haworth formula, where the carbon atoms are drawn in a ring, the hydroxyl group of carbon 1 is "down," below the ring, in the alpha form. Similarly, the beta form is written with the hydroxyl group on the left side in a Fischer formula and pointing "up" in a Haworth formula.

These two forms of the monosaccharide are isomers known as **anomers**. The carbon of the hemiacetal, carbon 1, is called the *anomeric carbon*. This dif-

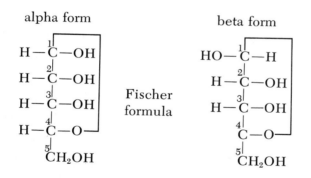

Figure 13-2. The closed forms of ribose. Note the difference in the orientation of the hydroxyl group at carbon 1 in the alpha and beta forms. The Fischer formula and Haworth formula are two different ways of representing carbohydrate molecules.

ference in the orientation of a hydroxyl group may seem slight, but you will learn how important it is when you study polysaccharides.

In solution, three different forms of the monosaccharide exist at the same time: the open-chain form and the two closed-chain or cyclic forms, alpha and beta. Even if you isolated one of the closed forms, it would partially convert to the other forms in water over several hours. An equilibrium or balance establishes itself, generally with most of the monosaccharide molecules existing in one of the closed arrangements. Conversion between the alpha and beta forms, called **mutarotation**, takes place by way of the open form:

alpha form of closed chain $\rightleftharpoons$ open chain $\rightleftharpoons$ beta form of closed chain

The presence of even small amounts of the open-chain form is vital in carrying out tests for monosaccharides since it makes available a reactive aldehyde (or ketone) group.

13.4 Glucose

The most important monosaccharides are hexoses, six-carbon molecules. Of these, only **glucose**, also called dextrose or blood sugar, plays a major role in your body. This molecule, which has the formula $C_6H_{12}O_6$, is shown below:

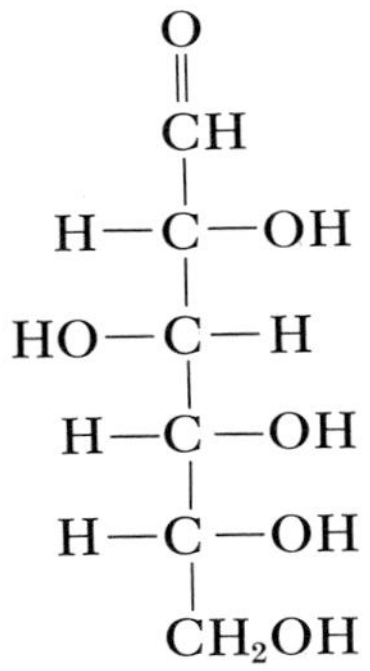

$$
\begin{array}{c}
\text{O} \\
\parallel \\
\text{CH} \\
| \\
\text{H}-\text{C}-\text{OH} \\
| \\
\text{HO}-\text{C}-\text{H} \\
| \\
\text{H}-\text{C}-\text{OH} \\
| \\
\text{H}-\text{C}-\text{OH} \\
| \\
\text{CH}_2\text{OH}
\end{array}
$$

glucose

The closed forms are presented in Figure 13-3. Notice that each cyclic form of glucose is a six-membered ring. Because it is similar to the oxygen heterocycle pyran, this monosaccharide is called a **pyranose**. Over 99% of glucose

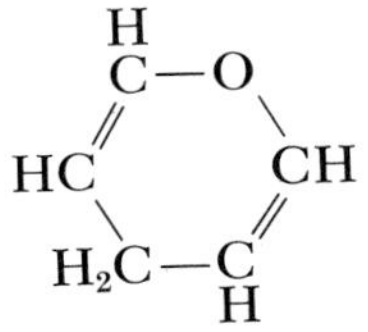

$$
\begin{array}{c}
\text{H} \\
\text{C}-\text{O} \\
\text{HC} \qquad \text{CH} \\
\text{H}_2\text{C}-\text{C} \\
\text{H}
\end{array}
$$

pyran

molecules exist in one of the closed forms in solution; of these, about two-thirds have the beta arrangement.

Glucose is the carbohydrate normally found in your bloodstream; it provides the major source of energy for life. The usual concentration is about 90 mg of glucose per 100 ml of blood. Glucose is administered intravenously to pa-

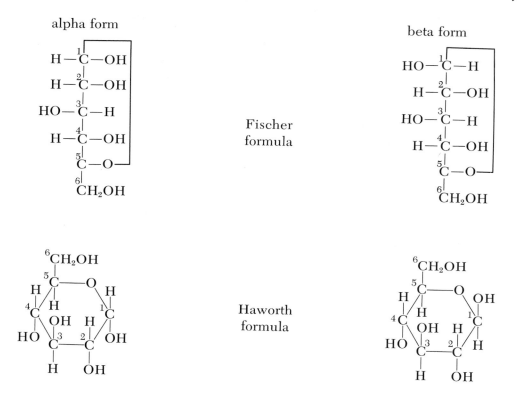

alpha form   beta form

Fischer formula

Haworth formula

Conformational structure

Figure 13-3. The closed forms of glucose. In solution, about two-thirds of the molecules exist in the beta form and one-third exists in the alpha form. The conformational structure is a more realistic picture of the shape of the molecule than given by the Fischer or Haworth formulas.

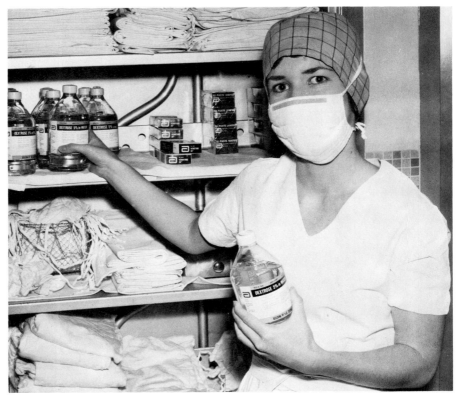

Figure 13-4. A surgical nurse with a bottle of dextrose (glucose).

tients who cannot take food by mouth. Since it requires no digestion, glucose can be used immediately by the body for energy. Intravenous glucose solutions are generally given at a concentration of 5% in order to be isotonic with the blood. This restriction limits the energy valued to only 200 kcal/liter of solution (Figure 13-4).

In certain conditions, such as the disease diabetes mellitus, the amount of glucose in the blood is above normal, and relatively large quantities of glucose appear in the urine. Various methods are used to test urine for the presence of glucose. Since some of the glucose molecules exist in the open form, a test for the aldehyde group can be performed.

As described in Chapter 11, Benedict's solution contains the cupric ion, Cu^{2+}, which gives it a blue color. The solution is basic and also contains sodium citrate to keep the cupric ion in solution. (Fehling's solution is similar but contains sodium tartrate instead and does not store as well.) If glucose is added to Benedict's solution, the aldehyde group is oxidized to a carboxylic acid group, while the cupric ion, Cu^{2+}, is reduced to the cuprous ion, Cu^{+}. The products are gluconic acid and an insoluble solid, a precipitate of cuprous oxide, CuO:

$$
\begin{array}{c}
\overset{\displaystyle O}{\underset{\displaystyle \|}{}} \\
CH \\
| \\
H-C-OH \\
| \\
HO-C-H \\
| \\
H-C-OH \\
| \\
H-C-OH \\
| \\
CH_2OH \\
\text{glucose}
\end{array}
\; + \; 2Cu^{2+} + 4OH^{-} \longrightarrow
\begin{array}{c}
\overset{\displaystyle O}{\underset{\displaystyle \|}{}} \\
C-OH \\
| \\
H-C-OH \\
| \\
HO-C-H \\
| \\
H-C-OH \\
| \\
H-C-OH \\
| \\
CH_2OH \\
\text{gluconic acid}
\end{array}
\; + \; Cu_2O + 2H_2O
$$

Benedict's solution

This same reaction takes place when Clinitest tablets are used instead of Benedict's solution (Figure 13-5). The color indicates the concentration of glucose

Figure 13-5. Testing urine for glucose. The color change indicates the glucose concentration of the urine specimen. (Photo by Al Green.)

present: green, a trace (symbolized +); yellow, up to 0.5% (+ +); orange, 0.5 to 1.5% (+ + +); red, over 1.5% (+ + + +). If the urine does not contain glucose, no change is observed.

In this reaction, glucose acts as a reducing agent; it is therefore called a **reducing sugar**. Any small carbohydrate is considered a reducing sugar if it can have a form in solution with an available aldehyde (or sometimes ketone) group. Thus, most monosaccharides, like glucose, fall into this category.

The oxidation of glucose to gluconic acid is the basis for another type of test, available commercially as Clinistix, Combistix, and Labstix. The test strip contains a substance (the enzyme glucose oxidase) that oxidizes any glucose in the urine sample to gluconic acid. Hydrogen peroxide, H_2O_2, which forms in this process, reacts (in the presence of another enzyme, peroxidase) with a dye, such as o-toluidine, changing its color to blue. The intensity of the color is a measure of the concentration of glucose present. Chapter 22 describes urine tests in more detail.

13.5 Other hexoses— galactose and fructose

Another aldohexose, **galactose**, is an isomer of glucose. The structural formulas of the open form and one of its closed forms (beta) are as follows:

open form   closed form (beta)

Galactose and glucose are called **epimers** because they differ only in the arrangement of groups at a single carbon atom, carbon 4. Galactose is found in brain and nervous tissue, and is used in a test of liver function. If the liver is damaged, the galactose that was administered appears unchanged in the urine.

The ketohexose **fructose**, or levulose, is a third isomer of glucose. Since the

open form   closed form (beta)   fructose

open form contains a ketone group, the closed form is called a cyclic hemi-ketal. As shown, fructose can exist both in the pyranose and furanose forms. Also known as fruit sugar, this monosaccharide is the most water-soluble as well as the sweetest sugar. It can be used for intravenous feeding as a 5 to 10% solution.

13.6 Disaccharides

Disaccharides consist of two monosaccharides joined together by an oxygen atom, as shown in **maltose**. Maltose consists of two glucose units, marked **A** and **B.** Carbon 1, the anomeric carbon atom of glucose A, in the alpha form, is connected by an oxygen "bridge" to carbon 4 of glucose B. Thus, the *hemi-*

glucose (A)   glucose (B)   maltose (beta form) *acetal* of one glucose molecule (A) has been converted to an *acetal* by reacting with the hydroxyl group at carbon 4 of another glucose. Carbon 1 of A now has two ether functional groups attached instead of one ether and one hydroxyl group. Carbon 1 of glucose B still forms a hemiacetal and can exist in either the alpha or beta form.

Maltose, or malt sugar, does not occur abundantly in nature, although it is found in sprouting grain. It is the main product of the hydrolysis of starch, a polysaccharide described in the next section. Maltose is used in formulas for feeding infants as well as in other beverages.

A second disaccharide, **lactose** or milk sugar, consists of glucose combined

lactose
(alpha form)

galactose   glucose

with galactose. The hemiacetal of a beta-galactose molecule (carbon 1) reacts with the hydroxyl group of a glucose molecule (carbon 4) to form an acetal linkage as in maltose. Lactose is found almost exclusively in milk; human milk contains about twice as much as milk from cows. When a woman lactates (secretes milk) after giving birth, small amounts of lactose appear in her blood and urine. Since lactose is a reducing sugar (the glucose end of the molecule contains a free hemiacetal) it gives a positive test with Benedict's solution, making the analysis for glucose misleading.

Many individuals, such as Americans of African descent, digest lactose poorly. Although milk is desirable for its nutritional value, it does not provide enough calories for these babies because the lactose cannot be used by their bodies. It may also cause gastrointestinal disturbances as the amount of undigested lactose increases. For these reasons, milk is not recommended as a food for all people.

The most common disaccharide is **sucrose**, formed from the linkage of glucose and fructose. The carbon that forms the hemiacetal in glucose (alpha form) is connected through an oxygen atom to the carbon that forms the hemiketal in fructose (beta form). The resulting bond cannot "open up" by mutarotation to produce an aldehyde or ketone group. Unlike other disaccharides, the anomeric carbons of both the monosaccharides are "tied up" in a new bond. As a result, sucrose does *not* have alpha and beta forms and is not a reducing sugar.

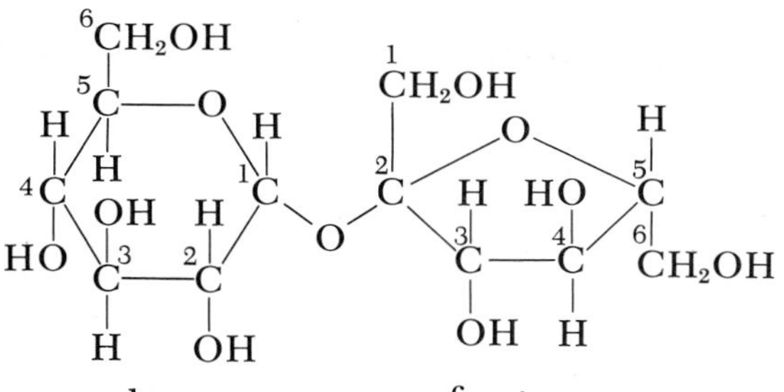

sucrose

glucose   fructose

Sucrose is found primarily in sugarcane and sugar beet, as well as in most fruits and vegetables. It is known as "table sugar" because of its use as a sweetener in the home. Sucrose is taken as the standard of sweetness, as shown in Table 13-1. Other sweeteners are compared to sucrose, which is given the value of 100 on this arbitrary scale. Tremendous quantities of this substance are consumed; each year, you probably eat about 100 pounds of sucrose. Many physicians think that such large amounts increase the risk of developing diabetes mellitus, in addition to other conditions, such as heart disease. Sucrose also supports dental caries, the decay or disintegration of the tooth structure.

As a disaccharide, sucrose cannot be used by the body if injected intravenously. Hydrolysis of sucrose, however, breaks the bond between the two monosaccharides, forming "invert sugar," a mixture of equal parts of glucose and fructose. This mixture can be administered directly into the bloodstream in place of glucose (dextrose) alone. "Invert sugar" is found naturally in honey.

Table 13-1 **Relative Sweetness of Sugars**[a]

Sugar (or sweetener)	Sweetness
saccharin	400
fructose	174
honey	120–170
molasses	110
sucrose	100
glucose	74
maltose	33
galactose	32
lactose	16

[a] Compared to sucrose as a standard, given the value 100.

13.7 Polysaccharides

Polysaccharides are polymers generally containing hundreds of monosaccharide molecules joined together through bridging oxygen atoms. Molecular weights range in value from the thousands to the millions. The most important polysaccharides are known as glucosans because they are formed from repeating glucose units.

Starch is a mixture of two types of polysaccharides—a straight-chain glucose polymer (20 to 30%) called **amylose** and a highly branched polymer (70 to 80%) called **amylopectin,** as illustrated in Figure 13-6. Amylose consists of 250 to 500 glucose molecules linked together in the alpha form, with a bond from carbon 1 of one unit to carbon 4 of its neighbor as in maltose. Amylopectin is mostly connected in the same way but contains about 1000 glucose molecules with a large number of branches, each consisting of about 25 monosaccharide units.

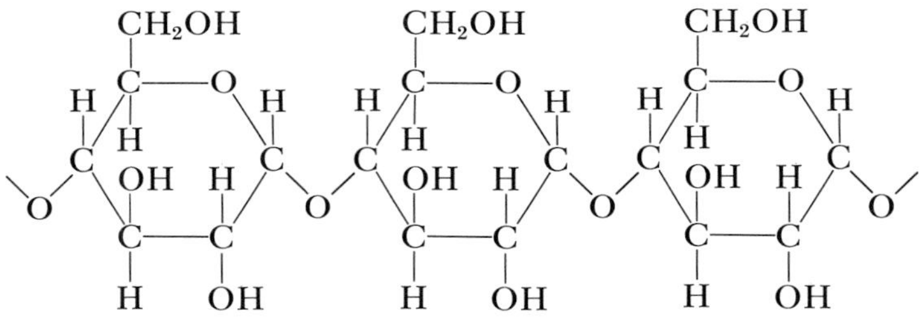

amylose

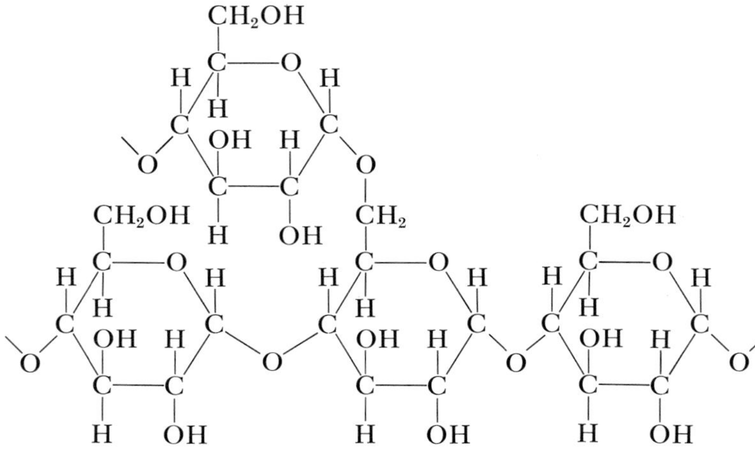

amylopectin

Figure 13-6. Starch is a mixture of 20 to 30% amylose and 70 to 80% amylopectin. As shown, amylose is a straight-chain polymer and amylopectin contains branches.

Starch is the most important carbohydrate that you eat. *It can be considered the basis for your diet,* since you ingest more carbohydrates than other types of food. Large amounts of starch are found in cereals (such as wheat, rye, and corn), potatoes, and vegetables (legumes). Starch is the form in which plants store energy obtained from sunlight. When you digest starch, it is broken down by hydrolysis into smaller pieces known as **dextrins**. Further hydrolysis results in maltose and finally glucose molecules.

When you toast bread, part of the starch is converted to dextrins, giving it a different taste and texture. Dextrins can form gummy masses and are therefore used in mucilage, such as on the back of stamps. A mixture of dextrins and maltose is commonly found in infant feeding formulas along with milk. They are digested easily and prevent the milk from forming large heavy curds in the infant's stomach.

Glycogen has a structure similar to the branched amylopectin form of starch, although it is larger, about 5000 monosaccharides, and has shorter chains, averaging 12 glucose units. This polysaccharide is the *storage form of carbohydrates in your body.* It is found primarily in the liver and muscle tissues and is drawn upon when needed as a source of energy. After you eat, part of the glucose produced by digestion is converted to glycogen (in a process called glycogenesis). Later, when the level of glucose in the blood-stream falls, glycogen is hydrolyzed back to glucose (through glycogenolysis) to be used by the body. Glycogen is well suited for storage because its large size prevents it from diffusing through the cell membranes.

Cellulose is the *most abundant organic substance found in nature;* it contains over 50% of all organic carbon atoms. Unlike starch and glycogen, cellulose consists of beta-glucose units, from 900 to 6000 per molecule, in a straight chain, as shown in Figure 13-7. Because the glucose monosaccharides are joined in their *beta* form, cellulose cannot be digested in your body. Only cer-

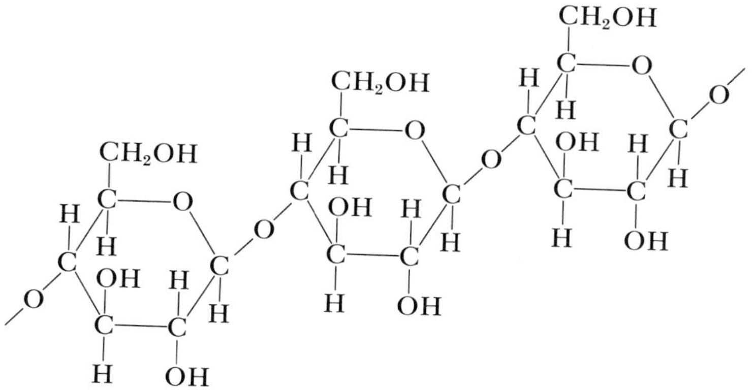

Figure 13-7. The structure of cellulose (compare with Figure 13-6). Cellulose consists of beta-glucose units while starch contains alpha-glucose units.

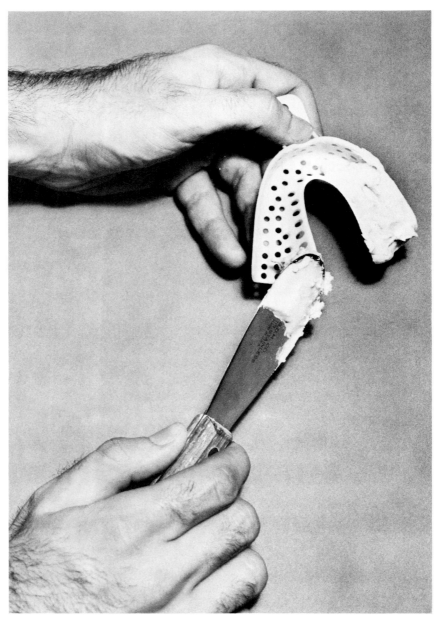

Figure 13-8. Hydrocolloid impression material (alginate). (Photo by Al Green.)

tain organisms, such as the cow and termite, can digest cellulose-containing materials for food because they contain microorganisms with the necessary enzymes. Since you do not digest it, cellulose forms part of the roughage of your diet. Its presence helps stimulate the movement of food (peristalsis) in the intestinal tract and aids in the formation of normal stools.

Cellulose is the main structural component of the cell walls of higher plants. Cotton consists almost entirely of this polysaccharide; wood is about 40 to 50% cellulose. Many commercially important products result from the chemical treatment of cellulose. "Permanent press" fabrics are made by introducing cross-links between the cellulose polymer chains, keeping the material in its desired shape. Rayon is made by treating cellulose with sodium hydroxide and carbon disulfide. Cellophane is prepared in a similar way. Cellulose nitrates, used in such products as movie celluloid (film) and explosives, are made by treatment with nitric acid. Collodion, which forms a protective film over the skin, consists of cellulose nitrate in a mixture of alcohol and ether. Cellulose acetate, another derivative, is used both as a plastic and as a fiber.

Dextrans (not to be confused with dextrins) are branched glucose polysaccharides obtained from certain microorganisms. Those dextrans with molecular weights from 25,000 to 75,000 are used clinically as blood "extenders" in such conditions as shock or blood loss. The dextran forms a colloid which increases the blood volume by osmosis. It is effective for periods longer than 2 hours.

Two polysaccharides that are found in seaweed are used in dentistry as impression materials for full or partial dentures (Figure 13-8). They are known as **hydrocolloids** since their large molecules form colloids when dispersed in water. Agar (or agar-agar) is a polysaccharide based on galactose; it forms a sol, a liquid colloid at high temperature (150° to 212°F), becoming an elastic semisolid gel as it cools (at 102°F), and then solidifies, making an impression of the surrounding teeth. In contrast to this process, which is reversible, alginic acid forms an irreversible hydrocolloid. A polysaccharide based on another

Table 13-2   Other Polysaccharides

Name	Source	Function or use
chitin	insects	skeletal support
gum arabic	vegetable	structural support
heparin	human tissue	prevent blood clotting
hyaluronic acid	higher animals	cement in connective tissue
inulin	plants	test kidney function
pectin	plants	cementing substance

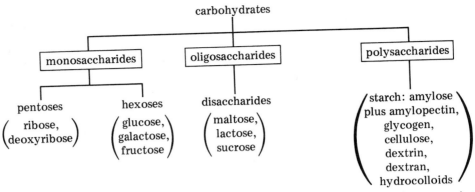

Figure 13-9. Summary chart of carbohydrates. Compounds in parentheses are typical examples of classes listed.

hexose, mannose, it reacts with calcium sulfate to form calcium alginate, a solid gel.

Other polysaccharides are listed in Table 13-2 (p. 311). Figure 13-9 contains a summary chart of carbohydrates.

13.8   Reactions of carbohydrates

Carbohydrates undergo various types of chemical reactions, depending on which functional groups are involved. For example, as described earlier, the aldehyde group can be oxidized to a carboxylic acid group by Benedict's solution and other means. Those sugars like glucose that can be oxidized in this way are *reducing sugars*. Table 13-3 compares the carbohydrates discussed in this chapter in terms of their reducing ability. The aldehyde group can also be reduced to an alcohol in the presence of an enzyme or with hydrogen and a metal catalyst. With glucose, the resulting alcohol is sorbitol.

$$
\begin{array}{ccc}
\underset{\text{C—OH}}{\overset{\overset{\textstyle O}{\|}}{}} & \underset{\text{CH}}{\overset{\overset{\textstyle O}{\|}}{}} & \text{CH}_2\text{OH} \\
\text{H—C—OH} & \text{H—C—OH} & \text{H—C—OH} \\
\text{HO—C—H} & \xleftarrow{\text{oxidation}}\ \ \text{HO—C—H} & \xrightarrow{\text{reduction}}\ \ \text{HO—C—H} \\
\text{H—C—OH} & \text{H—C—OH} & \text{H—C—OH} \\
\text{H—C—OH} & \text{H—C—OH} & \text{H—C—OH} \\
\text{CH}_2\text{OH} & \text{CH}_2\text{OH} & \text{CH}_2\text{OH} \\
\text{gluconic acid} & \text{glucose} & \text{sorbitol}
\end{array}
$$

Table 13-3 **Reducing Ability of Carbohydrates**

Reducing (positive test with Benedict's solution)	Nonreducing (negative test with Benedict's solution)
ribose	sucrose
glucose	starch
galactose	glycogen
fructose	cellulose
lactose	dextran
maltose	

In nature, most sugars other than glucose and fructose are found combined with nonsugars, usually compounds with a hydroxyl group or nitrogen heterocycle. These derivatives, formed by reaction with the hemiacetal group, are **glycosides**. If the reacting sugar is glucose, the product is called a *glucoside*. For example, methanol reacts with glucose in the presence of acid to form a methyl glucoside at carbon 1:

glucose   a methyl glucoside

Of course, when the reacting "alcohol" is the hydroxyl group of another monosaccharide, the resulting glycoside is a disaccharide. Glycosides containing a *nitrogen* atom bonded to the carbon instead of an alcohol are of great importance; the nucleic acids in the genes contain a nitrogen base, either a purine or pyrimidine, attached to a ribose or deoxyribose sugar. A medically significant glycoside is digitoxin, the main component of digitalis, used to treat heart failure by increasing the force of contraction of the heart muscle and slowing its rate.

The previously described reactions occur at the carbon of the aldehyde group; other important transformations involve the hydroxyl groups. They can react with phosphoric acid to form **phosphate esters,** just like the esters formed from an alcohol and a carboxylic acid. For example, a key step in the breakdown of glucose in the body is its phosphorylation, the formation of a phosphate ester, glucose 6-phosphate.

glucose   glucose 6-phosphate

Fermentation refers to the decomposition of a carbohydrate by the enzymes of a living organism. If yeast is added to certain sugars, for example, ethanol (ethyl alcohol) and carbon dioxide are generated. The equation for glucose fermentation is

$$C_6H_{12}O_6 \xrightarrow{\text{enzyme}} 2CH_3CH_2OH + 2CO_2$$
glucose   ethanol

In this way, wine is made from the sugars from crushed grapes and beer is made from those in sprouting malt. The process of fermentation continues until the alcohol content reaches about 14%.

In contrast to alcoholic fermentation, monosaccharides like glucose can also be broken down into two molecules of lactic acid:

$$C_6H_{12}O_6 \xrightarrow{\text{enzyme}} 2CH_3\overset{\displaystyle }{\underset{\displaystyle OH}{C}}H\overset{\displaystyle O}{\overset{\|}{C}}-OH$$
glucose   lactic acid

Cheese and sauerkraut, for example, result from lactic acid fermentation. Antibiotics are also produced by fermentation, as shown in Figure 13-10. In addition to taking place in microorganisms, a form of fermentation occurs in your body in a series of reactions called glycolysis, a process described in Chapter 17.

13.9 Optical isomers and carbohydrates

You have already come across several different types of isomers. Atoms that are bonded together in different ways but form molecules with the same molecular formula, like propanol and isopropanol, are structural isomers. The

Figure 13-10. The fermentation process. Antibiotics, such as penicillin, are produced by fermentation in the tanks shown. (Photo courtesy of Pfizer, Inc.)

cis and trans isomers of alkenes are known as stereoisomers because they have the same sequence of atoms bonded to each other but different arrangements of the atoms in space. Carbohydrates form special types of stereoisomers called **optical isomers**, molecules that are mirror images of each other. Optical isomers are related like your right and left hands.

When four different groups are attached to a carbon atom, a situation found in carbohydrates, the molecule is referred to as **chiral**. There exist two possible optical isomers, called **enantiomers**. As shown in Figure 13-11 for glyceraldehyde, these molecules are mirror images; the first is a mirror reflection of the second. *They cannot be superimposed*, one placed on top of the other.

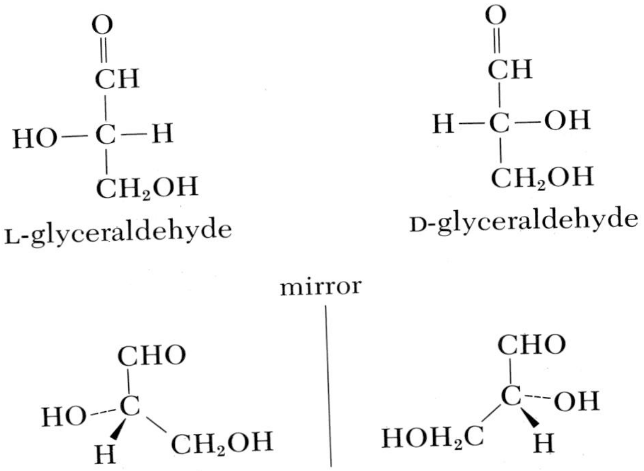

$$
\begin{array}{c}
\text{O} \\
\| \\
\text{CH} \\
| \\
\text{HO}-\text{C}-\text{H} \\
| \\
\text{CH}_2\text{OH}
\end{array}
\qquad
\begin{array}{c}
\text{O} \\
\| \\
\text{CH} \\
| \\
\text{H}-\text{C}-\text{OH} \\
| \\
\text{CH}_2\text{OH}
\end{array}
$$

L-glyceraldehyde   D-glyceraldehyde

Figure 13-11. The enantiomers of glyceraldehyde. The D and L forms are optical isomers; they are mirror images of each other, like right and left hands.

mirror

$$
\begin{array}{c}
\text{CHO} \\
\text{HO}-\text{C} \\
\diagup \quad \diagdown \\
\text{H} \qquad \text{CH}_2\text{OH}
\end{array}
\qquad
\begin{array}{c}
\text{CHO} \\
\text{C}\text{---OH} \\
\diagup \quad \diagdown \\
\text{HOH}_2\text{C} \qquad \text{H}
\end{array}
$$

You could interchange these enantiomers only by breaking and reforming chemical bonds to the carbon atom.

The two possible arrangements in space, or configurations, of glyceraldehyde are called D and L. Other molecules are labeled by comparing their configuration around the chiral carbon atom with D-glyceraldehyde and L-glyceraldehyde. The configuration for monosaccharides is based on the chiral carbon atom farthest from the carbonyl group. If the hydroxyl group is on the "right" side in the open-chain form with the carbonyl group at the "top," the configuration is D. *The common monosaccharides in nature exist in the D form.* Thus, glucose should be written D-glucose.

Light interacts with enantiomers in a special way. You can think of ordinary light in terms of vibrating waves; polarized light, produced when light passes through Polaroid sunglasses, for example, consists of waves vibrating in only one plane, such as just up and down. Enantiomers have the property of being able to rotate this plane of polarized light; they are called **optically active.** Those substances that rotate the plane in a clockwise direction are dextrorotatory, symbolized +, while those that cause rotation in a counterclockwise manner are levorotatory, symbolized −. (A solution containing equal parts of these two forms shows no optical rotation and is called a racemic mixture.) Even though all natural monosaccharides have the D configuration, some cause clockwise rotation of the plane of polarized light, symbolized D(+), and some cause a counterclockwise rotation, symbolized D(−).

Whether a molecule is in the D or L configuration has great biological importance. As you will see in a later chapter, the enzymes of your body are very specific; they recognize only one enantiomer of a molecule. Thus, *the form of the optical isomer determines the effect that a chiral molecule has on the body.* A drug, for example, must have the proper configuration to be able to carry out its designed role.

SUMMARY

Carbohydrates are one of your major sources of energy. They are organic molecules that contain carbon, hydrogen, and oxygen and can be defined as polyhydroxyl aldehydes or polyhydroxyl ketones or as substances that produce these compounds upon hydrolysis. Carbohydrates are classified as monosaccharides (single units), disaccharides (two units), or polysaccharides (many units). Low molecular weight carbohydrates that dissolve in water and taste sweet are known as sugars.

Monosaccharides can be further classified as aldoses or ketoses. They are also divided into groups depending on the number of carbon atoms. Glyceraldehyde, the simplest of all monosaccharides, is an aldotriose—it has three carbons, one of which is part of an aldehyde group.

Two important aldopentoses, ribose and deoxyribose, exist in open-chain and closed-chain forms. "Closing the chain" creates two possible cyclic hemiacetals, identified as alpha and beta. These isomers, known as anomers, differ only in the orientation of a hydroxyl group on one carbon atom. In solution they exist in equilibrium with a small amount of the open form.

The most important monosaccharide is glucose, an aldohexose known as dextrose or blood sugar. It is the carbohydrate normally found in your bloodstream that provides the major source of energy for life. Glucose is administered intravenously to patients who cannot take food by mouth. The amount of glucose in urine can be measured approximately by Benedict's test and related methods based on oxidation–reduction reactions. Galactose, another aldohexose, and fructose, a ketohexose, are isomers of glucose, also having the formula $C_6H_{12}O_6$.

Disaccharides consist of two monosaccharides joined together by an oxygen atom, forming an acetal linkage. Maltose consists of two glucose units. Lactose, or milk sugar, consists of glucose combined with galactose. The most common disaccharide, sucrose, or "table sugar," is formed from glucose and fructose. Hydrolysis of sucrose forms "invert sugar."

Polysaccharides are polymers generally containing hundreds of monosaccharide molecules joined together by bridging oxygen atoms. Starch is a mixture of two polysaccharides—straight-chain amylose (20 to 30%) and highly branched amylopectin (70 to 80%). It is the most important carbohydrate you eat and can be considered the basis of your diet. Glycogen, the storage form of carbohydrates in your liver and muscle tissue, has a structure similar to that of amylopectin. Cellulose is the most abundant organic substance found in nature; it contains over 50% of all organic carbon atoms. Because it consists of beta-glucose units, unlike starch and glycogen, which contain alpha-glucose units, cellulose cannot be digested by humans and forms part of the "roughage" of your diet.

Carbohydrates undergo various types of chemical reactions, depending on which functional groups are involved. Sugars that can be oxidized by Benedict's solution or similar means are known as reducing sugars because they act as reducing agents. Glycosides are sugar derivatives that contain a nonsugar part, formed by reaction with the hemiacetal group. Fermentation is the decomposition of a carbohydrate by the enzymes of a living organism.

Carbohydrates form special types of stereoisomers called optical isomers, molecules that are mirror images of each other. When four different groups are attached to a carbon atom, the molecule is referred to as chiral and can have two possible optical isomers, called enantiomers, designated D and L. Enantiomers are optically active—they can rotate the plane of polarized light. The configuration of an optical isomer has great biological importance.

Exercises

1. (13.1) How can you tell from its structure whether a molecule is a carbohydrate?
2. (13.1) What is a monosaccharide? disaccharide? oligosaccharide? polysaccharide? sugar?
3. (13.2) Identify the following according to the number of carbons and the functional group (such as aldotriose):

(a)
$$CH_2OH$$
$$|$$
$$C=O$$
$$|$$
$$HO-C-H$$
$$|$$
$$H-C-OH$$
$$|$$
$$CH_2OH$$

(b)
$$\overset{\overset{\displaystyle O}{\|}}{CH}$$
$$|$$
$$H-C-OH$$
$$|$$
$$H-C-OH$$
$$|$$
$$H-C-OH$$
$$|$$
$$HO-C-H$$
$$|$$
$$CH_2OH$$

(c)
$$\overset{\overset{\displaystyle O}{\|}}{CH}$$
$$|$$
$$HO-C-H$$
$$|$$
$$H-C-OH$$
$$|$$
$$CH_2OH$$

(d)
$$CH_2OH$$
$$|$$
$$C=O$$
$$|$$
$$CH_2OH$$

4. (13.2) How does ribose differ from deoxyribose? How are they similar?
5. (13.3) What are anomers?
6. (13.3) Describe what happens if you dissolve pure beta-ribose in water.
7. (13.4) Draw the straight-chain and cyclic forms of glucose.
8. (13.4) Why can glucose be administered intravenously?
9. (13.4) Why is glucose called a reducing sugar? How can this property be used to test for the presence of glucose in urine?
10. (13.5) How does galactose differ from glucose?
11. (13.5) How is fructose different from both glucose and galactose?
12. (13.6) Which disaccharide consists of (a) glucose and galactose? (b) two glucose units? (c) fructose and glucose?

13. (13.6) What is the nature of the chemical linkage between the units of a disaccharide?

14. (13.6) Why is milk not a recommended food for everyone?

15. (13.6) What is "invert sugar"? "table sugar"? "blood sugar"?

16. (13.7) Describe the composition of starch.

17. (13.7) What structural difference between starch and cellulose allows one but not the other to be digested?

18. (13.7) What is glycogen? What role does it serve?

19. (13.7) Describe the importance of cellulose.

20. (13.7) Explain the use of hydrocolloid impression materials.

21. (13.8) What is the product of (a) oxidation and (b) reduction of glucose?

22. (13.8) What is a glucoside? Describe the use of one glucoside.

23. (13.8) Describe the process of fermentation.

24. (13.9) Why can carbohydrates have optical isomers?

25. (13.9) Define chiral, enantiomer, optically active.

26. (13.9) What is the importance of optical isomers in the body?

Lipids

Lipids are a mixed group of organic compounds. They do not all have related structures like the carbohydrates; rather, lipids are substances found in plants and animals that dissolve in nonpolar organic solvents like ether, chloroform, and carbon tetrachloride. Lipids serve as vital components of all living matter. The group of lipids known as fats provides an important source of energy in your diet and a reserve supply stored in your body.

14.1 Fatty acids

Many lipids are special kinds of esters, formed from an alcohol and the class of carboxylic acids called **fatty acids**. Most fatty acids consist of a straight chain of carbon atoms with a carboxylic acid group at one end of the molecule. Those fatty acids found in nature contain an *even* number (2, 4, 6, 8, etc.) of carbon atoms, counting the carbon of the carboxyl group. They can be either saturated, having only single bonds between carbon atoms, or unsaturated with up to four double bonds in the molecule. (The possibility of cis and trans isomers exists in the fatty acids containing double bonds, but only the cis form is found in nature.)

Important fatty acids are listed in Table 14-1, and several structures are illustrated in Figure 14-1 (p. 322). Notice that the carboxylic acid functional group

$$\begin{matrix} & O \\ & \| \\ -&C-OH \end{matrix}$$

is abbreviated in the table as —COOH. Palmitic acid and stearic acid are the major saturated fatty acids, while oleic acid is the most common unsaturated fatty acid. An example not listed in the table, undecylenic acid, $CH_2{=}CH(CH_2)_8{-}COOH$, is used to treat fungus infections like athlete's foot.

Table 14-1 Common Fatty Acids

Name	Carbon atoms[a]	Formula	Source
saturated			
butyric acid	4	C_3H_7COOH	butter fat
caproic acid	6	$C_5H_{11}COOH$	butter fat
caprylic acid	8	$C_7H_{15}COOH$	coconut oil
capric acid	10	$C_9H_{19}COOH$	palm oil
lauric acid	12	$C_{11}H_{23}COOH$	coconut oil
myristic acid	14	$C_{13}H_{27}COOH$	nutmeg oil
palmitic acid	16	$C_{15}H_{31}COOH$	fats
stearic acid	18	$C_{17}H_{35}COOH$	fats
arachidic acid	20	$C_{19}H_{39}COOH$	peanut oil
unsaturated			
palmitoleic acid	16 (1)	$C_{15}H_{29}COOH$	butter fat
oleic acid	18 (1)	$C_{17}H_{33}COOH$	olive oil
linoleic acid	18 (2)	$C_{17}H_{31}COOH$	linseed oil
linolenic acid	18 (3)	$C_{17}H_{29}COOH$	linseed oil
arachidonic acid	20 (4)	$C_{19}H_{31}COOH$	nervous tissue

[a] Number of double bonds in parentheses.

Two unsaturated fatty acids, linoleic acid and linolenic acid, are known as **essential fatty acids**. You cannot synthesize them in your body and therefore must include these acids in your diet. They are needed to make other important molecules in the body; their absence may result in such conditions as growth failure in infants.

Prostaglandins are cyclic fatty acids; they contain 20 carbon atoms and form a five-membered ring in the middle of the molecule. All prostaglandins are derived from prostanoic acid. They are found in low concentrations in many

prostanoic
acid

tissues and body fluids. Prostaglandins have a wide range of physiological activities and are effective in very small amounts. They are known to influence the reproductive, respiratory, digestive, cardiovascular, and endocrine

Saturated

$$CH_3 \diagdown CH_2 \diagup CH_2 \diagdown CH_2 \diagup CH_2 \diagdown CH_2 \diagup CH_2 \diagdown CH_2 \diagup CH_2 \diagdown CH_2 \diagup CH_2 \diagdown CH_2 \diagup CH_2 \diagdown CH_2 \diagup CH_2 \overset{\displaystyle O}{\overset{\|}{C}}-OH$$

palmitic acid
($C_{16}H_{32}O_2$)

$$CH_3 \diagdown CH_2 \diagup CH_2 \diagdown CH_2 \diagup CH_2 \diagdown CH_2 \diagup CH_2 \diagdown CH_2 \diagup CH_2 \diagdown CH_2 \diagup CH_2 \diagdown CH_2 \diagup CH_2 \diagdown CH_2 \diagup CH_2 \overset{\displaystyle O}{\overset{\|}{C}}-OH$$

stearic acid
($C_{18}H_{36}O_2$)

Unsaturated

$$CH_3 \diagdown CH_2 \diagup CH_2 \diagdown CH_2 \diagup CH_2 \diagdown CH_2 \diagup CH_2 \diagdown CH=CH \diagup CH_2 \diagdown CH_2 \diagup CH_2 \diagdown CH_2 \diagup CH_2 \overset{\displaystyle O}{\overset{\|}{C}}-OH$$

oleic acid
($C_{18}H_{34}O_2$)

$$CH_3 \diagdown CH_2 \diagup CH_2 \diagdown CH_2 \diagup CH=CH \diagdown CH_2 \diagup CH=CH \diagdown CH_2 \diagup CH_2 \diagdown CH_2 \diagup CH_2 \overset{\displaystyle O}{\overset{\|}{C}}-OH$$

linoleic acid
($C_{18}H_{32}O_2$)

$$CH_3 \diagdown CH_2 \diagup CH=CH \diagdown CH_2 \diagup CH=CH \diagdown CH_2 \diagup CH=CH \diagdown CH_2 \diagup CH_2 \diagdown CH_2 \diagup CH_2 \overset{\displaystyle O}{\overset{\|}{C}}-OH$$

linolenic acid
($C_{18}H_{30}O_2$)

$$CH_3 \diagdown CH_2 \diagup CH_2 \diagdown CH_2 \diagup CH=CH \diagdown CH_2 \diagup CH=CH \diagdown CH_2 \diagup CH=CH \diagdown CH_2 \diagup CH=CH \diagdown CH_2 \diagup CH_2 \overset{\displaystyle O}{\overset{\|}{C}}-OH$$

arachidonic acid
($C_{20}H_{32}O_2$)

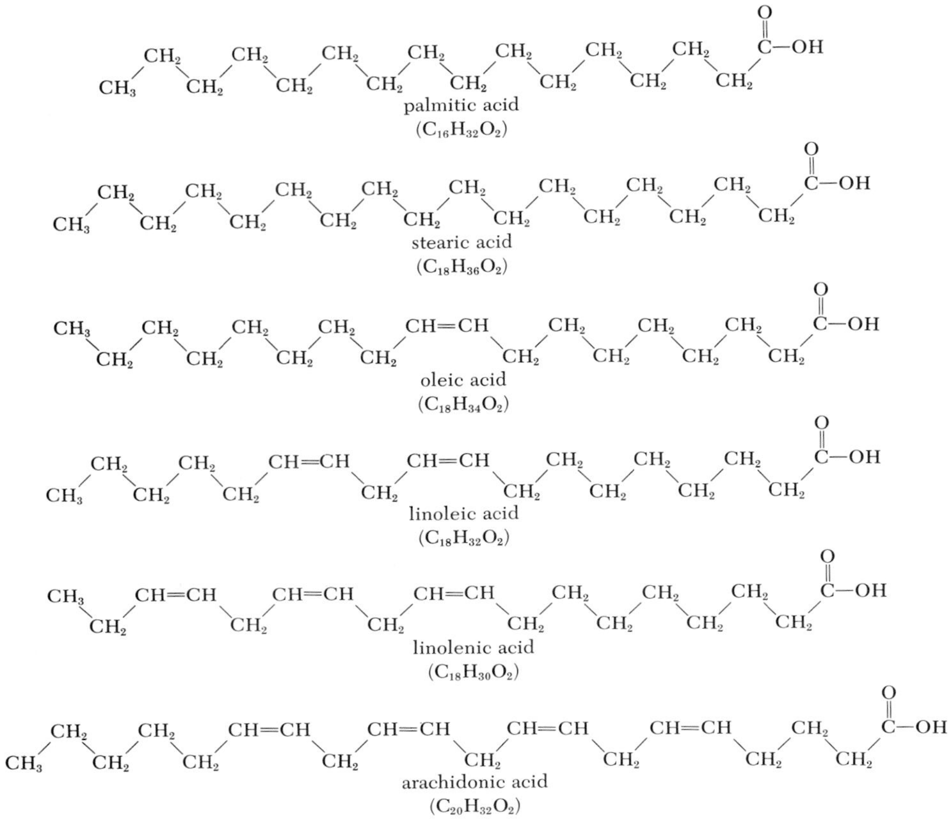

Figure 14-1. Structural formulas of major fatty acids. Linoleic acid and linolenic acid are known as essential fatty acids because they must be obtained from your diet.

systems, but their precise roles are not yet clearly understood. These fatty acids may have clinical applications in lowering blood pressure, controlling fertility, inhibiting blood clot formation, or preventing gastric ulcers.

14.2 Soaps and detergents

When a fatty acid reacts with a strong base, like sodium hydroxide or potassium hydroxide, a **soap** forms. A soap is simply the carboxylate salt of a fatty acid, such as sodium stearate, the salt of stearic acid. Soaps generally are

$$CH_3(CH_2)_{16}\overset{\displaystyle O}{\overset{\|}{C}}-OH + NaOH \longrightarrow CH_3(CH_2)_{16}\overset{\displaystyle O}{\overset{\|}{C}}-O^- \,{}^+Na + H_2O$$

stearic acid $\qquad\qquad$ sodium stearate
(a soap)

based on saturated fatty acids with 12 to 18 carbon atoms. Hard or solid soaps contain the sodium salts of fatty acids, while soft or liquid soaps consist of the potassium salts. Tincture of green soap consists of a solution of potassium salts of fatty acids in alcohol. Floating soaps are made by whipping air into them to make their density lower than that of water.

A soap, such as sodium stearate, contains a *nonpolar* end, the hydrocarbon chain of the fatty acid, and a *polar* end, the ionic carboxylate group:

$$\underbrace{CH_3CH_2CH_2CH_2CH_2CH_2CH_2CH_2CH_2CH_2CH_2CH_2CH_2CH_2CH_2CH_2CH_2}_{\text{nonpolar}}\underbrace{\overset{\overset{\displaystyle O}{\|}}{C}-O^-\ ^+Na}_{\text{polar}}$$

sodium stearate

Because "like dissolves like," the nonpolar part of the molecule can dissolve the greasy dirt present on your clothing or skin. The ionic end of the soap is attracted to water molecules and "pulls" the dirt away from the surface being cleaned, suspending it in water, as shown in Figure 14-2. Soaps thus act as **emulsifying agents**, dispersing nonpolar "dirt" molecules in water. In addition, soaps are surfactants (surface-active agents), which lower the surface tension of water, making it "wetter," allowing greater penetration and emulsification (Figure 14-3).

A major limitation of soap is the formation of insoluble curds instead of suds

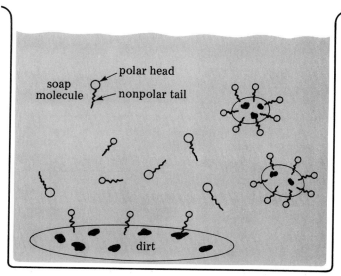

Figure 14-2.   The action of soap. The nonpolar "tail" of the soap molecule dissolves in the dirt, while the polar "head" is water soluble and "pulls" the dirt off the surface being cleaned.

Figure 14-3. Formation of an emulsion from two immiscible liquids. Before (left) and after (right) emulsification by addition of soap. (Photo by Al Green.)

in "hard" water. The soap reacts with the calcium or magnesium ions present to form a precipitate consisting of the calcium or magnesium salt of the fatty acid:

$$2CH_3(CH_2)_{16}\overset{\overset{\displaystyle O}{\|}}{C}-O^-\,{}^+Na\;+\;Ca^{2+}\longrightarrow\left[CH_3(CH_2)_{16}\overset{\overset{\displaystyle O}{\|}}{C}-O^-\right]_2Ca\;+\;2Na^+$$

insoluble in water

To overcome this problem, cleansers other than carboxylate salts have been developed. They are known as **synthetic detergents**, or "syndets" for short.

Detergents contain a surfactant that works in the same way as soap, such as sodium dodecyl sulfate or sodium dodecyl benzenesulfonate (derived from dodecylic, or lauric, acid). They differ from soaps in the nature of the ionic group attached to the hydrocarbon chain. Most detergents are anionic, like those shown below, containing either the negative sulfate or sulfonate groups.

$$CH_3(CH_2)_{10}CH_2-O-\overset{\overset{\displaystyle O}{\|}}{\underset{\underset{\displaystyle O}{\|}}{S}}-O^-\,{}^+Na$$

sodium dodecyl sulfate

$$CH_3(CH_2)_{10}CH_2-\!\!\!\left\langle\right\rangle\!\!\!-\overset{\overset{\displaystyle O}{\|}}{\underset{\underset{\displaystyle O}{\|}}{S}}-O^-\,{}^+Na$$

sodium dodecyl benzenesulfonate

Cationic detergents are based on positively charged quaternary ammonium groups; as you saw previously, such "quats" also act as disinfectants. An example is trimethyl dodecyl ammonium chloride. Nonionic detergents do not

$$CH_3(CH_2)_{10}CH_2-\overset{\overset{\displaystyle CH_3}{|}}{\underset{\underset{\displaystyle CH_3}{|}}{N}}-CH_3{}^+\ {}^-Cl$$

trimethyl dodecyl ammonium chloride

ionize in water but instead dissolve because they contain polar functional groups.

Although detergents solve the problem of cleaning in "hard" water, they can create other problems, such as water pollution (Fig. 14-4). The first detergents (such as Tide) were based on branched hydrocarbon chains. These were

Figure 14-4.   Pollution caused by sudsing. Early detergents were not "biodegradable" because they contained branched hydrocarbon chains. (Photo courtesy of U.S. Department of Agriculture.)

not "biodegradable": The microorganisms present in water could not break the carbon–carbon bonds, leaving the detergent to cause such problems as foaming in sewage disposal systems (Figure 14-4). The newer linear or straight-chain detergents are biodegradable and do not have this difficulty.

A second problem relates to the use of "builders," inorganic salts added to detergents to help soften the water and keep dirt particles suspended. The use of phosphates for this purpose (such as sodium pyrophosphate, $Na_4P_2O_7$) may result in a type of water pollution called **eutrophication**. Since phosphorus is an important nutrient and may be in short supply in a body of water, the addition of phosphates stimulates the growth of algae. The increased algae use up the dissolved oxygen needed by fish in the water, killing them and eventually creating a swamp. Phosphate builders are now banned in many areas and are being replaced by other salts such as sodium carbonate and sodium silicate.

14.3 Waxes

Simple lipids known as **waxes** are esters formed from a fatty acid and a high molecular weight alcohol. For example, the major component of beeswax is the ester formed by palmitic acid and myricyl alcohol:

$$CH_3(CH_2)_{14}\overset{\overset{\displaystyle O}{\|}}{C}{-}OH + CH_3(CH_2)_{28}CH_2OH \longrightarrow$$

palmitic acid   myricyl alcohol

$$CH_3(CH_2)_{14}\overset{\overset{\displaystyle O}{\|}}{C}{-}O{-}CH_2(CH_2)_{28}CH_3 + H_2O$$

myricyl palmitate

The natural waxes, listed in Table 14-2, are actually mixtures of waxes, containing a variety of these large esters.

Waxes are found in nature as protective coatings on the leaves of plants, feathers of birds, and fur of animals. Lanolin, from wool, is the most important wax medically; it is used as a base for many ointments. Beeswax is used in

Table 14-2 Common Waxes

Name	Source	Use
beeswax	honey comb	dental impressions
carnauba wax	palm leaves	floor wax, furniture polish
cerumen	ear wax	protect inner ear
lanolin	wool (sheep)	base for ointments and creams
spermaceti	sperm whale	cosmetics, candles

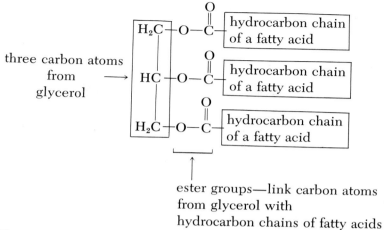

Figure 14-5. A triacylglycerol (triglyceride). The molecule is a triester formed from glycerol and three fatty acids.

dentistry to make impressions. Do not confuse paraffin wax, mentioned earlier with these waxes—it is not a lipid but a mixture of hydrocarbons.

14.4 Fats and oils

Fats, the most common lipids, are esters formed from fatty acids and the alcohol glycerol. Since this alcohol contains three hydroxyl groups, it can react

$$CH_2-CH-CH_2$$
$$\quad|\qquad|\qquad|$$
$$OH\quad OH\quad OH$$
$$glycerol$$

with three separate fatty acid molecules. An ester can form at each of the three carbon atoms of glycerol, producing a triester, known either as a **fat**, **triacylglycerol**, or **triglyceride**, as illustrated in Figure 14-5. For example, the fat triolein is produced from glycerol and three molecules of oleic acid by esterification:

$$
\begin{array}{l}
CH_2OH \\
| \\
CHOH \\
| \\
CH_2OH \\
glycerol
\end{array}
\;+\;
3\,C_{17}H_{33}\overset{\displaystyle O}{\overset{\|}{C}}-OH
\;\longrightarrow\;
\begin{array}{l}
CH_2-O-\overset{\displaystyle O}{\overset{\|}{C}}-C_{17}H_{33} \\
| \qquad\qquad O \\
CH-O-\overset{\|}{C}-C_{17}H_{33} \\
| \qquad\qquad O \\
CH_2-O-\overset{\|}{C}-C_{17}H_{33} \\
\text{triolein (a fat)}
\end{array}
\;+\;3H_2O
$$

oleic acid

Table 14-3
Typical Fatty Acid Composition of Fats

	Fatty acid composition (%)[a]							
	Saturated			Unsaturated		Iodine number	Melting point (°C)
Fat or oil	Myristic	Palmitic	Stearic	Oleic	Linoleic		
butter fat	11	29	9	27	4	36	32
tallow	5	25	31	36	4	40	42
lard	1	28	12	48	6	59	31
human fat	3	24	8	47	10	68	15
olive oil	—	7	2	84	5	81	−6
peanut oil	—	8	3	56	26	93	3
cottonseed oil	1	23	1	23	48	106	−1
corn oil	1	10	3	50	34	123	−20
soybean oil	—	10	2	29	51	130	−16
safflower oil	—	6	3	13	77	145	−18

[a] Total percentage may not equal 100% because of presence of small amounts of other fatty acids or round-off errors.

The three fatty acid molecules that combine with glycerol need not all be identical as in this case. In addition, most fats found in nature are *mixtures of triacylglycerols* (triglycerides). The fatty acid compositions of common fats are presented in Table 14-3.

As you can see from the table, fats come from both animal and vegetable sources. Animal fats include butter or milk fat from cows, lard from hogs, and tallow from beef and sheep. Vegetable fats, known as **oils**, are obtained from the fruit or seed. They include corn oil, olive oil, peanut oil, and safflower oil. The triacylglycerols (triglycerides) obtained naturally vary in composition depending on such factors as the diet of the animal or plant and the climate.

Notice from Table 14-3 that most *animal fats are solids* at room temperature (their melting points are higher than 20°C), while *vegetable oils are liquids* (the room temperature is higher than their melting points). This difference results from their fatty acid composition. The more double bonds in the fatty acids of a triacylglycerol (triglyceride), the greater its degree of unsaturation and the lower its melting point. Fats contain larger amounts of the saturated fatty acids and are solids. Oils, on the other hand, contain more unsaturated fatty acids and are liquids.

The **iodine number** measures the degree of unsaturation. This quantity is equal to the weight in grams of iodine, I_2, that can be added to 100 g of the fat. The more double bonds present in the fatty acid part of the molecule, the more iodine that can be added. In general, the higher the iodine number, the more unsaturated the fatty acid portion of the lipid. Oils, therefore, have higher iodine numbers than fats. Table 14-4 summarizes the differences between fats and oils.

You probably have heard the term **polyunsaturated**. It simply means that the fatty acid portion of a triacylglycerol (triglyceride) contains several double bonds. The degree of unsaturation of lipids in your diet appears to be related to the development of heart disease and certain types of cancer later in life. Leading medical authorities, such as the American Heart Association, recommend substituting unsaturated oils for saturated fats whenever possible (Figure 14-6). On this basis, you might consider using safflower oil, a highly unsat-

Table 14-4 **Fats and Oils**

Fat	Oil
mostly animal origin	mostly vegetable origin
solid at room temperature (high melting point)	liquid at room temperature (low melting point)
saturated	unsaturated (1–4 double bonds)
low iodine number	high iodine number

Figure 14-6. Copyright 1975, American Heart Association, Inc. Reprinted by permission.

urated (polyunsaturated) vegetable oil, instead of butter, a saturated animal fat, in your cooking.

Do not confuse the term "oil" as it is used here with either mineral oils or essential oils. **Mineral oils** are mixtures of hydrocarbons obtained from petroleum. **Essential oils** are aromatic substances from plants, such as oil of wintergreen. These liquids are also known as volatile oils because they evaporate readily. Neither mineral oils or essential oils, however, contain triacylglycerols (triglycerides), like the vegetable oils described in this section.

14.5 Properties of fats and oils

Pure fats have no color, odor, or taste; impurities give the naturally occurring fats these properties. This type of lipid feels slippery and leaves a transparent spot on a piece of paper. It is less dense than water and is insoluble in water. Oils can form an emulsion, however, in the presence of an emulsifying agent, such as a detergent. This process occurs during the digestion of fats, in which bile salts act to keep the fat globules dispersed.

The most important chemical reaction of triacylglycerols (triglycerides) is **hydrolysis**, the breakdown of the fat or oil by water (with a catalyst). This process is exactly the reverse of esterification, as you can see in the hydrolysis of triolein:

$$
\begin{array}{l}
CH_2-O-\overset{\overset{\displaystyle O}{\|}}{C}-C_{17}H_{33} \\[4pt]
CH-O-\overset{\overset{\displaystyle O}{\|}}{C}-C_{17}H_{33} \;+\; 3H_2O \longrightarrow \\[4pt]
CH_2-O-\overset{\overset{\displaystyle O}{\|}}{C}-C_{17}H_{33} \\[4pt]
\qquad\qquad \text{triolein}
\end{array}
\qquad
\begin{array}{l}
CH_2OH \\
CHOH \;+\; 3C_{17}H_{33}\overset{\overset{\displaystyle O}{\|}}{C}-OH \\
CH_2OH \\
\text{glycerol}\qquad\text{oleic acid}
\end{array}
$$

Glycerol and three fatty acids are the products of hydrolysis of a triacylglyc-erol (triglyceride). Hydrolysis takes place when your body digests fats.

Saponification is a reaction related to hydrolysis. Here the triacylglycerol (triglyceride) reacts with a strong base to form glycerol and three soap mole-cules, the salts of the fatty acids. Potassium oleate is formed in this way:

$$
\begin{array}{l}
CH_2-O-\overset{\overset{\displaystyle O}{\|}}{C}-C_{17}H_{33} \\[4pt]
CH-O-\overset{\overset{\displaystyle O}{\|}}{C}-C_{17}H_{33} \;+\; 3KOH \longrightarrow \\[4pt]
CH_2-O-\overset{\overset{\displaystyle O}{\|}}{C}-C_{17}H_{33} \\[4pt]
\qquad\qquad \text{triolein}
\end{array}
\qquad
\begin{array}{l}
CH_2OH \\
CHOH \;+\; 3C_{17}H_{33}\overset{\overset{\displaystyle O}{\|}}{C}-O^-\,{}^+K \\
CH_2OH \qquad\text{potassium oleate} \\
\text{glycerol}
\end{array}
$$

The weight in milligrams of potassium hydroxide required to completely sa-ponify 1 g of the fat is called its *saponification number.*

The addition reaction is a chemical change you already know about since it is the basis for determining the iodine number. Because the unsaturated fatty acid portions of fats have double bonds, they undergo the addition reactions characteristic of alkenes. Not only iodine, but also hydrogen, H_2, can add across a carbon–carbon double bond, converting it to a single bond. This process, **hydrogenation**, takes place at high temperatures using a catalyst. Under controlled conditions, a vegetable oil can be partially solidified through making it more saturated. Margarine is an emulsion prepared by churning partly hydrogenated vegetable oil with 15% milk by weight. Solid shortenings (like Crisco) consist of vegetable oils that have been hardened considerably by hydrogenation.

When allowed to stand in contact with the air, many fats become **rancid**—they develop unpleasant tastes and odors. Two chemical reactions cause this change. Hydrolysis of the ester linkages, as just discussed, pro-duces free fatty acids, many of which have strong odors. Caproic acid, for example, one possible product, is also found in human sweat. Oxidation of double bonds is the second reaction, forming short-chain aldehydes and car-boxylic acids, which also have strong smells. Rancidity of this kind can be inhibited by the presence of substances known as antioxidants. Also, keeping the fat cool slows down the rate of both of these two types of reaction.

14.6 Body fats

When you eat food containing more calories than your body needs, much of the excess is stored as fat in regions called **adipose tissue**. This tissue is located under the skin and around various organs. The stored fat, called **depot lipid**, is over 99% triacylglycerols (triglycerides); their fatty acids are about one-third saturated and two-thirds unsaturated (47% oleic acid, 24% palmitic acid, 10% linoleic acid, 8% stearic acid, 5% palmitoleic acid, 3% myristic acid, 3% gadoleic plus erucic acids).

Fats provide an excellent source of energy. They produce 9 kcal (38 kJ) per gram of fat, about twice as much as carbohydrates. Eskimos, whose source of carbohydrates is limited, eat a diet high in fats to supply the energy needed to maintain the normal body temperatures in their cold environment.

As depot lipid, *fats are a concentrated reserve supply of energy,* "mobilized" when needed by your body. Many people who are overweight have too much of this material. The added weight and strain on the heart resulting from the excess adipose tissue causes a reduced life span, on the average, and a greater risk of heart disease.

Body fat also serves other functions. It acts as insulation, slowing loss of heat through the skin and serving as a protection against the cold. Fat deposits around organs help support them and prevent injury. The triacylglycerols (triglycerides) that carry out this function have more saturated fatty acids and are therefore "harder" than the fats present under the skin.

14.7 Phospholipids

Phospholipids, or **phosphoglycerides**, differ from triacylglycerols (triglycerides) in that one of the hydroxyl groups of glycerol is esterified with phos-

$$
\begin{aligned}
&\quad\qquad\qquad \overset{O}{\overset{\|}{}} \\
&CH_2-O-C-(CH_2)_{16}CH_3 \\
&\quad\qquad\qquad \overset{O}{\overset{\|}{}} \\
&CH-O-C-(CH_2)_{16}CH_3 \\
&\quad\qquad\qquad \overset{O}{\overset{\|}{}} \\
&CH_2-O-P-OH \\
&\quad\qquad\qquad | \\
&\quad\qquad\quad OH
\end{aligned}
$$

a phosphatidic
acid

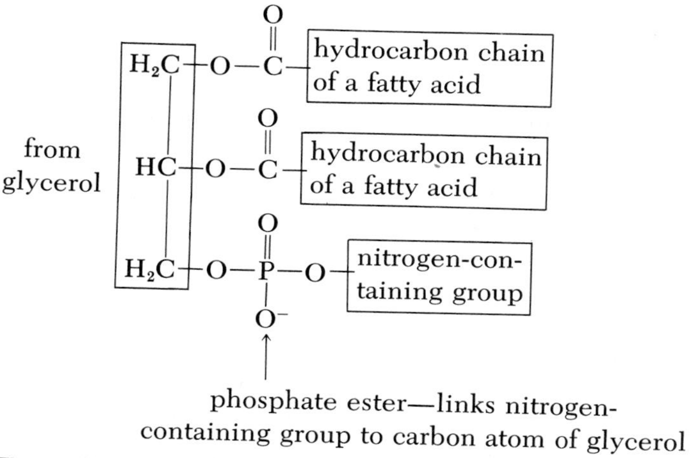

Figure 14-7. A phosphoglyceride. It consists of a phosphate ester with a nitrogen-containing group, along with two fatty acids joined to glycerol.

phoric acid, H_3PO_4, instead of a fatty acid. They are based on phosphatidic acid, a glycerol phosphoric acid. In phosphoglycerides, a nitrogen-containing group forms another ester at the free end of the phosphate, as shown in Figure 14-7. Such a group may come from amino alcohols such as choline or ethan-

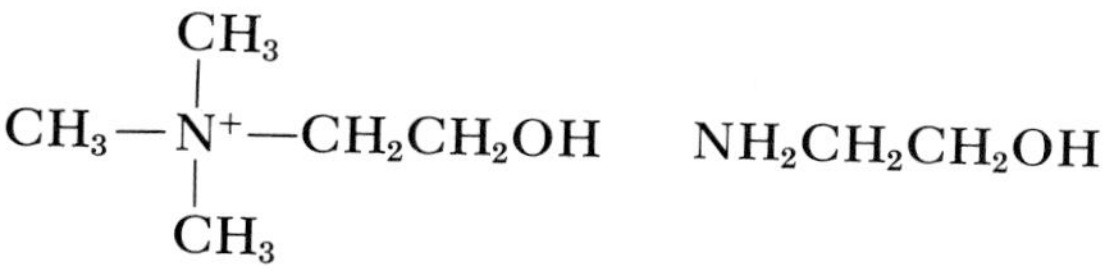

$$CH_3 - \overset{\overset{\displaystyle CH_3}{|}}{\underset{\underset{\displaystyle CH_3}{|}}{N^+}} - CH_2CH_2OH \qquad NH_2CH_2CH_2OH$$

choline   ethanolamine

olamine. The resulting phosphoglycerides are called *phosphatidylcholine* (choline phosphoglyceride) or *lecithin* and *phosphatidylethanolamine* (ethanolamine phosphoglyceride) or *cephalin*. These phosphoglycerides vary in

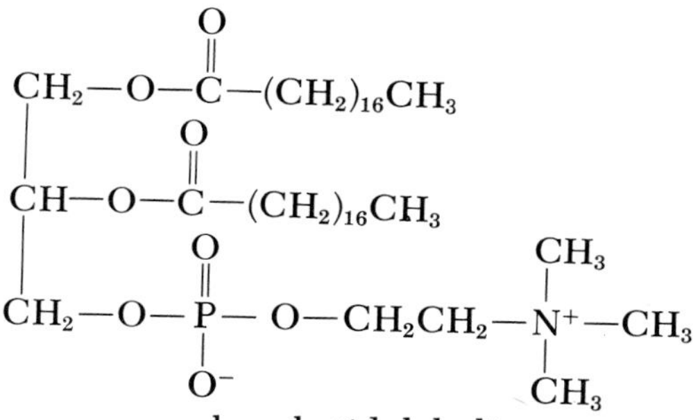

a phosphatidylcholine
(lecithin)

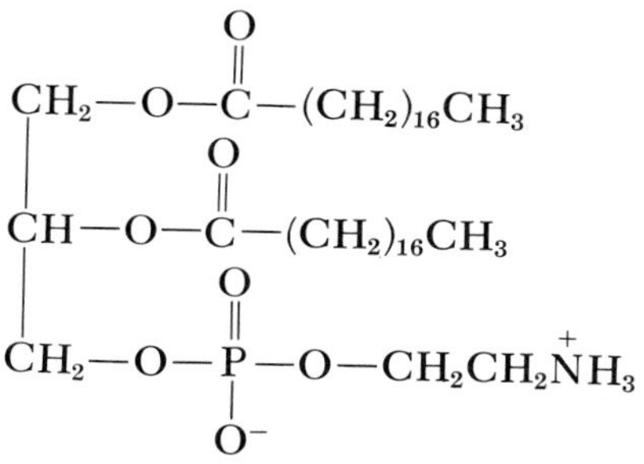

a phosphatidylethanolamine
(cephalin)

composition, depending on the types of fatty acids attached to the glycerol part of the molecule; often, one is saturated and the other is unsaturated. Phosphatidylcholines (lecithins) are the most common phosphoglycerides in your tissues. Phospholipids are found in cell membranes, the brain, the nervous tissue, and the liver.

Plasmalogens are another type of phosphoglyceride. In their molecules, one of the fatty acid esters has been replaced by an *ether* linkage to a long-chain alkyl group.

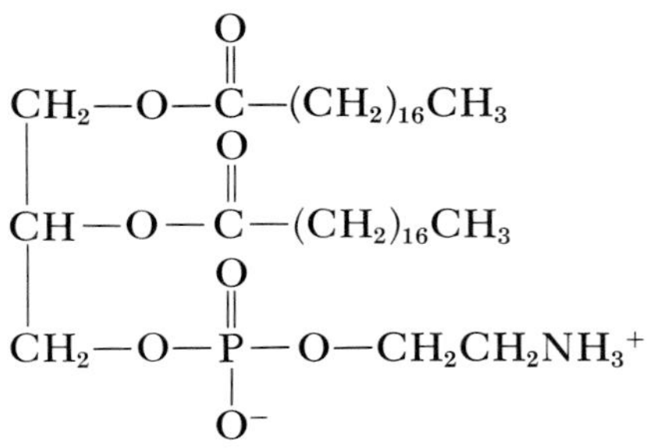

$$
\begin{array}{l}
\quad\quad\quad\quad \overset{\displaystyle O}{\overset{\|}{}} \\
CH_2-O-C-(CH_2)_{16}CH_3 \\
\quad\quad\quad\quad \overset{\displaystyle O}{\overset{\|}{}} \\
CH-O-C-(CH_2)_{16}CH_3 \\
\quad\quad\quad\quad \overset{\displaystyle O}{\overset{\|}{}} \\
CH_2-O-P-O-CH_2CH_2NH_3{}^+ \\
\quad\quad\quad\quad \overset{\displaystyle |}{O^-}
\end{array}
$$

a plasmologen

As you can see from their structures, phosphoglycerides have groups that ionize at the body pH. The result is shown in Figure 14-8. Phosphoglycerides consist of a polar "head," which contains the ionic groups, and a nonpolar "tail" consisting of the hydrocarbon chains. They are thus similar to detergents and in fact do act as emulsifying agents. At the surface of a cell, they bring water-insoluble lipids into contact with water-soluble substances like

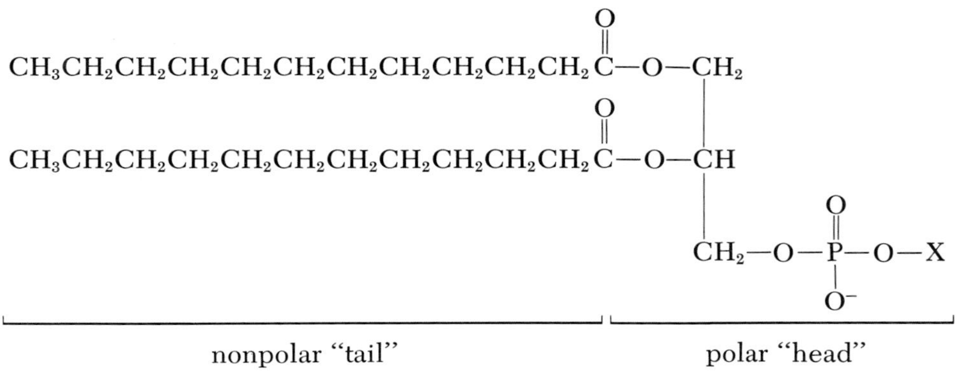

$$
\begin{array}{l}
\quad\quad\quad\quad\quad\quad\quad\quad\quad\quad\quad\quad\quad\quad\quad \overset{\displaystyle O}{\overset{\|}{}} \\
CH_3CH_2CH_2CH_2CH_2CH_2CH_2CH_2CH_2CH_2CH_2\,C-O-CH_2 \\
\quad\quad\quad\quad\quad\quad\quad\quad\quad\quad\quad\quad\quad\quad\quad \overset{\displaystyle O}{\overset{\|}{}} \\
CH_3CH_2CH_2CH_2CH_2CH_2CH_2CH_2CH_2CH_2CH_2\,C-O-CH \\
\quad\quad\quad\quad\quad\quad\quad\quad\quad\quad\quad\quad\quad\quad\quad\quad\quad\quad\quad\quad \overset{\displaystyle O}{\overset{\|}{}} \\
\quad\quad\quad\quad\quad\quad\quad\quad\quad\quad\quad\quad\quad\quad\quad\quad\quad CH_2-O-P-O-X \\
\quad\quad\quad\quad\quad\quad\quad\quad\quad\quad\quad\quad\quad\quad\quad\quad\quad\quad\quad\quad\quad\overset{\displaystyle |}{O^-}
\end{array}
$$

nonpolar "tail"   polar "head"

Figure 14-8. The structure of a phosphoglyceride. The nonpolar "tail" consists of the fatty acid hydrocarbon chains, and the polar "head" contains the ionic phosphate group (X = nitrogen-containing group.)

proteins. Phosphoglycerides are also involved in the transport of fatty acids in the blood and other processes, but their major role is in forming the cell membranes, as described in the following section.

14.8 Cell membranes and active transport

A membrane surrounds each cell, separating its "inside" from the outside. It is formed by two rows of phosphoglyceride molecules arranged as shown in Figure 14-9. The resulting "sandwich" or bilayer has the polar "heads" of the phosphoglycerides on the outside of the membrane and the nonpolar "tails" in the middle. This bilayer, about 0.00001 mm thick, is not homogeneous but contains a mixture of phosphoglycerides; the nature of the fatty acids present determines the properties of the membrane.

Water and neutral nonpolar molecules can pass through the membrane, but it is much less permeable to neutral polar molecules and nearly impermeable to charged molecules and ions. There are large molecules (proteins) embedded in the membrane which can carry across the membrane those molecules that otherwise could not pass through by diffusion. This arrangement allows the transfer of molecules from a region of lower concentration to one of greater concentration, as shown in Figure 14-10. Known as **active transport**, this energy-requiring process moves molecules across the membrane in the direction opposite to diffusion.

Active transport permits the cell to concentrate nutrients and "fuels" that are present in lower concentrations outside the membrane. This process is involved in keeping the relatively high concentration of potassium ion, K^+, and low concentration of sodium ion, Na^+, within the cell (see Chapter 16 for discussion of nerve impulse conduction). Through active transport, the cell can better adjust to a changing external environment and maintain its volume and osmotic pressure.

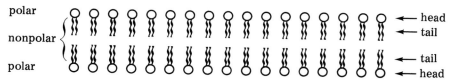

polar — head
nonpolar — tail
polar — tail — head

Figure 14-9. A model of membrane structure. The lipid bilayer is a "sandwich" formed by two rows of phosphoglyceride molecules. The outside is polar and the inside is nonpolar.

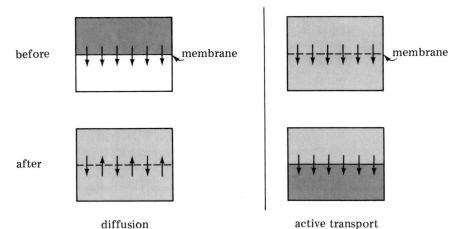

Figure 14-10. A comparison between the processes of diffusion and active transport. In diffusion, a substance moves from a region of higher concentration to a region of lower concentration. In active transport, energy is used to create a region of high concentration of a substance.

14.9 Sphingolipids

Sphingolipids do not contain glycerol but instead are based on the amino alcohol sphingosine. A lipid known as a ceramide results when a fatty acid

$$CH_3(CH_2)_{12}CH\!\!=\!\!CHCHCHCH_2OH$$
$$HO \quad NH_2$$

sphingosine

reacts with the amino group to form an amide. These molecules are widely

$$CH_3(CH_2)_{12}CH\!\!=\!\!CHCHCHCH_2OH$$
$$OH \quad \quad O$$
$$NH\!\!-\!\!C\!\!-\!\!(CH_2)_{16}CH_3$$

a ceramide

distributed in small quantities in plants and animals.

The most important sphingolipids are the *sphingomyelins*, which contain a phosphate ester of choline attached to the alcohol end of a ceramide. Certain

$$CH_3(CH_2)_{12}CH\!\!=\!\!CHCHCHCH_2\!\!-\!\!O\!\!-\!\!\overset{\displaystyle O}{\underset{\displaystyle O^-}{P}}\!\!-\!\!O\!\!-\!\!CH_2CH_2\!\!-\!\!\overset{\displaystyle CH_3}{\underset{\displaystyle CH_3}{N^+}}\!\!-\!\!CH_3$$
$$OH \quad \quad O$$
$$NH\!\!-\!\!C\!\!-\!\!(CH_2)_{16}CH_3$$

a sphingomyelin

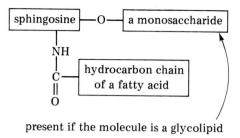

Figure 14-11. A sphingolipid. A fatty acid is attached to sphingosine by an amide linkage. If sphingosine also contains a monosaccharide, the molecule is a glycolipid.

conditions, such as Niemann–Pick disease and Gaucher's disease, cause the accumulation of sphingomyelins in the nervous system, often resulting in early death.

Another type of sphingolipid contains a carbohydrate attached to sphingosine, as shown in Figure 14-11. These molecules are known as **glycolipids** or **glycosphingolipids**. One kind of glycolipid is known as a cerebroside because it is found in the cerebrum of the brain. Cerebrosides generally contain the monosaccharide galactose combined with sphingosine along with a fatty acid.

$$CH_3(CH_2)_{12}CH{=}CHCHCHCH_2{-}O{-}C$$

a cerebroside

Milk is needed by infants not only for its nutritional value but also for the galactose (formed by hydrolysis from lactose) required to synthesize cerebrosides. The gangliosides are glycolipids found in significant amounts in nerve and spleen tissue. They contain a longer carbohydrate attached to a sphingosine-based ceramide.

14.10 Steroids

The lipids known as **steroids** are not esters, like the other molecules discussed so far. They consist of molecules built around a framework of four saturated hydrocarbon rings (labeled *A*, *B*, *C*, *D*) fused together:

$$\begin{array}{c} ^{12}CH_2 \quad ^{17}CH_2 \\ H_2C\,11 \quad 13\,CH \quad 16\,CH_2 \\ C \quad | \quad D\,15\,| \\ ^1CH_2 \quad ^9HC \quad 14\,CH - 15\,CH_2 \\ H_2C\,2 \quad 10\,CH \quad 8\,CH \\ | \quad A \quad | \quad B \quad | \\ H_2C\,3 \quad 5\,CH \quad 7\,CH - OH \\ ^4CH_2 \quad 6\,CH_2 \end{array}$$

steroid framework

Steroids are widespread in your body; very small amounts show marked biological activity. Small variations in the molecular structure result in great differences in their effects. The naturally occurring steroids include bile salts and many hormones, the regulators of chemical processes. These steroids are discussed in detail in later chapters.

Sterols, steroids containing a hydroxyl group, are the most abundant steroids. The most important sterol is *cholesterol*. Cholesterol is made by the

$$\begin{array}{c} CH_3 \\ | \\ CHCH_2CH_2CH_2CHCH_3 \\ H_3C \quad | \quad | \\ CH_2 \quad CH \quad CH_3 \\ H_2C \quad | \quad C \quad CH_2 \\ H_3C \quad | \quad | \\ CH_2 \quad HC \quad CH - CH_2 \\ H_2C \quad C \quad CH \\ | \quad | \quad | \\ CH \quad C \quad CH_2 \\ HO \quad CH_2 \quad CH \end{array}$$

cholesterol

body as well as obtained from the diet. It is used to synthesize molecules such as steroid hormones. This lipid is found in the brain and nervous tissue, where it forms part of myelin, the stable membrane that coats the nerve cells.

Heart disease and high blood pressure, hypertension, can result from deposits of cholesterol inside the walls of the arteries. This condition, atherosclerosis, is a type of arteriosclerosis or "hardening of the arteries" (Figure 14-12). Evidence shows that the level of cholesterol in the blood, and therefore the amount deposited, is related to the amount of saturated fats you eat. Cholesterol is also present in gallstones, the abnormal deposits from bile in the gallbladder (Figure 14-13).

Figure 14-14 presents a summary chart of lipids.

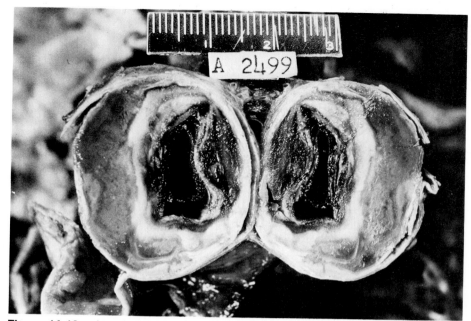

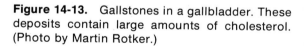

Figure 14-12. An atherosclerotic artery, gross view. (Photo by Martin Rotker.)

Figure 14-13. Gallstones in a gallbladder. These deposits contain large amounts of cholesterol. (Photo by Martin Rotker.)

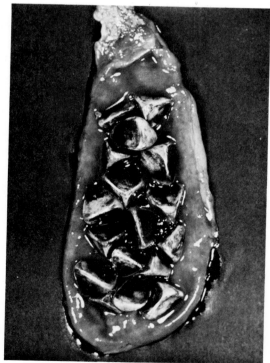

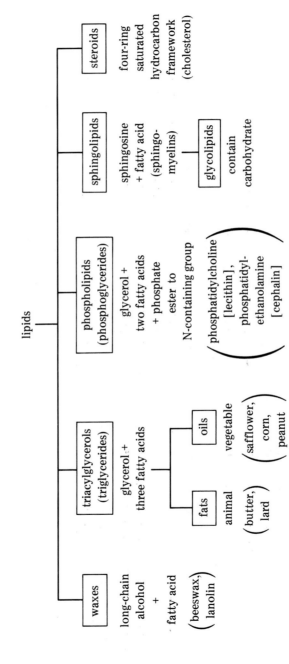

Figure 14-14. Summary chart of lipids. Compounds in parentheses are typical examples of classes listed.

SUMMARY

Lipids are a mixed group of organic compounds found in plants and animals that dissolve in nonpolar organic solvents like ether, chloroform, and carbon tetrachloride. Many lipids are special kinds of esters formed from an alcohol and the class of carboxylic acids called fatty acids. These acids consist of a straight chain with an even number of carbon atoms and from zero to four double bonds. The essential fatty acids, linoleic and linolenic acids, must be included in the diet. Prostaglandins are 20-carbon fatty acids with varied physiological effects.

When a fatty acid reacts with a strong base, a soap forms. It is simply the carboxylate salt of the fatty acid. Because it consists of a polar and nonpolar part, the soap molecule can remove greasy dirt and suspend it in water. Detergents, which contain an ionic group other than carboxylate, are effective in "hard" water, unlike soaps.

Waxes are esters formed from a fatty acid and a high molecular weight alcohol. Fats, the most common lipids, are triesters formed from three fatty acid molecules and the alcohol glycerol; they are triacylglycerols (triglycerides). Animal fats are saturated and exist as solids; vegetable fats, called oils, are unsaturated, containing one to four double bonds, and exist as liquids.

The most important chemical reaction of triacylglycerols (triglycerides) is hydrolysis, the breakdown of the fat or oil by water, a process that is the reverse of esterification. Saponification is a related reaction in which a strong base reacts with the triacylglycerol (triglyceride) to form glycerol and the salts of the fatty acids. Oils can be hardened, partially converted to fats, by an addition reaction called hydrogenation.

When you eat food containing more calories than needed by your body, much of the excess is stored as fat in regions called adipose tissue. Known as depot lipid, the fat serves as a concentrated reserve form of energy. Fat also provides insulation and internal support for various organs.

Phospholipids, or phosphoglycerides, differ from fats in that one of the hydroxyl groups of glycerol is esterified with phosphoric acid instead of a fatty acid. A nitrogen-containing group forms another ester at the free end of the phosphate, as in phosphatidylcholine (lecithin) and phosphatidylethanolamine (cephalin). These molecules have polar and nonpolar ends like detergents.

The membrane that surrounds a cell is formed by two rows of phosphoglycerides in a bilayer or "sandwich" with the polar "heads" on the outside and the nonpolar "tails" in the middle. Active transport is a process by which molecules such as nutrients and "fuels" are moved across the membrane in a direction opposite to diffusion.

Sphingolipids do not contain glycerol but instead are based on the amino acid sphingosine; the most important sphingolipids are the sphingomyelins. Glycolipids or glycosphingolipids contain a carbohydrate attached to sphingosine.

The class of lipids known as steroids consists of molecules built around a framework of four saturated hydrocarbon rings fused together. Cholesterol is the most important steroid; it is used to synthesize other molecules such as hormones. Heart disease can result from deposits of cholesterol inside the walls of the arteries.

Exercises

1. (Intro.) What are lipids?
2. (14.1) Describe the characteristics of a fatty acid.
3. (14.1) What are essential fatty acids?
4. (14.1) What are prostaglandins, and what effects can they have on the body?
5. (14.2) How is a soap formed?
6. (14.2) Explain how a soap or detergent works.
7. (14.2) What advantage do detergents have over soaps? How do their structures differ?
8. (14.2) What pollution problems can be caused by detergents?
9. (14.3) What is a wax?
10. (14.4) Draw the structure of the triacylglycerol (triglyceride) formed from the esterification of glycerol with three molecules of stearic acid.
11. (14.4) How do fats differ from oils chemically? How are their properties different?
12. (14.4) "A fat has a low iodine number." What does this statement mean?
13. (14.4) What is the meaning of the term "polyunsaturated"?
14. (14.5) Write an equation for the (a) hydrolysis and (b) saponification of tristearin (the triacylglycerol from stearic acid).
15. (14.5) How can oils be hardened?
16. (14.5) What causes rancidity? How can it be inhibited?
17. (14.6) What is depot lipid? What role does it play?
18. (14.6) Why can you eat more fats in cold weather?
19. (14.7) How do phosphoglycerides differ from triacylglycerols (triglycerides)?
20. (14.7) Which are the most common phospholipids found in your tissues?
21. (14.7) How are phosphoglycerides similar to detergents?
22. (14.8) Describe the structure of a cell membrane.
23. (14.8) How does active transport differ from diffusion? Why is this process necessary in the body?
24. (14.9) What are sphingolipids? glycolipids?

25. (14.9) Why is galactose needed for brain development in infants?

26. (14.10) Draw the steroid framework.

27. (14.10) What is cholesterol? How is it used in the body?

28. (14.10) What is the relationship between dietary fats, cholesterol, and atherosclerosis?

Proteins

Proteins are complex molecules essential to the structure and functions of the cells in your body. They are composed primarily of carbon, hydrogen, oxygen, and nitrogen. Because they contain about 16% nitrogen by weight, proteins in your diet are the main source of this element. Protein molecules usually contain sulfur and sometimes phosphorus or metals such as iron.

Their molecular weights vary from about 5000 to many millions. Because of their size, proteins are called macromolecules ("macro" means large). These molecules are so big because they are polymers, made by linking many simpler units called amino acids.

15.1 The amino acids

Amino acids contain both an amino group, —NH₂, and a carboxylic acid group, —COOH, in the same molecule. As shown in Figure 15-1, both groups are generally attached to the same carbon atom, called the alpha carbon. (The amino acids in proteins are therefore known as alpha-amino acids.) In addition to a hydrogen atom, another atom or group of atoms binds to the alpha

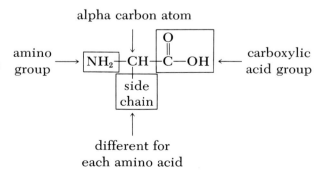

Figure 15-1. An amino acid. Each contains an amino group and a carboxylic acid group. The side chain makes one amino acid different from another.

carbon. This fourth group, the **side chain**, is what makes one amino acid different from another. *There are 20 major amino acids found in proteins.*

The simplest amino acid, *glycine,* abbreviated Gly, just has a second hydrogen atom bonded to the alpha carbon. Other amino acids contain nonpolar

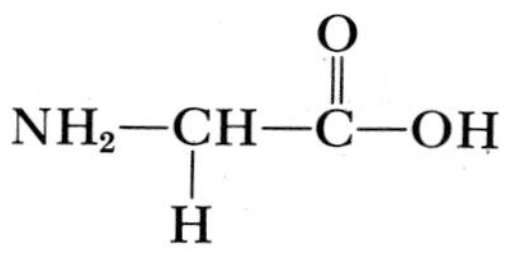

$$NH_2-CH-\overset{\overset{\displaystyle O}{\|}}{C}-OH$$
$$|$$
$$H$$

glycine (Gly)

hydrocarbon side chains, such as *alanine* (Ala), *valine* (Val), *leucine* (Leu), and *isoleucine* (Ile).

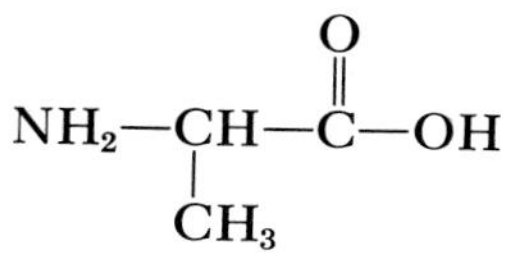

$$NH_2-CH-\overset{\overset{\displaystyle O}{\|}}{C}-OH$$
$$|$$
$$CH_3$$

alanine (Ala)

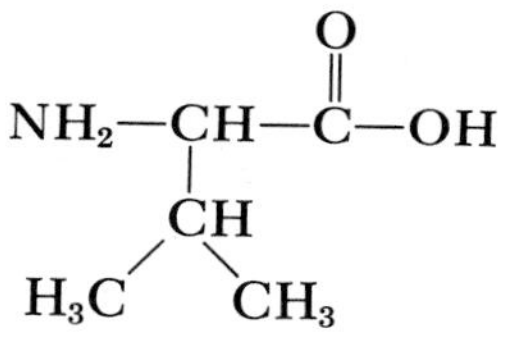

$$NH_2-CH-\overset{\overset{\displaystyle O}{\|}}{C}-OH$$
$$|$$
$$CH$$
$$H_3C \quad CH_3$$

valine (Val)

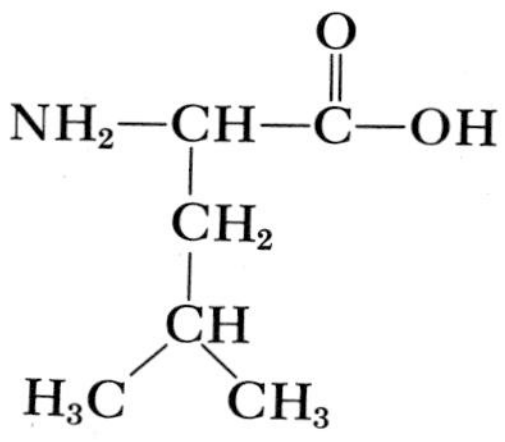

$$NH_2-CH-\overset{\overset{\displaystyle O}{\|}}{C}-OH$$
$$|$$
$$CH_2$$
$$|$$
$$CH$$
$$H_3C \quad CH_3$$

leucine (Leu)

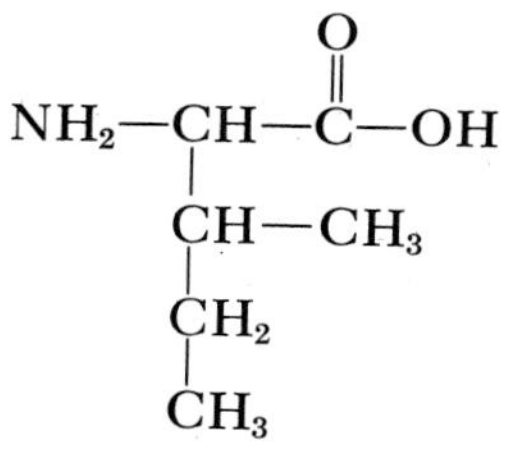

$$NH_2-CH-\overset{\overset{\displaystyle O}{\|}}{C}-OH$$
$$|$$
$$CH-CH_3$$
$$|$$
$$CH_2$$
$$|$$
$$CH_3$$

isoleucine (Ile)

Two amino acids contain a second carboxylic acid group—*aspartic acid* (Asp) and *glutamic acid* (Glu). You are probably familiar with the sodium salt

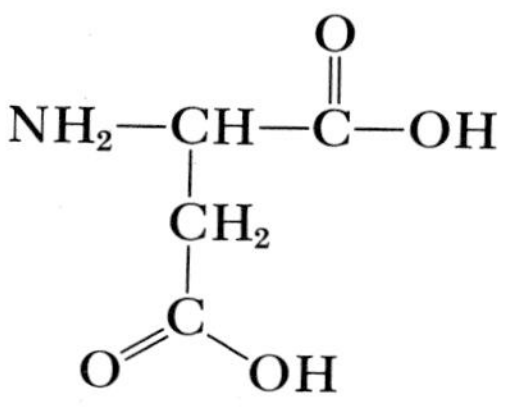

$$NH_2-CH-\overset{\overset{\displaystyle O}{\|}}{C}-OH$$
$$|$$
$$CH_2$$
$$|$$
$$\underset{O}{\overset{C}{\diagdown}}OH$$

aspartic acid (Asp)

$$NH_2-CH-\overset{\overset{\displaystyle O}{\|}}{C}-OH$$
$$|$$
$$CH_2$$
$$|$$
$$CH_2$$
$$|$$
$$\underset{O}{\overset{C}{\diagdown}}OH$$

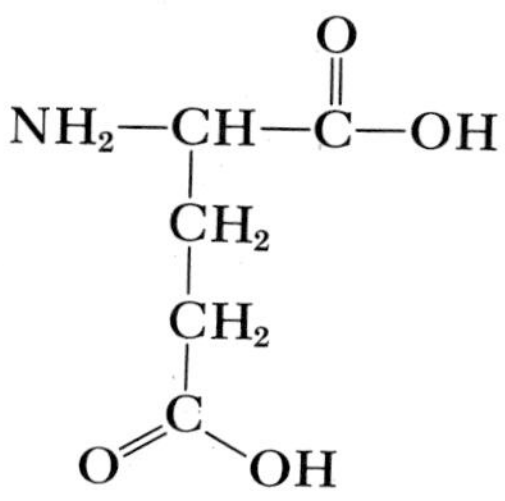

glutamic acid (Glu)

of glutamic acid, monosodium glutamate, or MSG. It is used to enhance flavor in food but may cause "Chinese restaurant syndrome," a tightness of the muscles in the face and neck, headache, and nausea. The amides formed from these two acids are *asparagine* (Asp) and *glutamine* (Gln).

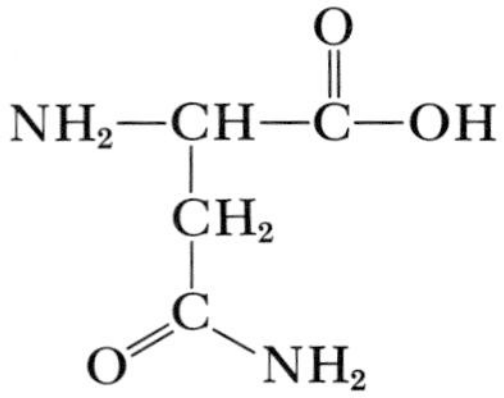

$$NH_2-CH-\overset{\displaystyle O}{\overset{\|}{C}}-OH \qquad NH_2-CH-\overset{\displaystyle O}{\overset{\|}{C}}-OH$$

asparagine (Asn)

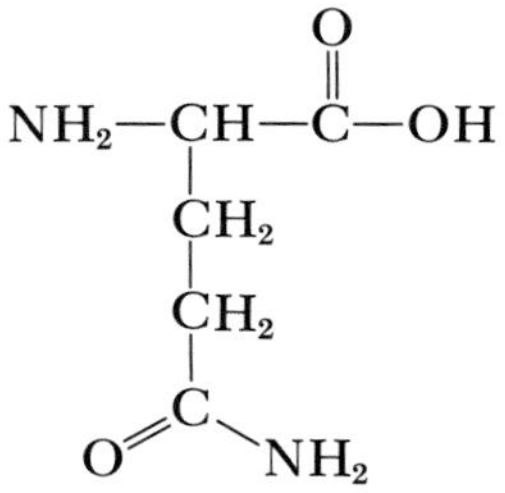

glutamine (Gln)

The two amino acids with additional amino groups are *lysine* (Lys) and the more complicated *arginine* (Arg).

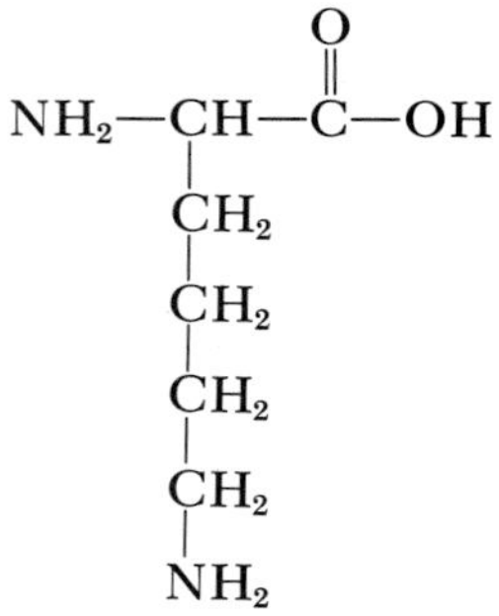

lysine (Lys)

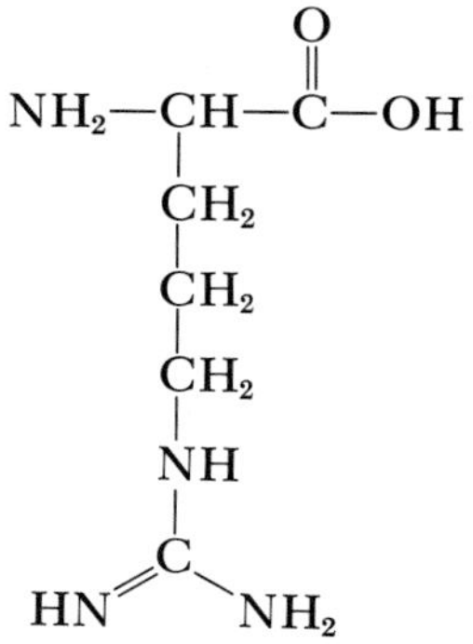

arginine (Arg)

Several amino acids have hydroxyl groups, such as *serine* (Ser) and *threonine* (Thr).

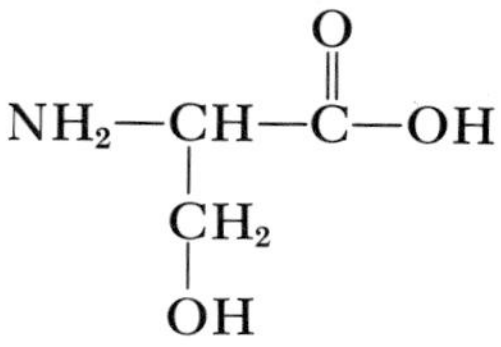

serine (Ser)

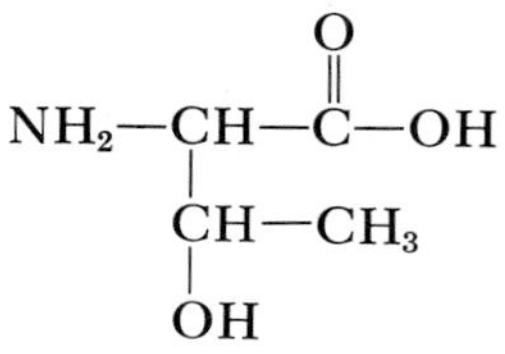

threonine (Thr)

Of the sulfur-containing amino acids, *cysteine* (Cys) has a thiol group, and *methionine* (Met) has a thioether group.

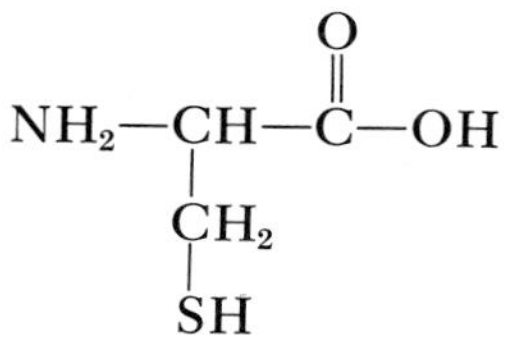

$$NH_2-CH-\overset{\overset{\textstyle O}{\|}}{C}-OH \qquad NH_2-CH-\overset{\overset{\textstyle O}{\|}}{C}-OH$$

cysteine (Cys)

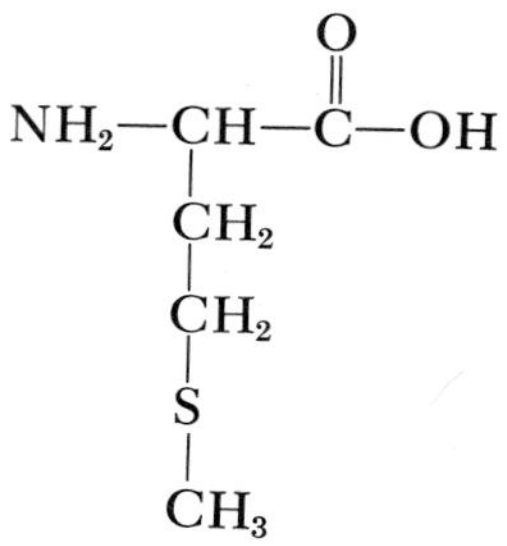

methionine (Met)

Amino acids with a phenyl group are the nonpolar *phenylalanine* (Phe) and polar *tyrosine* (Tyr), which has a hydroxyl group on the ring.

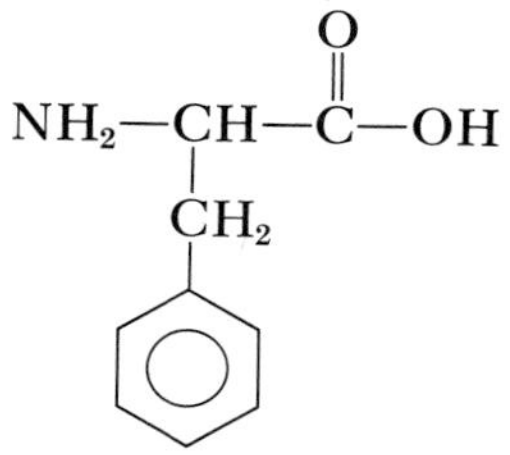

phenylalanine (Phe)

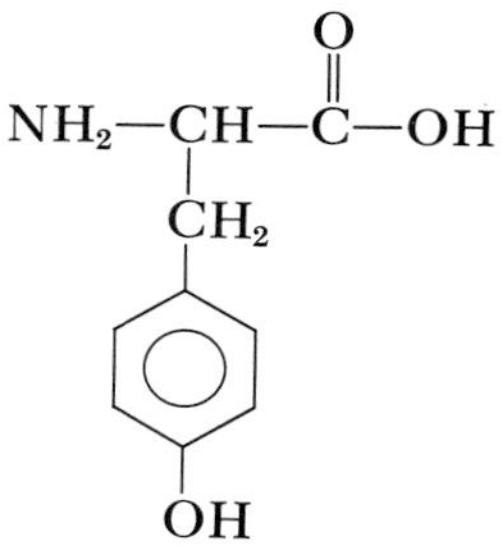

tyrosine (Tyr)

Other amino acids contain nitrogen heterocycles, basic *histidine* (His) and *tryptophan* (Trp).

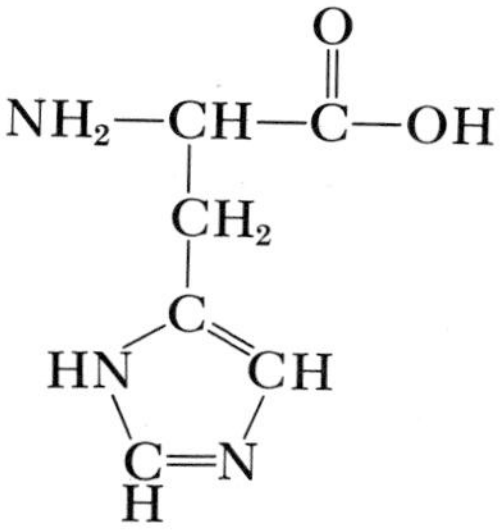

histidine (His)

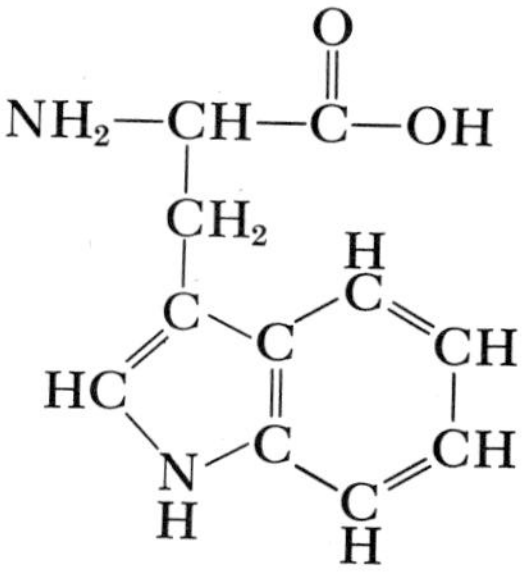

tryptophan (Trp)

Proline (Pro) forms a ring that includes both the amino nitrogen and the alpha carbon atoms.

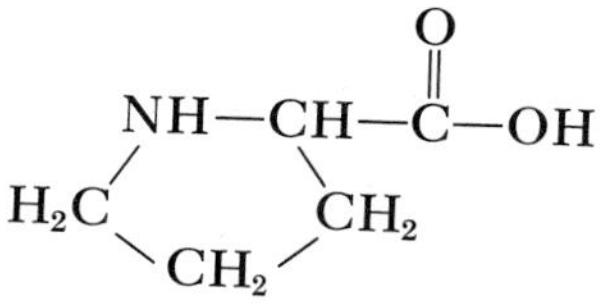

proline (Pro)

15.2   Properties of amino acids

All of the amino acids except glycine have optical isomers. Every chiral amino acid in your body has the L configuration, as you can see by comparing the arrangement of groups at the alpha carbon with L-glyceraldehyde:

L-glyceraldehyde   an L-amino acid   (R = side chain)

Bacteria can synthesize amino acids with the D configuration; some of these are used as antibiotics.

The charge of an amino acid depends on the pH of its solution. In a neutral solution, amino acids exist as doubly charged molecules called **zwitterions**. The carboxylic acid end loses its proton, becoming negatively charged. The amino part of the molecule gains a proton and is positive, as shown for glycine. The molecule as a whole remains neutral, however, because the posi-

$$\overset{+}{N}H_3-CH_2-\overset{\overset{\textstyle O}{\|}}{C}-O^-$$ the zwitterion form of glycine

tive and negative charges are opposite but equal. If the amino acid has other acidic or basic groups, as in aspartic acid or lysine, these too may become charged in solution.

Several of the amino acids are known as **essential amino acids**. They are isoleucine, leucine, lysine, methionine, phenylalanine, threonine, tryptophan, and valine. These amino acids cannot be synthesized in your body but are required for normal growth and maintenance of tissue. Therefore, you must include them in your diet. Histidine is an essential amino acid for children.

15.3   Primary structure of proteins

Two amino acids can react to form a chemical bond between the carboxylic acid group of one molecule and the amino group of the other. A molecule of water is lost (dehydration) and an *amide* results, as in the combination of glycine and alanine:

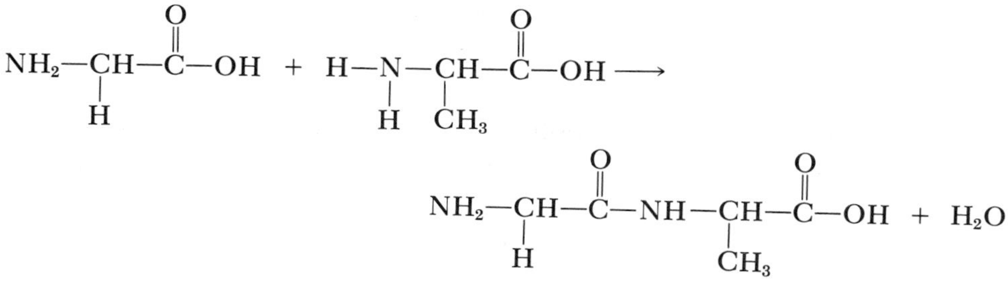

$$NH_2-CH-\overset{\overset{\textstyle O}{\|}}{C}-OH \ + \ H-N-CH-\overset{\overset{\textstyle O}{\|}}{C}-OH \longrightarrow$$

$$\underset{\substack{|\\H}}{} \qquad \underset{\substack{|\ \ \ |\\H \ \ CH_3}}{}$$

$$NH_2-CH-\overset{\overset{\textstyle O}{\|}}{C}-NH-CH-\overset{\overset{\textstyle O}{\|}}{C}-OH \ + \ H_2O$$

$$\underset{\substack{|\\H}}{} \qquad\qquad \underset{\substack{|\\CH_3}}{}$$

a dipeptide

The new bond, illustrated in Figure 15-2, is known as a **peptide bond**. The resulting molecule is called a dipeptide because it consists of two amino acids ("di-" means two).

Figure 15-2. The peptide bond. An amide linkage forms between the carboxylic acid end of one amino acid with the amino group of a second amino acid. One water molecule is released when the peptide bond forms.

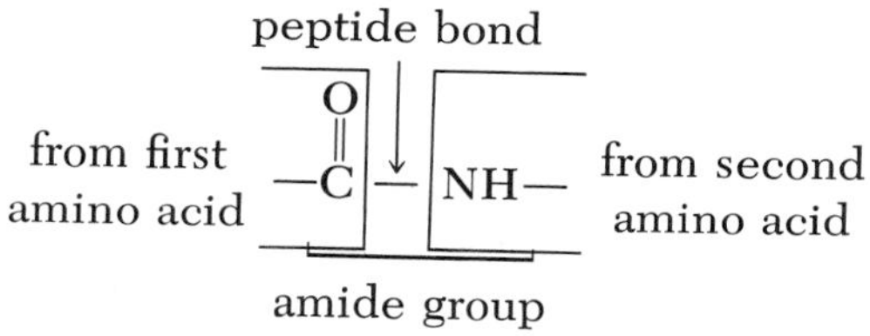

If this dipeptide reacts with a third amino acid, a tripeptide forms; these relatively small molecules are known as oligopeptides. When many units join in this manner, they make a **polypeptide**, a polymer of amino acids linked by peptide (amide) bonds, as shown in Figure 15-3. *Protein molecules consist of very large naturally occurring polypeptides.*

Think of a protein as a long chain, or several chains, of beads put together from only 20 different kinds of beads. Each protein has a definite, characteristic arrangement of the basic units, the amino acids. The **primary structure** of a protein is the *sequence* of its amino acids; it shows you exactly which are present and the order in which they are connected. You can also think of a pro-

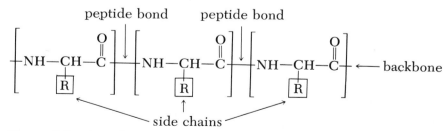

Figure 15-3. A portion of a polypeptide. Many amino acids (three of which are shown here) are joined by peptide bonds in a long chain.

tein as having a "backbone" which consists of nitrogen atoms, alpha carbon atoms, and carbonyl groups, forming a chain that has branches coming from it, the side chains. In this case, the primary structure describes the length of the backbone and its arrangement of side chains. As an example, the primary structure of one chain of hemoglobin is shown in Figure 15-4.

Because there are 20 different amino acids, and protein chains are long, a

Val—His—Leu—Thr—Pro—Glu—Glu—Lys—Ser—Ala—Val—Thr—Ala—Leu—Trp—Gly—Lys—Val—Asp—
 1   2   3   4   5   6   7   8   9 10 11 12 13 14 15 16 17 18 19

Val—Asp—Glu—Val—Gly—Gly—Glu—Ala—Leu—Gly—Arg—Leu—Leu—Val—Val—Tyr—Pro—Trp—Thr—
20 21 22 23 24 25 26 27 28 29 30 31 32 33 34 35 36 37 38

Glu—Arg—Phe—Phe—Glu—Ser—Phe—Gly—Asp—Leu—Ser—Thr—Pro—Asp—Ala—Val—Met—Gly—Asp—
39 40 41 42 43 44 45 46 47 48 49 50 51 52 53 54 55 56 57

Pro—Lys—Val—Lys—Ala—His—Gly—Lys—Lys—Val—Leu—Gly—Ala—Phe—Ser—Asp—Gly—Leu—Ala—
58 59 60 61 62 63 64 65 66 67 68 69 70 71 72 73 74 75 76

His—Leu—Asp—Asp—Leu—Lys—Gly—Thr—Phe—Ala—Thr—Leu—Ser—Glu—Leu—His—Cys—Asp—Lys—
77 78 79 80 81 82 83 84 85 86 87 88 89 90 91 92 93 94 95

Leu—His—Val—Asp—Pro—Glu—Asp—Phe—Arg—Leu—Leu—Gly—Asp—Val—Leu—Val—Cys—Val- Leu—
96 97 98 99 100 101 102 103 104 105 106 107 108 109 110 111 112 113 114

Ala—His—His—Phe—Gly—Lys—Glu—Phe—Thr—Pro—Pro—Val—Glu—Ala—Ala—Tyr—Glu—Lys—Val—
115 116 117 118 119 120 121 122 123 124 125 126 127 128 129 130 131 132 133

Val—Ala—Gly—Val—Ala—Asp—Ala—Leu—Ala—His—Lys—Tyr—His
134 135 136 137 138 139 140 141 142 143 144 145 146

Figure 15-4. The primary structure of a protein chain; a beta chain of hemoglobin.

large variety of combinations can exist. For example, consider a tripeptide, which is formed from three amino acids, say valine, arginine, and leucine. The following arrangements are then possible: Val-Arg-Leu, Val-Leu-Arg, Arg-Val-Leu, Arg-Leu-Val, Leu-Val-Arg, Leu-Arg-Val. Thus, six isomers can be formed from just three amino acids. The number of combinations increases dramatically as you make the chain longer and use more amino acids. For a relatively small protein, the possible sequence of isomers are equal to 10 multiplied by itself 300 times (10^{300}), an almost unlimited quantity of combinations.

It is the primary structure, the sequence of amino acids of the polypeptide chains, that determines the function of a protein in the body. A change in a single amino acid out of many hundred can completely alter the properties of the protein and affect its biological role. For example, in sickle cell anemia, the protein of red blood cells, hemoglobin, is defective. Out of about 300 amino acids, just 1, glutamic acid, is replaced by another, valine, at one position in part of the protein of a person with this condition. (The sixth amino acid in Figure 15-4 is the one affected.) As a result, the red cells become sickle shaped and fragile at low oxygen pressure, causing hemolysis and thus anemia. This condition is discussed in greater detail in Chapter 20.

15.4 Secondary structure of proteins

The **secondary structure** of a protein describes the regular shapes taken by portions of the protein backbone. These arrangements result from *hydrogen bonding*, the attraction of a hydrogen atom by another atom on the same molecule or a different molecule. The most important hydrogen bonds involve a hydrogen atom bonded to a nitrogen or oxygen atom being pulled toward another oxygen atom, as illustrated in Figure 15-5. These bonds are weak; they can break and re-form at a rapid rate. When several hydrogen bonds exist between two regions of polypeptide chains, however, the attractive force is strong enough to form a definite structure.

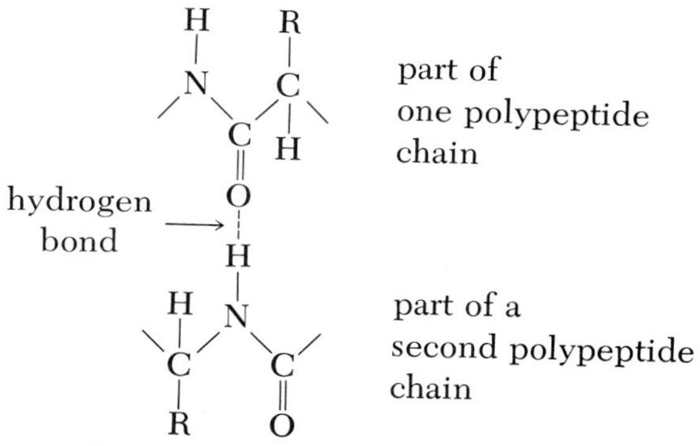

Figure 15-5. A hydrogen bond joining two strands of polypeptide. The hydrogen bonded to a nitrogen is attracted by the oxygen of a carbonyl group.

As shown in Figure 15-6, one type of secondary structure produced by hydrogen bonding is the **alpha helix**. It consists of the coiling of a polypeptide chain in a spiral arrangement. Like the threads of most screws, the helix is "right-handed." Each turn, which contains 3.6 amino acids, is held together by hydrogen bonding between the hydrogen atom of an amino nitrogen and the oxygen of a carbonyl group four amino acids away in the chain. The alpha helix is both flexible (can be bent or folded without breaking) and elastic (capable of recovering shape after stretching).

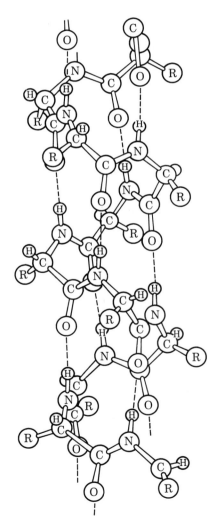

Figure 15-6. The arrangement in an alpha helix. The dashed lines indicate hydrogen bonds. (Redrawn from Linus Pauling, "The Nature of the Chemical Bond," 3rd ed. © 1939 and 1940, third edition 1960, by Cornell University, Ithaca, New York. Reprinted by permission of Cornell University Press.)

Another type of secondary structure involves hydrogen bonding between two strands of protein that run parallel to each other, as shown in Figure 15-7. This arrangement is called **pleated sheet** (or beta pleated sheet) because of the resulting folds or pleats in the sheetlike network. Pleated sheet structure is flexible but inelastic. Some types of protein, such as silk, consist primarily of pleated sheet; most others, however, are more complicated, as described in the following section.

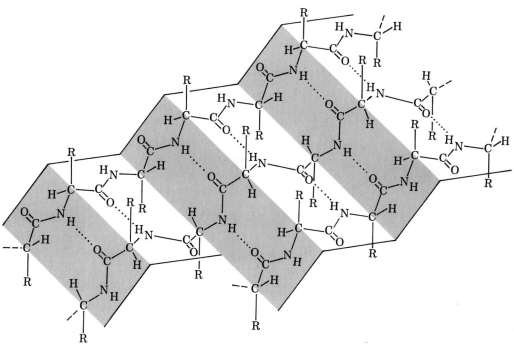

Figure 15-7. Pleated sheet formed from three parallel polypeptide strands. The dotted lines indicate hydrogen bonds; R stands for the side chain of an amino acid. (From P. Karlson, "Kurzes Lehrbuch der Biochemie," 9th ed. Georg Thieme Verlag, Stuttgart, 1974.)

15.5 Tertiary and quaternary structure of proteins

The **tertiary structure** of a protein is its complex three-dimensional shape, or conformation, resulting from the folding of the polypeptide chain. The interactions between amino acid side chains within the protein determine the linkages that hold the tertiary structure together.

Disulfide bonds form through the oxidation of the sulfhydryl (thiol) groups, —SH, of two cysteine side chains:

$$
\underset{\underset{\text{SH}}{\overset{\overset{\text{O}}{\parallel}}{\underset{|}{\overset{|}{\text{NH}_2-\text{CH}-\text{C}-\text{OH}}}}}{\overset{}{\underset{\text{CH}_2}{}}} + \underset{\underset{\text{SH}}{\overset{\overset{\text{O}}{\parallel}}{\underset{|}{\overset{|}{\text{NH}_2-\text{CH}-\text{C}-\text{OH}}}}}{\overset{}{\underset{\text{CH}_2}{}}} \xrightarrow{\text{oxidation}}
$$

$$
\text{NH}_2-\overset{|}{\underset{\text{CH}_2}{\text{CH}}}-\overset{\overset{\text{O}}{\parallel}}{\text{C}}-\text{OH} \quad \text{NH}_2-\overset{|}{\underset{\text{CH}_2}{\text{CH}}}-\overset{\overset{\text{O}}{\parallel}}{\text{C}}-\text{OH}
$$

$$
\text{CH}_2 - \text{S} - \text{S} - \text{CH}_2
$$

cystine

A disulfide linkage results, forming cystine. *This type of covalent bond creates cross-links that firmly connect different regions of the protein chain.* Disulfide bonds are about ten times stronger than the noncovalent bonds described below.

Hydrophobic bonds are extremely important in determining tertiary structure. The word "hydrophobic" literally means afraid of water. It refers to the clustering of nonpolar side chains of amino acids (like alanine, valine, leucine, isoleucine, and phenylalanine) with each other to form weak bonds. They stay away from polar water molecules, which also are highly associated (through hydrogen bonding). Because of these hydrophobic interactions, many proteins take folded shapes in which nonpolar groups are on the "inside." Polar side chains, such as those containing carboxyl groups, hydroxyl groups, and amino groups, stay on the "outside" surface of the protein. Here they can interact with water molecules through hydrogen bond formation.

Hydrogen bonds also form between side chains within the protein in ways other than the regular alpha helix or pleated sheet structures. For example, the hydrogen atom from a hydroxyl group may be weakly attracted by the oxygen of a carbonyl group of another amino acid in the protein. In addition, bonds called **salt bridges** form by ionic attraction between positively charged groups and negatively charged groups within the protein. These different types of forces within a protein molecule are summarized in Figure 15-8. Although they are all weak (except for the disulfide linkages), by acting together, many such bonds maintain the tertiary structure.

Because of these different kinds of interactions, proteins have very complex tertiary structures. *The shape of a protein is determined largely by its primary structure*, the sequence of amino acids. Each protein has a characteristic shape because of its unique composition and ordering of amino acids. By changing the arrangement of amino acids, you affect the types of bonds that can form within the protein and therefore change its conformation. The func-

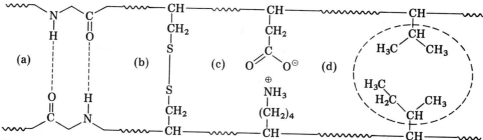

Figure 15-8. The types of bonds formed between different parts of a polypeptide chain. (a) Hydrogen bonds; (b) a disulfide bond between cysteines; (c) an ionic bond (between the side chains of aspartic acid and lysine); (d) a hydrophobic bond (between the side chains of valine and isoleucine).

tion of a protein, its job, depends on this shape and often on the presence of certain amino acids in particular places in the molecule. Since its function is based on its conformation, which in turn depends on the primary structure, it is actually the sequence of amino acids that ultimately determines how the protein works.

Some proteins are further complicated because they contain more than one polypeptide chain. Each separate chain of amino acids is called a subunit. The **quaternary structure** of a protein describes the way in which the subunits are arranged. An important example, hemoglobin, consists of four subunits. Its structure is described in the next section.

15.6 Examples of protein structure

Collagen is the most abundant protein in your body, making up about 30% of the total amount. It is the main component of your supportive and connective tissue—the skin and the organic part of the bone. The composition of this protein is unusual because such a large amount of three amino acids is present: glycine (33%), proline (and its derivative, hydroxyproline, 21%), and alanine (11%). Its structure is illustrated in Figure 15-9. The basic unit, called tropocollagen, is also unusual—a triple helix having a molecular weight of about 300,000. This structure is not the same as the alpha helix and is unique to collagen. It is formed by the twisting around each other of three polypeptide chains, which are held together by hydrogen bonds. The resulting "cables" are very long and thin but rigid. They line up alongside each other to form fibers; cross-linking gives the resulting combination extra strength.

Hemoglobin is the oxygen carrier of the red blood cells. Since its function is unlike that of collagen, you would expect its structure also to differ.

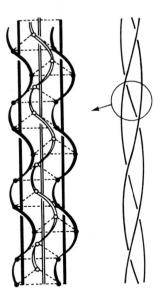

Figure 15-9. The structure of collagen. Three polypeptide chains form a triple helix in the cable-like tropocollagen molecules. They line up alongside each other, creating a fiber.

As you can see in Figure 15-10, the composition and shape of the protein are not at all similar to collagen. It consists of four separate polypeptide chains (labeled α_1, α_2, β_1, β_2), each consisting of about 150 amino acids. Also, each chain forms a "pocket" containing a nitrogen heterocycle called heme. An iron ion (Fe^{2+}) at the center of the heme group can bind a molecule of oxygen, carrying it from the lungs to the tissues. Notice that this protein is tightly folded and compact. Its chains, some of which contain regions of alpha helix, are held together by hydrogen bonding, salt linkages, and hydrophobic bonds. The quaternary structure of hemoglobin, the relation between the subunits, is closely tied to the ability of this protein to both bind and release molecular oxygen. As in the case of collagen, the structure of this protein is carefully "designed" to carry out its function.

15.7 Classification of proteins

Proteins can be classified, or divided into groups, in several ways. One set of categories is based on their shape, solubility in water, and composition.

Fibrous proteins are insoluble in water and resistant to digestion. They often consist of several parallel polypeptide chains that are coiled and stretched out, as in the case of collagen. Another example is elastin, the second major protein of connective tissue after collagen; it is found in elastic structures like large blood vessels. Keratins make up the fibrous protein of hair, nails, wool, and animal hoofs. They include a large amount of sulfur in the form of cystine (about 14%) and can exist in two forms, one consisting

(a)

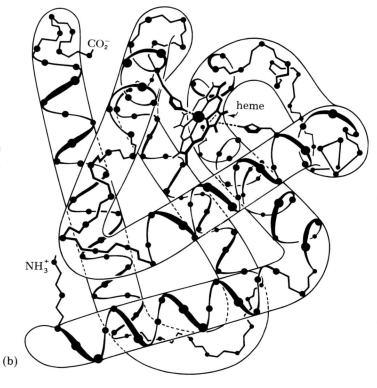

CO_2^-

heme

NH_3^+

(b)

Figure 15-10. The structure of hemoglobin: (a) the packing of its four polypeptide chains, each containing a heme group; (b) the backbone of one of its chains. (Courtesy of Dr. M. F. Perutz.)

largely of alpha helix (alpha form) and the other consisting mainly of pleated sheet (beta form).

Globular proteins consist of polypeptides tightly folded into the shape of a "ball" and are soluble in water. They include most of the proteins that carry out chemical processes in the body, such as enzymes. Albumins, such as egg albumin, are globular proteins that are coagulated, curdled, or solidified by heat and dissolve in pure water. Globulins, which include the gamma globulins of the blood, can also be coagulated but require the presence of small amounts of a salt to dissolve. Histones, the protein part of the genes, are water soluble and basic, containing large amounts of lysine or arginine. Protamines are basic globular proteins of low molecular weight.

Conjugated proteins contain another substance that is not made from amino acids in addition to the polypeptide portion. You have already studied one example, hemoglobin. The nonprotein part, heme in this case, is called the prosthetic group. As shown in Table 15-1, additional prosthetic groups include carbohydrates, lipids, and other molecules as well as simple metal ions.

You can classify proteins in another way, on the basis of their function—their effect or role in the body. **Enzymes** are probably the most important proteins because they allow all the chemical reactions inside of you to take place at a suitable rate. These biological catalysts are described in detail in the next chapter.

Structural proteins include collagen, the major protein of connective tissue and bone. This category also contains other fibrous proteins such as the keratins and the membrane-structure proteins, which are among the most abundant in the cell.

Contractile proteins, such as actin and myosin, are found in muscle. They stretch or contract as the muscle is relaxed or tensed through the successive breaking or making of cross-links between polypeptide chains.

Transport proteins, like hemoglobin, carry small molecules through the bloodstream. A second example is serum albumin, which brings fatty acids from adipose tissue to various organs. Other proteins transport lipids and iron.

Table 15-1   Prosthetic Groups of Conjugated Proteins

Class	Prosthetic group
glycoprotein	carbohydrate (small amount)
hemoprotein	heme
lipoprotein	lipids
metalloprotein	metal ion
mucoprotein (mucoid)	carbohydrate (large amount)
nucleoprotein	nucleic acid
phosphoprotein	phosphate

Hormones can consist of protein molecules. They are secreted in small amounts from the endocrine glands to regulate chemical processes in the body. Insulin, which controls the use of glucose, is an example. Chapter 21 deals with hormones in greater detail.

Storage proteins act as reservoirs for essential chemical substances. For example, ferritin stores iron, which is needed to make hemoglobin molecules in the spleen.

Protective proteins include the antibodies like gamma globulin. These proteins form complexes and thereby inactivate foreign proteins, called antigens. Other protective proteins are fibrinogen and thrombin, both involved in the process of blood clotting. Fibrin, formed from fibrinogen, is the fibrous insoluble protein that forms blood clots, trapping red blood cells, as shown in Figure 15-11. These proteins are described in Chapter 22.

Figure 15-11. A red blood cell (erythrocyte) enmeshed in fibrin. [Scanning electron micrograph (×20,500) by Emil Berstein and Eila Kairinen, Gillette Research Institute. Copyright 1971 by the American Association for the Advancement of Science. Reprinted with permission.]

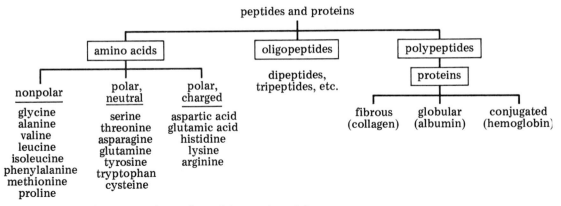

Figure 15-12. Summary chart of peptides and proteins.

Toxins are proteins harmful to the body; they are poisons, such as *Clostridium botulinum* toxin. This protein causes one type of bacterial food poisoning. Other examples are diphtheria toxin and various snake venoms.

Figure 15-12 presents a summary chart of peptides and proteins.

15.8   Properties of proteins

One of the most important properties of a protein molecule is its *large size*. Proteins therefore form colloids rather than solutions in water. They are too large to pass through the openings of the cell membranes and contribute to maintaining the osmotic pressure of body fluids. When membranes are damaged, however, as in kidney disease, proteins can pass from the blood plasma into the urine.

Because a protein contains many acidic or basic groups that ionize in water, the molecule as a whole may be charged. Whether the net charge is positive or negative depends on the pH and on the types of amino acids present in the protein. At a certain value of the pH, called its **isoelectric point** (pI), a protein is electrically neutral—it contains an equal number of positive and negative charges. (If placed in an electric field, as in the process of electrophoresis, it will be attracted to neither the positive terminal nor the negative terminal.) Because of their weak acidic and basic groups, proteins can act as buffers; those in the blood assist the carbonate and phosphate buffer systems in maintaining the pH of 7.4.

The most important chemical reaction undergone by proteins is **hydrolysis**, the breaking of peptide bonds by the addition of water molecules. Complete hydrolysis of a protein produces free amino acids. The hydrolysis of a dipeptide is shown in the following equation:

$$NH_2-\underset{\underset{H}{|}}{CH}-\underset{\overset{\displaystyle O}{\|}}{C}-NH-\underset{\underset{CH_3}{|}}{CH}-\underset{\overset{\displaystyle O}{\|}}{C}-OH \ + \ H_2O \ \xrightarrow{\ catalyst\ }$$

$$NH_2-\underset{\underset{H}{|}}{CH}-\underset{\overset{\displaystyle O}{\|}}{C}-OH \ + \ NH_2-\underset{\underset{CH_3}{|}}{CH}-\underset{\overset{\displaystyle O}{\|}}{C}-OH$$

Notice that this reaction is just the reverse of the one shown earlier in the chapter for the formation of a peptide linkage. Hydrolysis occurs when you digest proteins; the amino acids are then used by your body to synthesize its own proteins. They can also provide a source of energy, about the same amount as carbohydrates, 4 kcal/g (17 kJ/g).

15.9 Denaturation of proteins

Each protein in your body has a normal shape called its **native conformation**. This arrangement is required for the protein to perform its biological role. **Denaturation** is the disorganization of the protein structure, or destruction of the native conformation, as shown in Figure 15-13. This process, which is often reversible, generally involves breaking the weak noncovalent bonds that maintain the structure of the polypeptide chains.

Figure 15-13. The denaturation of a protein. Native conformation is lost, producing a random chain.

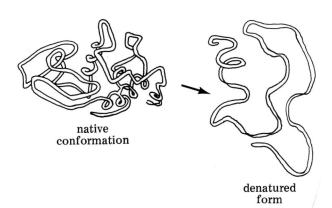

native
conformation

denatured
form

Although the peptide bonds are not broken as in hydrolysis, the secondary and tertiary structure is lost. The resulting random chains often clump together (coagulate) and precipitate (become insoluble). Because of denaturation, the protein no longer can carry out its intended function.

Heat readily denatures a protein. When you boil an egg, the protein of the egg white denatures and coagulates. By raising the temperature, you increase the motion of the atoms of the protein, causing the weak bonds that hold the tertiary structure together to break. Heat denaturation is used to destroy bacterial protein in the process of sterilization in an autoclave. A urine test for protein involves heating the sample and checking if any protein denatures and coagulates.

Alcohols and other *organic solvents* that dissolve in water denature protein by changing its normal aqueous environment. At a concentration of 50 to 60%, ethyl or isopropyl alcohol is an effective disinfectant. It destroys bacteria by penetrating them and denaturing their protein molecules. Tincture of iodine ($2\% I_2$ with 2.4% NaI in 50% alcohol) makes use of this effect in addition to the ability of the I_2 molecule to react with certain side groups on proteins.

Acids and *bases* cause denaturation by changing the distribution of positive and negative charges in the protein. The effect of changing the pH varies with the type of protein molecule; few can withstand exposure to strongly acidic (below pH 2) or strongly basic (above pH 12) conditions. The low pH of the gastric juice in the stomach aids digestion by denaturing ingested proteins. Their peptide bonds become more exposed, allowing hydrolysis to take place.

Metal ions, such as those of mercury, silver, and lead, bind strongly to certain groups on proteins, disrupting the normal weak bonds present. Insoluble "salts" of the protein may form, resulting in precipitation. Compounds of the heavy metals (those just mentioned) are poisonous because they destroy vital molecules like proteins.

Oxidizing and *reducing agents* denature proteins, usually by affecting the sulfhydryl groups of the amino acid cysteine. This effect is the basis for the "home permanent" because keratin, the protein in hair, contains a large amount of cysteine. A reducing agent (ammonium thioglycollate) first breaks the existing disulfide bonds in the protein. The hair is then given the desired shape with rollers and new disulfide bonds are formed using an oxidizing agent (potassium bromate). Keratin now keeps this conformation, but, of course, as new hair grows out, the "permanent" will have to be repeated.

Denaturation can be produced in other ways. Violent agitation, as in beating egg whites, produces denaturation (resulting in a meringue in this case). Certain compounds, like picric acid, cause precipitation of proteins; they have been used on burns to form protective coatings that exclude air.

Although denaturation has serious effects on the structure and function of a protein, it is not necessarily permanent. In many cases, if the conditions are returned to "normal," the protein goes back to its native conformation. This process is known as **renaturation**.

15.10 Protein and mineral—bones and teeth

Bone contains a unique combination of protein and mineral. The structural framework ("matrix"), making up much of the total volume, is provided by collagen, the fibrous protein described in Section 15.6. A complex salt called hydroxyapatite, having the composition $Ca_5(PO_4)_3OH$, forms crystals within the collagen fibers, creating a combination of great strength. The collagen provides toughness and springiness, while the hydroxyapatite provides hardness and rigidity.

The main part of a tooth (Figure 15-14) is made of dentine, a substance similar in composition to bone. It is much more dense, however, having a higher mineral content (75%), and is therefore harder. Enamel, which covers the ex-

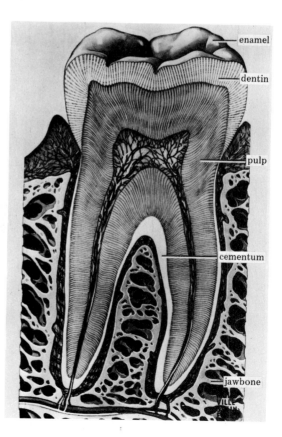

Figure 15-14. A tooth. (Copyright by the American Dental Association. Reprinted with permission.)

enamel

dentin

pulp

cementum

jawbone

posed outer portion of the dentine, is still harder (96% mineral). The root of the tooth, the part below your gum, is covered by cementum, a substance that helps hold the tooth in place; its structure is also based on hydroxyapatite.

Teeth and bones are formed by mineralization, the depositing of minerals in tissue. Specifically, the process is called **calcification**, since calcium is the basis for their structure along with phosphate. Vitamins A, C, and D are needed for tooth and bone formation. The hormones parathormone and calcitonin also play a role because they control calcium and phosphate metabolism (as described in Chapter 21).

The reverse process, **decalcification**, removal of calcium salts, results from our most common disease—*dental caries*. By fermenting carbohydrates, bacteria in the mouth produce lactic acid (pH 5.5 or lower), which supports this process, causing destruction of part of the tooth, as shown in Figure 15-15. Another dental problem related to calcium is the formation of *calculus* or "tartar," a deposit consisting of calcium phosphate (67%), calcium carbonate (13%), water with organic substances (20%), and traces of certain ions.

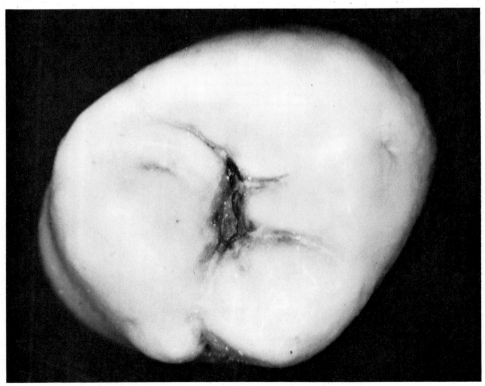

Figure 15-15. A tooth with decay in a fissure. (WHO photo.)

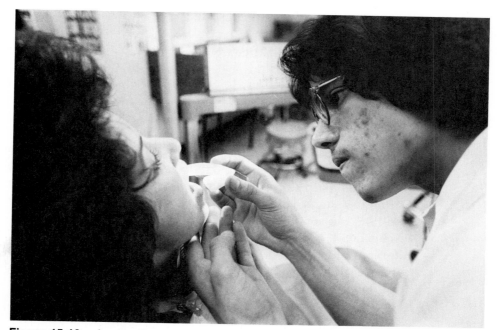

Figure 15-16. Application of fluoride to teeth. (Photo by Al Green.)

The demineralization of teeth is inhibited by the fluoride ion, F^-. By re-
placing hydroxide ion in the hydroxyapatite structure, it forms fluoroapatite,
with the composition $Ca_5(PO_4)_3F$. This material is less soluble and more resistant to attack by acid. When added to the water supply of a community (at a
concentration of about 1 ppm), fluoride generally reduces the number of decayed or missing teeth by 50 to 60% in children at the ages of 14 to 16. Topical
application of fluoride ion in the form of stannous fluoride, SnF_2, is believed
to convert hydroxyapatite to a new substance having the formula $Sn_3F_3PO_4$
(Figure 15-16).

SUMMARY

Proteins are complex molecules essential to the structure and functions of the
cells in your body. They are composed primarily of carbon, hydrogen, oxygen,
and nitrogen. Proteins in your diet are the main source of nitrogen.

Amino acids are the basic units of the huge protein molecules. They contain
both an amino group, —NH_2, and a carboxylic acid group, —COOH, in addition to another atom or group of atoms, the side chain, connected to the alpha
carbon. This side chain is what makes one amino acid different from another.
There are 20 major amino acids.

All amino acids in the body (except glycine, which does not have optical isomers) have the L configuration. In neutral solution, they exist as doubly charged molecules called zwitterions. Several are known as essential amino acids because they cannot be synthesized within the body but are required for health.

Two amino acids can react to form an amide linkage, or peptide bond, between the carboxylic acid group of one molecule and the amino group of the other. A molecule of water is lost and a dipeptide forms. Proteins consist of very large naturally occurring polymers of amino acids, polypeptides. The primary structure describes the sequence of amino acids connected together.

The secondary structure of a protein describes the regular shapes taken by portions of the protein backbone. These arrangements result from hydrogen bonding. Alpha helix consists of the coiling of a polypeptide chain in a spiral arrangement. Pleated sheet is formed from polypeptide strands running parallel to each other.

The tertiary structure of a protein is its complex three-dimensional shape resulting from the folding of the polypeptide chain through the interactions of the side chains. These interactions include the strong disulfide linkages and the weaker hydrophobic bonds, hydrogen bonds, and salt bridges. The function of a protein depends on its shape or conformation, which in turn depends on its primary structure.

Collagen is the most abundant protein in your body; it is the main component of supportive and connective tissue. The basic unit consists of a triple helix, formed by the twisting of three polypeptide chains. Hemoglobin, the oxygen carrier of the red blood cells, contains four separate chains, each with a heme group.

Proteins can be classified according to their shape, solubility in water, and composition as follows: fibrous, globular, and conjugated proteins. On the basis of function, they can be divided into enzymes, structural proteins, contractile proteins, transport proteins, hormones, storage proteins, protective proteins, and toxins.

One of the most important properties of a protein is its size; proteins therefore form colloids rather than solutions. The molecule as a whole can be charged because of its acidic and basic groups. The most important chemical reaction of proteins is hydrolysis, the breaking of the peptide bonds by addition of water molecules.

Each protein has a normal shape in solution called its native conformation, which is required for activity. Denaturation is the disorganization of this protein structure. It is caused by heat, organic solvents, acids and bases, metal ions, and oxidizing or reducing agents.

Bones and teeth are a unique combination of the protein collagen and the mineral hydroxyapatite. Dental caries produces decalcification, the removal of calcium salts from the tooth structure.

Exercises

1. (15.1) Describe an amino acid. What makes one different from another?
2. (15.1) Give an example of an amino acid with (a) a hydroxyl group; (b) a thiol group; (c) an acidic group; (d) a basic group; (e) a nonpolar group; (f) a heterocyclic ring.
3. (15.2) Explain how a zwitterion forms.
4. (15.2) What are essential amino acids?
5. (15.2) Write an equation for the formation of a dipeptide between valine and serine. Draw the structural formulas.
6. (15.3) What is a protein? What is its primary structure?
7. (15.3) Write five isomers of the following tetrapeptide: Gly-Phe-Arg-Ilu.
8. (15.4) What is secondary structure and what holds it together?
9. (15.4) How does alpha helix differ from pleated sheet in structure and properties?
10. (15.5) What is tertiary structure? quaternary structure?
11. (15.5) Identify the type of interaction that could take place between the side chains of the following pairs of amino acids: (a) lysine and aspartic acid; (b) valine and leucine; (c) two cysteines.
12. (15.5) What determines the function of a protein?
13. (15.6) Describe the structure and properties of (a) collagen; (b) hemoglobin.
14. (15.7) Describe how the following types of proteins differ: fibrous, globular, conjugated. Give an example of each.
15. (15.7) Describe five functions of proteins in the body.
16. (15.8) Why do proteins form colloids?
17. (15.8) What is the isoelectric point?
18. (15.8) Write an equation for the hydrolysis of the tripeptide Gln-His-Ilu, drawing all structural formulas.
19. (15.9) Describe the process of denaturation.
20. (15.9) Explain how each of the following causes denaturation: (a) acids and bases; (b) metal ions; (c) heat; (d) alcohol; (e) oxidizing and reducing agents.
21. (15.10) Describe the composition of bones and teeth.
22. (15.10) What are dental caries? What is the effect of fluoride on teeth?

Enzymes and digestion

The chemical changes in your body depend on **enzymes**. These molecules are *biological catalysts,* substances that increase the rate of a reaction but are not used up in the process. Without enzymes, the vital chemical reactions in the cells would take place too slowly to support life. Lack of even a single enzyme out of hundreds in the body can have very serious effects and possibly be fatal, because certain reactions cannot be carried out fast enough.

Every enzyme is a protein molecule. Therefore, your knowledge about proteins from the previous chapter will help you understand how enzymes work. As you will see, because enzymes are proteins, they are specific in their roles and highly efficient in carrying them out.

16.1 Mechanism of enzyme action

The molecule that is changed by a chemical reaction catalyzed by an enzyme is called the **substrate**. It binds to the enzyme, forming a temporary combination, the enzyme–substrate complex. At this stage, the chemical change takes place, converting the substrate molecule into the end product of the reaction. The product leaves the enzyme, making the enzymes available to bind another substrate molecule and repeat the process. Because an enzyme can catalyze so many such reactions in a short period of time, *only very small amounts of the enzyme are required* compared to the quantity of substrate.

Each enzyme has a special region in its protein structure into which the substrate molecule fits perfectly; it is known as the **active site**. You can think of the enzyme as a lock—only certain keys, the substrates, have the right shape to enter it (Figure 16-1). The factors involved in the enzyme–substrate interactions are similar to those that maintain the structure of a protein—hydrophobic bonds, ionic bonds, and hydrogen bonds.

The substrate has a bond or linkage that can be attacked by the enzyme. Groups on the enzyme help position the substrate molecule. Once it is in the active site, functional groups of the enzyme act on the substrate. Their effect

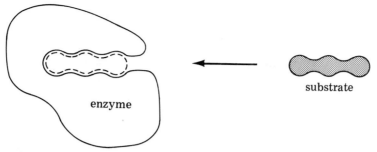

Figure 16-1. A simplified representation of an enzyme and its substrate. The substrate must have the proper structure to bind to and be acted on by the enzyme.

may be to weaken one of its bonds or to participate in a reaction by transferring an electron or proton.

As an example, look at the model of the enzyme lysozyme shown in Figure 16-2. You can see that the huge molecule has a well-formed crevice into which the substrate, a polysaccharide, fits. The backbone of 129 amino acids is illus-

Figure 16-2. A space-filling model of lysozyme. The arrow indicates the crevice in the molecule; this "opening" forms the active site. (Photo courtesy of Dr. John A. Rupley.)

trated in Figure 16-3, along with key functional groups that are part of the active site. Lysozyme, found in most body secretions, kills bacteria by breaking the bonds that hold together the polysaccharides of their cell walls. The hydrogen atom of a glutamic acid (amino acid number 35) "attacks" the oxygen atom that bridges two monosaccharide units in the substrate.

In general, for a chemical reaction to take place, the molecules involved must have a certain amount of energy, called the **activation energy**. *A catalyst*

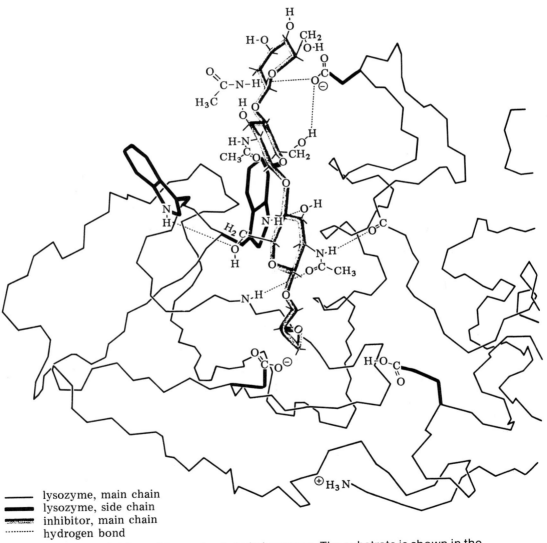

——— lysozyme, main chain
▬▬▬ lysozyme, side chain
▬▬▬ inhibitor, main chain
·········· hydrogen bond

Figure 16-3. The folding of the main chain in lysozyme. The substrate is shown in the crevice that forms the active site of the enzyme. (Courtesy of Dr. Nathan Sharon.)

such as an enzyme provides a reaction pathway that has a smaller energy of activation. As illustrated in Figure 16-4, by lowering the energy barrier or "hill" for the reaction, more molecules can react. Thus, the enzyme increases the rate of the reaction, producing more product in a shorter period of time. Enzymes are able to catalyze the reaction of millions of substrate molecules per minute.

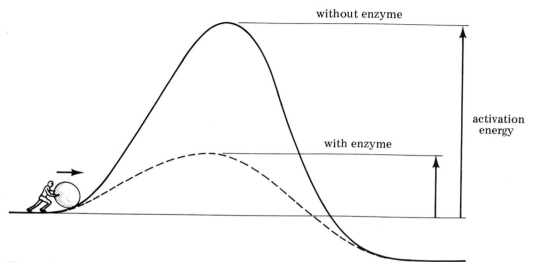

Figure 16-4.   A representation of the lowering of the activation energy by an enzyme. By lowering the energy of activation, the enzyme speeds up the rate of the reaction.

16.2   Classification of enzymes

Enzymes are named on the basis of either their substrate or the type of reaction they catalyze. In most cases, the name ends in "-ase." For example, a lipase is an enzyme whose substrate is a lipid. An oxidase is an enzyme that catalyzes a reaction involving oxidation of the substrate.

As shown in Table 16-1, enzymes are divided into six major categories. **Oxidoreductases** catalyze oxidation–reduction reactions, which involve the loss or gain of electrons (or hydrogen atoms). **Transferases** catalyze the transfer of a functional group, such as an amino group, from one molecule to another. **Hydrolases** catalyze hydrolysis, the reaction of a substrate with water. **Lyases** catalyze the breaking (or making) of double bonds, such as between two carbon atoms, a carbon atom and an oxygen atom (the carbonyl group), or a carbon atom and a nitrogen atom. **Isomerases,** as their name implies, catalyze isomerization reactions, such as the conversion of glucose to fructose. **Ligases** cat-

Table 16-1   **Classification of Enzymes**

Class	Examples	Function
oxidoreductases (catalyze oxidation–reduction reactions)	dehydrogenase oxidase	removes hydrogen adds oxygen
transferases (catalyze transfer of functional groups)	methyltransferase transaminase	transfers methyl group transfers amino group
hydrolases (catalyze hydrolysis reactions)	carbohydrase esterase lipase phosphatase protease	cleaves carbohydrates cleaves ester linkages cleaves lipids cleaves phosphate esters cleaves peptide bonds
lyases (catalyze breaking of double bonds)	decarboxylase aldolase	removes CO_2 breaks aldehyde bond
isomerases (catalyze isomerizations)	cis-trans isomerase mutase	converts cis and trans forms intramolecular transfer
ligases (catalyze bond formation)	synthetase carboxylase	combines 2 groups forms C—C bond

alyze the formation of a bond between a carbon atom and another atom—oxygen, sulfur, nitrogen, or a second carbon.

The name of the enzyme gives you an indication of how specific it is. For example, urease is an enzyme with only one substrate, urea; even though they may have similar structures, the hydrolysis of related molecules is not catalyzed by urease. Most enzymes are somewhat less specific. They catalyze certain types of reactions, such as those given as examples in Table 16-1. Some act on specific groups. A decarboxylase removes carboxyl groups, and a methyltransferase transfers methyl groups. Others catalyze reactions involving specific bonds, such as a peptidase, which breaks or cleaves peptide bonds. Enzymes can also recognize the difference between optical isomers; an enzyme specific for L-amino acids has no effect on the reactions of the D enantiomers. Thus, in addition to being very efficient, enzymes are selective for only certain substrates.

16.3 Factors affecting enzyme activity

Various factors influence the activity of an enzyme and therefore the rate of reaction. If you increase the *concentration of substrate* molecules, for example, the rate goes up. There is a limit however, at which point the enzyme becomes "saturated," causing the rate to level off at a maximum value. The rate also depends to a certain extent on the *concentration of enzyme*, even though only a small amount is required.

An enzyme catalyzes both the forward reaction, the conversion of substrate molecules into products, and the reverse reaction, the conversion of products back to the substrate, if the chemical change is reversible. When the rates of both these reactions are the same, a balance or equilibrium exists. The enzyme does not affect the relative amounts of unreacted substrate and product formed at equilibrium; it only speeds up the process of reaching this condition. But just as increasing the substrate concentration favors the formation of more product, the *buildup of product* causes an increase in the reverse reaction and a slowdown in the forward rate of reaction.

The *temperature* generally increases the rate of reaction by about 1.1 to 3 times for every rise of 10°C. But enzymes have an optimum temperature, at which their activity is greatest. The normal body temperature, 37°C (98.6°F), is near this "best" temperature for most enzymes. Since they are proteins, enzymes denature at high temperatures, such as at 60°C or above.

Enzyme activity also depends on the *pH* of the surrounding fluid. The activity is greatest at an optimum pH, which is often between pH 6 and 8. A notable exception is pepsin, the enzyme of the gastric juice in the stomach. Because the optimum pH of pepsin is about 2, it functions effectively in this acidic fluid.

16.4 Enzyme cofactors

Some enzymes need an extra nonprotein part, called a **cofactor**, for their activity. Neither the protein part alone (the apoprotein) nor the cofactor alone can act as catalyst in these cases; only their combination (the holoenzyme) can serve this function. When the cofactor is firmly bound to the protein, it is known as a **prosthetic group**.

If the cofactor is an organic molecule, it is called a **coenzyme**; as described in Chapter 21 many of the vitamins you eat become coenzymes. Some examples of coenzymes and their functions are listed in Table 16-2. Coenzymes often transfer atoms or electrons to or from the substrate molecule. They are discussed in greater detail in the next chapter.

Table 16-2 **Coenzymes**

Coenzyme	Function	Related vitamin
flavin adenine dinucleotide (FAD)	hydrogen carrier	riboflavin
flavin mononucleotide (FMN)	hydrogen carrier	riboflavin
nicotinamide adenine dinucleotide phosphate (NADP)	hydrogen carrier	nicotinic acid
pyridoxal phosphate	amino group carrier	pyridoxine
tetrahydrofolic acid	one-carbon carrier	folacin
thiamin pyrophosphate	aldehyde carrier	thiamin

Some enzymes cannot function without an **inorganic activator**, a metal ion. The cation, whether magnesium, zinc, manganese, cobalt, iron, copper, or molybdenum, is generally located at the active site. Enzymes requiring the presence of a metal ion are known as **metalloenzymes**; examples are listed in Table 16-3. Some enzymes are active only when a specific anion is present, such as the chloride ion in the case of amylase.

Table 16-3 **Metalloenzymes**

Enzyme	Metal	Reaction
alkaline phosphatase	Zn	phosphate ester hydrolysis
carboxypeptidase	Zn	peptide hydrolysis
cytochrome *c*	Fe	electron transfer
cytochrome oxidase	Cu, Fe	oxidation
glycol dehydrase	Co	rearrangement
pyruvate oxidase	Mn	oxidation
xanthine oxidase	Fe, Mo	oxidation

16.5 Inhibitors

Certain substances known as **inhibitors** prevent or slow down the action of an enzyme. **Competitive inhibitors** are molecules that have structures similar to that of a substrate. They bind *reversibly* at the active site, stopping the enzyme from catalyzing the reaction involving the substrate molecule as shown in Figure 16-5. For example, in cases of methyl alcohol poisoning, the victim is given ethyl alcohol, which has a closely related structure. The ethyl alcohol competes with the poison for the active site of the enzyme alcohol dehydrogenase, "tying" it up and preventing the formation of toxic products from methyl alcohol.

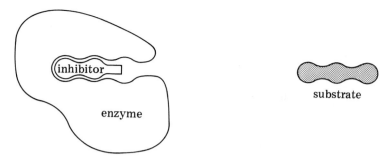

Figure 16-5. Competitive inhibition. Because the inhibitor has a shape similar to that of the substrate, it can bind (reversibly) to the active site, preventing the substrate from entering.

An important class of compounds, the **sulfa drugs**, acts through competitive inhibition of a key enzyme needed by bacteria for growth and reproduction. The chemical structures of these sulfonamides resemble that of a substrate of the bacterial enzyme, *p*-aminobenzoic acid; this molecule is a metabolite, an essential substance needed by the bacteria to produce folic acid. The sulfona-

$$H_2N-\!\!\!\bigcirc\!\!\!-\overset{\overset{\displaystyle O}{\|}}{C}-OH \qquad H_2N-\!\!\!\bigcirc\!\!\!-\overset{\overset{\displaystyle O}{\|}}{\underset{\underset{\displaystyle O}{\|}}{S}}-NH_2$$

p-aminobenzoic acid

a sulfonamide

mides are known as **antimetabolites** because they prevent the enzymes of the bacteria from synthesizing folic acid, which is required for their growth, by acting as inhibitors for the metabolite *p*-aminobenzoic acid. Examples of the sulfa drugs are shown in Figure 16-6; their basic structures are similar, but the different end groups make them active against different strains of bacteria.

Noncompetitive inhibitors bind reversibly to an enzyme at a place *other* than in the active site, changing the structure so that it cannot perform its function. The most common examples are heavy-metal ions, such as Hg^{2+}, Ag^+, or Cu^{2+}, which combine with the sulfhydryl groups, —SH, of the amino acid cysteine. Cyanide ion, CN^-, acts as a poison by binding to iron and thus inhibiting enzymes involved in respiration.

An **irreversible inhibitor** permanently modifies or destroys part of an enzyme. Certain organic compounds containing phosphorus are nerve poisons because they combine with and inactivate the enzyme acetylcholinesterase, which is needed for the functioning of the nervous system. Some of these irreversible inhibitors are used as insecticides.

One of the most important applications of inhibition in your body is the regulation of enzyme reactions. In many chemical processes, as you will see in the next chapter, a series of enzymes carries out several reactions, one after

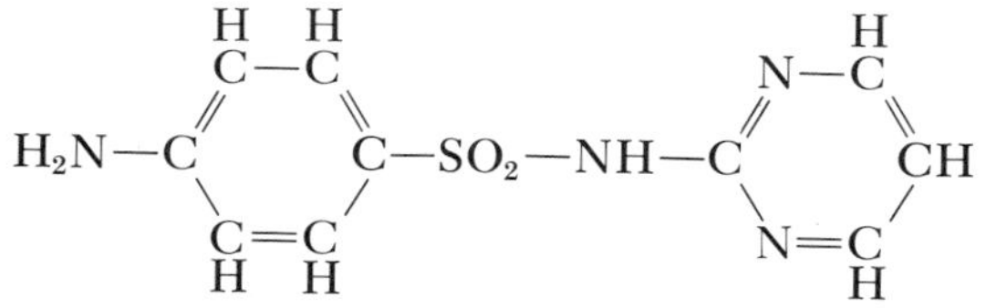

sulfadiazine

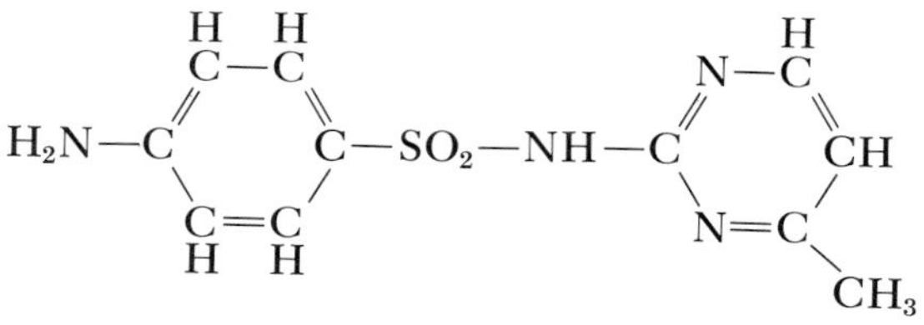

sulfamerazine

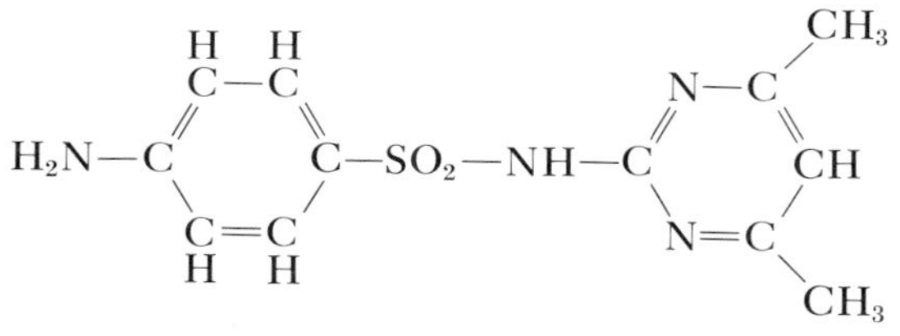

sulfamethazine

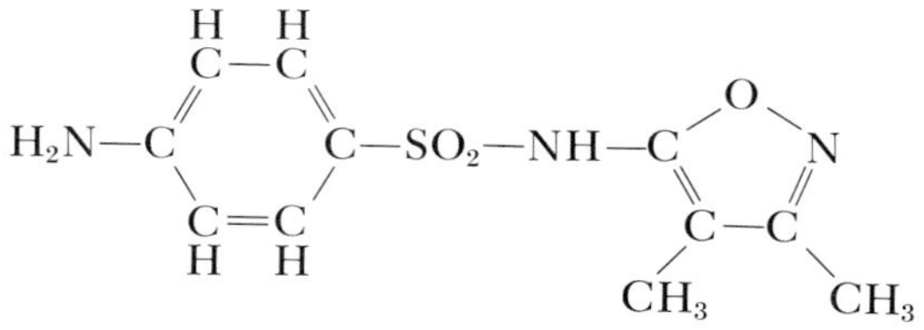

sulfisoxazole

Figure 16-6. Examples of sulfa drugs. These molecules are sulfonamides: they are competitive inhibitors for *p*-aminobenzoic acid, a bacterial metabolite.

the other. The product of the last enzyme in the chain acts in many cases as an inhibitor of the first enzyme, as shown in Figure 16-7. Thus, when enough product has been synthesized, the whole system "shuts down" because the initial step is inhibited. This process is also called **feedback inhibition**, and the enzyme being controlled is called a regulatory (or allosteric) enzyme. Of course, once your body has used up the supply of product, it is no longer present to act as an inhibitor, and the series of enzymes can begin functioning again.

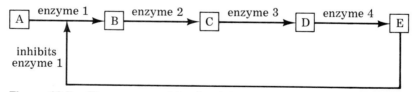

Figure 16-7. The process of feedback inhibition. In this series of reactions, A is converted to B, B to C, C to D, and D to E. The end product, E, inhibits the enzyme that catalyzes the initial step (A to B).

16.6 Applications of enzymes in medicine

Aside from their applications in industry and food processing, enzymes are used both for medical diagnosis and for treatment. *The presence of a specific enzyme in a patient's blood (or sometimes urine) may confirm the presence of disease.* If tissues are damaged, as in the heart or kidney, tissue enzymes are lost and appear in the blood. In certain conditions such as cancer, the production of enzymes may increase, causing them to "spill over" into the blood. Also, liver disease can block the normal elimination of proteins and cause a buildup in their concentration.

In hepatitis, damage of the liver causes an increase in the level of the enzyme glutamic-oxalacetic transaminase (and also glutamic-pyruvic transaminase). Analyzing a blood specimen for this enzyme is considerably easier and safer than examining a sample of liver tissue from the patient. In a heart attack (myocardial infarction), the blood level of the enzyme creatine phosphokinase, as well as glutamic-oxaloacetic transaminase, increases after 1 or 2 days because of destruction of heart tissue. These analyses are especially useful in cases in which interpretation of the patient's electrocardiogram (ECG) is difficult. Other enzymes and the possible conditions causing their increased concentration in blood are listed in Table 16-4.

Table 16-4   Enzymes Used in Diagnosis

Enzyme	Possible diagnosis
acid phosphatase	cancer of prostate
alkaline phosphatase	liver or bone disease
amylase	pancreatic or kidney disease, ulcer
glucose-6-phosphate dehydrogenase	blood hemolysis diseases
lactic dehydrogenase	heart attack, hepaptitis, cirrhosis of liver

The enzyme hyaluronidase is used medically as a spreading agent to help the absorption of injected drugs or fluids by hydrolyzing a complex polysaccharide of connective tissue, hyaluronic acid. Two enzymes, streptodornase and streptokinase, help to liquefy coagulated blood, remove clots, dissolve fibrous material, and eliminate pus accumulations after infection or injury. Trypsin, which hydrolyzes peptide bonds, acts on clotted blood and dead tissue to liquefy hematomas, such as bruises or black eyes. Chymotrypsin works in a similar way in the treatment of inflammations that cause pain and the escape of fluid from bruises and fractures. Because certain cancer cells need the amino acid asparagine for growth, the enzyme L-asparaginase is used in controlling some cancers, such as certain types of leukemia. It destroys this amino acid present from the diet, slowing the growth of the cancer cells; normal cells can make their own asparagine.

16.7 Enzymes and nerve impulses

Because enzymes carry out a special function called **active transport**, your cells contain a high potassium ion concentration and a low sodium ion concentration. This process is referred to as the **sodium–potassium pump** because potassium ions are "pumped" in and sodium ions are "pumped" out by enzymes in the cell membrane. The resulting difference in the concentration of the two cations creates what is known as an electrical "potential" (difference in electrical energy capable of doing work) across the membrane, giving the inside a negative charge compared to the outside, as shown in Figure 16-8.

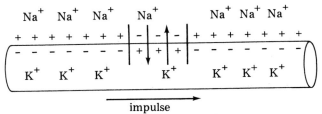

Figure 16-8. A representation of the process of conduction along a nerve cell. See text for details.

This charge difference (polarization) is the basis for **conduction**, the passage of impulses along a nerve cell, or neuron. When a nerve is "excited," the membrane changes, briefly allowing sodium ions to enter, which alters the electrical potential. This local change takes place many times along the nerve cell, carrying the signal (action potential) along its entire length, as each region "excites" the next in this way. Thus, changes in the permeability of the neuron membrane and the resulting movement of ions between the fluids on either side conduct the nerve impulse.

Chemical substances called **neurotransmitters** send on the electric signal after it reaches the end of a neuron. These molecules bring the "message" across the gaps, called synapses, between nerve cells and muscle cells, for example. A molecule, such as acetylcholine, is released from storage by the process of "excitation." It crosses the synapse to interact with the cells on the other side. After the signal is transmitted, the neurotransmitter is destroyed by an enzyme in the fluid, acetylcholinesterase in this case, restoring the "resting" condition. Compounds that inhibit this enzyme, such as phosphorus-containing insecticides, are very potent poisons.

16.8 Enzymes of carbohydrate digestion

Digestion is the process by which the complex molecules of the food you eat are broken down into simpler molecules. It takes place at various points in the digestive system (Figure 16-9). The end products of carbohydrate, lipid, and protein digestion are absorbed through the intestinal walls into your blood. Some substances, such as inorganic salts and vitamins, require no digestion; those substances that cannot be digested, like cellulose, are excreted by the bowels in the feces.

Enzymes are the basis for digestion. They catalyze hydrolysis of the food molecules, the reaction with water that breaks chemical bonds, forming smaller molecules. The juices secreted by organs along the digestive tract—saliva, gastric juice, pancreatic juice, and intestinal juice—all contain enzymes.

About 60% of your diet, about 400 to 500 g/day, consists of carbohydrates. Of this amount, approximately two-thirds consists of starch (and dextrins), and one-third is sucrose. The digestion of the large carbohydrate molecules begins in your mouth. Saliva contains the enzyme alpha-amylase, or ptyalin, which hydrolyzes many of the acetal bonds of the amylopectin and amylose molecules that make up starch. The products of this first step in digestion are lower molecular weight polysaccharides and maltose. When you chew on a cracker, its taste becomes sweet because of the sugars formed by this partial hydrolysis of starch.

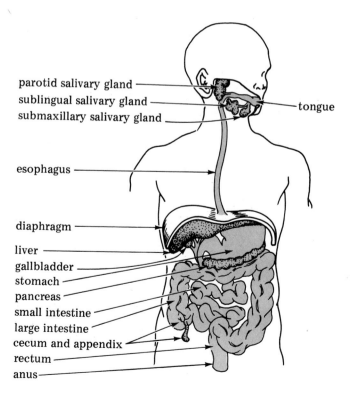

Figure 16-9. A diagram of the digestive system.

parotid salivary gland

sublingual salivary gland

submaxillary salivary gland

tongue

esophagus

diaphragm

liver

gallbladder

stomach

pancreas

small intestine

large intestine

cecum and appendix

rectum

anus

Little digestion of carbohydrates takes place in the stomach. Instead, the enzyme pancreatic amylase or amylopsin continues to break down starch in the small intestine. Final digestion is accomplished by sucrase, lactase, maltase, and isomaltase. The major end products are glucose (80%), fructose (10%), and galactose (10%), which are then absorbed into the bloodstream at different rates, with galactose and glucose being absorbed most rapidly. In most persons, the digestion and absorption of carbohydrates that can be used by the body is 95% complete.

16.9 Enzymes of lipid digestion

You eat about 200 g of lipids daily, mostly in the form of triacylglycerols (triglycerides). They pass unchanged through the mouth and stomach. Their presence in the stomach, however, slows down the rate at which it empties, making you feel more "full."

The major site of lipid digestion is the small intestine, which contains a pancreatic lipase, steapsin. This enzyme hydrolyzes at least 50 to 75% of the dietary fat, breaking the ester linkages that connect the fatty acids to the first and third carbons of glycerol. The resulting monoacylglycerol (monoglyceride), with a fatty acid at the middle carbon atom of glycerol, is cleaved by an esterase. Phospholipids are also digested in the intestine by such enzymes as phosphatases and phospholipases.

Before they can be digested by these enzymes, lipids must first be emulsified by **bile**, a secretion from the gallbladder. This fluid contains the bile salts sodium glycocholate and sodium taurocholate. They are salts formed by the steroid cholic acid coupled by an amide linkage to glycine, as shown below, or to taurine ($NH_2CH_2CH_2SO_3H$). These salts lower the surface tension of the intestinal fluid, helping to emulsify lipids like a detergent. Because bile is

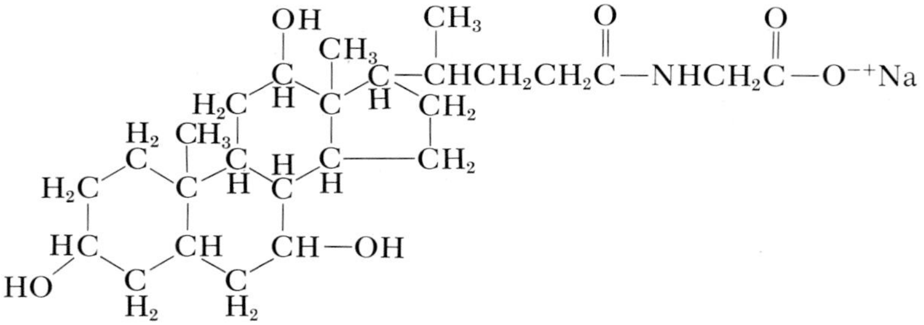

sodium glycocholate

basic, it neutralizes chyme, the acidic mixture of partially hydrolyzed food and gastric juice from the stomach, aiding the digestive processes of the small intestine.

The final products of lipid digestion are primarily free fatty acids (70%) and monoacylglycerols (monoglycerides, 25%). They are absorbed and resynthesized into triacylglycerols (triglycerides), which then reach the bloodstream by way of the lymph system.

16.10 Enzymes of protein digestion

Proteins make up about 100 g of your diet each day. Unlike carbohydrates and lipids, they are digested to a large degree in the stomach. Gastric juice, the digestive fluid of the stomach, contains 0.5% (0.1 M) hydrochloric acid, having a pH of about 1. The presence of a strong acid aids the digestion of proteins by causing them to denature, exposing their peptide bonds.

Hydrochloric acid also activates pepsinogen. This molecule is a **zymogen**, the inactive form of an enzyme, pepsin. The acid promotes removal of 42 amino acids from pepsinogen, freeing the active site. If pepsin were produced initially in an active form, the enzyme would digest the glands that released it. Pepsin hydrolyzes peptide bonds, forming polypeptides of various lengths from the protein molecule. An important feature of pepsin is its ability to digest collagen, the protein of connective tissue.

The major digestion of proteins then takes place in the small intestine. The pancreatic juice provides a zymogen, trypsinogen. It is converted into its active form, trypsin, by another enzyme, enterokinase (from the intestinal mucous membranes). Trypsin further hydrolyzes the products of protein digestion from the stomach into smaller polypeptides. The enzyme chymotrypsinogen is activated by trypsin. This enzyme produces still shorter polypeptides and amino acids. Carboxypeptidase, a zinc metalloenzyme, hydrolyzes dipeptides and splits the last amino acid (at the carboxyl end) from a longer polypeptide chain. Elastase, also released as a zymogen, attacks elastin as well as other types of protein. The roles of each of these enzymes are complementary—those bonds not hydrolyzed by one are split by another enzyme.

The process of protein digestion is completed by enzymes secreted by the small intestine. Dipeptidases and aminopeptidases produce the end products, amino acids, which are then absorbed into the blood. The digestion of proteins into amino acids is 95% complete in most individuals. Figure 16-10 presents a summary of the digestive process.

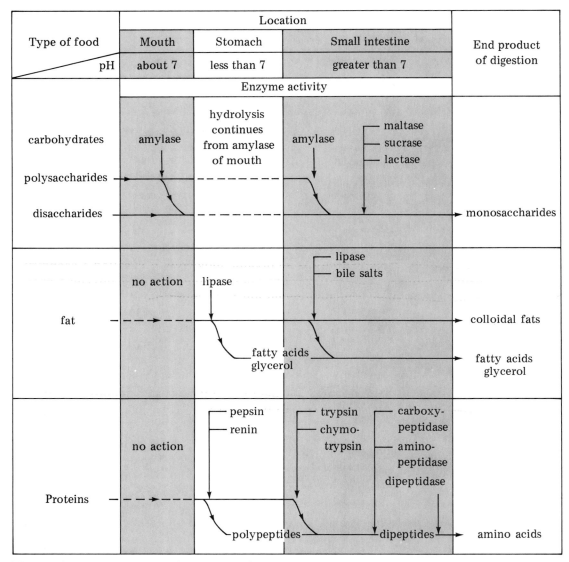

Type of food	Location			End product of digestion
	Mouth	Stomach	Small intestine	
pH	about 7	less than 7	greater than 7	

Figure 16-10. A summary of digestion. (From V. L. Parsegian, P. R. Shilling, F. V. Monaghan, and A. S. Luchins, "Introduction to Natural Science, Part 2: The Life Sciences." Academic Press, New York, 1970. Reprinted with permission.)

SUMMARY

Enzymes are biological catalysts, molecules that increase the rate of a reaction but are not used up in the process. Without enzymes, the vital chemical reactions in the cells would take place too slowly to support life.

The molecule that is changed by a chemical reaction catalyzed by an enzyme is the substrate. It binds to the enzyme at the active site, where a chemical reaction takes place. The product is released, freeing the enzyme to repeat the process. Enzymes provide a way for reactions to take place with a lower activation energy.

Enzymes are divided into six major categories: oxidoreductases, transferases, hydrolases, lyases, isomerases, and lipases. The name of the enzyme, based on either its substrate or type of reaction catalyzed, gives you an indication of how specific its activity is. Enzymes are very selective in addition to being efficient.

Various factors influence the activity of an enzyme and therefore the rate of reaction. They include concentration of substrate, concentration of enzyme, temperature, and pH. Enzymes catalyze both the forward and reverse reactions of a reversible process.

Some enzymes require an additional nonprotein molecule, a cofactor, for their activity. If the cofactor is an organic molecule, it is called a coenzyme; many vitamins become part of coenzymes. A large number of enzymes, known as metalloenzymes, contain a metal ion as an inorganic activator.

Certain substances called inhibitors prevent or slow down the action of an enzyme. Competitive inhibitors are molecules that have structures similar to that of the substrate and bind reversibly to the active site. The sulfa drugs resemble a substrate of bacterial enzymes and prevent growth of the bacteria by competitive inhibition. Noncompetitive inhibitors bind reversibly at places other than the active site; irreversible inhibitors permanently modify or destroy part of an enzyme. One of the most important applications of inhibition is the regulation of enzyme reactions by a feedback process.

Enzymes are used for both medical diagnosis and treatment. The presence of an enzyme in the blood may confirm the presence of disease. Certain enzymes help to absorb injected fluids and to promote the healing process after injury.

Enzymes are involved in the process of conducting nerve impulses. Through active transport, they maintain the electrical potential of nerve cells. In addition, enzymes are required to inactivate the neurotransmitter molecules that carry the impulse across the synapses between cells.

Digestion is the process by which the complex molecules of the food you eat are broken down into simpler molecules. Enzymes are the basis for digestion. They catalyze hydrolysis of the food molecules, forming smaller molecules that can be absorbed into the blood through the intestinal walls.

Digestion of carbohydrates occurs mainly in the small intestine, producing glucose, fructose, and galactose. Lipid digestion also occurs there; it is aided by the bile salts and results in free fatty acids and monoacylglycerols (monoglycerides). The digestion of proteins takes place in the stomach and small intestine, forming amino acids. The inactive forms of several enzymes, called zymogens, must first be activated for protein digestion to occur.

Exercises

1. (Intro.) What is an enzyme? Why are enzymes so important?
2. (16.1) Briefly describe how an enzyme works.
3. (16.1) Why are only small amounts of an enzyme required for catalysis?
4. (16.2) Describe the substrate or reaction catalyzed by the following enzymes: (a) sucrase; (b) dehydrogenase; (c) phosphatase; (d) lipase.
5. (16.2) What are the six major categories of enzymes?
6. (16.3) Describe the factors that influence the activity of an enzyme.
7. (16.4) Define cofactor, prosthetic group, coenzyme, inorganic activator.
8. (16.5) For each of the following, give an example and explain: (a) competitive inhibitor; (b) noncompetitive inhibitor; (c) irreversible inhibitor.
9. (16.5) How do the sulfa drugs work?
10. (16.5) Describe the process of feedback inhibition.
11. (16.6) A patient has an abnormally high level of glutamic-oxaloacetic transaminase in the blood. What might this mean? Explain.
12. (16.6) Describe the use of two enzymes in medical treatment.
13. (16.7) Describe the role of enzymes in nerve impulse conduction.
14. (16.7) What is a neurotransmitter?
15. (16.8) In general terms, what takes place during digestion?
16. (16.8–16.10) For each of the following describe the starting molecules and final molecules formed by complete digestion: (a) carbohydrates; (b) lipids; (c) proteins.
17. (16.8–16.10) Describe the reactions that take place during digestion in the (a) mouth; (b) stomach; (c) small intestine.
18. (16.9) What is the role of bile salts during digestion?
19. (16.10) Why is it important that pepsin, trypsin, and chymotrypsin are secreted as zymogens?

Energy and carbohydrate metabolism

In a real sense, you are what you eat. The atoms that make up your molecules come from food: carbohydrates, lipids, proteins, and other nutrients. Each year you ingest about 1400 pounds of food; part ends up as the macromolecules of your cells and part serves as "fuel" to run your chemical reactions. Through food you obtain both matter and energy from the surroundings. The series of complex chemical steps by which your enzymes perform this transfer is called **metabolism**. It includes all of the chemical processes that take place in your body.

17.1 Biochemical energy

The sun is the primary source of energy for all life. Through the process of **photosynthesis**, plants use solar energy to convert carbon dioxide, CO_2, and water into organic molecules such as the carbohydrate glucose, $C_6H_{12}O_6$:

$$6CO_2 + 6H_2O \xrightarrow{\text{light}} C_6H_{12}O_6 + 6O_2$$

In a complex series of reactions, *the energy from light is converted to chemical energy in the form of new bonds in a carbohydrate molecule.* Because energy is absorbed in this process, it is called **endergonic**. In general, energy is required to synthesize a more complex molecule like glucose from a simpler molecule such as carbon dioxide.

The chemical energy present in carbohydrates produced by plants can be used by higher organisms that eat them. Carbohydrates release the energy contained in their bonds in a series of chemical reactions. This process is **exergonic**—energy is given off. The carbohydrates serve as "fuels"; they are chemically converted to simpler molecules, releasing their energy.

You know that one way to obtain energy is by combustion, burning a fuel in the presence of oxygen. The energy that is liberated by this process exists mainly as heat. Combustion, however, cannot take place in your body. The heat produced by the release of a large amount of energy in one step would destroy the cells. Instead, the energy must be given off in a "flameless" process in a series of small steps.

Imagine a rock on top of a steep hill; it has much potential energy, as does a glucose molecule. If you push it over the edge, the rock releases a large amount of energy at the instant it hits the ground. On the other hand, if it rolls down in steps, as shown in Figure 17-1, energy is given off in smaller units. The total amount of energy released is the same in both cases, but the stepwise process provides energy in a more useful manner as far as your body is concerned.

Figure 17-1. Glucose as a high-energy compound. The energy from sunlight converts carbon dioxide and water to glucose in plants. This energy is released when glucose is broken down in the human body.

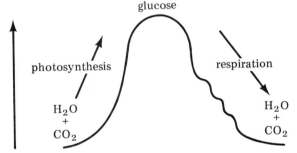

17.2 Electron transfer

Energy is released from carbohydrates by the process of **oxidation**. In a series of reactions, each catalyzed by an enzyme, *electrons are removed from the bonds of carbohydrate molecules.* The oxidizing agent, the final acceptor of electrons, is oxygen, O_2. The process by which electrons are transferred from one compound to another until they reach oxygen is called **electron transport**.

Certain coenzymes, the nonprotein substances that enzymes require for activity, serve as **electron carriers** in this process. One of the most important is nicotinamide adenine dinucleotide (NAD) (Figure 17-2). It can exist in two forms—an oxidized form, in which it can accept an electron, and a reduced form, from which it can donate an electron. They differ only in the pyridine ring of the nicotinamide portion of the molecule. By gaining two electrons (as well as a hydrogen ion) from another molecule, NAD is converted to NADH.

Nicotinamide adenine dinucleotide acts as a coenzyme for oxidation–reduction enzymes called dehydrogenases. These enzymes catalyze the transfer of electrons from the substrate to the coenzyme, the electron carrier. The substrate becomes oxidized in the process and energy is released.

Other coenzymes are employed for electron transport in oxidation–reduction reactions. Nicotinamide adenine dinucleotide phosphate (NADP), a mol-

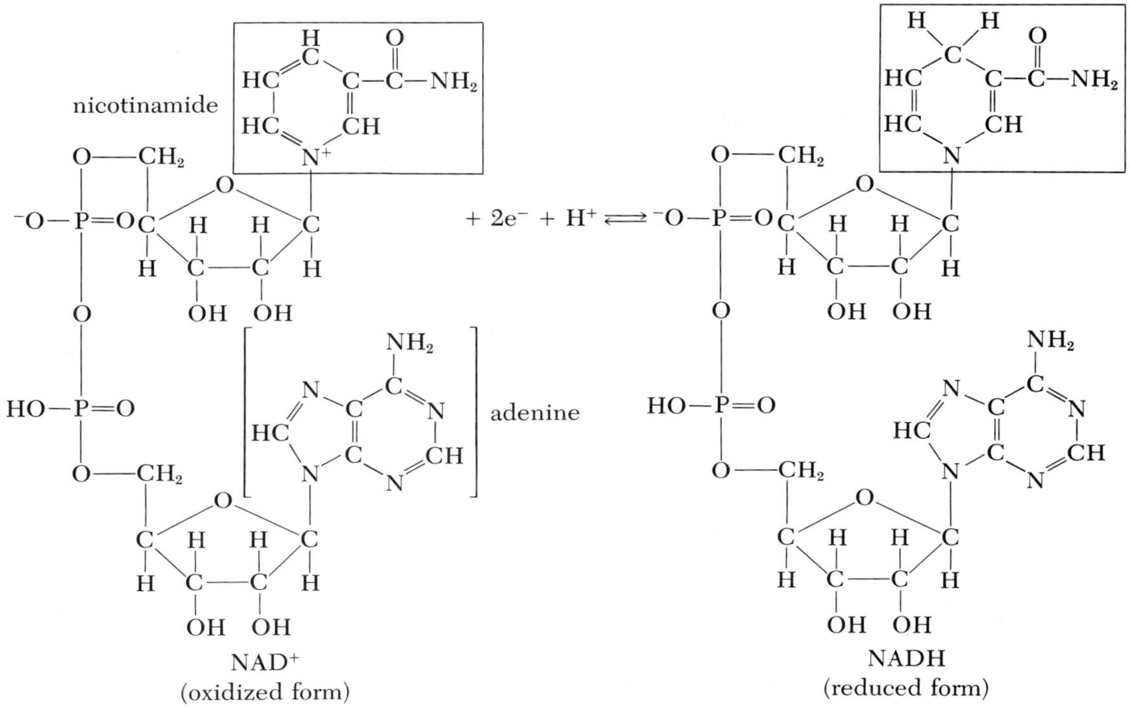

Figure 17-2. The structure of nicotinamide adenine dinucleotide (NAD). The oxidized form (NAD^+) and the reduced form (NADH) differ only in the nicotinamide portion of the molecule. (The $2e^-$ and H^+ can be thought of as H^-, a hydride ion.)

ecule similar to NAD, flavin adenine dinucleotide (FAD), and flavin mononucleotide (FMN) also work with dehydrogenases. These coenzymes are summarized in Table 17-1. A group of enzymes known as cytochromes can also transfer electrons. They contain iron in a heme group, which can readily pick up electrons (as Fe^{3+}) or lose them (as Fe^{2+}). All of these carriers eventually transport the electrons picked up during oxidation to molecular oxygen, O_2, which gets reduced to water.

Table 17-1   Oxidized and Reduced Forms of Coenzymes

Oxidized form (can accept electrons and hydrogen)	Reduced form (can donate electrons and hydrogen)
NAD^+	NADH
$NADP^+$	NAPH
FAD	$FADH_2$
FMN	$FMNH_2$

17.3 Storage and transfer of energy

The energy released by the oxidation of carbohydrate molecules must be stored in a form that can be used by the cells when needed. **Adenosine triphosphate** (ATP), illustrated in Figure 17-3, serves this purpose. It consists of

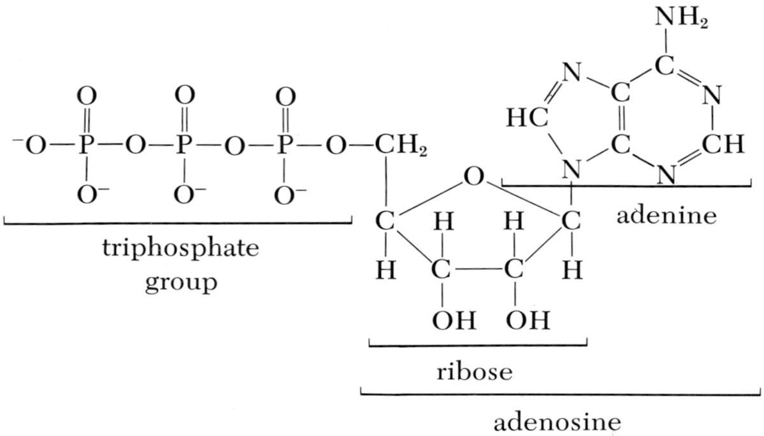

Figure 17-3. The structure of adenosine triphosphate (ATP).

a monosaccharide, ribose, to which a nitrogen heterocycle, adenine, is attached, forming a glycoside. The ribose also contains a triphosphate group attached by an ester linkage as shown.

Energy is stored when ATP forms from the addition of inorganic phosphate, $H_2PO_4^-$ (P_i), to adenosine diphosphate (ADP):

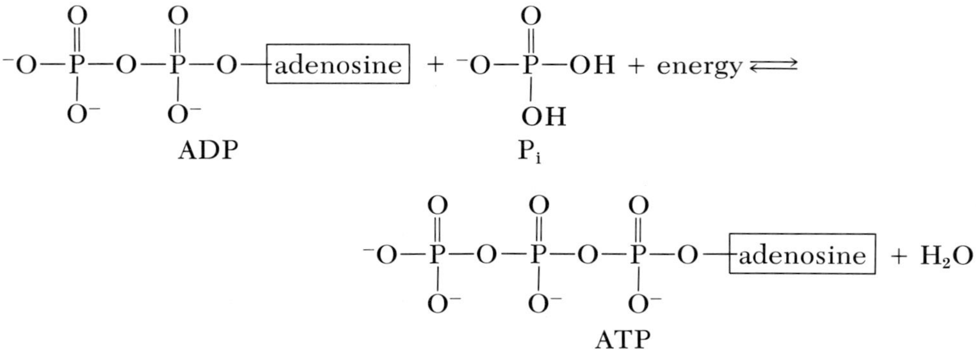

Adenosine triphosphate is a high-energy molecule; the energy absorbed in this endergonic reaction is stored in the chemical bonds of ATP. This process is called **oxidative phosphorylation**. When the energy is needed, the last phosphate group can be hydrolyzed in the reverse reaction:

$$ATP + H_2O \longrightarrow ADP + P_i + energy$$

The energy stored in ATP is then released as the bond connecting the last phosphate is broken. The ATP serves as a "go-between," picking up the energy given off during oxidation and transferring it to the processes in the cell that require work to be done (Figure 17-4).

Figure 17-4. The relationship between adenosine triphosphate (ATP) and adenosine diphosphate (ADP). Energy is stored when ATP is formed; it is released as ATP is converted to ADP.

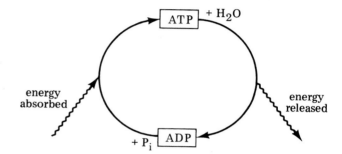

In general, the formation of a phosphate bond requires energy, and its hydrolysis releases energy. This principle is applied in many metabolic reactions in a process called **energy coupling**. Key endergonic reactions, which need energy, are linked with exergonic reactions, which give off the energy. Energy coupling can be accomplished by forming the phosphate ester of a substrate, using ATP to provide the energy and the phosphate group. This substrate is a high-energy compound as is ATP. Now its reaction to form products is "driven" by the energy released from breaking the new phosphate bond.

17.4 Uses of energy in the body

So far, you have learned about only one-half of the metabolic process, **catabolism**, which is the breaking down of complex molecules into simpler ones, releasing energy. The other closely related part is **anabolism**, the synthesis of large molecules needed by your cells using the small molecules and energy provided by catabolism. Thus, one of the major uses of ATP is to "run" the reactions of anabolism.

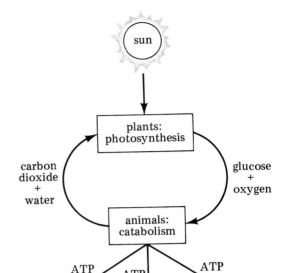

Figure 17-5. The source and uses of ATP.

In addition to this chemical work, you need energy to do mechanical work, such as contracting muscles. Transport work, like conducting nerve impulses and secreting urine, also requires energy, transferred by means of ATP molecules. The source and uses of ATP are summarized in Figure 17-5. Excess energy not used in any of these processes is given off as heat, part of which is used to keep your body temperature constant.

The amount of heat you give off 12 to 15 hours after the last meal under resting conditions is your **basal metabolism**. The basal metabolism accounts for about one-half the total energy output over a 24-hour period for a relatively inactive person. The heat released reflects the energy needs of the body's continuing processes, such as cell metabolism and the muscle contraction of breathing and blood circulation. It varies with the size of the body (its surface area), as well as with a person's sex, age, diet, and physical condition. An increased *basal metabolic rate* (BMR) may result from pregnancy, hyperthyroidism, infection, leukemia, congestive heart failure, or bronchial obstruction. Conditions such as malnutrition, shock, kidney disease, or hypothyroidism cause a decrease. One way of measuring the basal metabolic rate is shown in Figure 17-6; another method is based on a blood test (PBI, protein-bound iodine).

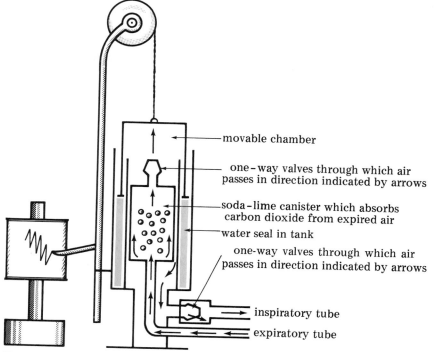

movable chamber

one-way valves through which air passes in direction indicated by arrows

soda-lime canister which absorbs carbon dioxide from expired air

water seal in tank

one-way valves through which air passes in direction indicated by arrows

inspiratory tube

expiratory tube

Figure 17-6. A respirometer for measuring the basal metabolic rate (BMR).

17.5   Glycolysis

The first steps in the catabolism of carbohydrates consist of a process that occurs in the *absence of oxygen,* anaerobic fermentation. It is called **glycolysis,** meaning the breakdown of sugar, and is a pathway or series of reactions for obtaining energy from fuels like glucose. Each step in glycolysis is catalyzed by a different enzyme.

This pathway (also known as the Embden–Meyerhof pathway) is outlined in Figure 17-7. All compounds exist in their normal ionized form. The sugar molecules have the naturally occurring D configuration and are drawn as the alpha arrangement (where anomers are possible). The details of each reaction, including the enzymes involved, are described in Appendix D.1. In this section, only the major changes are described.

In the first part of glycolysis (steps 1 to 4), a six-carbon glucose molecule is broken into 2 three-carbon molecules of glyceraldehyde. This process uses up two molecules of ATP in order to make high-energy phosphate esters of the sugar molecules. The most important reaction (step 5) is the oxidation of each glyceraldehyde to an acid, giving two electrons to an NAD coenzyme.

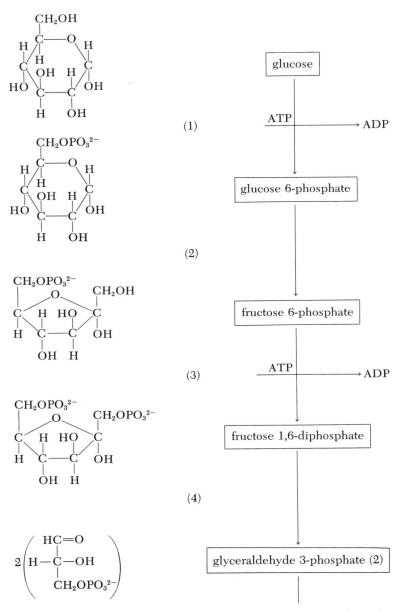

Figure 17-7. The glycolysis pathway. In this process each molecule of glucose is broken down into two molecules of lactic acid. See Appendix D.1 for a detailed description of each reaction.

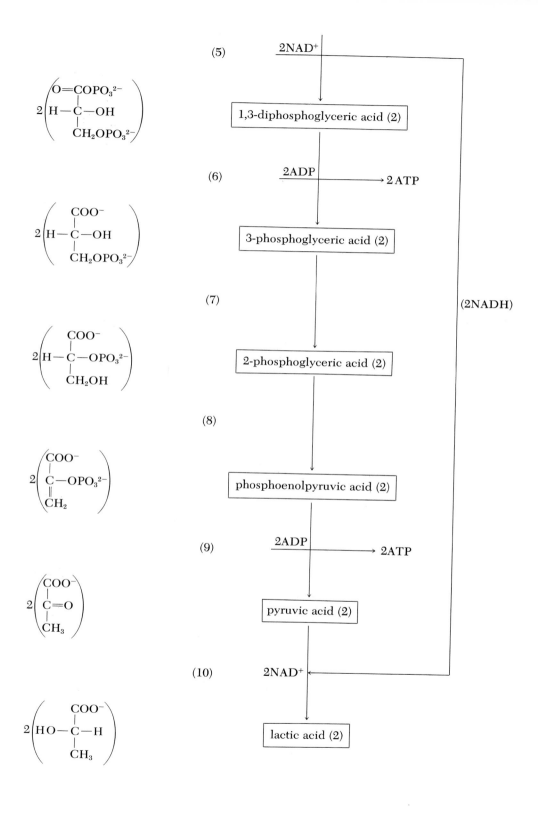

The large amount of energy released is stored by the formation of a phosphate bond. This energy is then used in two reactions (steps 6 and 9) to convert molecules of ADP to ATP. The last step, the reduction of pyruvic acid to lactic acid, oxidizes the NAD coenzyme back to its original form, so that glycolysis can continue.

This series of changes, in which the product of one reaction is the reactant for the next one, takes place in the cytoplasm of cells, such as in skeletal and cardiac muscle. Two ATP molecules are generated (four are formed but two were used up initially), and the overall reaction can be summarized by the following equation:

$$\text{glucose} + 2\text{ADP} + 2\text{P}_i \longrightarrow 2\text{lactic acid} + 2\text{ATP} + \text{H}_2\text{O}$$

In the absence of oxygen, lactic acid is a waste product, excreted by the cell.

17.6 Citric acid (Krebs) cycle

When oxygen is present, much more energy is released by the aerobic process called **respiration**. The need to generate NAD by the reduction of pyruvic acid to lactic acid disappears. Instead, pyruvic acid can itself now be oxidized and electrons can be transferred to oxygen.

The pyruvic acid produced by glycolysis moves from the cytoplasm to the mitochondria, the "power plants" of the cell. Here, the **citric acid cycle**, or **Krebs cycle**, takes place. The sequence of chemical reactions is presented in Figure 17-8; the details are given in Appendix D.2. Again, only the most important aspects of this complicated process are described here.

In the preparatory step before the citric acid cycle begins, pyruvic acid loses one carbon atom in the form of carbon dioxide. The remaining two-carbon fragment, the acetyl group

$$\text{CH}_3-\overset{\overset{\displaystyle O}{\displaystyle \|}}{\text{C}}-$$

becomes attached to **coenzyme A**, abbreviated CoA and pronounced "ko" A. ("Co" stands for coenzyme in this case, not the element cobalt.) The energy of this oxidation reaction is stored by the formation of a thioester (ester based on sulfur) bond; NAD is also reduced in this reaction. The resulting molecule, acetyl-CoA, is shown in Figure 17-9.

The citric acid cycle begins when the two-carbon acetyl group from acetyl-CoA joins with four-carbon oxaloacetic acid to form six-carbon citric acid. The energy for this combination comes from breaking the thioester bond. In the sequence of reactions that follows, two molecules of carbon dioxide form (in steps 3 and 4) and other transformations take place, converting citric acid back

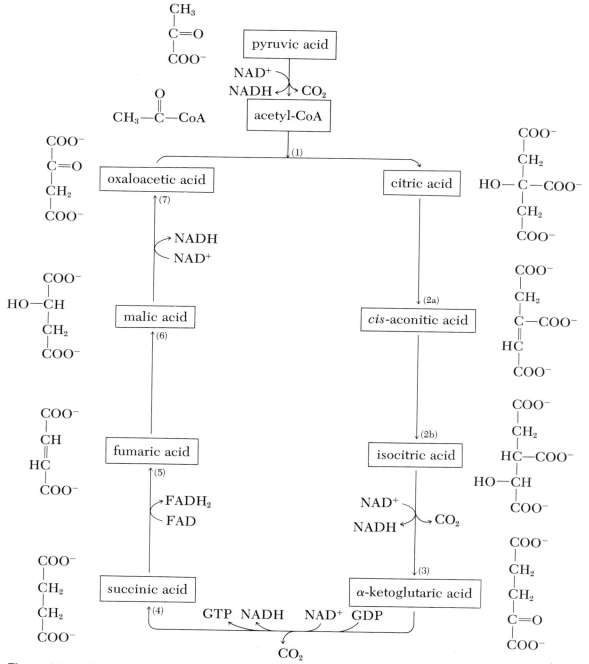

Figure 17-8. The citric acid (Krebs) cycle. See Appendix D.2 for a detailed description of each reaction.

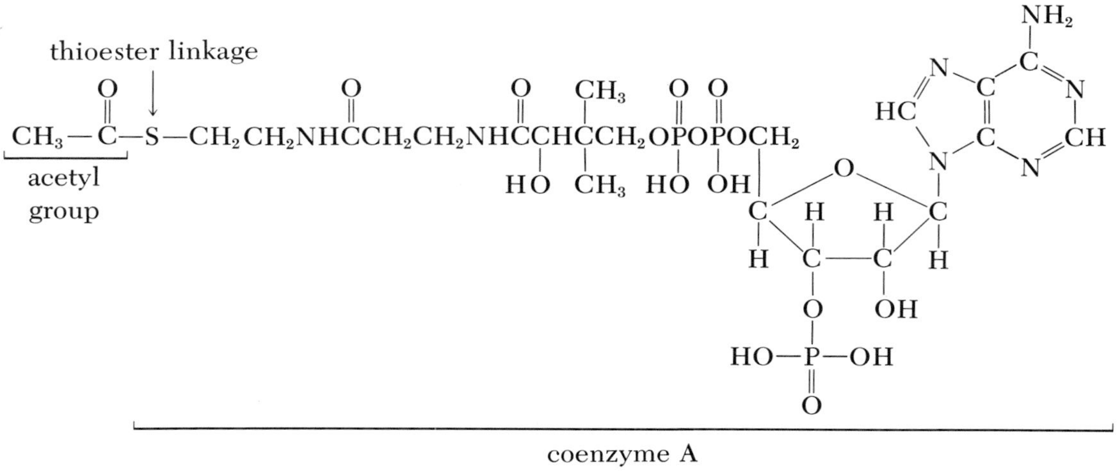

thioester linkage

$$CH_3-\overset{\overset{\displaystyle O}{\|}}{C}-S-CH_2CH_2NHCCH_2CH_2NHCCHCCH_2OPOPOCH_2$$

acetyl group

coenzyme A

Figure 17-9. The structure of acetyl coenzyme A (acetyl-CoA).

to oxaloacetic acid. Because the original molecule is synthesized again, this process is called a cycle.

17.7 The respiratory chain

At several stages in the citric acid cycle, oxidation occurs, resulting in electrons being transferred to, and reducing, coenzymes. Three molecules of NADH and one of $FADH_2$ form in this way. In a process of electron transport, the electrons carried by the coenzymes are transferred to oxygen. They "flow" through a series of enzymes called the **respiratory chain** (Figure 17-10).

In the respiratory chain, the transfer of electrons is coupled with the synthesis of ATP. The following equation represents this process of oxidative phosphorylation, or respiratory-chain phosphorylation:

$$NADH + H^+ + 3ADP + 3P_i + \tfrac{1}{2}O_2 \longrightarrow NAD^+ + 4H_2O + 3ATP$$

Thus, energy used to form each NADH during the citric acid cycle is channeled into three high-energy molecules of ATP.

The three molecules of NADH produced by each "turn" of the citric acid cycle, plus the one formed in the preparatory step, result in a total of 4×3, or 12, molecules of ATP. The $FADH_2$ enters the chain later than NADH, producing only two ATP molecules. In addition, one molecule of guanosine triphosphate, GTP, is formed in the cycle (step 4); this molecule also generates one molecule of ATP, for a total of $12 + 2 + 1 = 15$ molecules. Thus, one cycle of aerobic catabolism can be summarized by the following equation:

$$\text{pyruvic acid} + 2\tfrac{1}{2}O_2 + 15ADP + 15P_i \longrightarrow 3CO_2 + 15ATP + 17H_2O$$

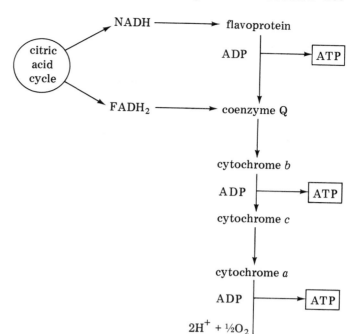

Figure 17-10. The respiratory chain. Oxygen is the final electron acceptor in this process of electron transfer.

By giving up their electrons to oxygen through the respiratory chain, the coenzymes are regenerated, producing their oxidized forms, which are ready to be used for another cycle.

17.8 Products of glucose catabolism

Through the citric acid cycle, the carbon from glucose is converted to the completely oxidized product, carbon dioxide, CO_2. Oxygen, the electron acceptor, is reduced, ending up as water, H_2O. The starting materials for glucose catabolism are the sugar and oxygen, both products of photosynthesis in plants. The end products of glucose catabolism, carbon dioxide and water, are the starting materials for photosynthesis. This relationship between plants and animals forms part of the **carbon cycle** (Figure 17-11), which represents the conversion between inorganic forms of carbon (such as CO_2) and organic forms of carbon (such as glucose).

By adding the equations for both glycolysis and the citric acid cycle, remembering that each glucose molecule forms two molecules of pyruvic acid, you can calculate the energy yield of glucose catabolism:

$$\text{glucose} + 6O_2 + 36ADP + 36P_i \longrightarrow 6CO_2 + 36ATP + 42H_2O$$

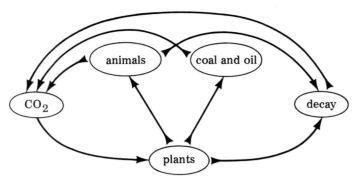

Figure 17-11. The carbon cycle.

(Also added in this equation is the oxidation of NADH produced by aerobic conversion of glucose to pyruvic acid.) Thus, *a total of 36 molecules of ATP forms through respiration,* 18 times the amount produced by the anaerobic process of glycolysis.

Another way you can look at glucose catabolism is in terms of the energy produced by the "burning" of this fuel. The reaction that takes place consists of the complete oxidation of glucose, the same change as occurs during combustion:

$$C_6H_{12}O_6 + 6O_2 \longrightarrow 6CO_2 + 6H_2O + energy$$

This equation is identical to the previous one except that the formation of ATP is represented simply as "energy." The reaction above liberates 686 kcal (2870 kJ) per mole (180 g) of glucose. (As it is used here, the calorie represents a unit of energy and not necessarily heat.) The reaction that forms ATP from ADP and inorganic phosphate requires 7.3 kcal/mole (30.5 kJ/mole). By producing 36 moles of ATP, glucose catabolism makes use of 36 moles × 7.3 kcal/mole, or 263 kcal, for an overall efficiency of about 39% (the 263 kcal divided by the maximum possible energy, 686 kcal).

The energy contained in ATP is then used by the cells to do work. The mitochondria (see Appendix E), in which respiration takes place, are often located near structures that require ATP or in some cases, like liver cells, are able to move around. In this sense, the mitochondria are not only the "power plants" of the cell but also the "distribution centers" for ATP.

17.9 Carbohydrate anabolism

In anabolism, products of carbohydrate catabolism are used to make more complex molecules with ATP as the energy source. The steps involved are *not* simply the reverse of the catabolic sequence. The path you take to go "uphill," that is, energy-requiring anabolism, is not always the same as that taken in "downhill" energy-producing catabolism. A path may be fine for going down but might be too "steep" for the upward climb, and an alternate path may have to be used. Thus, although the product of an anabolic sequence may be the same as the starting material for a catabolic sequence, the reactions and enzymes may be different. In addition, the cellular location of anabolism is often different from that of catabolism, and different regulatory procedures are generally employed.

Figure 17-12 summarizes the central pathway in the biosynthesis of carbohydrates. Most of the steps in the pathway from pyruvic acid to glucose occur by reversal of the sequence in glycolysis. However, three of them (steps 1, 3, and 9 in Figure 17-7) are irreversible and are replaced by "bypass" reactions more favorable energetically for synthesis. It is also possible for products of the citric acid cycle to be converted to glucose through oxaloacetic acid.

The overall process is called **gluconeogenesis**, the formation of new glucose from simpler molecules. Starting with pyruvic acid, reactions can be represented by the following equation:

$$2\text{pyruvic acid} + 4\text{ATP} + 2\text{GTP} + 2\text{NADH} + 2\text{H}^+ + 6\text{H}_2\text{O} \longrightarrow$$
$$\text{glucose} + 2\text{NAD}^+ + 4\text{ADP} + 2\text{GDP} + 6\text{P}_i$$

Six high-energy molecules (ATP and GTP) are consumed and two NADH

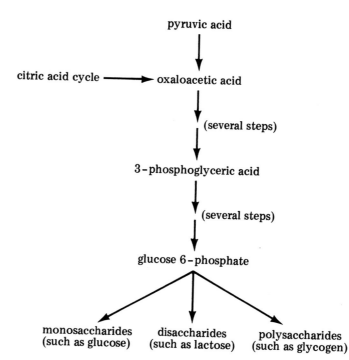

Figure 17-12. The central pathway in carbohydrate anabolism. The reactions are not simply the reverse of the catabolic reactions.

coenzyme molecules are needed for the production of one molecule of glucose. As you can see, this equation is quite different from the one representing the conversion of glucose to pyruvic acid in the catabolic sequence.

17.10 Glycogen metabolism

In another important anabolic pathway, glucose units are linked together to form glycogen. This process, **glycogenesis**, produces the polysaccharide that serves as your storage form of glucose. The reserve supply of glycogen is necessary because you eat only occasionally yet require fuel all the time.

Glycogen produced by anabolism is "potential" glucose, which can be drawn on, like a checking account in the bank. The process of converting glycogen back to glucose, by hydrolyzing the acetal linkages, is **glycogenolysis**. Glycogen is stored mainly in muscle, where it can be used to supply extra energy for contraction, and in the liver, from which the glucose can be added to the blood. An adult male weighing about 150 pounds (70 kg) has 120 g of glycogen in the muscle tissue, 70 g in the liver, and 2 g in the extracellular fluid. The total energy reserve from glycogen is equal to 192 g (120 g + 70 g + 2 g) times 4 kcal/g (the energy yield of carbohydrates) or 768 kcal (3213 kJ).

Consider a specific example of glycogen metabolism. During vigorous athletic exercise, the muscles require large amounts of energy in a short time. The glucose supplied by respiration is quickly used up; not enough oxygen can reach the muscle to produce sufficient glucose aerobically. So the muscle must draw on its reserve supply of glycogen and convert it to glucose.

The glucose undergoes the anaerobic process of glycolysis, producing the ATP for muscle contraction. Lactic acid, the end product of this pathway, circulates to the liver, where most of it is converted first back to glucose by gluconeogenesis, the anabolic pathway, and then to glycogen by glycogenesis. Vigorous exertion is said to create an "oxygen debt" because oxygen is needed to regenerate through respiration the glycogen from lactic acid. The interrelationship between glycogen and glucose in the muscle and liver, sometimes referred to as the Cori cycle, is shown in Figure 17-13.

Figure 17-13. The metabolism of glycogen (the Cori cycle). Do not confuse the terms glycolysis, gluconeogenesis, glycogenesis, and glycogenolysis.

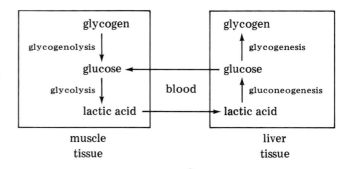

17.11   Blood glucose

The normal amount of glucose in your blood 8 to 12 hours after a meal is 70 to 90 mg/100 ml. Its concentration depends on several factors. As pointed out in the previous section, glycogenolysis, the breakdown of glycogen, provides part of the continuing supply of glucose, as does gluconeogenesis, the biosynthesis of glucose from simpler molecules. On the other hand, glucose is used up during glycolysis, the breakdown of glucose to supply energy (usually followed by the citric acid cycle), and glycogenesis, the synthesis of glycogen. By adjusting these processes, summarized in Figure 17-14, the body keeps the glucose concentration constant. This property is an example of homeostasis, the ability to maintain a stable state despite changing conditions. The regulation of glucose metabolism by hormones is discussed in Chapter 21.

When a malfunction in glucose regulation occurs, such as in diabetes mellitus, the concentration of blood glucose changes from the normal level, a state known as normoglycemia. In this case, the overproduction of glucose caused by a deficiency of the hormone insulin results in **hyperglycemia**, excess glucose in the blood. Glucose gets transferred to the urine when the renal threshold, the limiting concentration above which the kidney can no longer filter

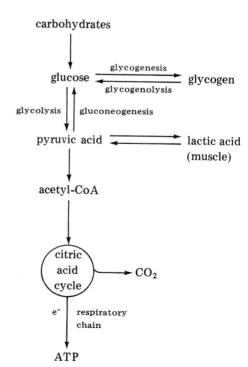

Figure 17-14.   A summary of carbohydrate metabolism.

and reabsorb blood glucose, is exceeded. This condition results in **glucosuria**. Instead of containing only 10 to 20 mg of glucose per 100 ml, the urine of a diabetic person contains a much larger amount (Figure 17-15).

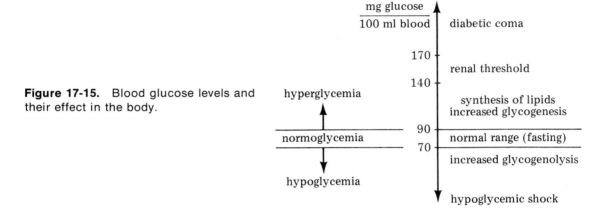

Figure 17-15. Blood glucose levels and their effect in the body.

If the glucose level drops below normal, **hypoglycemia** results. This condition may occur as a result of fasting, extreme activity, convulsions, or hormonal irregularities. Its most serious effect is on the central nervous system, causing coma and death in severe cases. This condition occurs because the brain obtains nearly all of its energy requirements from glucose but contains almost no glycogen as a reserve supply for times when the blood glucose level drops. Immediate treatment consists of ingestion of glucose; long-term therapy varies but generally involves diet regulation.

To test glucose metabolism, the **glucose tolerance**, the capacity of the body to dispose of glucose given either by mouth or by vein, is measured. In the oral glucose tolerance test, the concentration of blood glucose rises from about 90 mg to as high as 180 mg/100 ml of blood in 1 hour after the dose of 100 g of glucose is administered. In a normal individual, the blood glucose returns to its original values by the end of the second hour because of increased glycolysis and glycogenesis, which are stimulated by the added glucose and resulting increase in insulin secretion.

In a diabetic individual, the already high blood sugar level initially increases even further after the administration of oral glucose. The renal threshold may be exceeded, resulting in glucosuria. The decline in the concentration of the blood sugar level with time is slow. This patient is said to have a decreased glucose tolerance (Figure 17-16). A hypoglycemic individual shows an abnormally high glucose tolerance. The initial level is low and falls even further after the glucose has been administered because of insulin secretion, often to levels as low as 40 mg/100 ml.

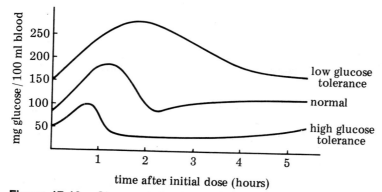

Figure 17-16. Glucose tolerance curves. They indicate the blood glucose level for several hours after an initial dose of glucose.

SUMMARY

Through food you obtain both matter and energy from the surroundings. The complex chemical steps by which your enzymes perform this transfer are called metabolism. It includes all of the chemical processes that take place in your body.

The sun is the primary source of energy for all life. In a series of reactions, the energy from light is converted by photosynthesis in plants to chemical energy in the form of new bonds in carbohydrate molecules. This chemical energy can be used by higher organisms that eat the carbohydrates.

Energy is released from carbohydrate molecules by oxidation, removal of electrons. The process by which electrons are transferred from one compound to another until they reach oxygen, O_2, is called electron transport. Certain coenzymes, such as NAD, serve as electron carriers.

The energy obtained from oxidation is stored in molecules of adenosine triphosphate, ATP. It is stored and then released by first making and then breaking the bond connecting the last phosphate group to the molecule. Energy coupling is the linking of key endergonic steps, which need energy, with exergonic reactions, which release energy, such as the hydrolysis of ATP.

The process of catabolism, the breakdown of complex food molecules, produces energy for anabolism, the synthesis of large molecules needed by the cells. In addition to this chemical work, energy is required for mechanical work (contracting muscle) and transport work (conducting nerve impulses and secreting urine). Basal metabolism reflects the energy needs of the body's continuing processes.

The first steps in carbohydrate catabolism occur in the absence of oxygen, a process of anaerobic fermentation known as glycolysis. The net result is the breakdown of a glucose molecule into two molecules of lactic acid. Two molecules of ATP are generated by this pathway.

When oxygen is present, much more energy is released by the aerobic process called respiration. For the citric acid or Krebs cycle part of pyruvic acid (from glycolysis) is attached to coenzyme A, forming acetyl-CoA. This molecule enters the cyclic series of reactions, undergoing transformations but synthesizing the starting molecule at the end.

At several stages in the citric acid cycle, oxidation occurs, resulting in electrons being transferred to coenzymes. These electrons "flow" through a series of enzymes called the respiratory chain until they are transferred to oxygen, the final electron acceptor. Electron transfer is coupled with the synthesis of ATP.

The overall reaction for carbohydrate catabolism is the conversion of glucose plus oxygen to carbon dioxide and water. A total of 36 ATP molecules is produced by the breakdown of 1 glucose molecule.

In carbohydrate anabolism, the chemical pathways are not simply the reverse of the catabolic ones. The formation of glucose from simpler molecules, called gluconeogenesis, requires the use of ATP (and GTP) molecules for energy.

In glycogenesis, glucose units are linked together to form glycogen, the storage form of carbohydrates. The reconversion of glycogen back to glucose when needed is glycogenolysis. This process (part of the Cori cycle) takes place in the muscle and liver tissue.

The normal amount of glucose in your blood (8 to 12 hours after a meal) is 70 to 90 mg/100 ml. Hyperglycemia is an abnormally high blood glucose level, as found in diabetes mellitus. This condition may be accompanied by glucosuria, relatively large amounts of glucose in the urine. The other extreme, hypoglycemia, is an abnormally low blood glucose level. Glucose metabolism can be tested by measuring glucose tolerance, the capacity of the body to dispose of sugar.

Exercises

1. (Intro.) What is the function of food?
2. (Intro.) What is metabolism?
3. (17.1) How do you indirectly use the energy from the sun?
4. (17.2) Describe the role of coenzymes in electron transfer.
5. (17.3) Explain the function of ATP in metabolism.
6. (17.3) Write an equation for the hydrolysis of ATP. Is energy given off or absorbed?
7. (17.3) Relate these three terms: endergonic, exergonic, energy coupling.
8. (17.4) What processes in the body require energy?
9. (17.4) Describe the relationship between catabolism and anabolism.
10. (17.4) What is basal metabolism? What does the BMR measure?

11. (17.5) What happens in glycolysis?

12. (17.6) Describe three differences between the citric acid cycle and glycolysis.

13. (17.6) What is acetyl-CoA? How is it formed?

14. (17.7) What is the respiratory chain?

15. (17.7) What role does oxygen play in aerobic catabolism?

16. (17.8) Write an overall equation for the complete oxidation of glucose. Explain it in words.

17. (17.8) How much energy is released by the oxidation of glucose?

18. (17.9) How does carbohydrate anabolism differ from catabolism?

19. (17.9) What is gluconeogenesis?

20. (17.10) Describe the relationship between glycogenesis and glycogenolysis.

21. (17.10) Explain what happens in muscle tissue during vigorous exercise.

22. (17.11) Define homeostasis. How is this term related to the blood glucose level?

23. (17.11) What is hyperglycemia? hypoglycemia?

24. (17.11) Compare the effect of an oral dose of glucose in a normal person and a diabetic.

Metabolism of lipids

Of all the foods you eat, lipids provide the most concentrated source of energy. Per gram they produce 9 kcal (38 kJ) compared to 4 kcal (17 kJ) for either carbohydrates or protein. But the real biological importance of lipids is that they make up the *main reserve supply of chemical energy in your body*. Any food digested which is not immediately needed is converted to fat and deposited for future use. Then, as required, it is drawn upon as a source of energy.

18.1 The fatty acid "cycle"

Most lipids, which come either from diet or from storage, exist as triacylglycerols (triglycerides). Before being metabolized, they must be broken down into fatty acids and glycerol. This process of hydrolysis occurs primarily in the gastrointestinal tract (as described in Section 16.9). The resulting glycerol enters the glycolysis pathway and is then used either to form glycogen or to generate ATP.

The fatty acids released are oxidized in the mitochondria by a series of reactions in which the long hydrocarbon chains are shortened two carbons at a time. This catabolic sequence is the **fatty acid** "cycle." It produces acetyl-CoA, which can then enter the citric acid cycle to generate ATP. The cycle is illustrated in Figure 18.1

In the first step, the fatty acid is activated by forming a thioester with coenzyme A, using up a molecule of ATP. Two phosphate groups are lost from ATP in this reaction, leaving adenosine monophosphate (AMP) and liberating pyrophosphate (PP_i). Next, dehydrogenation occurs; that is, two hydrogen atoms are removed, forming the reduced coenzyme $FADH_2$ and leaving a double bond (enol) in the fatty acid. Step 3 involves the addition of water, hydration, creating an alcohol group. The alcohol is then oxidized to form a ketone functional group, reducing NAD^+ to NADH. In the last step, a new CoA molecule enters and causes cleavage to produce acetyl-CoA and the CoA ester

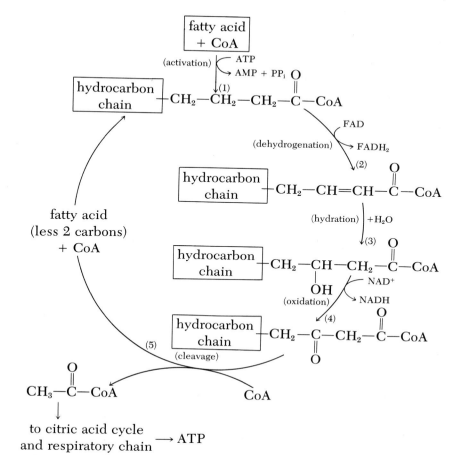

Figure 18-1. The fatty acid "cycle." A two-carbon fragment is removed during each complete turn of the "cycle."

of the fatty acid now shortened by two carbon atoms. This shorter fatty acid can then run through the cycle again.

With every passage through the "cycle," or "spiral," *two carbon atoms are cut off the fatty acid,* producing a molecule of acetyl-CoA:

$$\boxed{\text{hydrocarbon chain}}\!-\!CH_2CH_2CH_2\overset{\displaystyle O}{\overset{\|}{C}}\!-\!CoA + FAD + NAD^+ + CoA \longrightarrow$$

$$\boxed{\text{hydrocarbon chain}}\!-\!CH_2\overset{\displaystyle O}{\overset{\|}{C}}\!-\!CoA + acetyl\text{-}CoA + FADH_2 + NADH + H^+$$

The acetyl-CoA produced by fatty acid oxidation mixes with the acetyl-CoA generated by other reactions, such as reactions you learned about in carbohydrate catabolism. It can then be channeled into the citric acid cycle and the respiratory chain to generate ATP molecules.

18.2 Energy released by lipid catabolism

The total energy yield from one fatty acid molecule, such as 16-carbon palmitic acid, is tremendous. By going through the fatty acid "cycle" seven times, one molecule of palmitoyl-CoA (the thio ester of palmitic acid with coenzyme A) is cut into two-carbon pieces seven times to produce eight molecules of acetyl-CoA:

$$\text{palmitoyl-CoA} + 7\text{CoA} + 7\text{FAD} + 7\text{NAD}^+ + 7\text{H}_2\text{O} \longrightarrow$$
$$8\text{acetyl-CoA} + 7\text{FADH}_2 + 7\text{NADH} + 7\text{H}^+ \qquad (1)$$

Looking back to the respiratory chain, you will notice that each FADH_2 generates two ATP molecules, and each NADH produces three, for a total of five molecules of ATP formed by oxidative phosphorylation after each cycle. After seven cycles, $7 \times 5 = 35$ molecules of ATP are produced:

$$\text{palmitoyl-CoA} + 7\text{CoA} + 7\text{O}_2 + 35\text{ADP} + 35\text{P}_i \longrightarrow$$
$$8\text{acetyl-CoA} + 35\text{ATP} + 42\text{H}_2\text{O} \qquad (2)$$

In addition, the eight molecules of acetyl-CoA enter the citric acid cycle, generating another $8 \times 16 = 96$ molecules of ATP:

$$8\text{acetyl-CoA} + 16\text{O}_2 + 96\text{ADP} + 96\text{P}_i \longrightarrow$$
$$8\text{CoA} + 96\text{ATP} + 104\text{H}_2\text{O} + 16\text{CO}_2 \qquad (3)$$

The overall reaction can be obtained by adding equations (2) and (3):

$$\text{palmitoyl-CoA} + 23\text{O}_2 + 131\text{ADP} + 131\text{P}_i \longrightarrow$$
$$\text{CoA} + 16\text{CO}_2 + 146\text{H}_2\text{O} + 131\text{ATP}$$

Remembering that 1 molecule of ATP was required initially to make the palmitoyl-CoA, you see that *a total of 130 molecules of ATP is produced through the oxidation of 1 molecule of fatty acid,* compared to 36 from a molecule of glucose. The efficiency of converting the energy of palmitic acid to ATP is about 40% compared to the total available from complete oxidation.

18.3   **Lipogenesis**

The synthesis of fatty acids and triacylgylcerols (triglycerides) is an important process. It is through this anabolic pathway, **lipogenesis**, that excess glucose is stored as fat and that the lipids of membranes are replaced. Instead of being taken apart two carbons at a time as in catabolism, fatty acids are synthesized by joining together two carbon units. The anabolic reactions, however, occur in the cell cytoplasm, not in the mitochondria, where fatty acid oxidation takes place.

The overall reaction for the formation of palmitic acid is as follows:

$$8\text{acetyl-CoA} + 14\text{NADPH} + 14\text{H}^+ + 7\text{ATP} + 7\text{H}_2\text{O} \longrightarrow$$
$$\text{palmitic acid} + 8\text{CoA} + 14\text{NADP}^+ + 7\text{ADP} + 7\text{P}_i$$

Thus, 8 two-carbon pieces from acetyl-CoA are put togther using the energy of ATP and NADPH molecules. Once palmitic acid is formed, it can be lengthened by other enzymes, if necessary. The degree of unsaturation, the number of double bonds in the fatty acid, is determined by reactions with oxygenase enzymes.

Triacylglycerols (triglycerides) can then be synthesized by the reaction of three appropriate fatty acids (attached to CoA) with glycerol (as glycerol 3-phosphate). Phosphoglycerides or phospholipids, components of cell membranes, are formed from glycerol, two fatty acids, and a nitrogen-containing base such as choline. Sphingolipids, used for nerve and brain membranes, are synthesized from one molecule of fatty acid and the long-chain amino alcohol sphingosine, instead of glycerol.

Cholesterol (see structure in Section 14.10) is synthesized by a complex series of reactions beginning with acetyl-CoA, but by a pathway very different from the one in which fatty acids are formed. This molecule then serves as the starting point for the synthesis of fecal sterols, bile acids, and steroid hormones.

18.4   **Role of acetyl-CoA**

It should now be obvious to you that *acetyl-CoA plays a central role in the metabolism of both lipids and carbohydrates.* As you have seen in the previous chapter, this coenzyme can be formed from pyruvic acid and can serve as the entry point into the citric acid cycle. You now also know that it is the main product of fatty acid oxidation and that it is involved in the synthesis of fatty acids and cholesterol. Thus, *acetyl-CoA acts as a link between carbohydrates and lipids;* it is through this molecule that excess glucose is converted to fat in your body. After glucose is broken down to pyruvic acid through glycolysis, the acetyl-CoA is then used to make fatty acids and triacylglycerols (triglycerides) if energy is not immediately needed from the citric acid cycle.

In addition to the citric acid cycle and the synthesis of fatty acids, a third possible fate exists for acetyl-CoA. Two molecules of this coenzyme can react to form another coenzyme (acetoacetyl-CoA), which is converted in the liver to still another form (3-hydroxy-3-methylglutaryl-CoA, or HMG-CoA.) This modified coenzyme can then be used for the synthesis of cholesterol and steroids. The amount that can be used for this purpose is generally limited. Therefore, the modified coenzyme (HMG-CoA) in another reaction forms acetoacetic acid, which in turn generates β-hydroxybutyric acid and to a smaller extent acetone. These three molecules, which can then accumulate in the blood, are called **ketone bodies**.

$$CH_3-\overset{\overset{O}{\|}}{C}-CH_2-\overset{\overset{O}{\|}}{C}-OH \qquad CH_3-\underset{\underset{OH}{|}}{CH}-CH_2-\overset{\overset{O}{\|}}{C}-OH \qquad CH_3-\overset{\overset{O}{\|}}{C}-CH_3 \qquad \text{the ketone bodies}$$

acetoacetic acid $\qquad\qquad$ β-hydroxybutyric acid $\qquad\qquad$ acetone

Figure 18-2 summarizes the role of acetyl-CoA in metabolism. As you can see, acetyl-CoA can also be formed from the breakdown of proteins as well as carbohydrates or lipids. Thus, acetyl-CoA can be made from any of the major nutrients during catabolism. It is then channeled into the synthesis of fatty acids or cholesterol in anabolism, or the generation of ATP for energy through the citric acid cycle and respiratory chain, or the formation of the ketone bodies. Figure 18-3 summarizes the major features of lipid metabolism.

Figure 18-2. The role of acetyl-CoA in metabolism. Note its central position, relating carbohydrates, lipids, and proteins.

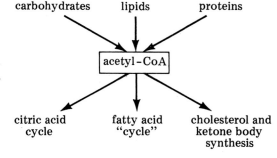

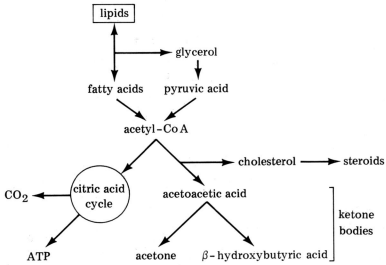

Figure 18-3. A summary of lipid metabolism.

18.5   Ketosis

The ketone bodies are normally present in small amounts in the blood. In fact, the muscles can use acetoacetic acid for a sizable fraction of their total energy needs. For this to happen, the molecule must first be converted to acetyl-CoA. Acetone also can be metabolized, either by being converted to pyruvic acid or by being broken down to a two-carbon (acetyl) and a one-carbon (formyl) fragment. Certain conditions, however, may lead to an overproduction of the ketone bodies. This event occurs when the other pathways for acetyl-CoA are restricted or when excessive amounts of acetyl-CoA are formed. If the ketone bodies are produced in large amounts, **ketosis** results.

An abnormally high concentration of the three ketone bodies in the blood is **ketonemia.** When the blood level becomes so high that it exceeds the renal threshold, the ketone bodies appear in the urine, resulting in **ketonuria.** It is also likely that, when both of these conditions exist, the smell of acetone will appear on the breath. *The term ketosis describes this combination of ketonemia, ketonuria, and acetone odor of the breath.*

One of the possible causes of ketosis is starvation. Because of the absence of carbohydrates, the main source of energy from the diet, glycogen is quickly used up, and the body must then draw upon its reserve of fat. Fatty acids are released and oxidized in the liver, resulting in a high concentration of lipids in the blood, known as **lipemia.** Production of acetyl-CoA increases and thus the ketone bodies form in large amounts. A recently popular diet causes ketosis by this mechanism because it is rich in lipids and low in carbohydrates ("Dr. Atkin's diet").

The most important clinical cause of ketosis is diabetes mellitus. In a diabetic individual, glucose is present but cannot be used normally because of a lack of the hormone insulin. The effect on lipid metabolism is similar to not eating any carbohydrates, as in the case of starvation. The patient is hyperglycemic, yet the muscles and liver cannot make use of the glucose; this situation has been referred to as "starvation in the midst of plenty." Therefore, storage lipids must be used for energy, generating acetyl-CoA in large amounts and resulting in ketosis. In severe cases, the blood ketone body concentration may reach 90 mg/100 ml compared to less than 3 mg/100 ml in a normal person. The urinary excretion may be 5000 mg in 24 hours instead of approximately 100 mg.

18.6   Ketoacidosis

One of the serious consequences of ketosis results from the acidic nature of two of the ketone bodies, acetoacetic acid and β-hydroxybutyric acid. Their presence in the blood leads to a decrease in its pH from the normal value of 7.4, causing acidosis, or **ketoacidosis**. This condition is the most serious acute complication of diabetes. It results in an increase in the symptoms of ketosis—weariness, discomfort, loss of appetite, thirst, excessive urine secretion—as well as nausea, vomiting, dizziness, deep breathing, and eventually coma.

The pH of the blood falls because the ketone bodies are more acidic than the carbonic acid in the blood's buffer system. The following reaction occurs, as shown for acetoacetic acid:

$$CH_3-\overset{\overset{\displaystyle O}{\|}}{C}-CH_2-\overset{\overset{\displaystyle O}{\|}}{C}-OH + HCO_3^- \longrightarrow H_2CO_3 + CH_3-\overset{\overset{\displaystyle O}{\|}}{C}-CH_2-\overset{\overset{\displaystyle O}{\|}}{C}-O^-$$

The acid reacts with the bicarbonate part of the buffer, decreasing its concentration. More carbonic acid is formed, as well as the anion of the acid, acetoacetate ion. The net result is an increase in the acidity of the blood, reflected by the lowering of its pH. This change is of great consequence since the proper functioning of the parts of the cell, such as its enzymes, depends on maintaining the proper pH.

In an attempt to counteract the acidosis, the kidney excretes the acids, which exist largely as anions in the urine. Since their negative charge must be balanced, sodium ions, Na^+, are eliminated at the same time. In addition, large quantities of fluid are lost in the urine, complicating the acidosis and electrolyte loss with dehydration.

18.7 Body lipids

Over 10% of your body weight is lipid, mostly triacylglycerols (triglycerides). Lipids are located in all organs in addition to special storage areas, the adipose tissue. In this type of connective tissue, much of the cell cytoplasm is replaced by droplets of lipid. These areas contain the **depot lipid** or depot fat; *they are the storehouses of chemical energy which can be used by the body when needed.*

The depot lipid is in a liquid state and is as saturated as possible without becoming solid at body temperature. (The greater the degree of saturation, the more single bonds rather than double bonds in the fatty acid and the larger the amount of energy that can be produced by oxidation.) Much is located under the skin and acts as an insulator against excessive loss of body heat. The depot lipid also serves as protection from mechanical blows by acting as a sort of cushion.

The fetus contains little depot lipid. While in the uterus, it continuously receives nourishment across the placenta from the mother's blood circulation. Furthermore, the fetus is in a well-protected location inside the body of its mother and does not require additional means to guard against temperature changes or mechanical shock. Thus, the fetus has no real need for depot lipid and gets it only shortly before birth. The adult, on the other hand, eats only a few times during the day but uses energy all the time. Therefore, the depot lipid is an important energy reserve as well as a "shield" against the environment.

The adipose tissue absorbs and stores lipid almost entirely in the form of triacylglycerols (triglycerides). After being absorbed during digestion, fats are transported from the liver to the adipose tissue by the blood as lipoproteins. Here they are broken down to fatty acids and glycerol, and the triacylglycerol (triglyceride) is resynthesized and deposited. In addition, fats are formed from carbohydrates through lipogenesis since, as you have seen, any substance that produces acetyl-CoA can serve as the starting point for fatty acid synthesis. This ability is of great importance since carbohydrates make up the bulk of your diet, and the capacity to store glycogen, the energy reserve formed from glucose, is limited. *Lipids are a better storage form since they provide more ATP during oxidation.* Table 18-1 shows the huge amount of energy available from fat in adipose tissue compared with other sources in the body.

When needed, the depot lipid can be mobilized, like the army reserve being called up and sent into action. It enters the bloodstream as fatty acids bound to a protein, serum albumin. Each protein molecule carries about 20 fatty acids to the liver for their breakdown to acetyl-CoA before entering the citric acid cycle. When excessive amounts of lipid are transported to the liver, as in starvation and diabetes, (as well as alcoholism), a condition known as "fatty liver" develops (Figure 18-4).

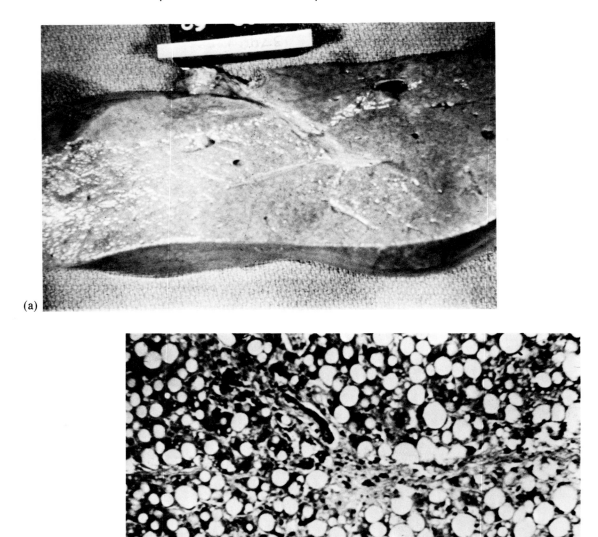

Figure 18-4. A fatty liver. (a) Gross view; (b) microscopic view. Note the droplets of fat. (Photos courtesy of Dr. Frank A. Seixas, National Council on Alcoholism.)

Table 18-1 Fuel Reserve Supply[a]

Fuel reserve	Supply (g)	Supply (kcal)	Supply (kJ)
fat (adipose tissue)	15,000	141,000	590,000
protein (muscle)	6,000	24,000	100,000
glycogen (muscle)	120	480	2,000
glycogen (liver)	70	280	1,200

[a] For an average 70-kg male.

18.8 Obesity

Under normal conditions, new lipids are continuously being deposited in the adipose tissue, and depot lipids are being mobilized. Because these two opposing processes balance each other, the total amount of body fat usually remains constant. If, however, the diet supplies more calories than are needed by the body for its basal metabolism and other types of work, the excess will form extra depot lipid that cannot be used up.

The condition of having a large excess of body fat is known as **obesity**. An extreme case is illustrated in Figure 18-5. It can also result from metabolic or other disorders, but the most common cause is overeating. Generally, each additional 9 kcal from food creates about 1 g of depot lipid.

Aside from the social and physical disadvantages of obesity, this state decreases a person's life expectancy because of the strain on the body. Among obese individuals, the chance of developing diabetes, or gallbladder, liver, and cardiovascular diseases is much higher than in others, as is the risk when surgery is performed. A weight that is 10% or more than normal for your height and build can be a medical danger. Unfortunately, millions of Americans suffer from this condition. Table 18-2 lists suggested weights according to height (Figure 18-6).

The most effective diet for an overweight individual involves "cutting down on calories," so that the amount taken in is less than what the body needs. This decrease leads to a loss of body lipid. Table 18-3 lists the recommended caloric intake for children and adults at different ages.

Another medical condition exists at the other extreme. When so few calories are ingested that adipose tissue completely disappears, cachexia results. It occurs in malnutrition, certain diseases, and metabolic disorders. Cachexia may be psychologically caused and is then called anorexia nervosa.

Figure 18-5. The heaviest human being of all time (Robert Earl Hughes). Here he is shown when he weighed only 700 pounds; his top weight was 1069 pounds. (From "The Guinness Book of World Records," © 1975 by Sterling Publ. Co. Inc., New York, 10016.)

Figure 18-6. © 1975, American Heart Association, Inc. Reprinted with permission.

Table 18-2   Suggested Weights by Height[a]

Height (inches)	Weight (pounds)[b]	
	Men	Women
60		100–118
62		106–124
64	122–144	112–132
66	130–154	119–139
68	137–165	126–146
70	145–173	133–155
72	155–182	140–164
74	160–190	
76	166–198	

[a] National Academy of Sciences, National Research Council (1974).
[b] Weight without clothing or shoes; based on college-age men and women.

Table 18-3   Recommended Caloric Intake[a]

Category	Age	Weight (pounds)	Height (inches)	Energy (kcal)
Children	1–3	28	34	1300
	4–6	44	44	1800
	7–10	66	54	2400
Males	11–14	97	63	2800
	15–18	134	69	3000
	19–22	147	69	3000
	23–50	154	69	2700
	over 51	154	69	2400
Females[b]	11–14	97	62	2400
	15–18	119	65	2100
	19–22	128	65	2100
	23–50	128	65	2000
	over 51	128	65	1800

[a] National Academy of Sciences, National Research Council (1974).
[b] Pregnant, add 300 kcal; lactating, add 500 kcal.

18.9   Blood lipids and atherosclerosis

Normally, the blood plasma, the liquid portion of whole blood, contains about 500 mg of total lipid per 100 ml. About 25% is triacylglycerol (triglyceride),

32% is phosphoglyceride, 36% or more is cholesterol, and the remainder includes other lipids and fatty acids. The composition and concentration of lipids in the blood is of great significance because certain disorders, particularly cardiovascular disease, are related to these factors.

Over half of all deaths in the United States result from cardiovascular disease. Of these about 85% are linked to **atherosclerosis**. This condition is a special kind of arteriosclerosis ("hardening of the arteries") in which the inner layer of the artery walls is thickened by lipid deposits called plaques, made largely of cholesterol, as shown in Figure 18-7. Atherosclerosis cuts down the

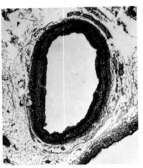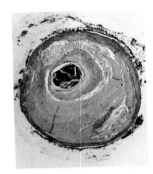

Figure 18-7. A normal artery (left); an artery with atherosclerotic deposits in the inner lining (center); and an artery narrowed by atherosclerotic deposits and now blocked by a blood clot, the dark inner circle (right). See Figure 14-12 for a gross view. (© 1975 American Heart Association, Inc. Reprinted with permission.)

circulation of blood in the same way that an old waterpipe, which may be partially blocked by mineral deposits, prevents the normal flow of water. High blood pressure, known as **hypertension**, develops when the blood must be pumped harder to make it circulate through the clogged arteries. Hypertension can result in damage to the body, especially to the kidneys. It is an important monitor of atherosclerosis since the blood pressure is easy to measure (Figure 18-8).

Atherosclerosis is the main factor involved in most heart attacks, strokes, and other cardiovascular diseases, such as angina pectoris. In the case of heart attack, a heart artery is blocked and, in a stroke, a brain artery is blocked, causing damage to the heart or brain, respectively, and possible death. Angina

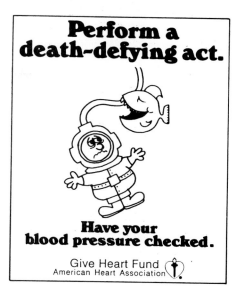

Perform a death-defying act.

Have your blood pressure checked.

Give Heart Fund
American Heart Association

Figure 18-8. © 1975, American Heart Association, Inc. Reprinted with permission.

pectoris results from a decreased flow of blood to the heart, creating pains in the chest. An intensive care cardiac unit is shown in Figure 18-9 (p. 420).

Various drugs, called **antilipemic** or **antihyperlipidemic**, are useful in maintaining normal lipid levels in the blood. Some are steroids that inhibit cholesterol biosynthesis or interfere with its absorption. One of the most widely used drugs is clofibrate. It appears to act by inhibiting synthesis of triacylglycerols (triglycerides) in the liver.

clofibrate

18.10   Lipids and diet

Evidence indicates that **hyperlipemia**, a very high concentration of blood lipid, and atherosclerosis are related to the diet. For example, people living in underdeveloped areas who eat vegetable oils as their major source of lipids have a low blood lipid concentration and little atherosclerosis. On the other

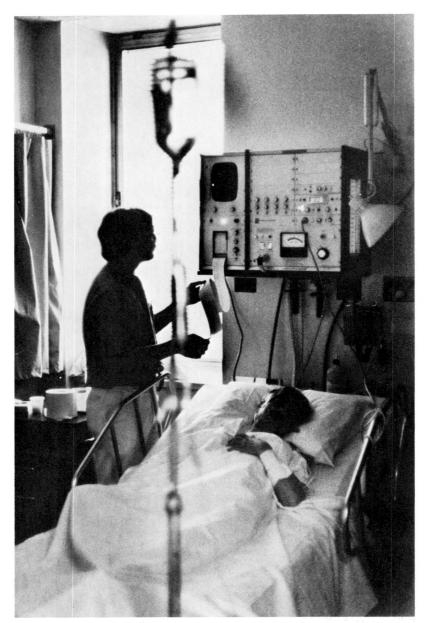

Figure 18-9. An intensive care cardiac unit. (Photo courtesy of St. Luke's Hospital Center, New York.)

hand, people in the United States and western Europe, who eat much more animal fat, have a higher level of blood lipids and a greater amount of athero- sclerotic disease. Of course, there are other risk factors involved in cardiovas- cular disease, such as lack of exercise, genetic tendencies, smoking, and emo- tional stress. But the food you eat plays a part, especially since a high lipid concentration in the blood is associated with the deposit of lipid plaques in the arteries.

The amount of lipid in the blood reflects the relative contribution of satu- rated and unsaturated fatty acids to your diet. Studies show that hyperlipemia and atherosclerosis may occur because of too much saturated lipid and not enough unsaturated or polyunsaturated lipid. This relationship may result from the fact that the more saturated the fatty acids you eat, the higher the rel- ative cholesterol concentration in the blood. Remember it is cholesterol, the important molecule needed for steroid synthesis, that forms the plaques that block the arteries.

Although the origin of cardiovascular disease is not yet completely under- stood, the American Heart Association gives the following advice:

> The reduction or control of fat consumption under medical supervision, with reasonable substitution of polyunsaturated fats, is recommended as a possible means of preventing atherosclerosis and decreasing the risk of heart attacks and strokes.

Vegetable oils, like safflower, corn, peanut, and cottonseed oils, are the most effective in keeping a low blood lipid concentration.

Controlling the total amount of lipids is important in maintaining the proper weight and thereby also lowering the risk of heart disease. In the present average diet in the United States, lipids account for 42% of the food as a source of energy. The American Heart Association recommends reducing this value to 35%.

SUMMARY

Lipids are the most concentrated source of energy of all of the foods you eat. Lipids are also important because they make up your main reserve supply of chemical energy as body fat.

The fatty acids, produced by hydrolysis of triacylglycerols (triglycerides), are oxidized in the mitochondria by a series of reactions in which the long hy- drocarbon chains are shortened two carbons at a time. Called the fatty acid "cycle," this process produces acetyl-CoA and reduced coenzymes, which can enter the citric acid cycle and respiratory chain. Palmitic acid, containing 16 carbons, generates a total of 130 molecules of ATP when completely oxidized to carbon dioxide and water by the fatty acid "cycle."

The synthesis of fatty acids and triacylglycerols (triglycerides) is known as lipogenesis. It is through this important anabolic pathway that glucose can be stored as fat. Two-carbon units are joined together in the cytoplasm using the energy of ATP molecules to form the fatty acid. Triacylglycerols (triglycerides), phosphoglycerides, and sphingolipids can then be synthesized using the appropriate fatty acids. Cholesterol is also made using acetyl-CoA but by a different pathway than that used for the fatty acids.

Acetyl-CoA plays a major role in both carbohydrate and lipid metabolism. It is formed during catabolism and then used in the citric acid cycle to generate ATP, to synthesize fatty acids or cholesterol, and to form the ketone bodies.

The ketone bodies are acetoacetic acid, β-hydroxybutyric acid, and acetone. An abnormally high concentration of these molecules in the blood is ketonemia. This condition in the urine is known as ketonuria. In addition, the odor of acetone appears on the breath. These three conditions are referred to as ketosis. It is caused by starvation and diabetes mellitus; because carbohydrates are either missing or cannot be used normally, lipids are drawn upon to provide energy. A high concentration of lipids in the blood, lipemia, results and causes an increased production of acetyl-CoA and the ketone bodies. The most serious consequence of ketosis is the lowering of the pH of blood from its normal value of 7.4, leading to acidosis.

The storehouse of chemical energy, the depot lipid, is contained in the adipose tissue, mostly under the skin. It also serves as an insulator against heat loss and mechanical shock. Under normal conditions, new lipids are continuously being deposited in the adipose tissue, and depot lipids are being mobilized. Because these two opposing processes balance each other, the total amount of body fat remains constant.

If the diet supplies more calories than the body needs, the excess is stored as body fat. If a large amount forms, obesity results. This condition leads to a decrease in life expectancy and an increased risk of cardiovascular disease.

The composition and concentration of the lipids in the blood are of great significance because certain disorders, particularly heart disease, appear to be partially determined by these factors. Atherosclerosis is a condition in which the inner layer of the artery walls is thickened by lipid deposits called plaques, made largely of cholesterol. They cut down the circulation of the blood and are the main factor involved in most heart attacks, strokes, and other cardiovascular diseases such as angina pectoris and high blood pressure (hypertension.)

Evidence indicates that hyperlipemia and atherosclerosis are linked to the diet. The American Heart Association recommends a reduction of fat consumption and reasonable substitution of polyunsaturated fats for saturated ones.

Exercises

1. (Intro.) Why are lipids important?

2. (18.1) What happens in the fatty acid "cycle"? Why is it not really a cycle like the citric acid cycle?

3. (18.2) Compare the energy produced from one molecule of glucose and one molecule of fatty acid during catabolism.

4. (18.3) What is lipogenesis? What is its biological role?

5. (18.4) How does acetyl-CoA provide a link between carbohydrate and lipid metabolism?

6. (18.4) Draw the structural formulas of the ketone bodies.

7. (18.5) Describe the condition of ketosis.

8. (18.5) Explain how starvation and diabetes mellitus can result in ketosis.

9. (18.6) What is ketoacidosis? How is it caused?

10. (18.7) What is depot lipid? Where is it located?

11. (18.7) What is the function of depot lipid?

12. (18.7) Why does the fetus need little depot lipid?

13. (18.7) Describe the storage and mobilization of depot lipid.

14. (18.8) How does overeating cause obesity?

15. (18.8) Why is obesity harmful?

16. (18.9) What is atherosclerosis?

17. (18.9) Describe the possible medical consequences of atherosclerosis.

18. (18.10) What is the relationship of diet and atherosclerosis?

19. (18.10) What is an antilipemic drug?

20. (18.10) What dietary recommendations are made by the American Heart Association?

Metabolism of proteins

Part of the protein that you eat each day goes into producing energy, in the same way as carbohydrates and lipids. The main job of protein, however, is to provide the amino acids needed to make your own proteins and many other important molecules. Of all the nutrients, *protein is the body's only major source of nitrogen atoms* (in the form of amino groups).

19.1   The nitrogen cycle

Nitrogen occurs in the environment in the form of N_2, the nitrogen molecule that makes up 80% of the air you breathe, and NO_3^-, nitrate ion, which is produced by lightning. The N_2 is converted by electrical discharge to NO_2, which forms nitric acid, HNO_3, and thus nitrate ions, when dissolved in water in the soil. You are not able to use these forms of nitrogen. Through a process called "nitrogen fixation," N_2 must first be reduced to ammonia, NH_3, by microorganisms in the soil. The ammonia and nitrates are converted to simple organic compounds which are used by plants to make amino acids.

You then eat the proteins made by the plants or those produced by animals who ate the plants. After metabolizing these proteins, you excrete nitrogen-containing waste products, which eventually return to the soil to be used again. This series of events is part of the **nitrogen cycle**, which is illustrated in Figure 19-1.

After eating plant or animal protein, you break the peptide linkages during the process of digestion in the gastrointestinal tract. The free amino acids produced by this process, **proteolysis**, are then rapidly absorbed by the small intestine and enter the bloodstream to be carried to the tissues and organs. Most of the amino acids end up in the liver, the main site of their metabolism. It is here that many important nitrogen compounds are formed. The liver is also responsible for sending a balanced mixture of amino acids to other organs and for disposing of any surplus amounts.

424

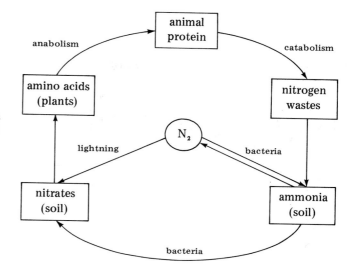

Figure 19-1.   The nitrogen cycle relates the metabolism of animals, plants, and microorganisms.

19.2   Amino acid catabolism

Most of the amino acids generated by proteolysis are used for synthesis. But if more amino acids are present in the liver than are needed for this purpose, they are broken down by 20 enzymatic pathways, one for each of the major amino acids. Just as in the breakdown of glucose and fatty acids, energy is released by this process. In addition, many of the intermediates of the catabolic sequences play important roles in other metabolic pathways.

The amino acids were originally divided into classes on the basis of the product formed by their catabolism. Some, capable of generating glucose, were labeled "glycogenic." Others, the "ketogenic" amino acids, could produce ketone bodies under certain conditions. A third category consisted of those that could form either glucose or ketone bodies. The results are summarized in Table 19-1; formulas of the amino acids can be found in Section 15.1.

Since this classification was made, more has been discovered about the products of the breakdown of the amino acids. Those that are transformed into one of the intermediates of the citric acid cycle are generally glycogenic, while those that are converted to acetyl-CoA or acetoacetyl-CoA during catabolism are ketogenic. Figure 19-2 summarizes the entry points of each of the amino acids into the pathways of carbohydrate and lipid metabolism.

The first step in the catabolic pathways for the amino acids is **transamination** (a reaction also found in anabolism). This process involves removal of the

Table 19-1 Glycogenic and Ketogenic Amino Acids

Glycogenic (produce glucose)	Ketogenic (produce ketone bodies)	Glycogenic and ketogenic
alanine	leucine	isoleucine
arginine		lysine
asparagine		phenylalanine
aspartic acid		tyrosine
cysteine		
glutamic acid		
glutamine		
glycine		
histidine		
methionine		
proline		
serine		
threonine		
tryptophan		
valine		

alpha-amino group and its transfer to another molecule, generally alpha-ketoglutaric acid, as follows:

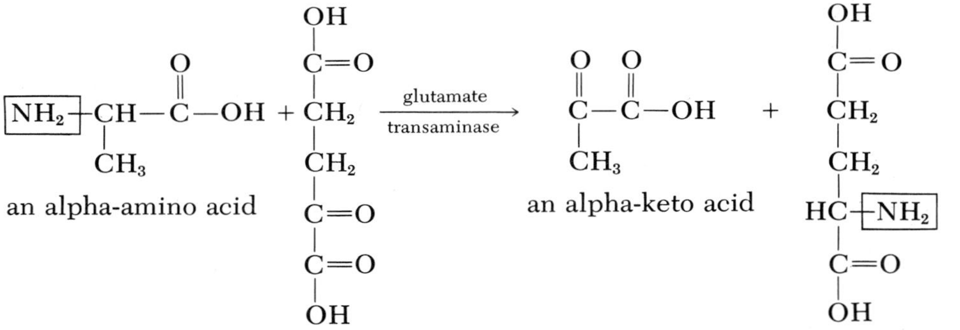

an alpha-amino acid; alpha-ketoglutaric acid; an alpha-keto acid; glutamic acid

Alpha-keto acids, such as alpha-ketoglutaric acid and one of the products of the above reaction, have a carbonyl group at the carbon atom next to the carboxylic acid group (the alpha carbon).

The enzyme for this step, a transaminase, requires the coenzyme pyridoxal

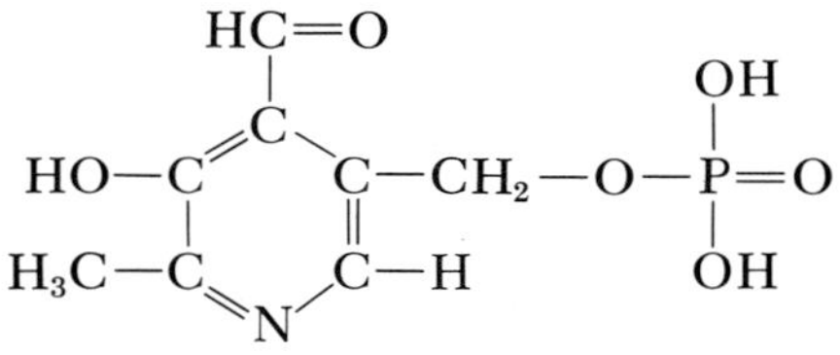

pyridoxal phosphate

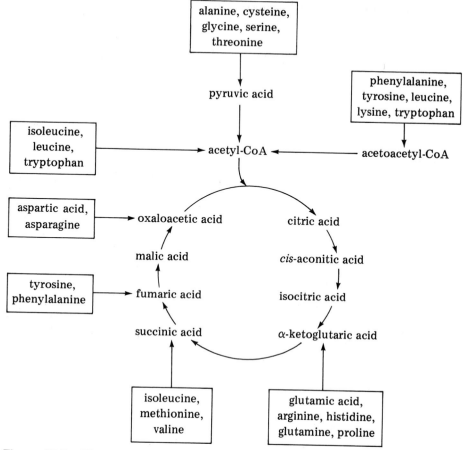

Figure 19-2. The catabolism of amino acids. They enter the citric acid cycle as indicated.

phosphate, a derivative of vitamin B_6. This molecule acts as the carrier of the amino group from the amino acid to the keto acid. The function of transamination is to transfer the amino groups of the different amino acids to only one, glutamic acid, which can then be channeled into the final series of reactions that form the nitrogen-containing end products.

The carbon skeletons that remain as the alpha-keto acids can enter the citric acid cycle after several other sets of reactions that further break down each of the amino acids. Most amino acids enter the cycle as acetyl-CoA; the remainder enter as alpha-ketoglutaric acid, succinic acid, fumaric acid, or oxaloacetic acid. These sequences then generate ATP as oxidation occurs and electrons are funneled into the respiratory chain.

19.3 The urea (ornithine) cycle

If no more nitrogen is needed by the body, glutamic acid undergoes a series of reactions in the liver called the **urea cycle**, or ornithine cycle. Here, the final end product of amino acid catabolism, urea, is produced for secretion in the urine from two amino groups, as shown in Figure 19-3. The first amino group is formed from glutamic acid in a process called oxidative deamination:

$$\text{glutamic acid} + NAD^+ + H_2O \xrightarrow[\text{dehydrogenase}]{\text{glutamate}}$$

$$\text{alpha-ketoglutaric acid} + NH_3 + NADH + H^+$$

This reaction is a way of "unloading" the nitrogen of amino acids; the alpha-ketoglutaric acid formed can enter the citric acid cycle.

The ammonia thus produced reacts with carbon dioxide to form a high-energy compound, carbamyl phosphate:

$$CO_2 + NH_3 + 2ATP + H_2O \xrightarrow[\text{synthetase}]{\text{carbamyl phosphate}}$$

$$\text{carbamyl phosphate} + 2ADP + 2P_i$$

This compound then transfers its carbamyl group

$$\overset{\overset{\textstyle O}{\|}}{-C-NH_2}$$

to ornithine to make citrulline:

$$\text{carbamyl phosphate} + \text{ornithine} \xrightarrow[\text{transcarbamylase}]{\text{ornithine}} \text{citrulline} + H_3PO_4$$

A second amino group now enters the cycle as aspartic acid, also formed from glutamic acid:

$$\text{glutamic acid} + \text{oxaloacetic acid} \xrightarrow[\text{transaminase}]{\text{aspartate-glutamate}}$$

$$\text{alpha-ketoglutaric acid} + \text{aspartic acid}$$

In a two-step reaction, the amino group of aspartic acid reacts with the citrulline from the first part of the cycle to form argininosuccinate, which then breaks down to make arginine and fumaric acid (which returns to the citric acid cycle):

$$\text{citrulline} + \text{aspartic acid} + ATP \xrightarrow[\text{synthetase}]{\text{argininosuccinate}} \xrightarrow{\text{argininosuccinase}}$$

$$\text{arginine} + \text{fumaric acid} + AMP + PP_i$$

(In this reaction, ATP does not lose one phosphate to form ADP, but loses two,

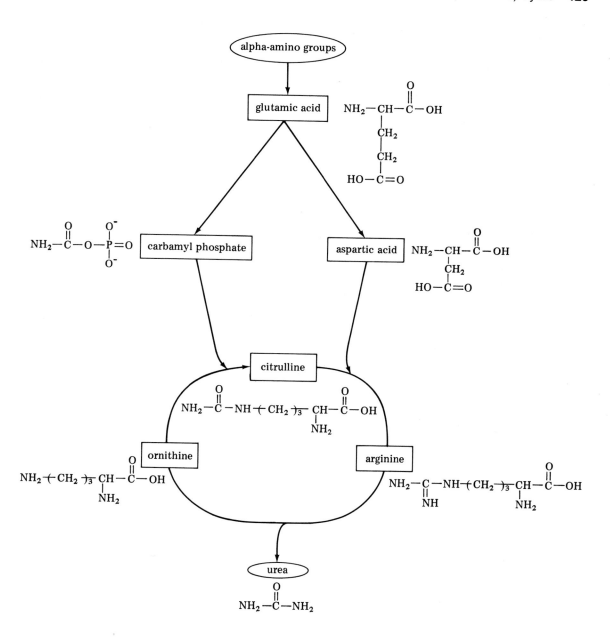

Figure 19.3 The urea (ornithine) cycle. Through this process, amino groups end up as urea, which is excreted in urine.

as pyrophosphate, PP_i, leaving behind adenosine monophosphate, AMP.) The arginine is then finally split by water to form urea and regenerate ornithine:

$$\text{arginine} + H_2O \xrightarrow{\text{arginase}} \text{urea} + \text{ornithine}$$

The overall equation for the urea cycle is as follows:

$$2NH_3 + CO_2 + 3ATP + 2H_2O \longrightarrow \text{urea} + 2ADP + 2P_i + AMP + PP_i$$

Because another ATP is needed to hydrolyze pyrophosphate to two phosphate groups, a total of four ATP molecules is needed for the urea cycle. Each time one molecule of urea is excreted, two molecules of toxic ammonia are removed from the body.

19.4 Amino acid anabolism

During evolution, human beings lost the ability to make certain amino acids. Since you must therefore include them in your diet, they are called **essential amino acids**. The others, which your body can synthesize, are called nonessential; this term does not mean they are unimportant, however. The amino acids are divided into these two categories as follows:

Essential: histidine (for children), isoleucine, leucine, lysine, methionine, phenylalanine, threonine, tryptophan, valine
Nonessential: alanine, arginine, asparagine, aspartic acid, cysteine, glutamic acid, glutamine, glycine, proline, serine, tyrosine

Ammonia is the raw material for making the nonessential amino acids. It adds to alpha-ketoglutaric acid in a reaction that is really the reverse of the step that starts the urea cycle:

$$NH_3 + \text{alpha-ketoglutaric acid} + NADPH + H^+ \xrightarrow[\text{dehydrogenase}]{\text{glutamate}}$$

$$\text{glutamic acid} + NADP^+$$

The formation of glutamic acid is the only significant pathway for the production of alpha-amino groups directly from ammonia. Glutamic acid then serves as an amino group donor, transferring it to other alpha-keto acids through transamination. This process represents the major pathway for the addition of the alpha-amino group during the synthesis of most other amino acids.

Glutamine, for example, is made from the addition of a second ammonia to glutamic acid:

$$\text{glutamic acid} + NH_3 + ATP \xrightarrow[\text{synthetase}]{\text{glutamine}} \text{glutamine} + ADP + P_i$$

Proline is also derived from this amino acid. Alanine and aspartic acid are produced by transamination to pyruvic acid and oxaloacetic acid, respectively:

$$\text{glutamic acid} + \text{pyruvic acid} \xrightarrow[\text{transaminase}]{\text{glutamate-pyruvate}}$$

$$\text{alanine} + \text{alpha-ketoglutaric acid}$$

$$\text{glutamic acid} + \text{oxaloacetic acid} \xrightarrow[\text{transaminase}]{\text{glutamate}}$$

$$\text{aspartic acid} + \text{alpha-ketoglutaric acid}$$

Asparagine is then formed from the reaction of ammonia with aspartic acid:

$$\text{aspartic acid} + NH_3 + \text{ATP} \xrightarrow[\text{synthetase}]{\text{asparagine}} \text{asparagine} + \text{ADP} + P_i$$

The addition of an —OH group, hydroxylation, converts the essential amino acid phenylalanine to tyrosine:

$$\text{phenylalanine} + \text{NADPH} + H^+ + O_2 \xrightarrow[\text{hydroxylase}]{\text{phenylalanine}}$$

$$\text{tyrosine} + \text{NADP}^+ + H_2O$$

Serine is generated in a series of reactions from 3-phosphoglyceric acid, an intermediate of glycolysis. Glycine is made from serine by the removal of a carbon, along with its hydroxyl group, by the coenzyme tetrahydrofolic acid. The reaction of serine with methionine, an essential amino acid, occurs in several steps and leads to cysteine, replacing the hydroxyl group by a sulfhydryl group (—SH).

Besides the synthesis of nonessential amino acids, further anabolic pathways are available for producing many important nitrogen-containing molecules. The amino acids available from either diet or synthesis are modified through a series of reactions until the desired product is made. One type of process that can take place is transamidination, the transfer of the guanidine part of arginine

$$\overset{\displaystyle NH}{\underset{\displaystyle}{\overset{\|}{-C}}}-NH_2$$

Another possible reaction, transmethylation, is the transfer of a methyl group from methionine to the sulfur, nitrogen, carbon, or oxygen of another molecule. A third possible change is decarboxylation, the removal of a carboxyl group as carbon dioxide, CO_2, to form an amine. It is through these and other reactions that amino acids are turned into other molecules such as hormones, coenzymes, alkaloids, pigments, neurotransmitters, and porphyrins. Table 19-2 lists a few of the many compounds made from amino acids.

The main fate of amino acids is protein synthesis, formation of the specific proteins needed by your body. This process is much more complicated than either carbohydrate synthesis, in which identical mono- or disaccharides are linked together, or lipid synthesis, in which fatty acids are joined to an al-

Table 19-2 **Some Compounds Synthesized from Amino Acids**

Amino acid	Compounds
aspartic acid	pyrimidines
glycine	purines, creatine
serine	sphingosine
tyrosine	epinephrine
tryptophan	skatole, serotonin

cohol. Even the synthesis of a complex lipid like cholesterol is simpler than the making of a protein. The reason is that proteins must form from 20 different building blocks, the amino acids, in a very specific order. Hundreds of these amino acids must somehow be directed to make peptide bonds according to a fixed "recipe" for each particular protein. This procedure, in which proteins are made from the "recipes" in the genes, is discussed in detail in the next chapter.

19.5 Metabolism of other nitrogen compounds

The nucleic acids, the huge molecules that carry your genetic information, contain two types of nitrogen-containing molecules—purines and pyrimidines. Neither of these are required in your diet since they can be synthesized in the body in anabolic pathways from the materials shown in Figure 19-4. More is said about the structure of the nucleic acids in the next chapter.

The catabolism of *purines* leads to the formation of uric acid, which is excreted in the urine. A metabolic defect, which causes the overproduction of

uric acid and its deposit in cartilage, results in gout (Figure 19-5). The *pyrimidines* are broken down to ammonia and urea.

Another important nitrogen-based molecule, *heme*, is the prosthetic group of the protein hemoglobin, the oxygen carrier of the red blood cells. Heme is based on a class of compounds called porphyrins, whose skeleton is illustrated in Figure 19-6a. It is synthesized in a complex series of reactions

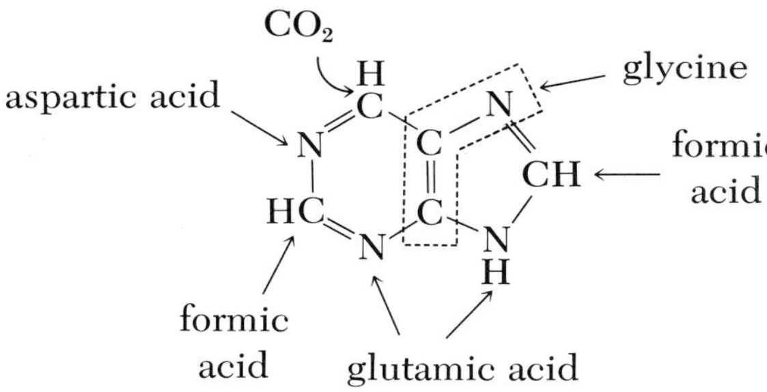

Figure 19-4.   The sources of the atoms in the purine ring and pyrimidine ring.

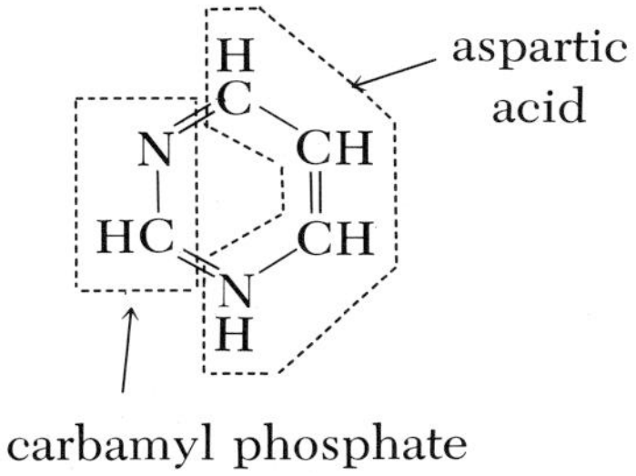

forming first a substituted pyrrole, the five-membered nitrogen heterocycle. Four of these molecules, called porphobilinogens, join together to make protoporphyrin III, a porphyrin with specific organic groups attached to the ring carbons. The addition of an iron atom to the center of the molecule results in heme (Figure 19-6b).

Heme catabolism begins when the red blood cells break down at the end of their approximately 126-day life span. First a bond is broken at one of the carbons that bridge the pyrrole rings (making choleglobin). Removal of the protein part, globin, leaves verdohemochrome, which is then converted to biliverdin. The iron is removed and stored for later reuse; the globin portion is broken down to its amino acids, which may then also be used by the body. Biliverdin is reduced to bilirubin and transported from the spleen to the liver and then the intestine. Here it is converted to products (urobilinogen and stercobilinogen) that are oxidized in air to urobilin and stercobilin. These orange-red bile pigments are excreted and contribute to the color of the urine and feces, respectively. When those pigments accumulate in the blood plasma giving the skin a yellowish tint, jaundice develops. This condition may result from increased destruction of red blood cells (hemolytic jaundice), liver disease such as infectious hepatitis and cirrhosis, or obstruction of the biliary passages, the pathway to the bowel ("surgical jaundice").

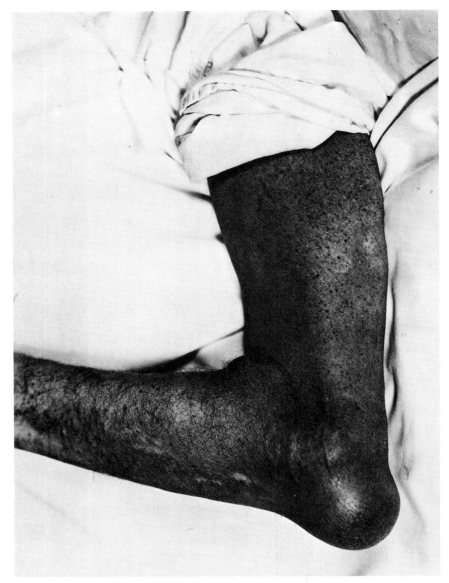

Figure 19-5. Elbow of person with gout. (Photo by Leonard D. Dank.)

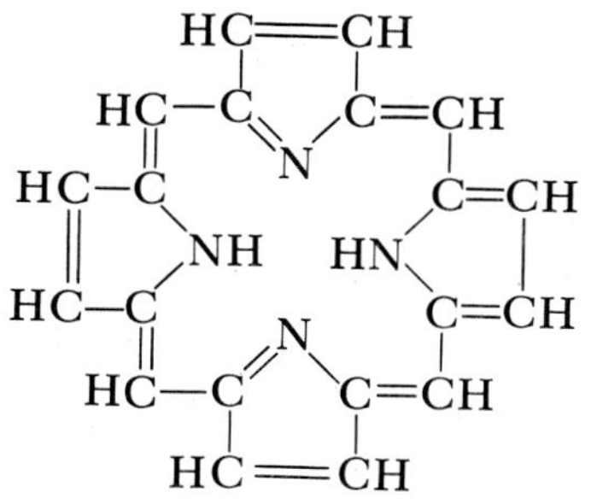

porphyrin ring

(a)

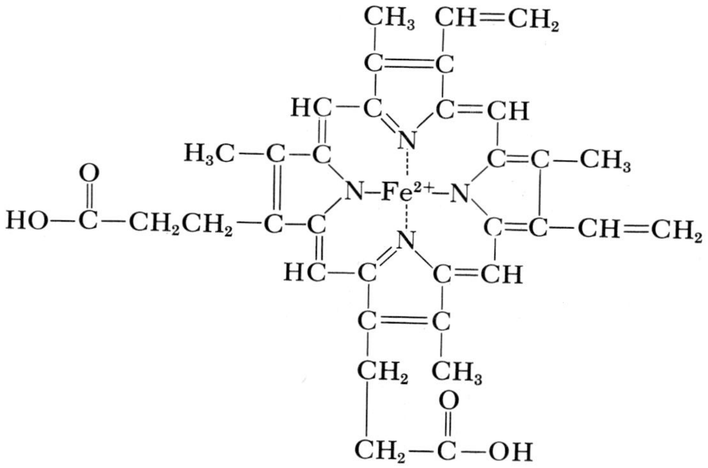

heme group

(b)

Figure 19-6. The structures of the (a) porphyrin ring and (b) heme group. Oxygen binds to the iron at the center of the heme groups in hemoglobin.

19.6 Nitrogen balance

Nitrogen compounds are constantly being metabolized by your body. Their main sources are the amino acids of the proteins you eat. The resulting "pool" of absorbed amino acids can be drawn from to make your proteins or to form other nitrogen-based molecules. The amino acids can further break down, forming urea, with the carbon part entering the citric acid cycle. Figure 19-7 shows the possible sources and fates of the amino acids.

The continuing entry and loss of nitrogen compounds creates a **nitrogen balance** in the body. In adults with adequate protein in the diet, the amount of

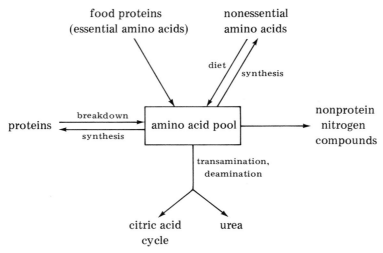

Figure 19-7. The amino acid "pool." Amino acids cannot be stored in the body but are constantly being drawn upon or excreted.

nitrogen ingested, mostly as amino acids, is equal to the amount of nitrogen excreted, primarily as urea. When the intake of nitrogen is greater than the output, as occurs during infancy, childhood, convalescence from a severe illness, and pregnancy, a *positive* nitrogen balance results. The opposite condition, a greater excretion of nitrogen than input, which occurs in old age and starvation, results in a *negative* nitrogen balance. The amount of imbalance in either of these two cases is only a small fraction of the total amount of nitrogen metabolized.

Most of the proteins of the body are constantly undergoing **turnover**, a process of being broken down and re-formed. Since the total amount of protein stays the same, these two processes must take place at the same time. The rate of turnover for different proteins can vary from several days as in the case of blood proteins to many months for muscle protein. Some types of protein such as the collagen that makes up your connective tissue do not appear to undergo turnover at all. An adult (weighing about 150 pounds) breaks down and resynthesizes almost 1 pound (400 g) of protein every day. Of this amount about 33 g of protein are lost daily; extreme stress, such as infection, fever, or surgery, increases this loss. There is no overall master control of protein metabolism; instead, synthesis occurs for each protein independently, at a rate that meets the needs of the body.

Because the fixation of molecular nitrogen is a process that is limited to a relatively small number of organisms, the forms of nitrogen usable by the body are scarce compared to the other nutrients. Therefore, your body practices economy in protein metabolism to get the most out of the roughly 100 g of amino acids you eat each day. Amino groups can be "salvaged" or spared for reuse by trapping them with alpha-ketoglutaric acid during catabolism. Also, the synthesis of the nonessential amino acids is under feedback control by the regulatory enzymes that make them. The synthesis of these enzymes is further controlled by the amount of amino acid already present. If enough is available from the diet, the enzyme needed to synthesize that particular amino acid will not be made. In this way, your body produces only those nitrogen compounds that it immediately needs.

The precise amount of protein you eat varies from day to day. Yet a nitrogen balance exists because the amount of urea that is formed can also change. It is therefore the excretion of urea that serves as a "leveling device" to maintain your nitrogen balance. In this sense, the amount of amino acids that you eat controls their rate of breakdown; this mechanism is the biological answer to the problem of a variable diet.

19.7 Dietary protein

To receive the greatest nutritional value, you should eat *a balanced mixture of the essential amino acids.* Your diet must not necessarily contain an equal amount of each, however. Table 19-3 lists the number of grams of essential amino acids needed each day by a person weighing about 150 pounds. The total recommended protein allowance is 56 g/day for a 150-pound man and 46 g/day for a 130-pound woman, with an additional 30 g/day during pregnancy and 20 g/day extra during lactation.

Therefore, it is important to not eat just any protein, but those that contain the proper balance of amino acids. This balance is measured by the **biological value** of the protein, which is defined as the ratio of nitrogen retained to the

Table 19-3 **Estimated Amino Acid Requirements (g/day)[a]**

histidine	?
isoleucine	0.84
leucine	1.12
lysine	0.84
methionine[b]	0.70
phenylalanine[c]	1.12
threonine	0.56
tryptophan	0.21
valine	0.98

[a] From National Research Council, National Academy of Sciences (1974); based on a 150-pound (70-kg) adult.
[b] Includes cystine.
[c] Includes tyrosine.

nitrogen absorbed by the body. A protein that lacks one of the essential amino acids has no biological value compared to "complete" proteins such as casein and lactoglobulin found in milk. Animal proteins in general have a greater biological value than plant proteins. Plants have proteins that are lower in lysine, methionine, and tryptophan and contain a lower concentration of protein in a less digestible form. Digestibility (or net protein utilization) is the ratio of nitrogen retained to the nitrogen intake from diet.

By combining protein from different sources, it is possible to make up for the inadequacies of each by itself. Most grains are low in lysine and most legumes are low in methionine. The traditional Latin combination of rice and beans has a higher biological value than each food eaten by itself. To be effective, however, the proteins must be eaten in the same meal, since amino acids cannot be stored.

A vegetable mixture that has a biological value almost identical to cow's milk consists mainly of cottonseed meal, whole corn, and whole sorghum. Another mixture contains corn meal, soy flour, and nonfat dried milk. These combinations are used to supplement the diets in countries where malnutrition is present.

19.8 Protein deficiency—kwashiorkor

Protein deficiency can have serious consequences because of the shortage of one or more essential amino acids. Such symptoms as anemia, edema, peptic ulcers, low basal metabolism, growth failure, and diminished brain activity may result. It is generally more harmful to eat plenty of an unbalanced mixture of amino acids than an insufficient amount of balanced protein.

The world's food problem is really a protein problem. The most serious nutritional disorder is **kwashiorkor** (pronounced kwa-shee-or'-kor), a disease found in the children of economically depressed areas in central and south Africa, India, and Latin America. It is caused by severe protein deficiency (with lack of most of the essential amino acids) in the presence of enough or even excess calories from carbohydrates. Children are affected most because growth requires continual synthesis of new protein. The child who gets this disease, such as the one shown in Figure 19-8, has often been weaned from its mother's breast to a diet based on starchy food like yams or cassava. The symptoms are edema, a "flaky paint" skin condition, thinning hair, discoloration, enlarged "fatty" liver, and apathy. Mortality is high, from 15 to 40%. Treatment consists of a diet high in good-quality protein. There are many programs under way to improve the protein value of diets. One involves developing cereals with a greater amount of protein and a higher concentration of lysine, since kwashiorkor results in part from a deficiency of this essential amino acid.

Figure 19-8.   A two-year-old with kwashiorkor; the child cannot stand up. (FAO photo by F. Botts.)

19.9 Metabolism of carbohydrates, lipids, and proteins

Your body can be thought of as a food-processing factory. The input consists of food, but it must be the "right" food, that is, containing the essential raw materials. The food molecules are modified by a large number of reactions which occur in small steps, each catalyzed by an enzyme. These chemical changes are the basis of metabolism. They serve two purposes: to provide energy and to supply intermediates for the synthesis of larger molecules. The waste products from these processes are excreted from the "factory."

Figure 19-9 is a very simplified drawing of the major pathways of metabolism. A complete chart would be extremely complex since many of the intermediates formed in one sequence may enter into two or more other pathways

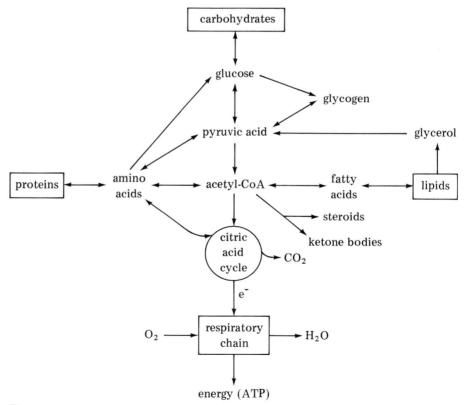

Figure 19-9. A summary of the major metabolic pathways. Note the central role of acetyl-CoA and the citric acid cycle. This is a highly simplified diagram of the actual metabolic pathways involved.

depending on the needs of your body. Notice in this diagram the close inter-connections between carbohydrate, lipid, and protein metabolism. This rela-tionship allows you to make the most efficient use of your food.

Now you should understand why you are what you eat. But you must also realize that your body is not just a collection of atoms obtained from food. Liv-ing organisms use energy to remain organized systems; they have a high de-gree of order. In this important sense, you are much more than the food you eat.

SUMMARY

Although part of the protein you eat is used to produce energy, most is used to provide the amino acids needed to make your own proteins and other impor-tant molecules. Dietary protein is the body's major source of nitrogen.

Nitrogen occurs in the environment in the form of N_2. The nitrogen mole-cule is chemically converted to different forms through the nitrogen cycle. This series of changes relates nitrogen metabolism in microorganisms, plants, and animals.

During catabolism, certain amino acids called "glycogenic" are transformed into intermediates of the citric acid cycle; others called "ketogenic" are con-verted to acetyl-CoA or acetoacetyl-CoA. The first step in the catabolism of any amino acid is transamination, removal of the alpha-amino group and its transfer to an alpha-keto acid such as alpha-ketoglutaric acid. The carbon skel-etons can then enter the citric acid cycle to generate ATP.

The urea (ornithine) cycle removes nitrogen from the body when an excess is present. The two amino groups of urea are derived from glutamic acid. In the cycle, arginine is converted to ornithine, and urea is split off and excreted in the urine.

Amino acids that must be obtained from the diet are called "essential." The others, the "nonessential" amino acids, can be synthesized in the body. Nitrogen-containing molecules can be synthesized by such reactions as trans-amidination (transfer of a guanidine group), transmethylation (transfer of a methyl group), and decarboxylation (loss of a carboxyl group.) Hormones, vit-amins, coenzymes, alkaloids, and other compounds are made through anabo-lism.

The purines and pyrimidines are the nitrogen heterocycles present in the nucleic acids. They can be synthesized from amino acids. The breakdown of purines forms uric acid; an excess leads to gout. Heme, the prosthetic group of hemoglobin, is synthesized in a complex series of reactions forming a substi-tuted pyrrole and then the porphyrin ring. When the red blood cells break down after 126 days, heme is catabolized to urobilin and stercobilin, the orange-red pigments that contribute to the color of urine and feces.

Nitrogen compounds are constantly being metabolized by the body. An amino acid "pool" is formed from the diet and the synthesis of amino acids, as well as the breakdown ("turnover") of proteins. The amino acids either are used to make new molecules or are broken down to urea, with the carbon skeleton entering the citric acid cycle. A nitrogen balance results from the continuing entry and loss of nitrogen compounds. When the intake of nitrogen is greater than the amount excreted, a positive nitrogen balance results; the opposite condition is a negative nitrogen balance.

To provide maximum nutritional value, a balanced mixture of the essential amino acids must be provided by the diet. The biological value of a protein measures the ratio of nitrogen retained to the nitrogen absorbed by the body. Animal protein generally has a greater biological value than plant protein. Proteins can be combined from different sources and eaten together to make up for inadequacies of each by itself.

Protein deficiency can have serious consequences resulting from the absence of one or more of the essential amino acids. The world's most serious nutritional problem is kwashiorkor, a disease of children in economically depressed areas, usually caused by an exclusive diet of a starchy cereal.

The process of metabolism can be summarized as a modification of food molecules by a large number of reactions which occur in small steps, each catalyzed by an enzyme. The purpose of metabolism is to provide energy and to supply intermediates for the synthesis of larger molecules. The pathways of carbohydrate, lipid, and protein metabolism are closely interconnected, allowing the body to make efficient use of the food you eat.

Exercises

1. (Intro.) What does protein supply that carbohydrates and lipids do not?
2. (19.1) Describe the nitrogen cycle.
3. (19.2) What is the first step in amino acid catabolism? Describe its function.
4. (19.2) How is amino acid catabolism related to the citric acid cycle?
5. (19.3) Describe what happens in the urea cycle.
6. (19.4) What is the difference between essential and nonessential amino acids?
7. (19.4) What is transamidination? transmethylation? decarboxylation?
8. (19.4) What makes protein synthesis different from the synthesis of other nitrogen compounds?
9. (19.5) What is gout? How is it caused?
10. (19.5) Summarize heme metabolism.
11. (19.5) Describe the cause of jaundice.
12. (19.6) What is the amino acid "pool?"
13. (19.6) Explain the term "nitrogen balance."
14. (19.6) What is a positive nitrogen balance? negative nitrogen balance?

15. (19.6) Describe the process of protein turnover.

16. (19.6) Why is biologically usable nitrogen scarce? Relate this scarcity to protein metabolism.

17. (19.7) What is the importance of the biological value of protein?

18. (19.8) Describe the origin of kwashiorkor.

19. (19.9) How is the body like a food-processing factory?

20. (19.9) In a short paragraph, summarize the most important aspects of metabolism.

20

Heredity and protein synthesis

In the previous chapters, you learned why you are what you eat. The processes of metabolism explain how atoms of your food become parts of the molecules that make up your body. But this series of events cannot be the whole picture, because food is only the raw material for your chemical reactions. Many of the same foods are ingested by very different types of organisms. There is something else, a set of instructions, which determines whether you grow up to become a man (or woman) or a mouse.

The genes contain the information that decides your characteristics. They determine what proteins to make and therefore which chemical processes take place in the body. Some of these biochemical reactions result in observable traits like brown hair or being right-handed. But more importantly, the type of proteins, and especially enzymes, synthesized from "orders" contained in the genes makes you a member of *Homo sapiens*, the human species.

20.1 Nucleotides

The genes are located on the chromosomes along with a protein component (histones). Each chromosome is thought to consist of a single threadlike molecule of a large polymer called **deoxyribonucleic acid**, or **DNA**. To understand its structure, you must first look at the separate parts of this complex macromolecule.

One component is the five-carbon sugar ribose. Deoxyribonucleic acid contains a ribose that has no oxygen on carbon 2′ as shown below (in the beta form); it is therefore called deoxyribose. (The symbol 2′, read "two prime," is used to indicate the carbon number on the sugar.) Ribose is found in another very important molecule, similar to DNA, called **ribonucleic acid**, or **RNA**.

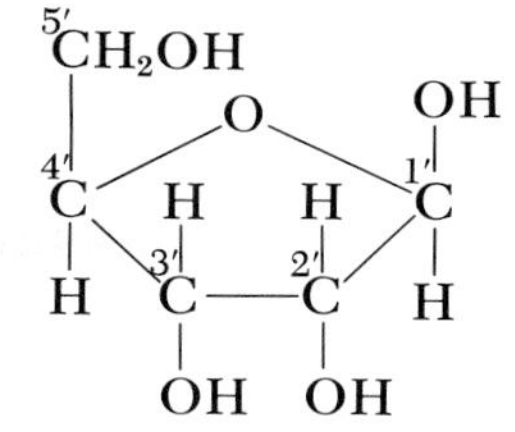

ribose (in RNA) 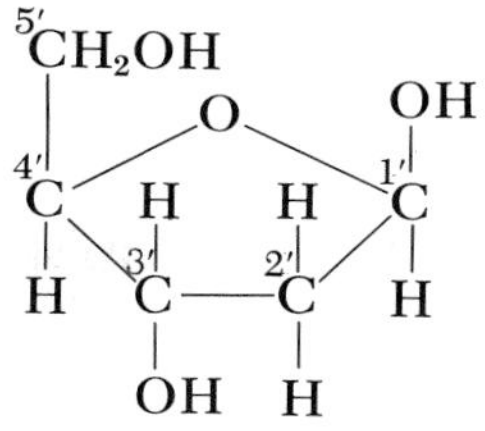   deoxyribose (in DNA)

Attached to the sugar is a nitrogen heterocycle referred to as the **base**. It can be either one of the purines or one of the pyrimidines (Figure 20-1). Each type

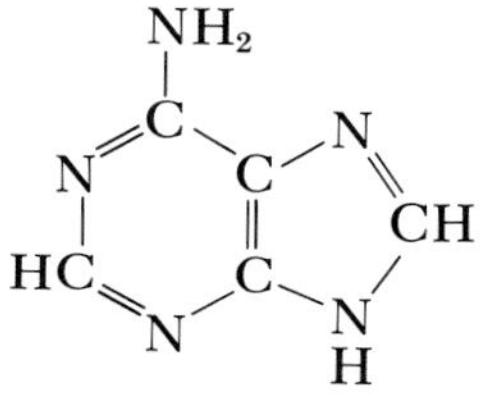

adenine (A)
(found in DNA and RNA)

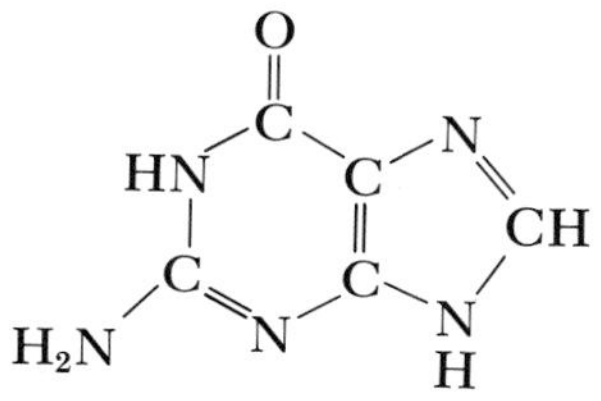

guanine (G)
(found in DNA and RNA)

purines

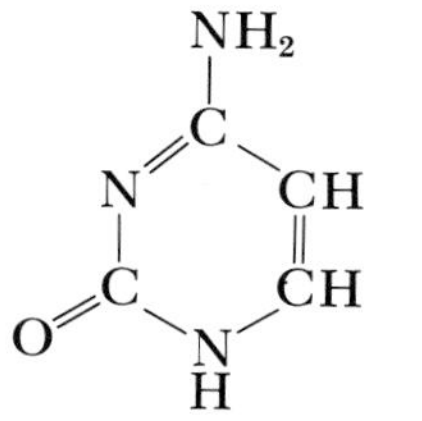

cytosine (C)
(found in DNA and RNA)

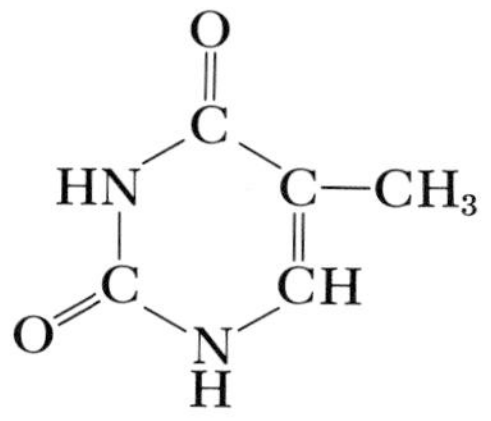

thymine (T)
(found in DNA only)

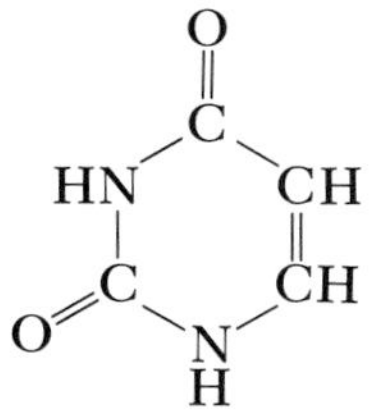

uracil (U)
(found in RNA only)

pyrimidines

Figure 20-1. The bases of DNA and RNA: DNA contains A, G, C, and T, while RNA contains A, G, C, and U.

of nucleic acid contains only four different nitrogen bases: DNA has adenine, guanine, cytosine, and thymine, while RNA has adenine, guanine, cytosine, and uracil. When one of the bases is attached to ribose, the resulting glycoside is called a **nucleoside**. The nucleosides are named adenosine, guanosine, cytidine, thymidine, and uridine.

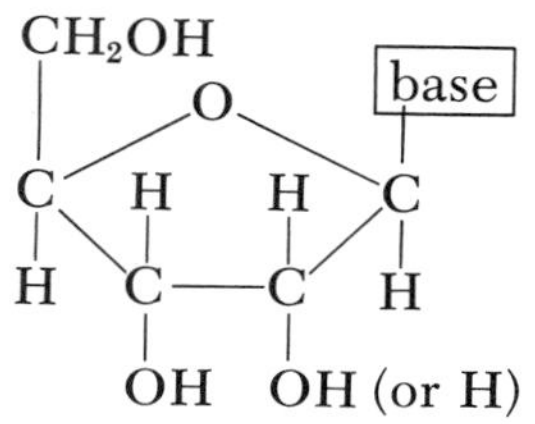

a nucleoside

The repeating unit in a nucleic acid polymer is a nucleoside with a phosphate group attached at carbon 5'; this molecule is called a nucleoside monophosphate, or **nucleotide**. *The nucleotides are the basic building blocks of the nucleic acid polymers,* just as amino acids make up proteins, and monosaccharides combine into polysaccharides. These fundamental units are illustrated in Figure 20-2a,b. Notice that ATP (adenosine triphosphate) is made

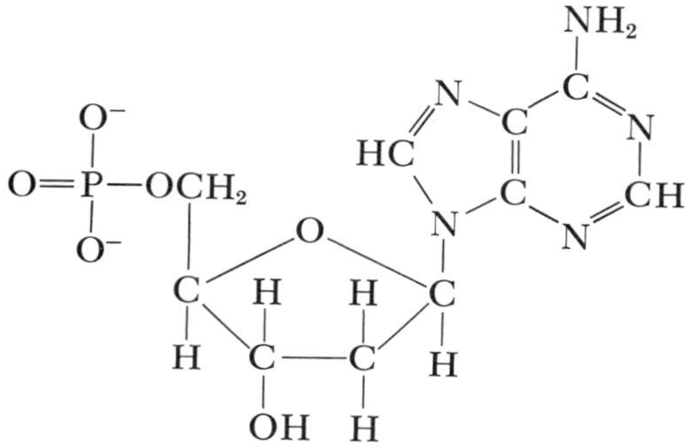

deoxyadenosine
monophosphate
(dAMP)

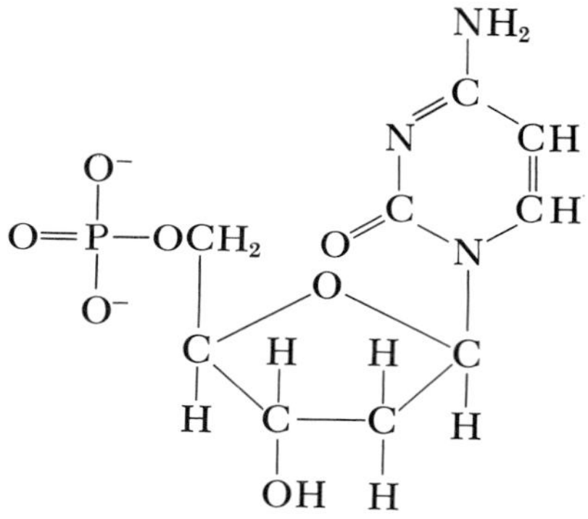

deoxycytidine
monophosphate
(dCMP)

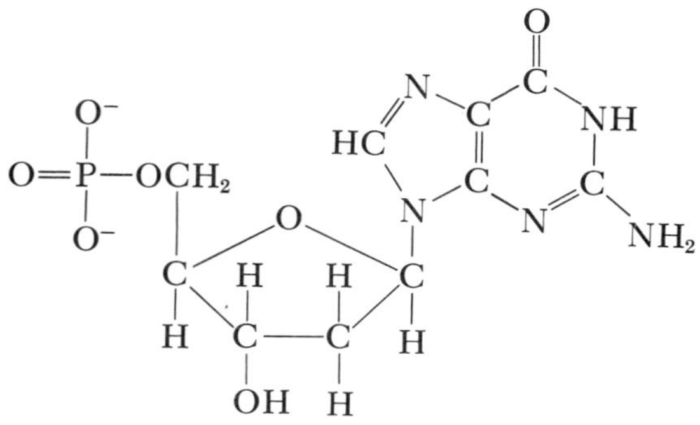

deoxyguanosine monophosphate
(dGMP)

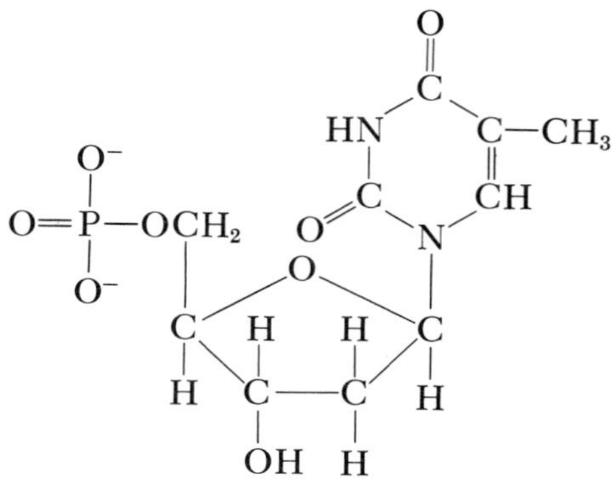

deoxythymidine
monophosphate
(dTMP)

Figure 20-2a. The nucleotides of DNA

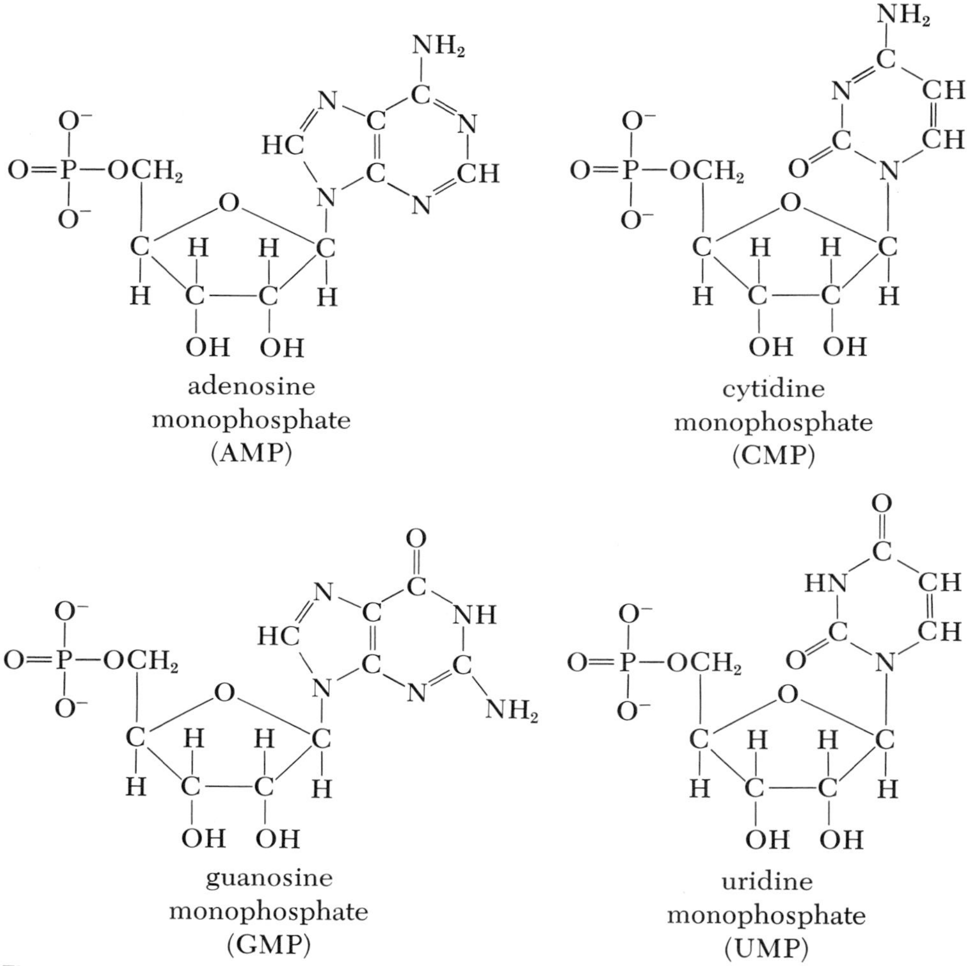

adenosine
monophosphate
(AMP)

cytidine
monophosphate
(CMP)

guanosine
monophosphate
(GMP)

uridine
monophosphate
(UMP)

Figure 20-2b. The nucleotides of RNA.

from AMP (adenosine monophosphate) by adding two more phosphate groups. Similarly, other molecules you know about, such as NAD (nicotinamide adenine dinucleotide) and FMN (flavin mononucleotide), are also related to nucleotides, the basic units of DNA and RNA.

20.2 Polynucleotides and base pairing

In each of the polymers, the nucleotides are linked together by a **phosphodiester bond**; that is, they are joined through the phosphate group to form a polynucleotide. The phosphate group connects carbon 3' on one nucleotide with carbon 5' on another, as shown in Figure 20-3. The polynucleotide chain

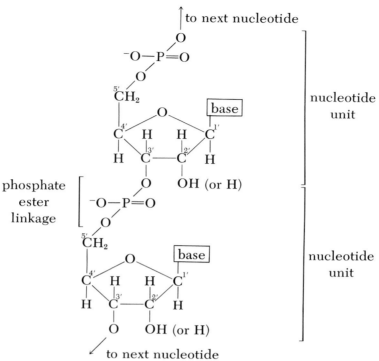

Figure 20-3. The phosphate ester linkage in a polynucleotide chain joins carbon 3′ of one nucleotide with carbon 5′ of the next nucleotide.

can be symbolized in the following way:

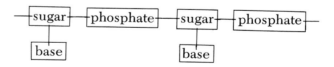

It consists of a "backbone" of sugars linked by phosphate groups with the bases coming off the chain. These polymers are tremendously long; DNA molecules contain several million nucleotides joined together in this way. *The size of this polymer is thus much, much larger than the longest protein or carbohydrate molecule.*

The actual structure of a DNA molecule is related to the way the bases interact with each other. A purine can form hydrogen bonds with a pyrimidine; however, because of their shape, *adenine can interact only with thymine (or uracil)* and *guanine can interact only with cytidine*, as shown in Figure 20-4. This special type of hydrogen bonding is called **base pairing**. Adenine is said to be "complementary" to thymine (or uracil); guanine and cytosine are also

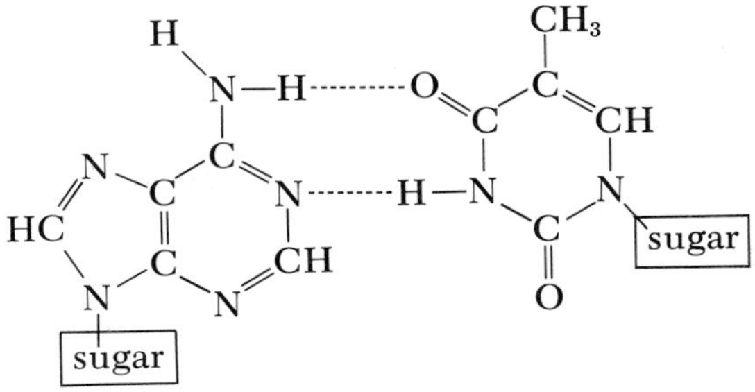

A ::::::::: T

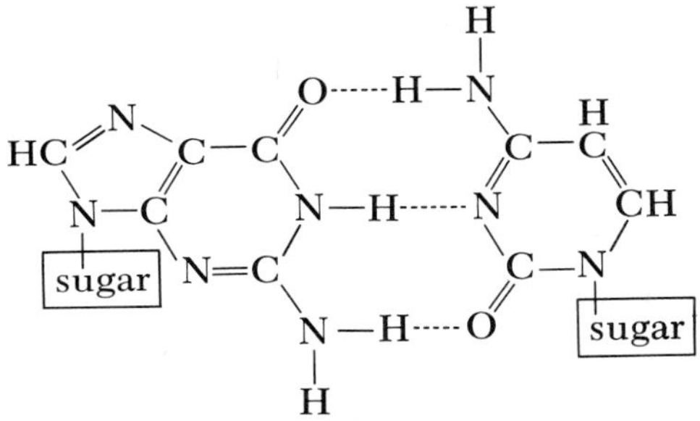

Figure 20-4. Base pairing. The only possible base pairs are between adenine and thymine (or uracil) and between guanine and cytosine

G :::::::: C

complementary base pairs. As you will see, these weak but very *specific* interactions between only certain pairs of bases provide a basis for the functioning of DNA and RNA.

The DNA molecule consists not of a single chain of nucleotides, but of two such separate strands; the molecule is "double stranded." The two strands run parallel to each other and wind around in the form of a right-handed helix. This "double helix" looks somewhat like a spiral staircase, as seen in Figure 20-5. Notice that the two chains interact through hydrogen bonding between parallel base pairs; there are ten such pairs per turn.

One strand must be exactly complementary to the other; when a guanine appears on one side, a cytidine must be on the other chain, and similarly for adenine and thymine. Thus, if the helix were unwound, the two strands would be related as outlined in Figure 20-6. If a part of one chain has the sequence of bases shown, pTpCpGpA (where the small "p" represents the phosphate linkage between the deoxyribose units containing the bases), the other strand must have the complementary arrangement, pApGpCpT.

In addition, notice that one chain runs in the opposite direction from the other, as shown by the arrows; that is, if the phosphate diester bonds are from carbon 3′ of one nucleotide to carbon 5′ of the next nucleotide (3′ ⟶ 5′) on

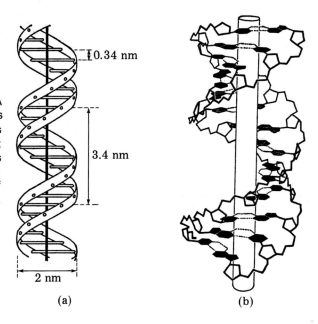

Figure 20-5. Representations of the DNA molecule. (a) The phosphate–sugar chains are shown as ribbons and the base pairs as horizontal rods. (b) A model of the DNA helix; dotted lines represent the hydrogen bonds between bases. (Reprinted with permission from J. N. Davidson, "The Biochemistry of Nucleic Acids," 7th ed. Chapman and Hall, London, 1972.)

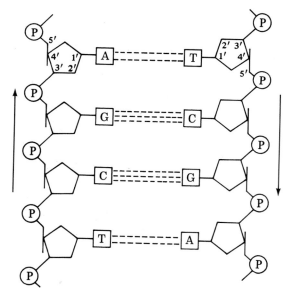

Figure 20-6. A representation of part of a DNA molecule, showing complementary base pairs along the two strands. (Reprinted with permission from J. N. Davidson, "The Biochemistry of Nucleic Acids," 7th ed. Chapman and Hall, London, 1972.)

the first strand, the linkages run $5' \longrightarrow 3'$ on the other chain. The double helix is stabilized by hydrophobic interactions between the flat aromatic rings of the bases which are stacked on top of each other in the center of the helix. The helix can be denatured by heat, changes in pH, or various chemicals (such as urea or formamide). In contrast to this arrangement in DNA, RNA molecules are usually single stranded and linear.

20.3   Transcription

At the beginning of this chapter, you learned that the genes contain the instructions for the synthesis of protein. Since the genes consist of DNA, you must look at the structure of this nucleic acid to see how these instructions are stored. The only differences between various DNA molecules lie in their sequence of nucleotides, the order in which the bases are arranged in each strand. *It is this sequence that determines what protein is synthesized by controlling the order of its amino acids.*

This discovery, that the genes through the sequence of bases in their DNA form a code, the **genetic code**, was one of the most important scientific events of this century. Along with the unraveling of the structure of the DNA molecule, these advances revolutionized the science of biology, forming a new branch called "molecular biology." For the first time, the chemical basis of the storage of genetic information was understood at the molecular level.

The genetic code consists of three-letter words, or *triplets*. Every three nucleotides ("letters") is a signal. It may say, "Use this particular amino acid next," or "Stop, the protein is finished." A three-letter code of four different bases can make 64 "words" ($4 \times 4 \times 4 = 64$). Thus, there are more than enough triplets of nucleotides to specify all 20 of the amino acids.

But how is the information contained in the DNA code used to make proteins? The first step must involve carrying the instruction from the nucleus of the cell, where the DNA is located, to the cytoplasm, where protein synthesis takes place. A special molecule, **messenger RNA**, symbolized **mRNA**, serves this purpose. It is formed by an enzyme (RNA polymerase) in a process referred to as **transcription**, since the DNA code is copied (transcribed) into an RNA molecule.

The procedure is illustrated in Figure 20-7. The DNA strands must first separate or "unzip." One of them then serves as a template or guide for the formation of the mRNA by means of *complementary base pairing*. The RNA nucleotides (in the form of the triphosphates) line up next to their complementary bases on the "unzipped" strand of DNA and are attached to each other by the enzyme. (The conversion of the triphosphate to the monophosphate provides the energy for this linkage.) The mRNA thus formed is an exact copy of the DNA except that it is a sort of "mirror image" because it consists of the complementary bases and not the same bases as the DNA strand it is made from.

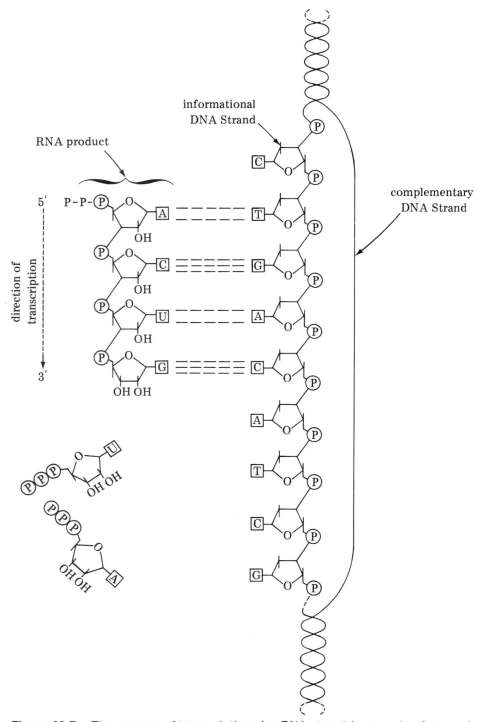

Figure 20-7. The process of transcription. A mRNA strand forms using free nucleotides by base pairing along one chain of DNA. Note that "P" in the figure represents a phosphate group. (Reprinted with permission from J. N. Davidson, "The Biochemistry of Nucleic Acids." 7th ed. Chapman and Hall, London, 1972.)

20.4 Translation of the genetic code

Messenger RNA carries the information that it has faithfully transcribed from the DNA into the cytoplasm of the cell, where it attaches to a ribosome, the site of protein synthesis (see Appendix E). The ribosome itself contains another kind of ribonucleic acid, **ribosomal RNA** or **rRNA**, whose function is not well understood, in addition to protein. It is here that the process of **translation** occurs, taking the code from the mRNA and interpreting it into the language of the proteins.

Since it came directly from the DNA, the mRNA code is also written in groups of three nucleotides. Each such triplet is called a **codon**; it is a three-letter "code word." Table 20-1 lists the meaning of each possible three-letter codon. As you can see, it is possible for more than one triplet to code for the same amino acid. (The genetic code is thus said to be "degenerate.")

Significantly, *the code is universal*: In every known organism, these codons correspond to the same amino acid. For example, part of the mRNA may contain the sequence AAA/CAC/UUU (short for pApApA/pCpApC/pUpUpU).

Table 20-1 The Genetic Code

mRNA codon	Translation
AAA, AAG	lysine
AAC, AAU	asparagine
ACA, ACC, ACG, ACU	threonine
AGA, AGG, CGA, CGC, CGG, CGU	arginine
AGC, AGU, UCA, UCC, UCG, UCU	serine
AUA, AUC, AUU	isoleucine
AUG	methionine (START)
CAA, CAG	glutamine
CAC, CAU	histidine
CCA, CCC, CCG, CCU	proline
CUA, CUC, CUG, CUU, UUA, UUG	leucine
GAA, GAG	glutamic acid
GAC, GAU	aspartic acid
GCA, GCC, GCG, GCU	alanine
GGA, GGC, GGG, GGU	glycine
GUA, GUC, GUG, GUU	valine
UAC, UAU	tyrosine
UGC, UGU	cysteine
UGG	tryptophan
UUC, UUU	phenylalanine
UAA, UAG, UGA	STOP

The resulting portion of the protein will consist of lysine–histidine–phenyl-alanine because AAA = lysine, CAC = histidine, and UUU = phenylalanine, whether the translation occurs in a person or a mouse.

Translation is made possible by a third type of RNA, **transfer RNA** or **tRNA**, which contains about 70 to 85 nucleotides. This molecule has two main parts: one end that can "read" the code of the mRNA and another that can attach it-self to a particular amino acid. There are thus at least 20 different kinds of tRNA, one to carry each of the amino acids. The tRNA picks up its "brand" of amino acid in a process that involves ATP and activating enzymes.

The other side of the tRNA molecule contains a group of three nucleotides called an "anticodon" because they are complementary to a mRNA codon for the amino acid carried by the tRNA. For example, the tRNA that carries phen-ylalanine has the anticodon AAG, which is complementary to the codon UUC on the mRNA chain. (The match between the tRNA anticodon and the mRNA codon need not be exact for all three nucleotides; thus, phenylalanine tRNA can also bind to the codon UUU on the mRNA.) Transfer RNA serves as an "adaptor" into which an amino acid is plugged in order to be adapted to the genetic code on the mRNA.

20.5 Protein synthesis

The actual process of protein synthesis is illustrated in Figure 20-8. Part A shows a DNA segment and the mRNA produced from one of the strands. No-tice in part B that the ribosome contains binding sites for two tRNA mole-cules—one for the growing polypeptide chain (the peptidyl-tRNA site) and one for the new amino acid being added (the aminoacyl-tRNA site). The codon AUG of the mRNA is the starting signal for protein synthesis; it codes for a special amino acid, formylmethionine, abbreviated fMet. The tRNA with this amino acid finds its way to the first site on the mRNA.

Alanine-tRNA occupies the second site; alanine will become the next amino acid in the chain. The first two amino acids are thus brought together in the order specified by the mRNA. By step C the ribosome has moved to the right by one codon in a process called translocation (which requires GTP, guanosine triphosphate, as a source of energy). Through an enzyme, alanine has formed a peptide bond with the carboxyl end of the first amino acid, making a dipeptide which is now attached at codon 1. The first tRNA, now without its amino acid, is ejected from the ribosome. The next codon brings serine-tRNA into the aminoacyl-tRNA site.

Part D carries the synthesis one step further; a tripeptide has been formed and the ribosome has moved down another codon. By E, seven amino acids have been joined, with an eighth, serine, ready to be added. In each step, the

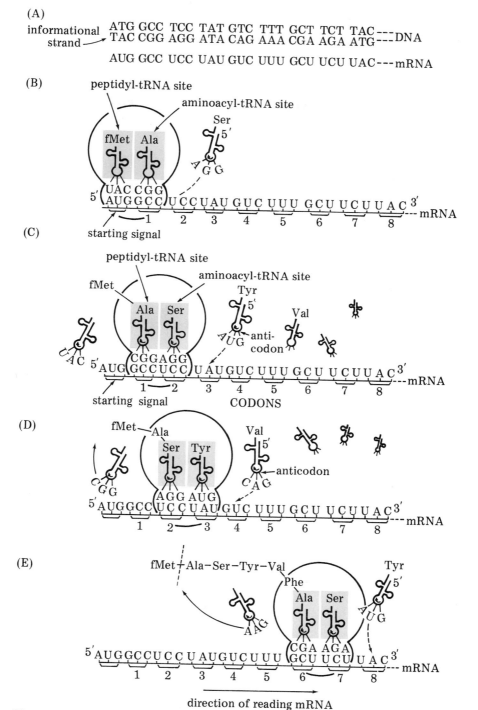

Figure 20-8. Protein synthesis. See text for details. (Reprinted with permission from J. N. Davidson, "The Biochemistry of Nucleic Acids," 7th ed. Chapman and Hall, London, 1972.)

tRNA anticodon recognizes the mRNA codon and carries the next amino acid into place on the ribosome for linkage to the growing polypeptide chain. When one of the termination codons (UAA, UAG, UGA) is read, protein synthesis stops and the polypeptide is released.

When protein synthesis occurs in the cell, several ribosomes are moving along each strand of mRNA at the same time, each at a different stage. This collection, called a polyribosome or polysome, is illustrated in Figure 20-9. This situation is similar to an assembly line in a factory.

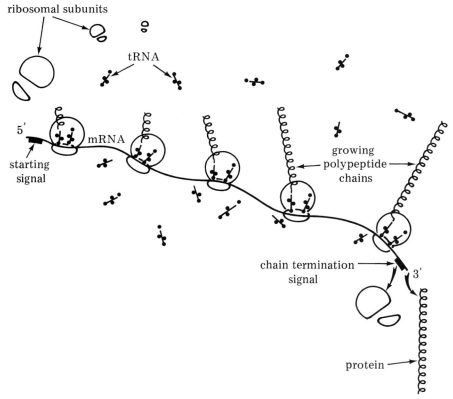

Figure 20-9.   A representation of the polysome (polyribosome) showing five ribosomes passing down a strand of mRNA. A protein chain is synthesized at each ribosome. (Reprinted with permission from J. N. Davidson, "The Biochemistry of Nucleic Acids," 7th ed. Chapman and Hall, London, 1972.)

20.6   Regulation of synthesis

Every cell contains all of the genetic information about you. Your cells are thus able to synthesize thousands of proteins. But not all possible proteins that can be made by a DNA molecule are being synthesized at the same time. The normal state for much of the DNA must be one of **repression**.

A special protein, the repressor, made by a "regulatory" gene, binds at a control point on the DNA (the operator site), thereby preventing production of mRNA and protein synthesis. The group of genes whose activity is controlled together by the same regulator is known as an operon. The operon consists of a number of "structural" genes (cistrons), each of which makes one polypeptide chain. The cistrons are found next to each other on the DNA because the proteins they code for are needed at the same time by the body.

When the repressor is inactivated (a process called "induction") protein synthesis begins. After enough protein has been made, the repressor protein becomes active again and binds at the operator gene, causing repression of the operon and a stop to the synthesis. Thus, *the repressor acts like an on–off switch* in controlling the manufacturing of protein; the process is summarized in Figure 20-10.

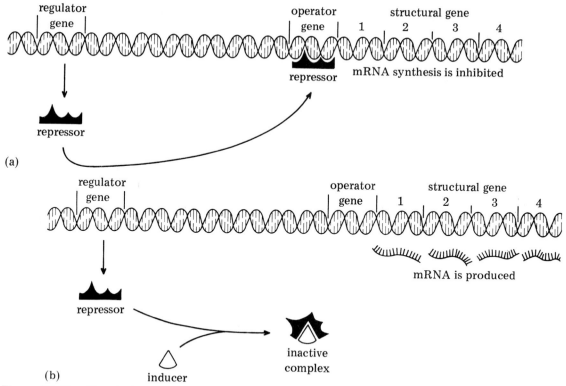

Figure 20-10.   Regulation of protein synthesis. (a) The normal state—repression; no synthesis. (b) If the repressor is inactivated, synthesis begins. (From P. Karlson, "Kurzes Lehrbuch der Biochemie," 9th ed. Georg Thieme Verlag, Stuttgart, Germany, 1974.)

In addition to these normal controls, the translation process can be inhibited by certain drugs. Streptomycin, erythromycin, tetracyclines, and others bind to the ribosomes of bacteria, preventing protein synthesis. Thus, they are effective as antibiotics.

20.7 DNA replication

You can recognize in yourself certain traits such as eye color which have been passed down from your parents. The inborn ability to develop characteristics of your ancestors is known as heredity. Traits are passed on because the fertilized egg from which you grew contained 23 chromosomes from your mother and 23 from your father. Furthermore, as you developed, more cells were generated through division, each having this identical set of 46 chromosomes. It is the chromosomes that carry the genetic information; they contain the genes, which consist of DNA.

The process by which a DNA molecule makes a copy of itself is called **replication**, the production of a copy or replica. This process transfers information from an existing cell to a new one. The obvious mechanism for making the copy is through base pairing. As seen in Figure 20-11, the double helix separates and two new strands start forming, using the original ones as templates in much the same way as mRNA is produced. The deoxyribonucleotides at-

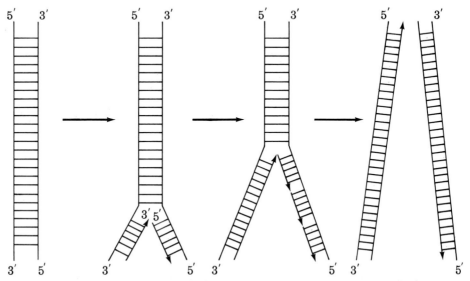

Figure 20-11. The process of DNA replication. A new strand forms by base pairing along each of the two original DNA strands. (Reprinted with permission from J. N. Davidson, "The Biochemistry of Nucleic Acids," 7th ed. Chapman and Hall, London, 1972.)

tach themselves to the separated strands and are then linked together by an enzyme. When the process is finished, *two complete double helices are formed, identical with the original DNA*. Each consists of one strand from the original helix and one new strand, complementary with the first.

A similar process is used by the **virus** particle, a structure consisting of a core of nucleic acid surrounded by a protein covering. It is not "alive" in the sense of having an independent metabolism. Yet a virus can infect almost any organism, including bacteria. The virus takes over a "host" cell, forcing it to make more viruses instead of the normal cell contents. The production of new virus particles usually kills the infected cell, causing such diseases as the common cold, poliomyelitis, chicken pox, mumps, measles, smallpox, and rabies. Viruses are essentially "portable" genes which make copies of themselves by replication. An example is shown in Figure 20-12.

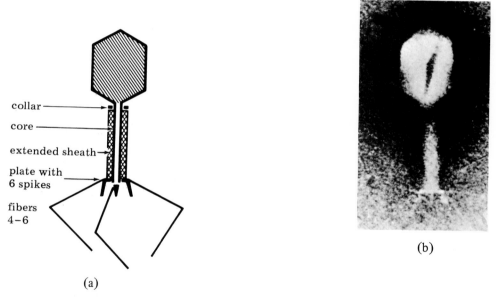

collar
core
extended sheath
plate with
6 spikes
fibers
4–6

(a)

(b)

Figure 20-12.   A virus particle. (a) Line drawing; (b) photo. (Photo courtesy of Carl Zeiss, Inc.)

20.8   Mutations

Replication faithfully copies the base sequence of DNA. If the DNA has been altered in some way, changing its order of nucleotides, that modification will also be transmitted. Thus, replication results in permanently inherited changes. The altered DNA will, of course, be reflected in the amino acid sequence of the protein for which it codes.

Any change in the sequence of nucleotides in a DNA molecule is known as a **mutation**. It may consist of replacement of one pair of complementary bases by another pair. More serious mutations involve the addition (insertion) or removal (deletion) of a nucleotide pair. These last types are known as frame-shift mutations since they cause the DNA to be read incorrectly; all the triplets after the mutation are read out of register. For example, if the original sequence was AAT/GCA/TCA and one nucleotide in the first triplet, say the T, is deleted, the strand will be read as AAG/CAT/CA. This type of mutation is usually lethal because it results in the synthesis of a completely different protein. Fortunately, most mutations are of the first type, point mutations, resulting in a change of only one amino acid.

Mutations can take place spontaneously. In fact, they provide a basis for evolution, the modification of a species. They are the ultimate source of all variations. Certain mutations may result in proteins that provide a trait giving an individual a better chance to survive and reproduce. Through replication, the offspring will also carry that favorable adaptive trait. Mutations occur unpredictably since they are molecular accidents, but once they have taken place, the altered DNA will be mechanically reproduced into millions of copies—the chance event becomes a certainty.

20.9 Mutagens

Mutations can also result from artificial means. Ionizing radiation like x rays or gamma rays cause mutations; the greater the dosage, the higher the rate of genetic change. The deformed children born to survivors of the atomic bombs dropped in Japan resulted from genetic damage caused by radiation. Ultraviolet radiation also produces mutation.

Certain substances known as **chemical mutagens** cause mutation as well. These include alkylating agents such as compounds of the form $ClCH_2CH_2N(R)CH_2CH_2Cl$ (R = alkyl) [a nitrogen mustard used in warfare] which can react with the nitrogen of the DNA bases. Nitrous acid, HNO_2, removes amino groups, changing cytidine to uridine, for example. Molecules with structures similar to the bases like 5-bromouracil, an analog of thymine, can become part of a DNA but will not hydrogen bond in the same way as the base it is replacing. Certain dyes such as the acridines can slip between the base pairs (a process called intercalation), causing mistakes during translation. All of these mutagens result in modifications of different severity in the proteins produced from the DNA. Because you are such a complex creature dependent on the delicate interplay of many biochemical processes, *most mutations can cause something to go wrong and may therefore be harmful or lethal*. Figure 20–13 shows an example of a genetic deformity.

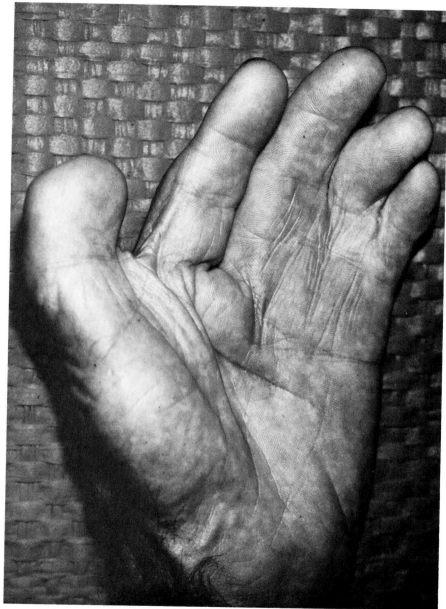

Figure 20-13.   A genetic deformity: polysyndactyly (webbed fingers). (Photo courtesy of Center for Disease Control, Atlanta, Georgia.)

20.10   **Molecular (genetic) diseases**

Certain mutations result in a specific defective protein, leading to a **molecular or genetic disease**. This hereditary condition results from a mistake in the DNA which is then transmitted by replication from parent to child. About 1500 such diseases exist. Molecular diseases tend to appear in particular racial groups because of intermarriage.

In *sickle cell anemia*, which occurs primarily in blacks, one portion of the DNA that controls the synthesis of hemoglobin may contain a single substitution, such as adenine for thymine, in one triplet. This defect causes valine to replace glutamic acid at a certain point in the protein chain. Because of the change in the hemoglobin, the red blood cells become sickle shaped at low oxygen pressures, as seen in Figure 20-14, from precipitation of the protein from the solution as a polymer. They become trapped in small blood vessels, impairing circulation and damaging organs, especially the kidney and bone. The sickled cells are also fragile and therefore hemolyze readily, resulting in anemia.

If a person has the defective gene from only one parent, the individual does not have anemia, but does have the sickle cell trait. It results in mild disturbances and a poor ability to adapt to conditions of low oxygen, as during exertion. About 10% of blacks in the United States have this trait. In tropical Africa, where the trait appears in approximately 25% of the native blacks, the sickled cell protects the carrier from malaria. (The protozoan that causes malaria must spend part of its life cycle within red blood cells but cannot grow in a sickled cell.)

Inborn errors of metabolism result from a missing or defective enzyme. *Phenylketonuria* (PKU) is caused by the lack of phenylalanine hydroxylase, the enzyme that converts phenylalanine to tyrosine. This disease is found in 1 out of 20,000 newborn. Other metabolic pathways, little used in the normal individual, become significant when the level of phenylalanine increases, as shown in Figure 20-15. As a result, severe mental retardation develops, and a large amount of phenylpyruvic acid (up to 1 to 2 g/day) is excreted in the urine. A victim of PKU is shown in Figure 20-16 (p. 465). Careful restriction of the intake of phenylalanine in the diet of the growing child having this disorder may prevent the symptoms from appearing. Babies with PKU are fed a milk substitute (Lofenalac) containing partially digested protein that is adequate in all amino acids except phenylalanine, in addition to lipids, carbohydrates, vitamins, and minerals.

Other hereditary metabolic disorders may be caused by the complete absence of the final product of a particular pathway, such as in albinism, in which the pigment melanin is not present because of a defective tyrosinase enzyme. An intermediate metabolic product may accumulate as in glycogen

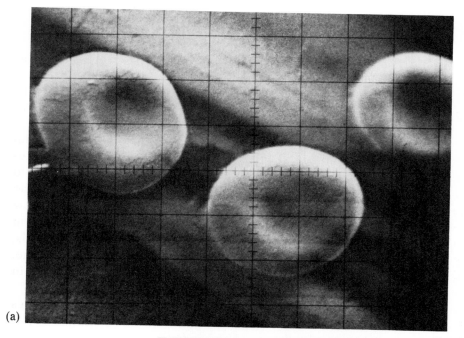

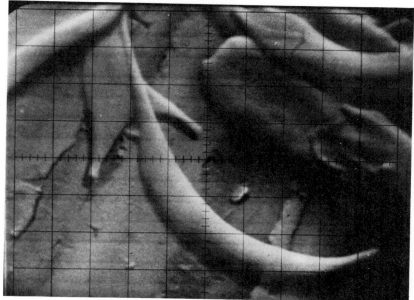

Figure 20-14. Comparison of (a) normal red blood cells with (b) sickle cells. [Micrographs (magnification ×5000); courtesy of Philips Electronic Instruments, Inc.]

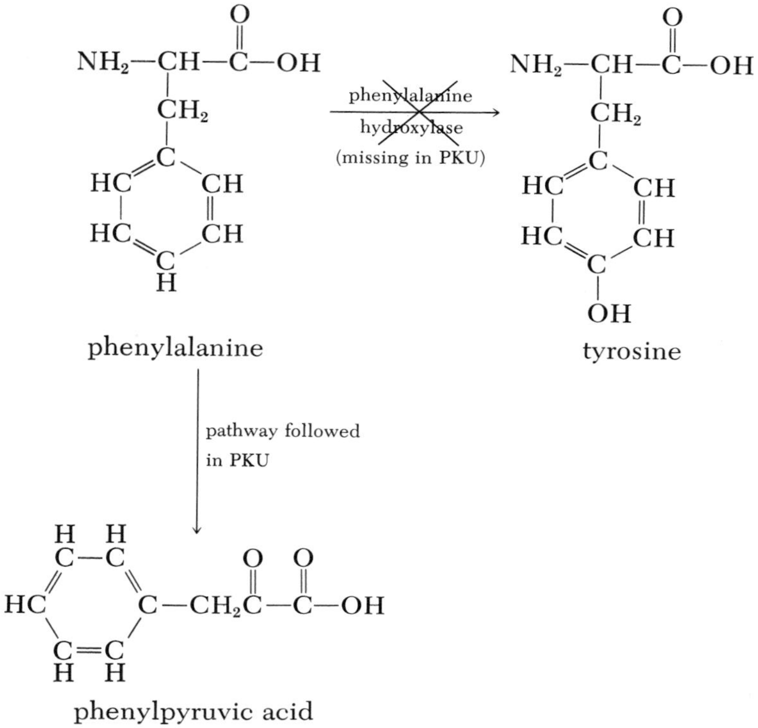

Figure 20-15. The metabolic change in phenylketonuria (PKU), a molecular (genetic) disease. The phenylpyruvic acid appears in the infant's urine. It can be tested for with ferric chloride.

storage disorders, which result from a lack of specific enzymes of glycogen metabolism. The defective enzymes or proteins for several other molecular diseases are presented in Table 20-2. In addition, Table 20-3 lists some

Table 20-2 Identity of Missing Protein in Molecular Diseases

Disorder	Missing enzyme or protein
galactosemia	galactose-1-phosphate uridylyltransferase
Gaucher's disease	glucocerebrosidase
goiter (familial)	iodotyrosine dehalogenase
hemophilia	antihemophilic factor
neonatal jaundice	glutathione peroxidase
pentosuria	xylulose dehydrogenase
pulmonary emphysema	antitrypsin
Tay-Sachs disease	hexosaminidase

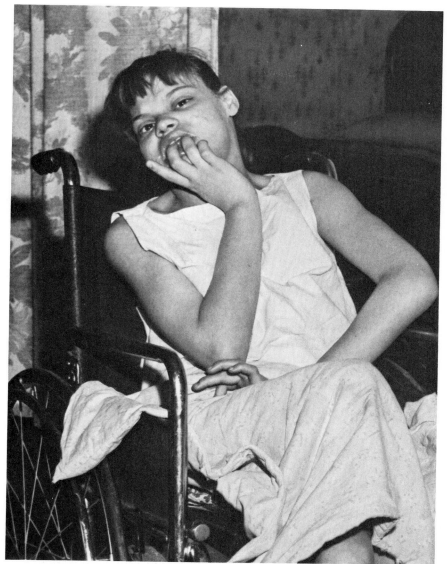

Figure 20-16. A victim of PKU. (Photo courtesy of National Foundation-March of Dimes.)

genetic diseases and their frequency in particular populations. Because each of these conditions comes from a mutant DNA, they can never be cured or eliminated; they can be controlled in certain cases.

Table 20-3 **Frequency of Molecular Diseases in Certain Populations**[a]

Condition	Population	Frequency (per 10,000)
porphyria variegata	South African whites	30
adrenogenital syn-drome	Yupik Eskimos	20
myotonic dystrophy	New Zealand	9
alpha-antitrypsin deficiency	Scandinavia	8
cystic fibrosis (pancreas)	Victoria, Australia	4.8
Gaucher's disease	Ashkenazi Jews, Israel	4
Duchenne muscular dystrophy	Northeast England	3.3
phenylketonuria	Scotland	2.0–2.8
Tay-Sachs disease	Ashkenazi Jews, United States	1.7
Huntington's chorea	Tasmania	1.7
tyrosinemia	Chicoutimi region, Canada	1.5–1.8

[a] From D. J. H. Brock and O. Mayo, eds., "The Biochemical Genetics of Man," Academic Press, New York, 1972. Reprinted with permission.

20.11 Information transfer

Deoxyribonucleic acid is the hereditary storage molecule. In the sequence of its nucleotides, it contains the "recipes" for making all proteins and enzymes needed for life. The flow of information travels from DNA to mRNA and then into a sequence of amino acids. This sequence is all that is needed because, as you already know, the order of amino acids completely determines the three-dimensional structure of a protein.

Once the information has been transferred to the protein, it cannot be recovered. A protein molecule does not carry genetic information; only DNA can serve this role in your body. And it is through the exact process of replication that DNA transfers its instructions to other cells. Figure 20-17 summarizes the major allowed general transfers of information. This simple picture represents what has been called the **central dogma** or doctrine of molecular biology.

Cell biologists have developed the technique of isolating a single cell and causing it to grow and divide. The resulting new cell or **clone** has DNA that is

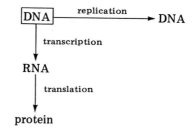

Figure 20-17. The central dogma of molecular biology. The flow of information begins with DNA.

identical to that of the original cell. If this technique of "cloning" could be applied to humans, a body (somatic) cell might be fused with an egg cell (whose nucleus was removed) to make an artifical embryo, which could then be carried by a foster mother in the normal way. The offspring would have exactly the same genes as the single body cell. Because it bypasses the normal means of transmitting genetic information, this asexual technique of reproduction raises the alarming possibility of producing large numbers of identical human beings.

SUMMARY

The genes contain the information that decides your characteristics. They contain molecules of a large polymer called deoxyribonucleic acid, or DNA. The repeating unit is a nucleotide. This molecule consists of three components: a five-carbon sugar, deoxyribose; one of four types of nitrogen heterocycles, called bases—either a purine (adenine and guanine) or a pyrimidine (cytosine and thymine); and a phosphate group. Deoxyribonucleic acid is a polynucleotide containing a repeating sequence of sugars and phosphate groups with a nitrogen base on each sugar molecule; the linkages in the chain are phosphodiester bonds.

The complete structure consists of two polynucleotide strands wound around each other in a double helix. The bases of one strand are complementary to those of the other; that is, an adenine always matches a thymine, and a guanine always pairs with a cytosine. These interactions, called base pairing, result from specific hydrogen bonds. The double helix of DNA is right-handed and stabilized by hydrophobic attractions between the aromatic rings of the bases. Molecules of RNA, ribonucleic acid, contain the sugar ribose and the base uridine instead of thymine; they are single stranded.

The order of the nucleotides in the DNA molecule determines the sequence of amino acids in proteins. The genetic code consists of groups of three nucleotides, each such "triplet" specifying a particular amino acid. Molecular biology deals with this chemical basis for the storage and transfer of genetic information.

The first step in protein synthesis is the formation of messenger RNA by the process of transcription. Through base pairing, it copies the sequence of the DNA and carries this information from the nucleus to the ribosomes, the site of protein synthesis. It is here that translation occurs, the process of interpreting the genetic code through molecules of transfer RNA. There is a separate tRNA for every amino acid; each contains an anticodon which recognizes the appropriate codon (nucleotide triplet) on the mRNA.

The actual process of synthesis occurs as two molecules of tRNA bind to the ribosome. One contains the growing peptide chain, while the other carries the next amino acid to be added. The ribosome moves over by one codon each time a peptide bond is formed. When a termination signal is read, synthesis stops, and the polypeptide is released.

Synthesis is regulated by repressor molecules, which can bind at a control point on the DNA and thereby prevent formation of mRNA. During induction, the repressor is inactivated and synthesis proceeds.

Heredity is the inborn ability to develop the characteristics of your ancestors. It is made possible by the replication of DNA, the formation of an exact copy through the mechanism of base pairing.

Any change in the sequence of nucleotides in DNA is known as a mutation. It may consist of substitution, deletion, or addition. Mutations occur naturally and provide a basis for evolution. Mutations can be caused by ionizing radiation, by ultraviolet radiation, or by chemical mutagens, which are substances such as alkylating agents, structural analogs, or dyes.

Molecular diseases result from specific defective proteins or enzymes and are hereditary since they result from a mutation. Examples include sickle cell anemia and phenylketonuria.

Information can be transferred from DNA to RNA and then to a protein. The DNA is the hereditary storage molecule that contains the genetic instructions. Through replication, DNA transmits the code to other cells. These relationships represent the "central dogma" of molecular biology.

Exercises

1. (Intro.) Why are the genes so important?
2. (20.1) Draw the structures of ribose and deoxyribose. Point out the differences.
3. (20.1) Which bases are found in DNA? in RNA? Identify which are purines and which are pyrimidines.
4. (20.1) What is a nucleoside? a nucleotide?
5. (20.2) What kind of bonds join the repeating units in a polynucleotide?
6. (20.2) What is base pairing? List the possible base pairs.
7. (20.2) Describe the double helix of DNA.

8. (20.2) What are complementary strands? Write the strand complementary to ACC-TACTAACCCTAG.

9. (20.3) What is the genetic code?

10. (20.3) Why must the genetic code consist of "words" of at least three nucleotides?

11. (20.3) What is transcription? Describe the function of mRNA.

12. (20.4) What is translation? Describe the role of tRNA.

13. (20.5) Write the sequence of amino acids that will be made from the following mRNA: AAC/AUG/UGU/GAU/GGG/CCC/AAG. (Refer to Table 20-1.)

14. (20.5) Write a possible base sequence of mRNA that could produce the following peptide: alanine–glycine–leucine–tryptophan–lysine. (Refer to Table 20-1.)

15. (20.5) Summarize in your own words the key steps in the process of protein synthesis.

16. (20.6) Describe how protein synthesis is regulated.

17. (20.7) What is heredity? How is it related to DNA?

18. (20.7) Describe replication.

19. (20.8) What is a mutation? Describe the different types of mutations and their effects.

20. (20.9) What is a mutagen? Give several examples.

21. (20.10) Why are molecular diseases not contagious?

22. (20.10) Describe the origin and effects of sickle cell anemia.

23. (20.10) What causes phenylketonuria?

24. (20.11) Describe the "central dogma" of molecular biology.

21

Vitamins and hormones

Your body needs fairly large amounts of the basic foods—carbohydrates, lipids, and proteins—to provide energy and synthesize important molecules. Certain other substances are equally necessary, but only in very small amounts. Some, called **hormones**, are made by specific glands in the body. Others, the **vitamins**, must be supplied by your diet.

21.1   Vitamins

The term vitamin comes from "vitamine" ("vital amine") because the first food factors that were found to be essential for life happened to be amines. This term is now used for any biologically active organic substance necessary in trace quantities for normal health and growth but not made by the body. Only small amounts are needed because *most of the vitamins serve as catalysts in the form of coenzymes.* They take part in a chemical reaction and are changed but are then restored by a second reaction. Thus, they do not get used up in the process.

The structures of the vitamins differ greatly. Some are polar molecules and are therefore water soluble. Others are lipid soluble and are absorbed from the intestine with the dietary lipids and can even be stored in the liver for various periods of time. Not all vitamins are present in food in the form that is used by your body. A few exist as provitamins, which must first be modified by a chemical reaction, similar to the way zymogens are converted to active enzymes.

Disease results when a vitamin is missing or deficient, conditions known, respectively, as avitaminosis and hypovitaminosis. Lack of some vitamins causes **deficiency diseases**; other vitamins have less specific effects. The skin appears to be especially sensitive to missing vitamins. Deficiencies occur mainly in populations where the choice of available food is limited.

Because they are stored in fat depots, the lipid-soluble vitamins can build up to toxic concentrations. Too much of a vitamin, hypervitaminosis, can

470

therefore be dangerous. The use of some vitamins in big doses to treat disease, megavitamin therapy, is currently being investigated. Orthomolecular psychiatry is a new field of study in which vitamins are used on an experimental basis to treat some mental disorders.

21.2   Water-soluble vitamins—thiamin

This vitamin, thiamin, is also known as B_1. Its active form, thiamin pyrophosphate, serves as a coenzyme for decarboxylases, which remove CO_2 from molecules in carbohydrate catabolism.

thiamin

Lack of sufficient thiamin results in *beriberi*. This disease is found in areas such as parts of Asia where polished rice, which contains very little of the vitamin, is the main food. Cardiovascular ("wet") beriberi leads to edema and acute cardiac symptoms, including heart failure. The neurological ("dry") effects involve weight loss, neuritis (nerve inflammation), loss of reflexes, wasting of muscles, and mental confusion or anxiety.

Thiamin is found in the outer layers of plant seeds in addition to animal tissues, especially pork. The amount of this vitamin in the diet can be increased by eating whole wheat bread rather than white bread because the portion of the grain that contains this vitamin is missing in white bread. Thiamin is also found in peas and beans but can be lost or destroyed by overcooking.

21.3   Water-soluble vitamins—riboflavin

Riboflavin is a member of the vitamin B_2 complex (along with nicotinic acid, folacin, and pantothenic acid). It is converted to the coenzymes flavin adenine dinucleotide (FAD) and flavin mononucleotide (FMN). The coenzyme FAD serves as a carrier of hydrogen atoms and electrons in the citric acid cycle. Flavin coenzymes are also needed for other oxidases and dehydrogenases in carbohydrate and amino acid metabolism. A deficiency in riboflavin leads to cheilosis (fissures of the corners of the mouth), dermatitis of the face and scrotum ("shark skin"), eye damage, and a magenta-colored tongue. Further-

$$CH_2-CH-CH-CH-CH_2OH$$

riboflavin

more, many liver enzymes lose their activity since they may require a coenzyme derived from riboflavin.

This vitamin is found in organ meats—kidney and liver. It is also present in yeast, wheat germ, milk, eggs, and green leafy vegetables. The vitamin is destroyed by sunlight and prolonged cooking.

21.4 Water-soluble vitamins—niacin

Niacin is the second member of the B_2 complex. It includes both nicotinic acid and its amide, nicotinamide. They are used to make two important coenzymes, nicotinamide adenine dinucleotide (NAD) and its phosphate derivative (NADP). As you have seen, these molecules play a major role in hydrogen transfer in metabolic reactions.

nicotinic acid   nicotinamide   niacin

Insufficient niacin results in *pellagra*, the disease of the "3D's": dermatitis, diarrhea, and dementia. It is characterized by dermatitis of the areas exposed to sunlight, stomatitis (inflammation of the mouth), inability to digest food, a sore magenta tongue, and disturbance of the central nervous system. During the first part of this century, as many as 170,000 cases were reported annually in the southeastern United States. Pellagra occurs in regions where corn is the major food (Figure 21-1).

Niacin is found in meat, especially organs and poultry, in addition to yeast and legumes (like beans and peas). This vitamin can be synthesized from tryptophan so sources of this amino acid such as milk and eggs supply niacin too.

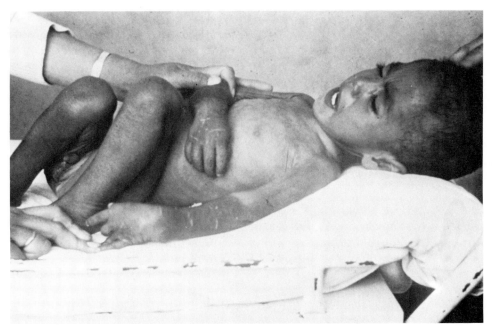

Figure 21-1. A child with pellagra. This condition results from insufficient niacin. (Photo courtesy of Center for Disease Control, Atlanta, Georgia.)

21.5   Water-soluble vitamins—folacin

The third member of the B_2 complex is folacin, or folic acid. Two double bonds are hydrogenated to make tetrahydrofolic acid, the metabolically active form of the vitamin. It serves as a carrier of one-carbon groups (such as formyl and hydroxymethyl) in pathways involving amino acids and nucleotides.

folacin

Since it is needed for cell growth and multiplication, drugs that interfere with folacin metabolism are used to control certain types of leukemia.

Lack of folic acid is probably one of the most common deficiencies in North America and western Europe. It results in a failure to make purines and thymine for DNA and decreased porphyrin synthesis. The symptoms that develop include a form of anemia, changes in the gastrointestinal tract, and growth failure.

The name "folacin" comes from the Latin for leaf, since it is found in spinach and other green leafy vegetables. It is also present in liver, yeast, asparagus, and wheat germ. Deficiencies are most common in those parts of the population on a subsistence level, whose diet usually does not include these foods.

21.6 Water-soluble vitamins—pantothenic acid

This vitamin is the last member of the B_2 complex. It is used to make coenzyme A (CoA). Look at the structure of CoA in Figure 17-9 and identify the part that comes from pantothenic acid. This coenzyme plays a central role in carbohydrate, lipid, and protein metabolism.

$$HO-CH_2-\underset{\underset{H_3C}{|}}{\overset{\overset{CH_3}{|}}{C}}-\underset{\underset{OH}{|}}{CH}-\overset{\overset{O}{\|}}{C}-NH-CH_2-CH_2-\overset{\overset{O}{\|}}{C}-OH$$

pantothenic acid

No specific disease results from a deficiency of pantothenic acid. Its absence may cause neuromotor disturbances, cardiovascular and digestive disorders, weakness, and susceptibility to infection.

Pantothenic acid is found along with the other B vitamins in yeast, liver, and eggs. Royal jelly, from bees, is an especially rich source.

21.7 Water-soluble vitamins—vitamin B₆

Three forms exist for vitamin B_6: pyridoxine, pyridoxal, and pyridoxamine. Any of these molecules can be used to make the coenzyme pyridoxal phosphate, which is the amino-carrying group in transamination. This coenzyme is very important in amino acid metabolism.

pyridoxine   pyridoxal   pyridoxamine

Vitamin B$_6$ deficiency does not cause a specific disease. It results in weight loss, anemia, and convulsions in infants and depression, confusion, and convulsions in adults.

Liver, nuts, wheat germ, brown rice, and yeast are good sources of vitamin B$_6$. It is also found in meats, vegetables, and egg yolks.

21.8 Water-soluble vitamins—vitamin B$_{12}$

As you can see from Figure 21-2, vitamin B$_{12}$ has a very complicated structure. Cobalt is located at the center of the ring, giving this vitamin the name cobalamin. Vitamin B$_{12}$ is a coenzyme for enzymes that catalyze reduction, dehydration, and transmethylation in carbohydrate and lipid metabolism as well as in DNA, RNA, and protein synthesis.

Persons with gastric disturbances who lack "intrinsic factor," a mucoprotein, cannot absorb this vitamin and develop *pernicious anemia*. This disease disturbs the maturation of red blood cells. Besides the symptoms of anemia, degeneration of the spinal cord may result.

Vitamin B$_{12}$ is the only vitamin manufactured in large amounts by microorganisms. It is made by soil and intestinal bacteria. The major sources from diet are animal tissue, especially kidney, liver, and brain. Only very small amounts are needed, but strict vegetarians, as well as persons who have had portions of their small intestine surgically removed, may develop a deficiency.

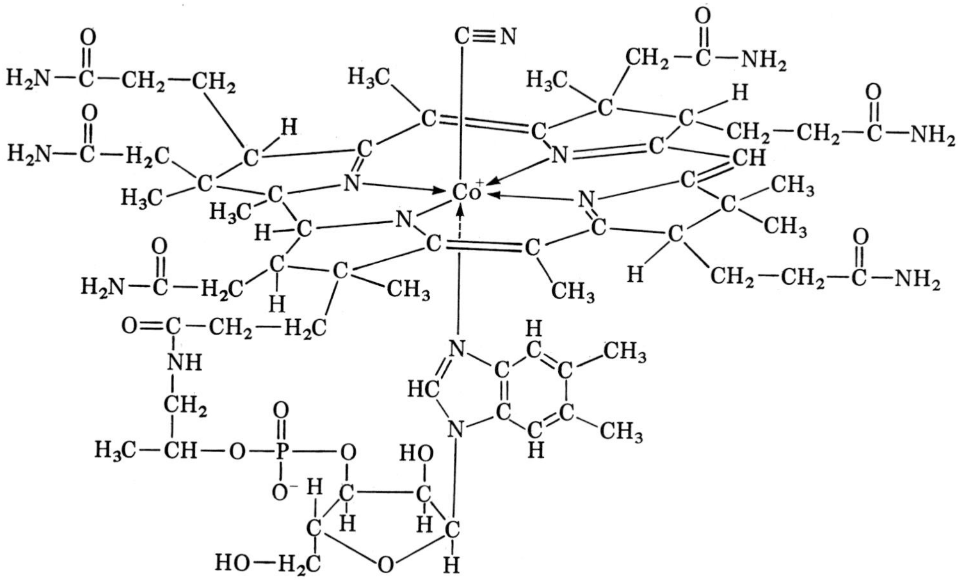

Figure 21-2. The structure of vitamin B$_{12}$.

21.9 Water soluble vitamins—ascorbic acid

Ascorbic acid is commonly known as vitamin C. This vitamin and its derivative, dehydroascorbic acid, take part in biological oxidation and reduction—the loss and gain of hydrogen, respectively. Ascorbic acid also serves as

$$HO-CH-CH_2OH$$

ascorbic acid

a coenzyme in hydroxylation reactions, the addition of hydroxyl groups, as in the formation of hydroxyproline from proline, a step in collagen synthesis.

Vitamin C deficiency results in *scurvy,* a disease that causes sore gums and teeth loosening, hemorrhages, edema, joint pain, anorexia (loss of appetite), and anemia. The symptoms of scurvy partially come from decreased abilities to form the structural protein collagen and to store iron (Figure 21-3).

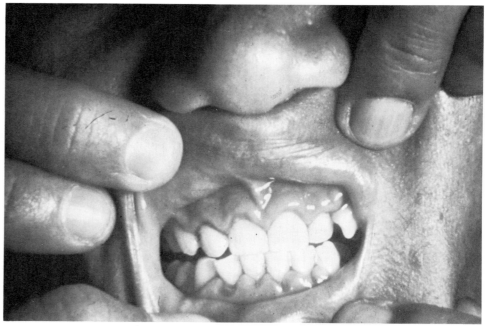

Figure 21-3.   The gums of a person with scurvy. This condition results from a vitamin C deficiency. (Photo courtesy of Center for Disease Control, Atlanta, Georgia.)

This vitamin is found in fresh fuits and vegetables, especially broccoli, brussel sprouts, collards, sweet peppers, and turnip greens. Ascorbic acid has become the subject of much discussion since the chemist Linus Pauling, a winner of two Nobel prizes, recommended the ingestion of large additional doses of vitamin C (over 250 mg daily) to lower the number and severity of colds. His hypothesis, as well as possible side effects of taking large doses of this vitamin, are currently being investigated in clinical studies.

21.10   Water-soluble vitamins—biotin

Biotin is also known as vitamin H. It is a coenzyme for carboxylase enzymes, which join CO_2 to organic molecules, and for transcarboxylases, which transfer the CO_2 group. Biotin is needed for carbohydrate, amino acid, and fatty acid synthesis.

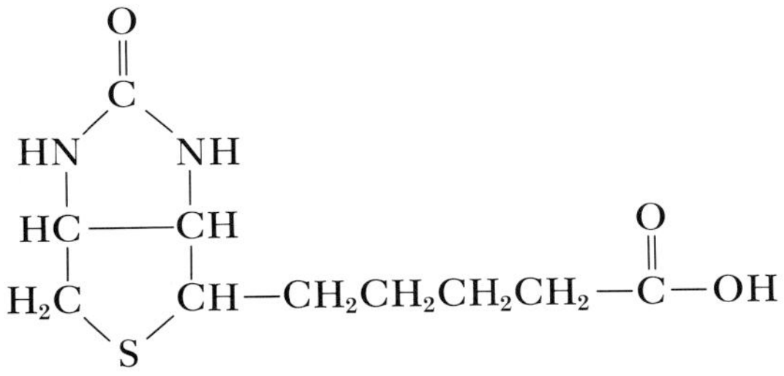

biotin

No specific diseases result from a deficiency of this vitamin. Its absence may cause dermatitis and weariness. Biotin is found in liver, yeast, grains, eggs, nuts, and chocolate. It is also made by intestinal bacteria.

21.11   Lipid-soluble vitamins—vitamin A

Vitamin A, also called retinol, is a lipid-soluble vitamin, in contrast to those discussed previously, which are water soluble. The primary role of this vitamin, after it has been converted to an aldehyde, retinal, is the regeneration

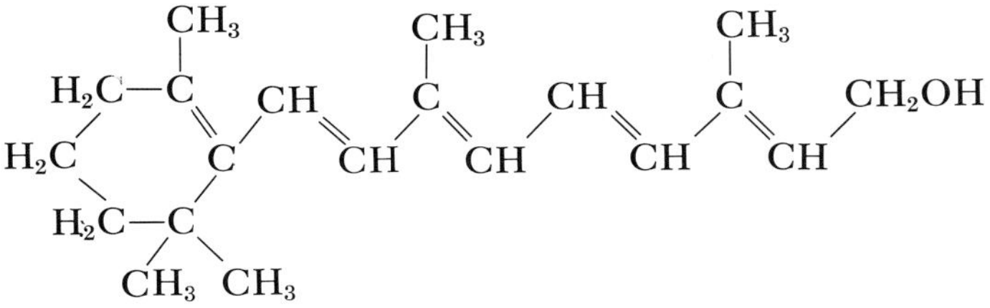

vitamin A

of rhodopsin (visual purple), the protein pigment that becomes bleached when light strikes the eye, an essential part of the process of seeing. The vitamin also acts as a coenzyme for several metabolic reactions.

Deficiency of vitamin A affects all organs. The first sign is "night blindness," the inability to see at low light levels. A more serious effect is *xerophthalmia*, a major cause of blindness in young children in parts of Asia, Africa, and Latin America. It results from the formation of keratin, the protein of the outer skin layer, in the conjunctiva, which is the mucous membrane of the eyelid and eye, causing damage to the cornea and loss of the lens. Vitamin A deficiency can also prevent growth and create a condition known as "toad skin."

The vitamin is found in liver, fish liver oils, fruits, and vegetables. Vitamin A is toxic in large amounts; overdoses cause irritability, fatigue, insomnia, and painful bones and joints.

21.12 Lipid-soluble vitamins—vitamin D

Vitamin D, calciferol, has two forms, D_2 (ergocalciferol) and D_3 (cholecalciferol). (In case you are wondering, D_1 was found to be a mixture, and this symbol is no longer used.) The steroidlike vitamin is involved in protein syn-

$(R = C_9H_{17}$ in D_2 and $R = C_8H_{17}$ in $D_3)$

vitamin D

thesis and calcium transport; it is a coenzyme for phosphatases and phosphorylases of phosphate metabolism.

The deficiency disease of this lipid-soluble vitamin is *rickets*, resulting from failure in the deposit of inorganic bone material, such as calcium. This condition may cause bowlegs, knock-knees, or "pigeon breast." Besides the malformation of bone, normal growth is retarded (Figure 21-4).

Vitamin D is found in few foods; fish liver oil is a notable exception. Fortunately, it can be made from the provitamins ergosterol (for D_2) from plants and 7-dehydrocholesterol (for D_3), which is found in the skin, by the action of sunlight. Rickets used to be common among children living in northern areas because the sun is less effective there in converting the provitamins to the vi-

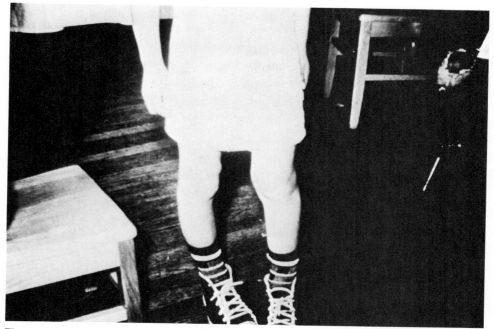

Figure 21-4. A child with the bowlegs and knock-knees of rickets. A vitamin D deficiency causes this condition. (Photo courtesy of Center for Disease Control, Atlanta, Georgia.)

tamin. Milk, an important source of calcium, is now fortified with vitamin D, the factor necessary for its utilization. Vitamin D is toxic in large amounts, causing loss of calcium from bone and such symptoms as nausea, diarrhea, muscular weakness, and joint pains.

21.13 Lipid-soluble vitamins—vitamin E

This vitamin is also called tocopherol; the most common and most active form is the type known as alpha-tocopherol. Its function is to prevent the oxidation

$$H_3C-C\overset{\overset{\displaystyle CH_3}{|}}{C}\overset{\displaystyle C}{=}\overset{O}{C}-\overset{\overset{\displaystyle CH_3}{|}}{C}-(CH_2)_3\overset{\overset{\displaystyle CH_3}{|}}{CH}(CH_2)_3\overset{\overset{\displaystyle CH_3}{|}}{CH}(CH_2)_3\overset{\overset{\displaystyle CH_3}{|}}{CH}CH_3$$

vitamin E

of unsaturated fatty acids. (Remember that because of double bonds, fatty acids are easily oxidized.) Vitamin E is also needed to preserve cell membranes and to maintain normal growth.

Lack of vitamin E shortens the life of red blood cells because of hemolysis. Its absence in rats causes loss of fertility, but no similar result is known for humans. A deficiency is often found in persons having a defective fat absorption and transport system. Tocopherol occurs in plant oils, especially cottonseed, corn, soybean, safflower, and wheat germ.

21.14 Lipid-soluble vitamins—vitamin K

This vitamin has two forms, K_1 (phylloquinone), which is present in plants, and K_2 (menaquinone) which is found in intestinal bacteria. Lack of vitamin K lengthens the time needed for the blood to clot, making minor injuries more serious through the possibility of shock and death from hemorrhage and loss of blood.

vitamin K_1

Sources of the vitamin are dark green leafy vegetables and some fruits and seeds. Deficiencies are rare because vitamin K can be made by intestinal bacteria. Because newborn infants do not have these bacteria, they may be given vitamin K by injection immediately after birth.

21.15 Recommended dietary allowances

Table 21-1 contains a list of the recommended daily quantities of the vitamins. Some of these numbers are estimates since it is often difficult to determine the exact amount required. The requirements of niacin, folacin, biotin, and vitamin K are either entirely or partially supplied by intestinal bacteria.

The quantities listed correspond to the level considered adequate to meet the known nutritional needs of almost everybody. Each person has slightly different requirements for the vitamins. The **recommended dietary allowances** (RDA) provide enough excess to cover this type of variation in the gen-

Table 21-1   **Recommended Daily Vitamin Allowances**[a]

Vitamin	Allowance, adult
thiamin	1.0–1.5 mg
riboflavin	1.1–1.8 mg
niacin	12–20 mg
folacin	0.4 mg
pantothenic acid	5–10 mg
vitamin B_6	2.0 mg
vitamin B_{12}	0.003 mg
ascorbic acid	45 mg
biotin	0.002 mg (?)
vitamin A	4000–5000 IU[b]
vitamin D	400 IU[b]
vitamin E	12–15 IU[b]
vitamin K	0.001 mg (?)

[a] Based on National Research Council, National Academy of Sciences (1974) recommendations.
[b] The abbreviation IU denotes international units.

eral population. In certain conditions, such as pregnancy, more than these amounts may be required.

The vitamins discussed in this chapter are summarized in Table 21-2 (p. 482).

21.16   Hormones

In contrast to the vitamins, hormones are all produced within the body by special ductless glands called the **endocrines** (Figure 21-5). Along with the nervous system, hormones traveling through the bloodstream are a means of carrying information from one organ to another. They integrate the various functions of your body to create a proper balance by serving as regulators.

Hormones are secreted by a gland when needed to influence the rate of a particular reaction going on in certain cells. Generally, they exert their effect by controlling the synthesis of specific proteins. They have a short life span. The hormone is released, carried by the blood to its target, performs its job, and is quickly inactivated. Some hormones are destroyed only a few minutes after they have been secreted by an endocrine gland.

A nucleotide, cyclic AMP, **cyclic adenosine monophosphate (cAMP)**, is involved in hormone action. Although hormones related in structure to steroids can pass directly into the cytoplasm of the target cell, polypeptide hormones cannot. They bind to the membrane, where they stimulate the activity of the

Table 21-2   Vitamin Summary Chart

Name	Function	Deficiency disease	Sources
thiamin (B_1)	decarboxylation (as pyrophosphate)	beriberi	seeds, pork, whole wheat bread
riboflavin (B_2)	dehydrogenation	cheilosis, "shark skin"	organ meats, yeast, wheat germ
niacin (B_2, nicotinic acid)	dehydrogenation (as NAD and NADP)	pellagra	meat, yeast, legumes
folacin (folic acid, B_2)	one-carbon carrier (as tetrahydrofolate)	anemia, gastrointestinal changes	green leafy vegetables, liver
pantothenic acid (B_2)	part of coenzyme A	neuromotor disturbance	yeast, liver, eggs
vitamin B_6	amino carrier (as pyridoxal phosphate)	skin lesions, anemia	liver, nuts, wheat germ
vitamin B_{12}	reduction, dehydration, transmethylation	pernicious anemia	bacteria, organs
ascorbic acid (vitamin C)	oxidation–reduction, connective tissue	scurvy	fruits, vegetables
biotin (vitamin H)	carboxylation	dermatitis	liver, yeast, grains
vitamin A (retinol)	regenerates pigment for vision	"night blindness," xerophthalmia	liver, fruits, vegetables
vitamin D (calciferol)	calcium transport	rickets	fish liver oil
vitamin E (tocopherol)	inhibits oxidation of fatty acids, hematopoiesis	hemolysis of blood cells	plant oils
vitamin K (phylloquinone)	blood clotting	loss of blood	green leafy vegetables

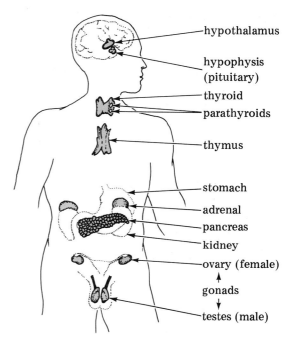

hypothalamus

hypophysis
(pituitary)

thyroid

parathyroids

thymus

stomach

adrenal

pancreas

kidney

ovary (female)
↑
gonads
↓
testes (male)

Figure 21-5. The location of the endocrine glands.

enzyme adenyl cyclase, which converts ATP to cyclic AMP. The cAMP has been called a "second messenger" since it transmits the instructions of the hormone from the surface to the inside of the cell. It also amplifies the message because the resulting cAMP concentration is about 10,000 times as great as the concentration of the hormone. Cyclic AMP then regulates the rates of enzymatic reactions or processes within the cell. A second nucleotide, cyclic guanosine monophosphate (cGMP), may also translate external stimuli into cellular activities.

cyclic AMP

21.17 Thyroxine (thyroid gland)

Tetraiodothyronine, or thyroxine, is made by the addition of iodine to the amino acid tyrosine in the thyroid gland. Along with triiodothyronine, which

$$HO - \text{(ring, I, I)} - O - \text{(ring, I, I)} - CH_2 - \underset{NH_2}{CH} - \underset{\underset{O}{\|}}{C} - OH$$

thyroxine

contains one less iodine atom, thyroxine is *the only molecule in the body that contains iodine.* Transported in the blood by carrier proteins, it affects all organs and tissues. It speeds up the cellular reactions and stimulates protein synthesis, influencing growth and the rate of basal metabolism.

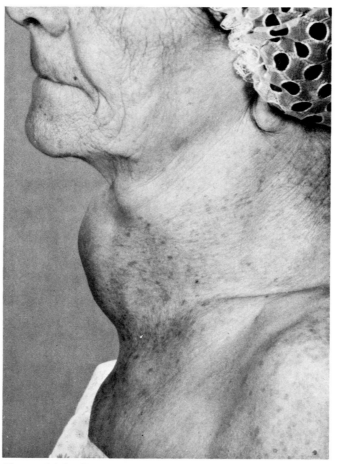

Figure 21-6 The neck of a person with goiter. This condition results from hyperthyroidism, an overactive thyroid gland. (Photo by Martin Rotker.)

Underactivity of the thyroid, *hypothyroidism,* leads to a slowing of the body processes because of a lack of thyroxine. When hypothyroidism occurs at birth, *cretinism* results. The cretin has little or no thyroid gland; physical and mental development is retarded. In the adult, lack of thyroid hormone may result in *myxedema,* characterized by puffy face and hands, slow speech, and mental apathy.

Hyperthyroidism, an overactive thyroid, produces an excess of thyroxine, increasing the metabolic rate by 30 to 60%. The resulting condition is called *Grave's disease* or toxic goiter. The term *goiter* refers to enlargement of the thyroid gland that occurs in this hyperactive state (Figure 21-6). Another prominent sign is the bulging of the patient's eyes (exophthalmia). Other symptoms include overactivity, restlessness, weakness, hyperglycemia, glucosuria, and a negative nitrogen balance. Treatment may involve the use of antithyroid agents. The addition of iodides to table salt (1:5000 to 1:200,000), forming "iodized salt," helps prevent goiter, particularly in mountainous areas far from oceans, where the iodine of seawater cannot enrich the soil and drinking water.

21.18 Parathormone (parathyroid gland)

Parathormone is a polypeptide of 84 amino acids. It acts mainly on the kidney, skeleton, and gastrointestinal tract. The secretion of parathormone is *inversely* related to the blood calcium level; that is, the lower the amount of Ca^{2+} in the blood, the more hormone is released. Its role is to increase reabsorption of calcium as well as magnesium by the kidneys, to release calcium from bone by a process called demineralization, and to increase calcium uptake by the cells of the intestine. It also influences phosphate metabolism.

21.19 Calcitonin (parathyroid gland)

Calcitonin is also a polypeptide, containing 32 amino acids. As with parathormone, the job of this hormone is to maintain the calcium balance in the body. Its secretion, however, is *directly* related to the calcium concentration of the blood; the greater the level of Ca^{2+}, the more hormone is given off by the parathyroid gland. Calcitonin opposes the effects of parathormone, inhibiting loss of calcium from bone; its function is to decrease the calcium level. These two hormones provide a dual feedback system, capable of immediate action if the blood calcium concentration either rises above or falls below its normal value (4.5 to 5.6 mEq/liter).

An underactive parathyroid, hypoparathyroidism, is rare. It leads to a lowering of the calcium level (hypocalcemia), resulting in tetany (a syndrome of spasms and convulsions) and death. Hyperparathyroidism, due to overactivity of the gland, causes *ostitis fibrosa cystica*. Calcium is removed from the skeleton, leading to bone destruction and an increase in the possibility of bone fractures. The blood calcium level is high and large amounts of the ion are excreted, forming calculi, solid deposits that may damage the kidney.

21.20   Insulin (pancreas)

Insulin consists of two polypeptide strands connected by disulfide bonds with a total of 51 amino acids. Its secretion by the pancreas is governed by the blood glucose level; more is released when the sugar concentration is high. Insulin is called the "hypoglycemic factor" because the hormone reduces the amount of glucose in the blood by increasing its entry into the cells. This hormone increases glycogenesis and decreases glycogenolysis and gluconeogenesis. It also increases the formation of protein from amino acids and stimulates the formation of lipids from carbohydrates. All of these processes decrease the level of blood glucose.

When too little insulin is secreted, or when its function is inhibited, *diabetes mellitus* results. As described in the chapters on carbohydrate and lipid metabolism, the body cannot properly make use of glucose and must employ its protein and lipid pathways to a greater extent. Gluconeogenesis increases and protein synthesis decreases since amino acids are being converted to glucose. Lipids are mobilized to make acetyl-CoA since the route from pyruvate is decreased, leading to lipemia, "fatty" liver, and ketosis.

Diabetes mellitus can be controlled with the help of daily injections of insulin. The hormone cannot be taken orally because insulin is a protein and would be digested. There are several kinds of insulin commercially available, obtained from either beef or pork. They are divided into slow-, intermediate-, and fast-acting types in strengths of 40 and 80 units/ml (1 unit = 0.0417 mg). The most common form ("NPH" insulin) lasts 28 to 30 hours and peaks after 8 to 12 hours. The amount needed for control depends on many factors: diet, time of injection, type of insulin, body weight, and severity of the disease (Figure 21-7).

If the patient takes too much insulin or does not eat enough food, **insulin shock** results. This condition is hypoglycemia, resulting in weakness, hunger, sweating, and nervousness caused by a low blood glucose level. Neural damage can result if the brain goes for a long time without enough glucose. A diabetic patient taking insulin should carry sugar or candy at all times so that it can be eaten immediately when the initial symptoms of faintness, dizziness, and blurred vision are felt.

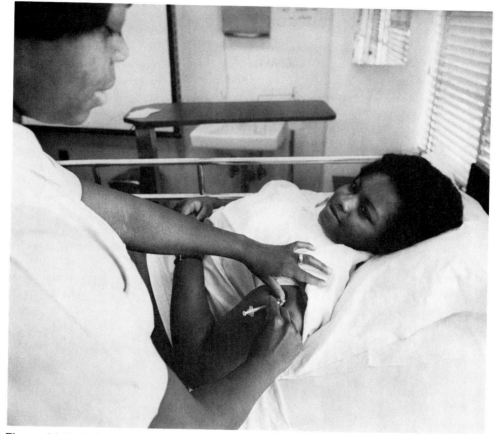

Figure 21-7. Injection of insulin. (Photo by Al Green.)

Milder cases of diabetes can be controlled through diet and oral hypoglycemic agents, such as drugs related to sulfonamides that stimulate the pancreas to secrete more insulin. One example is tolbutamide. Some appear to in-

$$H_3C-C \underset{\underset{H}{\overset{\displaystyle C=C}{|}}}{\overset{\overset{\displaystyle C-C}{\underset{H\ \ H}{}}}{}} C-\overset{\overset{\displaystyle O}{\|}}{\underset{\underset{O}{\|}}{S}}-NH-\overset{\overset{\displaystyle O}{\|}}{C}-NH-CH_2-CH_2-CH_2-CH_3$$

tolbutamide

crease the patient's risk of cardiovascular disease. The American Diabetes
Association recommends that these drugs be used only in moderate cases in
which the diabetes is poorly controlled by diet alone and the patient cannot or
will not take insulin.

21.21 Glucagon (pancreas)

This polypeptide of 29 amino acids works in the opposite direction from insulin. Glucagon is secreted when the blood sugar level is low. It thus helps maintain normal concentrations during fasting. Its presence increases the blood glucose level by stimulating glycogenolysis in the liver. The hormone also increases gluconeogenesis, inhibiting the formation of protein. In addition, fatty acids are released from adipose tissue and fewer fatty acids are synthesized.

21.22 Testosterone (male gonads)

Testosterone and dihydrotestosterone are androgens, male hormones, secreted by the testes. They are 19-carbon steroids made from cholesterol. The

testosterone

function of testosterone is to promote and maintain the male sexual organs and secondary sex characteristics. It also increases the growth of muscles, liver, and kidney by stimulating protein synthesis.

When too little testosterone is secreted as in hypogonadism (underactivity of the gonads) or castration (removal of the testes), *eunuchoidism* develops. Puberty does not take place, causing the individual to have a high-pitched voice, long, thin limbs, lack of body and facial hair, less muscular strength, and underdevelopment of the penis and sexual feelings. The effects of castration after puberty has occurred are less dramatic; treatment with androgenic compounds can reverse these changes.

Anabolic steroids are synthetic compounds that stimulate protein synthesis but with minimal increase of masculine characteristics. They lead to retention of nitrogen by the body and are used after surgery, disease, or malnutrition. Dangers involved in this therapy include loss of sex drive, edema, and damage to the liver and bone.

21.23   Estrogen (female gonads)

The female sex hormones are 18-carbon steroids based on estradiol. Estradiol is produced in the ovaries from testosterone, the male sex hormone. As you

estradiol

can see, the structures of the two molecules are quite similar, but the difference in their effects is tremendous. Secretion of estrogens produces the secondary sex characteristics of the female, the growth of axillary (armpit) and pubic hair, and changes in body shape as well as proper bone formation. Estrogens increase protein synthesis, especially in the target tissues, the uterus, vagina, and mammary glands. The hormones are also involved in regulating the menstrual cycle and female sex behavior. They have been used to treat the symptoms of menopause, although this use may be related to cancer.

Removal of the ovary, ovariectomy, is essentially castration of the female. When this happens before puberty, the menstrual cycle does not begin, lipids are not deposited in the mammaries, body hair does not grow, and the pelvis is not enlarged. If the ovariectomy is performed after puberty, menstruation stops, secondary sex characteristics disappear, and the uterus, vaginal mucous membranes, and fallopian tubes atrophy (reduce in size).

21.24   Progesterone (female gonads)

Progesterone, a steroid hormone, is found in the ovary (corpus luteum), placenta, and adrenals. It is secreted during the second half of the menstrual cycle. The hormone causes mucus formation in the ovary, which is necessary if an egg is to be implanted. Continued secretion is required during pregnancy if the egg is fertilized.

Progesterone prevents ovulation (release of the egg) when given during the fifth to twenty-fifth days of the menstrual cycle. Synthetic steroids similar to progesterone, such as those shown in Figure 21-8, provide the basis for birth control pills. The **oral contraceptives** are almost 100% effective in preventing

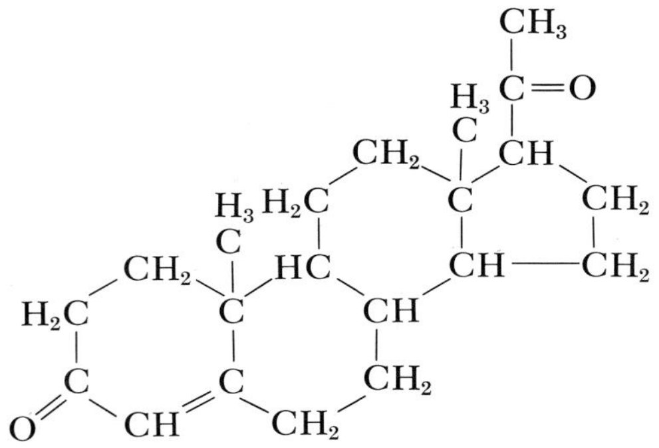

progesterone

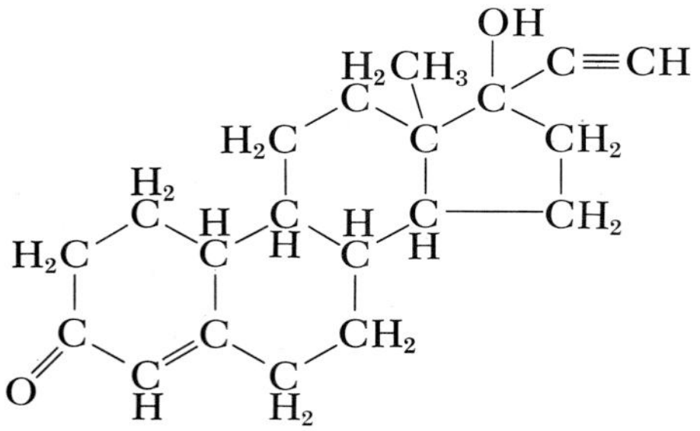

norethindrone

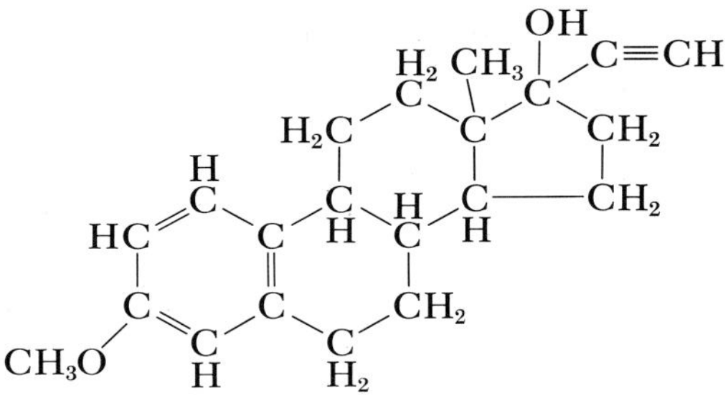

mestranol

Figure 21-8. Structural formulas of oral contraceptive drugs. Note the similarity in structure to progesterone.

conception because they trick the body into believing that progesterone is being secreted. But, in addition to other possible side effects, they increase the risk of blod clot formation.

21.25 Epinephrine and norepinephrine (adrenal gland)

Epinephrine (adrenaline) and norepinephrine (noradrenaline) are catecholamines, amines derived from catechol (*o*-dihydroxybenzene). Epinephrine increases the blood pressure, the force of the heart contraction, and the pulse rate. It is the "fright or flight" hormone that is secreted in times of stress. Epinephrine provides emergency energy in a hurry by stimulating glycogenolysis, gluconeogenesis, and mobilization of fatty acids from adipose tissue. Because of the increased blood flow, the glucose can be rushed to the brain, liver, and muscle tissues.

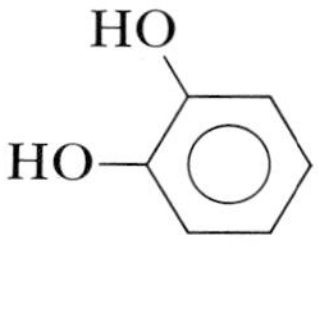

catechol 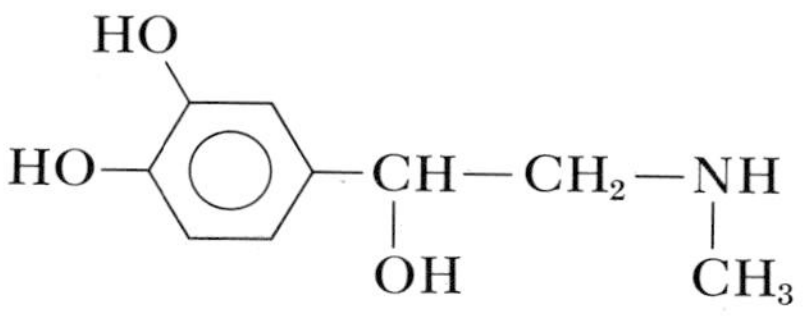   epinephrine 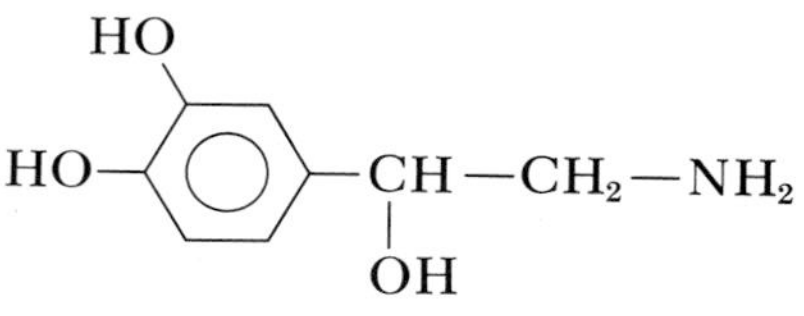   norepinephrine

Insufficient epinephrine is not fatal but slows down response to emergency situations, hard work, and emotional disturbances. Epinephrine is given in cases of bronchial asthma to widen the lung channels and in acute cardiac arrest as a stimulant injected directly into the heart. In local hemorrhage it checks bleeding by constricting blood vessels; when administered with a local anesthetic, epinephrine prevents diffusion of the drug from the site of injection by the same mechanism.

The effects of norepinephrine are somewhat similar to those of epinephrine. Norepinephrine also increases the blood pressure but does not change the heart output or rate. It causes constriction of blood vessels as does epinephrine when given intravenously (moderate doses cause dilation). A very important role for norepinephrine is to conduct impulses at nerve endings; a deficiency therefore results in poor nerve condition. Norepinephrine is used in the treatment of shock.

21.26 Cortisol and aldosterone (adrenal gland)

The hormones of the adrenal cortex are 21-carbon steroids. They consist of cortisol (hydrocortisone), corticosterone (which has one less hydroxyl group than cortisol), and aldosterone. Cortisol is the main secretion. It is responsible for increasing the blood glucose level and liver glycogen by stimulating both

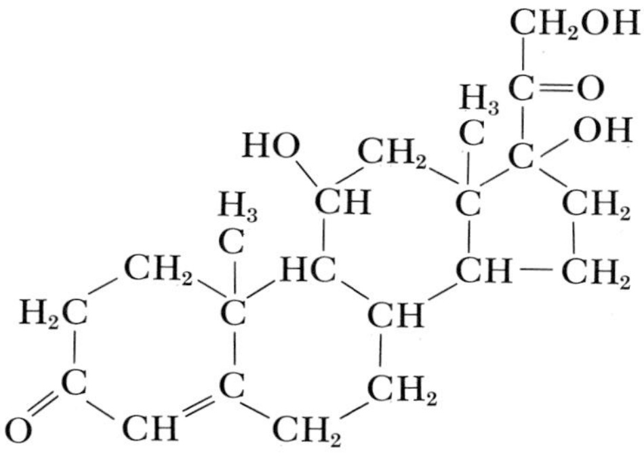

cortisol
(hydrocortisone)

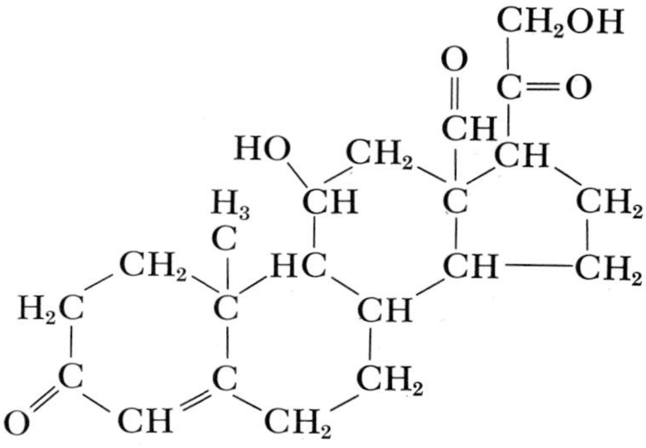

aldosterone

gluconeogenesis and glycogenesis. Cortisol promotes synthesis of liver protein and triacylglycerols (triglycerides) and provides resistance to harmful stimuli like infection, trauma, or hemorrhage.

Aldosterone, one of the most essential hormones, has similar metabolic effects, such as accelerating gluconeogenesis, but mainly serves to keep constant the electrolyte balance of the blood. Secretion depends on the amount of sodium in the diet, the quantity of potassium in the body, and the volume of the blood plasma. Aldosterone is sometimes called a "mineralcorticoid" because of its effect in maintaining a mineral balance, in contrast to cortisol, a "glucocorticoid," because its primary role involves glucose metabolism.

If not enough of the adrenal cortex hormones are being produced, *Addison's disease* develops. Lack of cortisol results in hypoglycemia, weakness, reduced cardiac output, and lowered resistance. Deficiency of aldosterone causes severe dehydration, low blood pressure (hypotension), and circulatory collapse. Skin pigmentation changes also occur in Addison's disease (Figure 21-9).

Hyperfunction of the adrenal cortex, on the other hand, causes *Cushing's syndrome* with obesity, muscle weakness, and hypertension due to excess

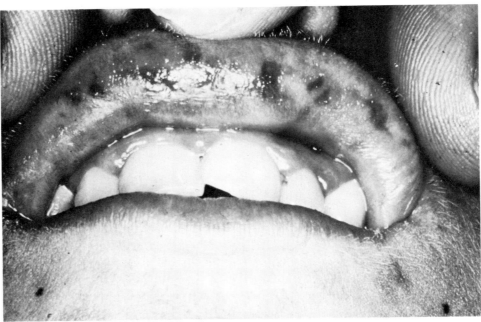

Figure 21-9. Lip pigmentation in an 11-year-old boy with Addison's disease. This condition is caused by an insufficient adrenal cortex hormone supply. (Photo courtesy of Center for Disease Control, Atlanta, Georgia.)

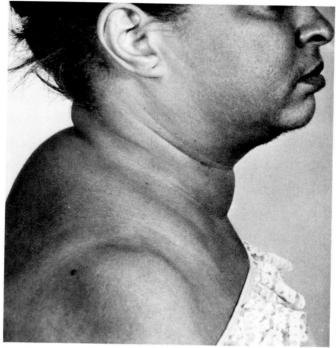

Figure 21-10.   A person with Cushing's syndrome. This condition, which includes the "hump" shown, results from overproduction of the adrenal cortex hormones. (Photo by Martin Rotker.)

cortisol (Figure 21-10). Too much aldosterone results in edema and hypertension with possible heart failure. If accompanied by hypersecretion of the androgens, *adrenal virilism* may result. In children, this appears as premature sexual development, in women as ending of menstruation, lowering of voice, and growing of body hair, and in men as increasing sexual desire and hair growth.

Synthetic molecules similar to the corticosteroids (such as prednisone and prednisolone) are many times more powerful than the natural hormones and have more specific effects. Some are used for their ability to reduce inflammation, some for their ability to retain sodium ions, and others for their influence on carbohydrate metabolism. Arthritis, rheumatic fever, allergies, and skin conditions are treated with these drugs. In all cases, **steroid therapy** must be carefully monitored. Undesirable side effects include weight gain, ulcers, edema, hypertension, infection, hyperglycemia, and psychiatric abnormalities.

21.27 Vasopressin (hypophysis or pituitary)

Secretion of vasopressin, a nonapeptide (nine amino acids), is stimulated by emotion, pain, and trauma. Its rate of release is controlled by the volume and osmotic pressure of the blood. The job of this hormone is to increase the blood pressure by constricting the arteries. It also regulates the water balance by its antidiuretic effect—increasing the rate of water reabsorption by the kidneys, thereby decreasing the volume of urine excreted. Thus, the most important causes of vasopressin release are hemorrhage, dehydration, and excess salt intake.

Too little vasopressin causes *diabetes insipidus*, which results in polyuria, excretion of very large quantities of urine having low specific gravity (1.002 to 1.006). Generally, 4 to 5 liters/day are lost compared to the normal volume of 500 ml; in one unusual case, a patient excreted 56 liters of urine in 1 day.

21.28 Adrenocorticotropin (hypophysis)

Adrenocorticotropin is a polypeptide made up of 39 amino acids. Also known as corticotropin and ACTH, it is one of the most essential hormones. It stimulates growth of the adrenal cortex and the release of its steroid hormones—cortisol and aldosterone. Secretion is *inversely* proportional to the concentration of these adrenal cortex hormones in the blood. A low level thus stimulates ACTH release. Trauma, stress, drugs, and bacteria also influence secretion of the hormone. Adrenocorticotropin causes the responses of the adrenal hormones: gluconeogenesis, inhibition of protein synthesis (except in the liver), and increased lipid mobilization. Oversecretion of this hormone as in Cushing's disease results in obesity of the trunk, skin pigmentation, body hair growth, bone demineralization, hypertension, hyperglycemia, loss of sex functions, and acne. Its deficiency causes the diseases of an underactive adrenal cortex such as Addison's disease.

21.29 Gonadotropic hormones (hypophysis)

This group of polypeptide hormones is called gonadotropic because their targets are the gonads—the ovary in the female and the testes in the male.

Follicle-stimulating hormone (FSH) causes growth of the ovarian follicles in the female and spermatogenesis, sperm formation, in the male. *Luteinizing hormone* (LH), or interstitial-cell-stimulating hormone (ICSH), promotes estrogen and progesterone secretion, ovulation, and ripening of the ovarian follicles; it stimulates secretion of testosterone in men. Both of these hormones are required for reproduction and sexual development. They are released when the sex hormone level decreases. *Prolactin* acts with estrogen to promote mammary gland growth. It also stimulates lactation, the formation and secretion of milk. Prolactin is antigonadotropic—it inhibits release of FSH and LH, and vice versa.

Chorionic gonadotropin, made in the placenta, is similar to the hypophysial (pituitary) hormones. It affects the ovary, maintaining growth of the corpus luteum, which is formed at the site of a ruptured follicle during pregnancy. Since it appears in the urine at about the initial week after the first missed menstrual period, this hormone can be the basis for pregnancy tests. When it is injected into female mice, rats, or rabbits, the animal's ovaries become enlarged.

Oxytocin is not really a gonadotropic hormone at all, but it has related activity. It causes strong contractions of the uterus, such as during the start of labor. In addition, it stimulates the muscles of the intestine, gallbladder, ureter, and urinary bladder and causes milk ejection, also by muscle contraction. It may be used to induce labor at the terminal stages of pregnancy.

21.30   Growth hormone (hypophysis)

Growth hormone is a polypeptide of 190 amino acids. Also known as somatotropin, it affects the rate of growth of the skeleton and of gain in body weight. Growth hormone causes increases in protein synthesis, blood glucose level, urine glucose concentration, muscle glycogen, release of free fatty acids from adipose tissue, kidney size and function, and bone formation.

This hormone is not the only one that influences growth, but its absence results in *dwarfism*, premature stopping or stunting of skeletal growth (Figure 21-11). On the other hand, too much growth hormone results in *acromegaly*, with enlarging of bones in the face, hands, and feet, bowing of the spine (kypnosis), increases in body hair and in the size of organs, hyperthyroidism, hyperglycemia, and loss of sexual function. If the excess occurs before puberty, the individual may grow to a height of seven or eight feet, a condition known appropriately enough as *gigantism*.

Figure 21-11. Examples of dwarfism. The absence of growth hormone causes this condition. (Photo courtesy of National Foundation–March of Dimes.)

21.31 Thyrotropin (hypophysis)

Thyrotropin, which is also called thyroid-stimulating hormone (TSH), consists of two polypeptide chains having a total molecular weight of 28,300. As its name suggests, it is involved with the thyroid gland—increasing the rates of iodide removal from the blood by the thyroid, formation of thyroxine from iodide, and secretion of this hormone.

21.32 Hormonal regulation

Hormones are molecules that regulate various substances in your body. But their release is regulated by still other molecules, some of which are produced by the hypothalamus gland. The hormones of this gland, the **hypothalamic releasing factors**, serve as chemical messengers to another gland, generally the hypophysis (pituitary). They themselves are controlled by your nervous system, acting as a link between the body's two primary means of communication.

Thyrotropin-releasing hormone is typical of these releasing factors. It controls secretion of thyrotropin by the hypophysis gland. This hormone in turn regulates the release of still another hormone, tyroxine, as shown in Figure 21-12. Both of these processes involve stimulation, an increase in hormone secretion. Thyroxine, which now influences other organs and tissues, inhibits further release of the original hypothalamic hormone. This "feedback" mechanism maintains the proper amount of thyroxine in the blood. The overall relationships between your body's hormones are summarized in Figure 21-13.

In addition to these hormones, there are other molecules in the body that act as regulators. They cannot strictly be considered hormones because these substances are not secreted by distinct glands. The so-called **tissue hormones** are listed in Table 21-3. Histamine and antihistamines are discussed in the following chapter.

Figure 21-12. An example of hormone action. Hormone secretion is carefully regulated by the body as shown in the case of thyroxine.

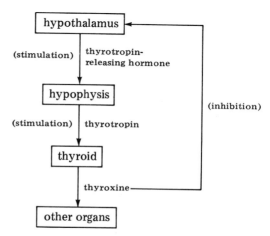

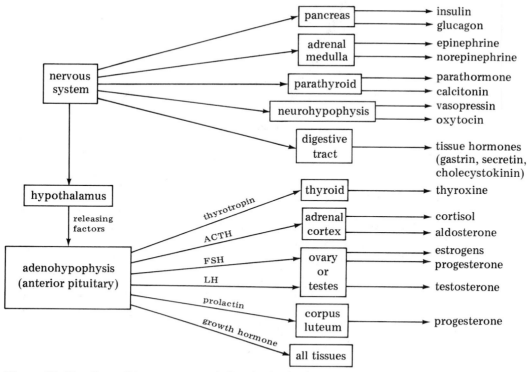

Figure 21-13. Overall hormone regulation in the body.

Table 21-3 Tissue Hormones

Hormone	Function
angiotensin	stimulates aldosterone production; increases blood pressure
bradykinin	lowers blood pressure; contracts smooth muscle
cholecystokinin—pancreozymin	contracts gallbladder; increases secretion of digestive enzymes by pancreas
enterogastrone	inhibits gastric secretion, slowing digestion
gastrin	promotes production of HCl in stomach and secretion of pepsin
histamine	controls local blood circulation; involved in allergic reactions
secretin	stimulates pancreas and flow of bile
serotonin	causes constriction of blood vessels; promotes peristalsis
tyramine	raises blood pressure; stimulates smooth muscle

21.33   Examples of hormone function

Using a mechanical model, Figure 21-14 summarizes the role of hormones in regulating the *blood sugar level*. Since this job is so vital, several hormones are involved in the process. Insulin is the most important, but growth hormone, glucagon, epinephrine, cortisol, and thyroxine also participate. Each of these hormones stimulates either a glucose-consuming process like glycogenesis, glycolysis, and fat formation, or a glucose-producing process like glycogenolysis or gluconeogenesis. Specifically, insulin lowers the blood sugar level by promoting the use of glucose, while other hormones have the opposite effect. It is through the complex interplay of these hormones that the concentration of glucose in the blood can be kept so constant.

The monthly *menstrual cycle* is another example of how hormones interact. The rhythmic sexual cycle of the female depends on the secretion of the gonadotropic hormones of the adenohypophysis. Follicle-stimulating hormone (FSH) with luteinizing hormone (LH) stimulates the development of a follicle in the ovary. This event causes estrogen to be released, allowing FSH to cause the follicle to ripen. The secretion of FSH is decreased by the presence of estrogen, but that of LH and prolactin now increases. These cause the rupture of the mature follicle, ovulation (release of the egg), and development of the corpus luteum ("yellow body").

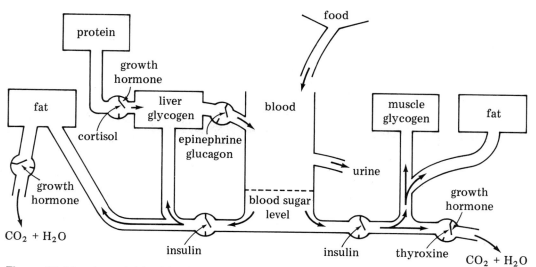

Figure 21-14.   A model for blood glucose regulation. The action of hormones is symbolized by the opening or closing of valves. (From P. Karlson, "Kurzes Lehrbuch der Biochemie," 8th ed. Georg Thieme Verlag, Stuttgart, 1967.)

Table 21-4
Hormone Summary Chart

Name	Location	Effect	Deficiency	Excess
thyroxine	thyroid	metabolic	cretinism	Grave's disease, goiter
parathormone	parathyroid	calcium and phosphate metabolism	low Ca²⁺ level	high Ca²⁺ level
calcitonin	parathyroid, thyroid	calcium and phosphate metabolism	high Ca²⁺ level	low Ca²⁺ level
insulin	pancreas	blood glucose	diabetes mellitus	hypoglycemia
glucagon	pancreas	blood glucose	hypoglycemia	hyperglycemia
testosterone	testes	male sex characteristics	eunuchoidism	–
estrogen	ovary	female sex characteristics	no sexual development	–
progesterone	ovary (corpus luteum)	uterine mucosa	–	–
epinephrine	adrenal medulla	pulse, blood pressure	slow response	pheochromo-cytoma (tumor)
norepinephrine	adrenal medulla	blood pressure, nerve con-dition	poor nerves	pheochromo-cytoma (tumor)

Hormone	Source	Function	Hyposecretion	Hypersecretion
cortisol	adrenal cortex	glucose metabolism	Addison's disease	Cushing's syndrome
aldosterone	adrenal cortex	mineral balance	dehydration	edema
vasopressin	neurohypophysis	blood pressure	diabetes insipidus	–
oxytocin	neurohypophysis	contraction of uterus	–	–
adrenocorticotropin	adenohypophysis	adrenal hormone stimulation	Addison's disease	Cushing's syndrome
follicle-stimulating hormone (FSH)	adenohypophysis	estradiol production	sterility	–
luteinizing hormone (LH)	adenohypophysis	sex hormone production	sterility	–
prolactin	adenohypophysis	milk secretion	–	–
chorionic gonadotropin	placenta	ovary development	–	–
growth hormone (somatotropin)	adenohypophysis	growth of bone/muscle	dwarfism	acromegaly, gigantism
thyrotropin	adenohypophysis	thyroid hormone formation	cretinism	Grave's disease, goiter
hypothalamic releasing factors	hypothalamus	control of adenohypophysis	–	–

Progesterone is secreted, which inhibits release of LH and prolactin. If an egg has not been fertilized, menstruation takes place, and a new cycle begins. This process is illustrated in Figure 21-15. It consists of the periodic matura-

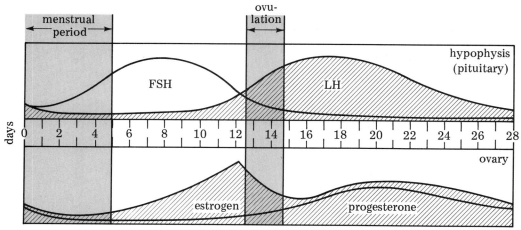

Figure 21-15. The ovarian cycle—variation of the hormone levels.

tion of follicles of the ovary accompanied by changes in the mucous membranes of the uterus. The cyclic process is regulated by action of the hormones of the hypophysis on the ovary followed by feedback to the gland. The critical timing of the cycle results from this interplay, which has been called a "symphony of hormones."

Table 21-4 (pp. 500–501) summarizes the important properties of each of the hormones mentioned in this chapter.

SUMMARY

In addition to the relatively large amounts of carbohydrates, lipids, and proteins which you need for energy and synthesis, other substances are equally necessary but only in very small quantities. Those called hormones are made within the body; vitamins must usually be provided by diet.

Vitamins are biologically active substances required in trace amounts for normal health and growth. Most serve as catalysts in the form of coenzymes, taking part in chemical reactions and undergoing changes but then being restored by other reactions.

Lack of some vitamins causes deficiency diseases. Too much of others can be toxic. Megavitamin therapy, the use of large doses of vitamins to treat disease, is being clinically investigated.

The structures of vitamins differ greatly. They can be either water soluble or lipid soluble. Some foods contain vitamins in the form of a provitamin, which must first be changed by a chemical reaction before it can be used by the body.

In contrast to the vitamins, hormones are all produced within the body by special ductless glands called endocrines. In addition to the nervous system, hormones traveling through the blood are a means of carrying information from one organ to another. By serving as regulators, they integrate the various functions of the body. Hormones are secreted by a gland when needed to influence the rate of a particular reaction of a cell; they are carried by the blood to their target, perform their job, and are quickly inactivated, often in a matter of minutes.

These regulating substances are generally either polypeptides or similar in structure to steroids. Their release is controlled by many factors including feedback, the hypophysis gland, and the releasing factors of the hypothalamus. Cyclic AMP serves as a "second messenger" for transmitting the information from the hormone to the cells.

The blood glucose level and the menstrual cycle are examples of processes controlled by hormones.

Exercises

1. (21.1) What is a vitamin?

2. (21.1) Why are only small amounts of the vitamins needed?

3. (21.1) What is a deficiency disease?

4. (21.1) What is megavitamin therapy?

5. (21.2–21.14) For each of the vitamins, give its function in the body, deficiency disease (if any), and foods in which it is found.

6. (21.2–21.14) Which vitamins are toxic in large doses?

7. (21.12) Why is milk fortified with vitamin D?

8. (21.15) Why do you think it is difficult to determine human vitamin requirements?

9. (21.16) What are hormones? How do they differ from vitamins?

10. (21.16) What could be the advantage of a short life span of a hormone?

11. (21.16) What is the role of cyclic AMP?

12. (21.17–21.31) For each of the hormones, give its function in the body, the effect of too little being secreted, and the effect of too much being secreted.

13. (21.20) What causes insulin shock?

14. (21.24) What hormone is the basis for birth control pills? Explain.

15. (21.25) Which is the "emergency hormone"? Explain.

16. (21.26) What are corticosteroid drugs?

17. (21.32) What are the hypothalamic releasing factors?

18. (21.32) What factors control hormone secretion?

19. (21.33) Describe how control of the glucose level and the menstrual cycle are examples of hormonal regulation.

Chemistry of the body fluids

Your body contains 60 to 70% water by weight. The exact amount depends on your sex, age, and fat content. Of the approximately 40 liters in a 150-pound adult, 25 liters are **intracellular fluid** (inside the cells) and 15 liters are **extracellular** (outside the cells). The extracellular part includes *plasma*, the liquid portion of your blood; *interstitial fluid*, in the spaces between the cells; *gastrointestinal secretions*, gastric juice, pancreatic juice, etc.; *cerebrospinal fluid*, surrounding the brain and spinal cord; *intraocular fluid*, aqueous humor of the eyes; and *fluids of the body spaces*, such as in the joints.

22.1 Blood plasma

One of the most important fluids, blood, is the *transport system of the body*. The main job of blood is to bring oxygen and food molecules to your tissues and to remove carbon dioxide and the waste products of metabolism. It also helps to maintain the fluid balance, the acid–base balance, and the normal body temperature, as well as to transport hormones and metabolic intermediates and to protect against infection.

Blood makes up about 8% of your body weight; an adult has 5 to 6 liters of the fluid. Most of the blood consists of a liquid portion, the plasma. Suspended in it are the cellular components ("formed elements"): the *erythrocytes* (red blood cells), the *leukocytes* (white blood cells), and *platelets* (thrombocytes).

Proteins are the most abundant part of the blood plasma, approximately 7%. The remainder consists of inorganic salts (1%) and organic compounds other than protein (2%), as shown in Table 22-1. Over half the plasma protein (59%) is **albumin**; its primary function is osmotic regulation (as described in Section 7.10).

504

Table 22-1   Organic (Nonprotein) Compounds in Blood Plasma

Component	Normal range (mg/100 ml)
fatty acids (total)	150–500
phosphoglycerides	150–250
cholesterol	100–225
triacylglycerols (triglycerides)	80–240
polysaccharides (as hexose)	70–105
glucose	70–105
amino acids	35–65
urea	20–30
sphingomyelin	10–50
lactic acid	8–17
fructose	6–8
glycogen	5–6
organic acids (other than lactic)	3–10
uric acid	3–8
pentoses	2–4
creatinine	0.6–1.1
bilirubin	0.2–1.2
creatine	0.2–0.9

A second function of plasma albumin is to transport molecules in the blood that are only slightly soluble in water. When a substance binds to the protein, its solubility greatly increases. Fatty acids and other metabolic products are examples of molecules carried by albumin.

22.2   Blood antibodies

Globulins are another class of proteins present in blood plasma. The most important are the gamma (γ) globulins, also known as immunoglobulins because these proteins provide immunity against disease. They form a defense system by acting as **antibodies**, destroying or inactivating microorganisms and foreign protein molecules known as **antigens**. Included in the category of foreign protein molecules are toxins, the poisonous substances produced by bacteria, plants, and animals; the antibodies formed to counteract these antigens are known as antitoxins. (When the toxin is a snake venom, the antitoxin is called an antivenin. It is used to overcome the effects of snakebite.)

Antibodies are very specific; gamma globulins formed in response to an antigen react only with that specific molecule, whether it is a toxin, virus, or bacteria. They can neutralize the antigen, cause it to precipitate, make several clump together, or rupture its cell membranes, in all cases making the invading particle harmless. Figure 22-1 illustrates the antigen–antibody interaction.

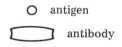

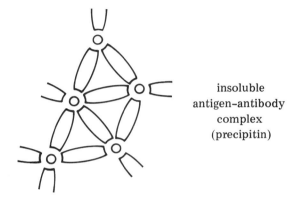

Figure 22-1. The antigen–antibody interaction. By combining with an antigen, the antibody can inactivate it.

insoluble
antigen–antibody
complex
(precipitin)

Artificial immunity can be produced by injecting a nonpathogenic antigen, such as the killed organism. Antibodies develop which are then ready to defend the body against the live form. The poliomyelitis vaccine is an example. The Salk type consists of inactivated poliovirus, while the Sabin type contains live but weakened virus particles.

Allergies are a side effect of immunity. Certain types of allergy, such as hay fever, result from an abnormal immunity to common substances like plant pollens. An allergic person forms antibodies that react with these antigens, releasing the substance histamine. This molecule causes blood vessel dilation

$$N-C-CH_2-CH_2-NH_2$$
$$HC \quad CH$$
$$N$$
$$H$$

histamine

and loss of fluid through capillary membranes, resulting in the symptoms of the allergy. **Antihistamines** (such as diphenhydramine or chlorpheniramine) are drugs that block the activity of histamine. Other types of allergy include asthma and uticaria (hives), as well as serum sickness (a reaction to foreign proteins producing edema and fever).

22.3   Blood clotting

The third major protein of blood plasma is **fibrinogen**, which is needed to form a blood clot (thrombus). **Clotting** is the chemical defense against loss of blood; it results from the formation of a jellylike solid from liquid blood. Normal human blood, when "shed," clots in 5 to 8 minutes at 37°C.

The mechanism of blood clotting (Figure 22-2) is complex. The major

Figure 22-2.   The mechanism of blood clotting. See Figure 15-11, a photograph of an erythrocyte enmeshed in fibrin.

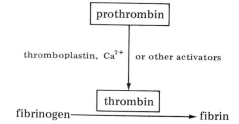

change that takes place is the conversion of fibrinogen to fibrin, a fiberlike insoluble polymer that forms the hard clot. (Refer to Figure 15-11 for a close-up view of fibrin fibers.) The enzyme that catalyzes this change is thrombin. It is not present in the blood normally; if it were, blood could not flow through your body. Thrombin is generated from an inactive zymogen, prothrombin, in the presence of calcium ion and another protein, thromboplastin. Other factors are also known to activate prothrombin. Vitamin K is needed for its biosynthesis.

The blood platelets (thrombocytes) play an important role in clotting. They contain thromboplastin as well as other activating factors which are released when the platelet ruptures. In addition, after a blood vessel is injured, the platelets gather at the site, sealing the wound and acting as a center around which the fibrin strands can form. Thrombocytopenia, a deficiency of platelets, results in a prolonged bleeding time.

Other factors lengthen clotting time. Hemophilia is a hereditary disease in which one of the activating molecules is missing; minor injuries can cause fatal hemorrhaging (bleeding). In contrast to the absence of some factor, the presence of certain compounds prevents coagulation of the blood. One such molecule is heparin, a polysaccharide found in many tissues. It interferes with the conversion of prothrombin to thrombin and prevents the clumping together of platelets. Dicumarol (bishydroxycoumarin) interferes with prothrombin formation in the liver, thus also preventing clot formation. The related compound sodium warfarin is poisonous to rodents—it causes death by internal bleeding from impaired clotting. These anticoagulants are sometimes used after surgery to inhibit the formation of clots. A blood clot that

dicumarol

forms and breaks loose can block a major blood vessel, causing serious damage or death.

Freshly drawn blood can be prevented from clotting by the addition of a compound that "traps" calcium ion, preventing it from entering the clotting process. An example is potassium citrate, which removes the ion as slightly soluble calcium citrate. This salt is added to blood that will be used for a transfusion or donation. Sodium oxalate and sodium fluoride also serve as plasma anticoagulants.

A system exists in your body for the dissolving of blood clots; this process normally occurs after a few hours or days. The enzyme that catalyzes the dissolving is a protease, plasmin. It normally exists as the zymogen plasminogen. The conversion from the inactive form is stimulated by various factors such as emotional stress and exercise, as well as normal recovery from injury.

22.4 Erythrocytes and gas transport

The major job of the red blood cells, the erythrocytes, is the *transport of oxygen and carbon dioxide*. These cells contain the most concentrated protein solution in the body; they are 34% hemoglobin (about 1,000,000 molecules per cell). The normal blood range for hemoglobin is 13–17 g/100 ml (2.0–2.6 mmol/l). Erythrocytes permit the circulation of this large amount of protein without the high viscosity (resistance to flow) that would result if it were free in solution. The cells are formed in the bone marrow and have a lifetime of about 126 days. There are 4.2 to 5.9 million erythrocytes per cubic millimeter (mm^3) of blood (4.2 to 5.9 × 10^{12}/liter).

The structure of hemoglobin is illustrated in the chapter on protein (see Figure 15-10) and that of heme, its prosthetic group, is shown in the protein metabolism chapter (see Figure 19-6). Each day, hemoglobin molecules carry about 600 liters of O_2 from the air to the farthest parts of the body in a matter of seconds. The oxygen molecule binds reversibly to the iron atom of heme, converting the deoxygenated form (symbolized Hb or HHb$^+$) to the oxygenated form, oxyhemoglobin (HbO$_2$). The conformation of the four subunits of hemoglobin changes when oxygens bind, and the color of the blood becomes redder.

As you remember from your study of the gas laws, gases move because of differences in their partial pressures. Thus, in the lungs, where the partial pressure of O_2 is high, 96% of the hemoglobin of the erythrocytes is converted to oxyhemoglobin. When the blood reaches the tissues, which have a lower partial pressure of oxygen, the gas is released, leaving only about 64% of the hemoglobin in the oxygenated form. In addition, the oxygen affinity of hemoglobin, its tendency to bind the gas, is lower inside the red blood cell than in free solution, allowing oxygen to be "unloaded" at the capillaries. The difference, 32%, has been transferred to the tissues—over 6 ml of O_2 per 100 ml of blood. During exercise, more oxyhemoglobin is converted to hemoglobin, delivering an even greater amount of oxygen.

The carbon dioxide produced as a by-product of metabolism has a higher partial pressure in the tissues than in the erythrocytes. About one-third of the CO_2 binds directly to the amino groups of all four chains of hemoglobin, forming carbaminohemoglobin, which contains the carbamate group:

$$\boxed{\text{hemoglobin chain}}\text{—NH}_2 + CO_2 \longrightarrow \boxed{\text{hemoglobin chain}}\text{—NH—}\overset{\displaystyle O}{\overset{\|}{C}}\text{—O}^- + H^+$$

carbamate
group

The large remainder of CO_2 is hydrated to carbonic acid, H_2CO_3, by the erythrocyte enzyme carbonic anhydrase:

$$CO_2 + H_2O \xrightarrow[\text{anhydrase}]{\text{carbonic}} H_2CO_3$$

The resulting acid then dissociates to form bicarbonate ion and a hydrogen ion:

$$H_2CO_3 \longrightarrow HCO_3^- + H^+$$

(The proton is accepted by the side chains of the amino acid histidine in hemoglobin, so the pH remains constant.) This process, the **isohydric shift**, results in a major part of the CO_2 being transported away from the tissues as HCO_3^- in the erythrocytes. Much of the newly formed bicarbonate ion escapes from the erythrocytes into the blood plasma and is replaced by chloride ions moving in the other direction, into the red blood cell; this effect is known as the **chloride shift**. The net result is that 60% of the total carbon dioxide is transported away from the tissues as plasma bicarbonate ion, and the remainder is transported as bicarbonate ion in the erythrocytes or as carbaminohemoglobin.

When the blood returns to the lungs through the veins, the reverse processes occur. Carbon dioxide is re-formed by the dehydration of bicarbonate ion by carbonic anhydrase, and it is also released from the carbaminohemoglobin. You then exhale the gaseous CO_2. Oxygen once again oxygenates hemoglobin and the cycle repeats.

22.5 Abnormal number of erythrocytes

Under certain conditions, the erythrocytes are present in much larger or smaller amounts than normal. An overproduction of red blood cells is known as **polycythemia**. (When caused by a tumor, the condition is polycythemia vera or erythremia.) The red cell count may reach as high as 11 million per cubic millimeter, far above the normal range of 4.2 to 5.9 million. The viscosity of the blood increases greatly, from the normal value of 3 times that of water to about 15 times as viscous as water.

The total blood volume may also double in this disease. Circulation time increases from about 60 seconds to as much as 120 seconds. A larger amount of hemoglobin loses oxygen and is reduced to its blue form, resulting in cyanosis, a bluish-purple discoloration of the skin. A less extreme form of polycythemia is produced after a person stays at high altitudes for several weeks. More erythrocytes are formed by the body to make up for the lower oxygen pressure.

Anemia results from a deficiency of red blood cells. Blood loss after hemorrhage causes anemia. Although the body replaces plasma in 1 to 3 days, the erythrocyte concentration does not return to normal until 3 or 4 weeks. *Aplastic anemia* results from destruction of the bone marrow that produces red blood cells; it may be caused by exposure to gamma or x rays and certain chemicals or drugs (such as the antibiotic chloramphenicol). A defect in the production of "intrinsic factor," a mucoprotein of the stomach, leads to an inability to absorb vitamin B_{12} from food in the intestinal tract, resulting in *pernicious anemia*. Lack of vitamin B_{12} prevents red blood cells from developing normally ("maturation failure"). *Hemolytic anemias* result from fragile red blood cells, which easily break as they pass through the capillaries. You already know about one such example, sickle cell anemia, the hereditary disease (Section 20.10).

22.6 Blood groups

The membranes of the erythrocytes contain two different glycosphingolipids known as **blood group substances**. They act as antigens when mixed with the blood of another group containing gamma globulin antibodies that react against these glycosphingolipids in the plasma. Four different blood types exist, as shown in Table 22-2. During a transfusion (Figure 22-3), you react

Table 22-2 **The ABO Blood Groups**

Blood group	Erythrocyte antigen	Plasma antibody
A	A	anti-B
B	B	anti-A
AB	A and B	none
O	none	anti-A and anti-B

against the blood of another group as if it were an invading organism. You release antibodies, which cause the "foreign" blood to "agglutinate" (form clumps), making it useless as a blood replacement (Figure 22-4).

The **Rh factors** are a second type of special group of erythrocyte components that can act as antigens. The term "Rh-positive" means that a person has one of these factors; "Rh-negative" means that these factors are absent from the blood. Unlike the case of the ABO groups, a person does not have antibodies against the Rh antigen unless the body has been previously exposed to blood containing it. Problems may result when an Rh-negative mother gives birth to

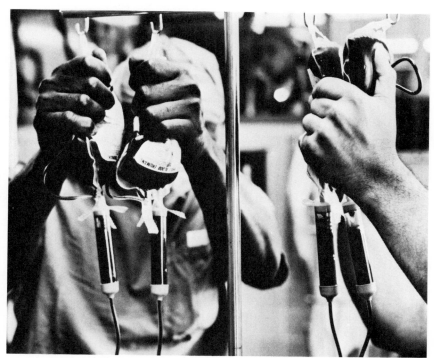

Figure 22-3. Blood infusion. (Photo courtesy of St. Luke's Hospital Center, New York.)

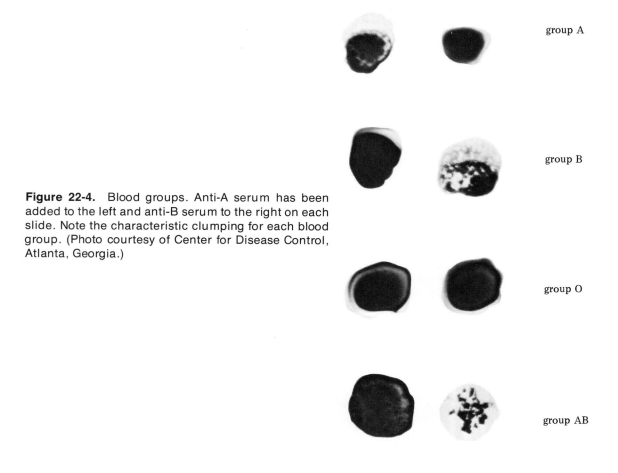

group A

group B

group O

group AB

Figure 22-4. Blood groups. Anti-A serum has been added to the left and anti-B serum to the right on each slide. Note the characteristic clumping for each blood group. (Photo courtesy of Center for Disease Control, Atlanta, Georgia.)

an Rh-positive baby because she forms antibodies against the infant's erythrocytes (a condition known as erythroblastosis fetalis). Transfusions and injection of antibodies may be necessary.

22.7   Leukocytes

The white blood cells or leukocytes (Figure 22-5) are part of the *body's protective system*. After formation in the bone marrow (granulocytes) or lymph nodes (lymphocytes), they are carried to areas of inflammation. Here, some of the white blood cells (phagocytes) surround and destroy bacteria and fragments of damaged tissue, forming the mixture known as pus. Other types of leukocytes form the plasma cells which make the antibodies that inactivate

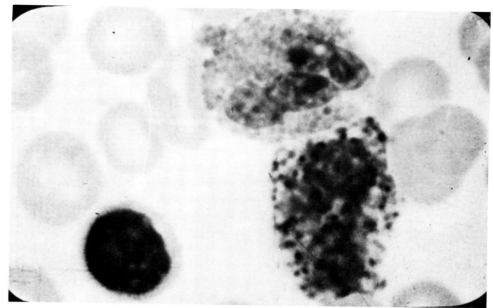

Figure 22-5. Leukocytes. The smaller cells surrounding the three leukocytes are erythrocytes. (Photo courtesy of Center for Disease Control, Atlanta, Georgia.)

antigens. This second stage of the defense system provides immunity against the infection on a long-term basis.

The normal adult has about 5000 to 10,000 white blood cells per cubic millimeter of blood. The number rises during infection. Uncontrolled production of leukocytes is a form of cancer known as leukemia. Leukemic cells are produced in great quantities, using up the body's supply of nutrients and metabolites but serving no useful function. The white cell count drops as a result of typhoid, tuberculosis, measles, flu, rubella, mumps, and radiation exposure.

22.8 Blood analysis

Various analyses can be performed on whole blood, on plasma, or on the blood serum, which is plasma with the fibrinogen removed. Table 22-3 lists the most important tests, the normal ranges, and the conditions that may cause either an increase in or a decrease from these values. In the SI system, the units are generally given as millimoles per liter (mmol/liter) or micromoles per liter (μmol/liter) as indicated. A clinical laboratory is shown in Figure 22-6.

Table 22-3 Common Blood Tests

Name of test	Normal range (per 100 ml)	Normal range (SI units)	When increased	When decreased
albumin	4–5 g	540–770 μmol/liter	—	kidney and liver disease
bilirubin	0.5–1.4 g	3–21 μmol/liter	red cell destruction, liver damage (jaundice)	—
cholesterol	150–250 mg	2.6–5.9 mmol/liter	diabetes, lipemia; pregnancy	severe infection, pernicious anemia, epilepsy
creatinine	0.7–1.5 mg	53–106 μmol/liter	kidney disease, intestinal or urinary obstruction	—
glucose	70–100 mg	3.9–5.8 mmol/liter	diabetes mellitus, hyperthyroidism, pregnancy, stress	starvation, hyperinsulinism, hyperthyroidism, liver disease
inorganic phosphate	3–4.5 mg	1.0–1.5 mmol/liter	starvation, kidney disease, hypoparathyroidism	rickets, myxedema, hyperparathyroidism
total protein	6–8 g	60–80 g/liter	multiple myeloma	infectious hepatitis, nephrosis, cirrhosis of liver
urea nitrogen (BUN)	8–25 mg	2.1–7.1 mmol/liter	metal poisoning, kidney disease, dehydration	pregnancy, liver failure
uric acid	3–7 mg	0.18–0.48 mmol/liter	gout, leukemia, liver and kidney disease, infection	liver atrophy, salicylate therapy

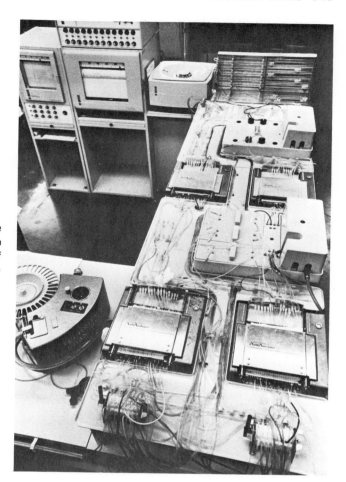

Figure 22-6. A clinical laboratory. The automated equipment shown is used to perform blood tests. (Photo courtesy of St. Luke's Hospital Center, New York.)

22.9 Other extracellular fluids

In addition to the blood plasma, other specialized fluids are needed for the normal functioning of your body. A large fraction of the total extracellular liquid is taken up by the **interstitial fluid**, which contributes about 15% of the body weight. It lies in the spaces between the cells; most of it cannot flow freely, but rather hydrates the molecules in the interstitial spaces. A large part is present in a gel composed mainly of hyaluronic acid, a polysaccharide that acts as a cement holding cells together. Dissolved substances can diffuse through this gel on their way between blood and tissue cells.

Although the terms "interstitial fluid" and "lymph" are sometimes used to mean the same thing, **lymph** is more correctly defined as the fluid in the lymphatic ducts. These are part of a special circulatory system that includes lymph veins and capillaries but no arteries, as shown in Figure 22-7. The lymphatic system serves to return materials from the tissue to the blood, particularly protein and excess tissue fluid. This process is important for maintaining the normal osmotic balance between the blood and tissues. When it is not functioning properly, edema, an excessive accumulation of fluid in the tissue spaces, develops. Another important property of the lymph is its ability to absorb fats from the intestine. In addition, the lymph nodes, masses of connective tissue present in the lymphatic system, form certain types of white blood cells and act as filters, trapping dead cells and destroying bacteria.

Important extracellular fluids are produced by **secretion**, that is, formation or separation from the blood or interstitial fluid by an energy-requiring process. Examples are the aqueous humor of the eye and the cerebrospinal fluid, which surrounds the brain and spinal cord. Tears are a secretion that keeps the surface of the cornea wet, protecting the eye and improving its optical properties. This fluid contains the enzyme lysozyme to protect the cornea from infection by hydrolyzing the cell walls of bacteria.

Saliva is a secretion from several glands (parotid, submaxillary, sublingual) in the mouth. This tasteless fluid is over 99% water and has a pH between 6.4

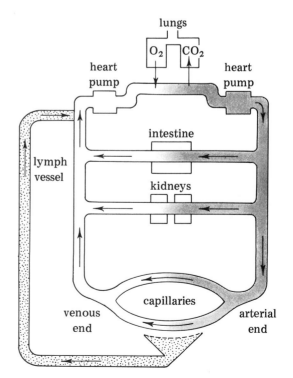

Figure 22-7. The lymphatic system and its relation to the circulatory system.

and 7.0. Approximately 1000 to 1500 ml are produced daily. Saliva contains the enzyme alpha-amylase (ptyalin), which begins the hydrolysis of starch to smaller polysaccharides. In addition, a mixture of glycoproteins known as mucin forms the basis of mucus, providing lubrication to aid swallowing. Saliva also serves as a cleansing agent for food particles that stick to your teeth. Other secretions of the digestive tract include juices from the stomach, pancreas, and small intestine (see Sections 16.8, 16.9, and 16.10).

Sweat is secreted to cool the body through evaporation. It is a means of keeping a heat balance in the body and is the only way you can lose heat when the temperature of your surroundings is greater than the body temperature. Table 22-4 lists the ways heat is gained and lost. The production and composi-

Table 22-4   Heat Production and Loss by the Body

Heat production	Heat loss
basal metabolism	evaporation of sweat
muscle activity	radiation
increased metabolism—	conduction
hormones or temperature	

tion of sweat depends on individual as well as environmental factors. At high temperatures, the maximum rate of sweat production can reach 1.5 to 4 liters/hour, depending on how well adjusted (acclimatized) you are to the hot surroundings. This loss can amount to a reduction in body weight of 8 pounds/hour. Unreplaced loss of large volumes of perspiration results in shrinkage in both the cellular and extracellular parts of the body, causing central nervous system changes. The resulting disorders include heat prostration and heat cramp; they can be prevented by taking sodium chloride tablets (1 to 2 g with water, four or more times daily).

22.10   Milk

Milk is secreted by the mammary glands of the mother at the end of pregnancy. *This fluid is one of the most complete natural foods;* milk contains an abundant supply of nutrients most needed for growth and development of the infant. The composition of human milk is compared in Table 22-5 with that of cow's milk. The main difference is the higher carbohydrate content of human milk and the greater protein content of cow's milk. The white color of milk comes from emulsified lipid and from the calcium salt of casein, milk's most important protein. The pH of milk is 6.6 to 6.8, but it becomes more acidic on

Table 22-5   **Comparison of Human Milk and Cow's Milk**

Component	Human milk	Cow's milk
water	87.5%	87%
total solids	17.5%	13%
carbohydrates	7.0–7.5%	4.5–5.0%
lipid	3.0–4.0%	3.5–5.0%
protein	1.0–1.5%	3.0–4.0%
ash	0.2%	0.75%

standing as a result of formation of lactic acid through fermentation of lactose by microorganisms.

Casein represents about 40% of the protein in human milk; it binds phosphate ions as well as calcium ions, making milk the best source of these two nutrients. The remainder of the milk protein is present in the "whey," the portion obtained by removal of the cream and acidification to pH 4.7, which causes the casein to precipitate. The principal protein here is beta-lactoglobulin. Together, these two proteins contain all the amino acids, including the essential ones.

The main carbohydrate in milk is the disaccharide lactose. Lipids are present as triacylglycerols (triglycerides); they contain all the saturated fatty acids and provide much of the caloric content of milk (1 cup = 160 kcal). Milk has vitamin A and riboflavin and, in smaller amounts, ascorbic acid, vitamin D, thiamin, pantothenic acid, and niacin. The ascorbic acid (vitamin C) is destroyed by pasteurization, the heat treatment (62°C for 30 minutes) that kills harmful microorganisms in milk. The important inorganic constituents of milk besides calcium and phosphorus include potassium, sodium, magnesium, and chlorine (these make up its mineral "ash"). Milk is low in iron; anemia develops in the infant if milk is its only food.

22.11 Urine

Urine is the fluid separated from the blood plasma by the kidney, the body's major organ of secretion (Figure 22-8). It contains waste products from metabolism as well as any other substances present in excess amounts that must be eliminated. By controlling the nature of urine excreted, the kidney helps regulate the volume and composition of the extracellular fluid of your body.

The volume of urine voided (discharged) by a normal adult can be anywhere from 600 to 2500 ml in a 24-hour period. The exact amount depends on such factors as the volume of fluid intake, the ingestion of certain foods or drugs, the presence of disease, and the environmental temperature. A chemi-

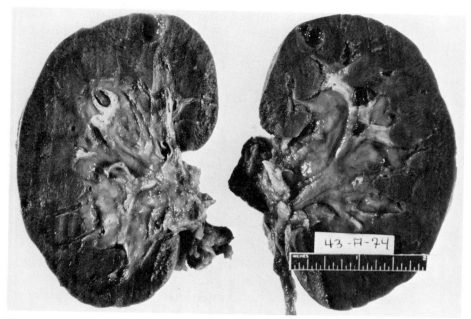

Figure 22-8. A human kidney. (Photo by Martin Rotker.)

cal substance that increases the volume of urine excreted is a **diuretic**. Caffeine, a purine derivative found in coffee and tea, is an example. Diuretic drugs (such as hydrochlorothiazide) are used to treat edema. The hormone vasopressin (also known as antidiuretic hormone) helps to regulate urinary volume.

In certain diseases, the volume of urine increases or decreases greatly. **Polyuria**, the excretion of excessive quantities of dilute urine, occurs in diabetes insipidus because of a deficiency of vasopressin. **Nocturia** is the passage of more than 500 ml of urine (with specific gravity below 1.018) during a 12-hour period at night; this condition occurs in the early stages of kidney disease and diabetes mellitus. A decrease in urine volume, **oliguria**, or a complete absence of urine, **anuria**, occurs in the later stages of kidney disease.

Normal urine has a specific gravity (see Section 6.6) of 1.005 to 1.030. An increased specific gravity results from diabetes mellitus and dehydration; a decreased specific gravity results from diabetes insipidus, nephritis, collagen diseases, and hypertension. Its color is amber because of the presence of the pigment urochrome, and it is clear. After standing, a sediment may form, consisting of proteins and phosphate salts.

The pH of urine is most commonly between 5.5 and 6.5, but it can vary from 4.8 to 8.0. Urine is generally acidic because the residue of an average diet contains sulfuric acid, H_2SO_4, from sulfur-containing amino acids, and phosphoric acid, H_3PO_4, from phosphorus-containing molecules like nucleic acids and phosphoglycerides. Alkaline urine results from a diet consisting of fruits and vegetables.

22.12 Normal composition of urine

The kidneys reabsorb molecules needed by the body, including 99% of the water, glucose, and amino acids, as well as sodium ion, chloride ion, and bicarbonate ion. The ions and molecules that are secreted and appear in the urine are listed in Table 22-6.

The most abundant positive ions in food, sodium ion, Na^+, and potassium ion, K^+, are the major cations in urine. The amount varies with diet, but the potassium ion excretion has a minimum value of about 1 g/day. Calcium, magnesium, and ammonium ions are found in smaller amounts (although large quantities of ammonium ion, NH_4^+, appear in severe acidosis). The major anion of the urine is chloride ion, Cl^-; the amount excreted is about the same as the quantity ingested. Phosphorus appears as phosphate ion, PO_4^{3-}; the

Table 22-6 **Composition of Urine**

Component	Amount[a]
sodium ion	2.4 g
potassium ion	1.5–2.0 g
magnesium ion	0.1–0.2 g
calcium ion	0.1–0.3 g
iron ion	0.2 mg
ammonium ion	0.4–1.0 g N
hydrogen ion	trace
chloride ion	9–16 g
bicarbonate ion	0–3 g
phosphate ion	0.7–1.6 g P
sulfate ion	0.6–1.8 g S
organic sulfate	0.06–0.2 g S
uric acid	0.08–0.2 g N
amino acids	0.08–0.15 g N
hippuric acid	0.04–0.08 g N
creatinine	0.3–0.8 g N
peptides	0.3–0.7 g N
urea	6–18 g N

[a] g N denotes grams nitrogen; g P, grams phosphorus; g S, grams sulfur.

concentration increases with acidosis or alkalosis and decreases with kidney damage, pregnancy, and diarrhea. Sulfur is found as sulfate ion, SO_4^{2-}.

The major organic part of urine is *urea*, the end product of nitrogen metabolism. Because of the nitrogen balance, the amount excreted depends directly

$$NH_2-\overset{\displaystyle O}{\overset{\|}{C}}-NH_2$$
urea

on the quantity of nitrogen-containing foods, particularly protein, that you eat.

Uric acid (shown in Section 19.5) is a waste product of purine metabolism; its concentration is related to the amount of nucleoprotein (present especially in glandular meats) in the diet. The diseases gout, leukemia, polycythemia, and hepatitis result in increased excretion of this molecule.

Creatine and *creatinine* are related molecules found in muscle tissues; creatinine, a normal constituent of urine, is the end product of creatine metabolism. The amount of creatinine in the urine is a measure of your muscle mass.

$$HO-\overset{\displaystyle O}{\overset{\|}{C}}-CH_2-\overset{\displaystyle H_3C}{\overset{\displaystyle |}{N}}-\overset{\displaystyle NH}{\overset{\|}{C}}-NH_2$$
creatine

creatinine

(The "creatinine coefficient," the milligrams excreted per 24 hours per kilogram of body weight, is 18 to 32 in men and 10 to 25 in women.) Creatine is excreted in larger amounts than creatinine during muscle wasting, as in starvation, fever, diabetes and muscular dystrophy.

Hippuric acid (benzoylglycine) is the form in which benzoic acid is excreted. (Benzoic acid is present in foods such as fruits and berries.)

hippuric acid

Other organic molecules found in urine include urobilinogen and small amounts of vitamins and hormones.

22.13   Abnormal compounds in urine

Under certain conditions, other molecules appear in urine. These are its abnormal constituents; they often indicate the presence of disease in the body.

When a sugar is present in unusually large amounts, the condition is **glycosuria**. Specific names are given depending on the type of sugar present. **Glucosuria** is an excess of glucose in urine, above the normal range of 10 to 20 mg/100 ml. The most common cause is diabetes mellitus; other causes are anesthesia, asphyxia, emotional states, and hyperthyroidism. **Pentosuria** is an excess of a pentose, as may occur after eating large amounts of fruit. Other possible conditions are lactosuria (excess lactose), galactosuria (excess galactose), and fructosuria (excess fructose).

Proteinuria, a large amount of protein in the urine, is found in kidney disease (such as nephritis). Albumin is generally the major constituent, so this condition is known as **albuminuria**. Other abnormal constituents include the ketone bodies present in ketosis. Porphyrias are diseases in which various substituted porphyrins are excreted. Bilirubin and urobilinogen are found in urine and liver diseases. Blood ("occult blood") is present after internal bleeding, and pus indicates an infection of the kidney or urinary tract. Commercial urine tests for these and other substances are listed in Table 22-7. A test for the approximate glucose concentration of urine is shown in Figure 22-9.

Table 22-7 **Commercial Urine Tests**

Name	Test
Acetest, Ketostix	acetoacetic acid, acetone
Albustix, Albutest, Bumintest	protein
Azostix	urea nitrogen
Bili-Labstix	pH, protein, glucose, ketones, bilirubin, blood
Clinistix, Tes-Tape	glucose
Combistix	protein, glucose, pH
Hema-Combistix	protein, glucose, pH, blood
Hemastix, Occultest	blood
Ictotest	bilirubin
Labstix	protein, glucose, pH, blood, ketones
Phenistix	phenylpyruvic acid
Uristix	protein, glucose
Urobilistix	urobilinogen

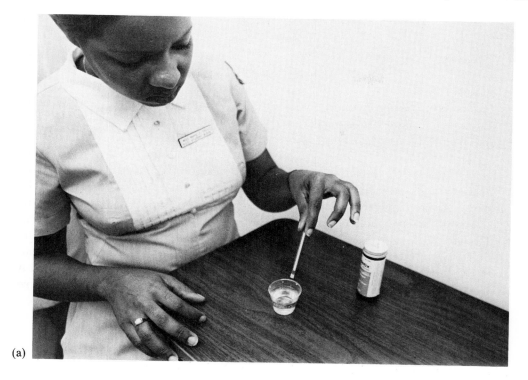

(a)

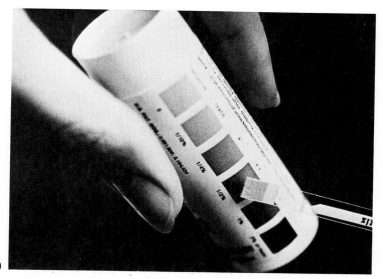

(b)

Figure 22-9. (a) A person performing a commercial urine test. (b) The color change
on the "dip-and-read" strip (Clinistix) indicates the approximate glucose concentration. [Photo (a) by Al Green; (b) courtesy of Ames Company.]

SUMMARY

Your body contains 60 to 70% water by weight. Of the approximately 40 liters in a 150-pound adult, 25 liters are intracellular fluid and 15 liters are extracellular fluid.

Blood, one of the most important fluids, is the transport system of the body. Its main job is to bring oxygen and food molecules to the tissues and to remove carbon dioxide and the waste products of metabolism. An adult has 5 to 6 liters of blood.

Proteins are the most abundant part of the blood plasma, the liquid portion of the blood. Over half of the plasma protein is albumin. This protein helps maintain the osmotic pressure of the blood.

The globulins, such as gamma globulin, provide immunity against disease. These proteins form a defense system by acting as antibodies. They destroy or inactivate microorganisms and foreign protein molecules known as antigens.

The protein fibrinogen in blood plasma is needed for blood clot formation. Clotting is the chemical defense against loss of blood. It results from the conversion of fibrinogen to fibrin, the substance that forms the jellylike clot. Numerous chemical substances affect the clotting time.

The major role of the erythrocytes, the red blood cells, is the transport of O_2 and CO_2. Oxygen is carried by binding to the iron of heme groups in the protein hemoglobin. Carbon dioxide is transported as bicarbonate ion in the plasma or as carbamate groups on the hemoglobin chains.

Certain conditions result in overproduction of the erythrocytes, known as polycythemia. Anemia results from a deficiency of red blood cells.

The membranes of the erythrocytes contain chemical groups that divide them into blood group substances. Four different blood types exist—A, B, AB, and O. Antibodies in the plasma (anti-A, anti-B) can react with antigens of another blood type.

The leukocytes, or white blood cells, are part of the body's protective system. They are formed in the bone marrow or lymph nodes and are carried to areas of inflammation. The number of leukocytes rises during infections.

In addition to plasma, other specialized fluids are needed for normal functioning of the body. These include the interstitial fluid, lymph, aqueous humor, cerebrospinal fluid, tears, digestive tract secretions such as saliva, and sweat. Milk, secreted by the mammary glands of a mother at the end of pregnancy, is one of the most complete natural foods.

Urine is the fluid separated from the blood plasma by the kidney. It contains waste products from metabolism as well as any other substances present in excess amounts that must be eliminated. The volume voided varies from 600 to 2500 ml in 24 hours. A diuretic increases the amount excreted.

Urea is the major organic part of urine. Under certain conditions, other molecules appear in the urine. The presence of these abnormal constituents may indicate the presence of disease in the body.

Exercises

1. (22.1) Describe the role of blood in the body.

2. (22.1) What function does plasma albumin play?

3. (22.2) How do the globulins provide immunity against disease?

4. (22.3) Describe the process of blood clotting.

5. (22.3) What is an anticoagulant? Give an example.

6. (22.4) Explain how the blood transports (a) oxygen; (b) carbon dioxide.

7. (22.5) What is polycythemia? anemia?

8. (22.6) Why must the type of a patient's blood be determined before a transfusion?

9. (22.7) Describe the function of the leukocytes.

10. (22.8) Give possible interpretations of the following blood test results: (a) 400 mg cholesterol; (b) 2 g albumin; (c) 10 mg uric acid; (d) 150 mg glucose; (e) 2.5 mg bilirubin, each per 100 ml. (Refer to Table 22-3.)

11. (22.9) What is the interstitial fluid?

12. (22.9) Describe the role of lymph.

13. (22.9) List three fluids produced by secretion.

14. (22.9) What is the function of sweat?

15. (22.9) Explain how saliva aids in the digestion of food.

16. (22.10) Why is milk considered one of the most complete natural foods?

17. (22.11) Why must the body secrete urine?

18. (22.11) Define polyuria, nocturia, oliguria, anuria.

19. (22.11) What is a diuretic? Give an example.

20. (22.12) What are the major inorganic and organic components in normal urine?

21. (22.13) Describe three possible abnormal constituents of urine and the conditions that would cause them to be present.

Drugs and poisons

Drugs are chemical substances that affect your body. They are either administered internally or applied externally for purposes of medical diagnosis, preventing or curing disease, treating symptoms, or birth control. The drugs used in a health center are called ethical drugs; they require a prescription. Others, that can be bought "over the counter" without prescription, are known as proprietary drugs. With certain notable exceptions, the latter category of drugs, which includes such preparations as mouthwashes and cold remedies, has relatively little medical importance. Many drugs have already been mentioned, as summarized in Table 23-1; those in this chapter are the most important drugs, which have far-reaching effects in the body.

23.1　The action of drugs

Most drugs act by stimulating or depressing certain activities in the cell, replacing a deficient substance, killing or weakening a foreign organism, or causing irritation. In many cases, the action of a drug results from its influence on enzymes, on cell membranes, or on other "receptors," molecules with which the drug interacts.

After a drug is administered, the first event that takes place is absorption, its transfer to the blood. Then it is distributed to various parts of the body, depending on the size and solubility properties of the drug molecule. Drugs may undergo a process called biotransformation, a chemical change catalyzed by enzymes, commonly in the liver. Finally, the drug is excreted, generally by the kidneys into the urine.

The patient's response to a particular drug depends on many factors, including age, sex, weight, genetic makeup, means and time of administration, emotional state, and previous drug background (Figure 23-1). In addition, *no drug produces a single effect*—a drug given to treat a certain condition causes other changes in the body, which must be taken into account. The guiding

Table 23-1 Index to Drugs in Previous Chapters

Drug class	Section in text	Drug class	Section in text
alkaloid	12.7	cardiac glycoside	13.8
anabolic steroid	21.22	diuretic	22.11
antacid	8.7	fluoride	15.10
antiangina	12.8	hypoglycemic	21.20
anticoagulant	22.3	inorganic salt	3.13
antihistamine	22.2	"quat"	12.4
antilipemic	18.9	radioisotope	9.10
antiseptic	15.9	steroid	21.26
antitoxin	22.2	sulfa	16.5
birth control	21.24	vaccine	22.2
cancer therapy	16.6		

Figure 23-1. Administration of a drug. (Peace Corps photo by Paul Conklin. Courtesy of National League for Nursing.)

principle in prescribing drugs, and in fact for all medical intervention, is that if nothing else at least the patient should not be harmed.

23.2 Antibiotics

Antibiotics are chemical substances used to treat infectious diseases; this application of drugs is referred to as **chemotherapy**. Antibiotics are based on a principle known as "selective toxicity": they are designed to be more harmful to the invading organism than to the patient. Their function is to either *kill or inhibit the growth of a microorganism.* Antibiotics are generally produced or derived from living cells by the process of fermentation. (The sulfa drugs, described in Section 16.5, are not considered antibiotics because they do not come from living organisms.) This class of compounds is one of the most widely prescribed group of drugs.

Bacteria are the greatest single cause of disease. They are divided into two main groups depending on their ability to be stained by a method devised by a Danish physician named Gram. Those bacteria that can be stained are known as "gram-positive" and those that cannot are known as "gram-negative." These two classes of bacteria respond differently to antibiotics. Some drugs act on bacteria from one group but not the other. Certain antibiotics are called **broad spectrum** because they treat diseases caused by both types, as well as some other kinds of foreign organisms.

The most potent antibacterial drugs are the *penicillins* (Figure 23-2). They are effective mainly against gram-positive bacteria, such as those that cause staphylococcus, streptococcus, pneumococcus, meningococcus, and gonococcus infections. They inhibit the synthesis of bacterial cell walls, causing the cell to burst as water flows inside. Penicillin G, or benzyl penicillin, is the most important form of this antibiotic. Newer derivatives, which contain groups other than the benzene ring, are less potent but overcome the disadvantages of

penicillin G
(benzyl penicillin)

penicillin G—its breakdown in the stomach and poor absorption after oral administration. They include penicillin V (phenoxymethyl penicillin), which is more acid stable, methicillin (2,6-dimethoxyphenyl pencillin), which is resistant to the penicillinase enzyme, and ampicillin (6-aminobenzyl penicillin),

Figure 23-2. A penicillin mold. A mutant form of this green mold, *Penicillin chrysogenum,* produces most of the world's supply of the antibiotic penicillin. (Courtesy of Pfizer, Inc.)

which is also effective against some gram-negative bacteria. Although penicillins are among the least toxic drugs, they cause allergic reactions in about 5 to 10% of the adults in the United Sates—the most common drug allergy. For these patients, the antibiotic *erythromycin* is often prescribed because of its similar activity (it is also effective against certain bacteria resistant to penicillin).

Tetracyclines are the most widely used class of antibiotics, largely because of their broad spectrum of activity. They are effective against streptococcus and staphylococcus infections, pneumonia, gonorrhea, syphillis, typhus, urinary infections, and acne. This group of antibiotics inhibits protein synthesis in the bacteria at its ribosomes by blocking the binding of tRNA molecules carrying amino acids. The structure of the parent molecule of the group, tetracycline is as follows:

tetracycline

Derivatives, such as aureomycin (chlorotetracycline), have similar activity. Side effects include gastrointestinal irritation and alteration of the normal microbial population of the intestine, possibly resulting in a secondary infection (superinfection).

Table 23-2   Common Antibiotics

Name	Clinical application
ampicillin	urinary tract infections, salmonella infections, bacterial meningitis
bacitracin	skin infections (topical use only)
cephalosporin	gram-negative bacterial infections
chloramphenicol	typhoid and salmonella infections
erythromycin	penicillin substitute
gentamicin	gram-negative bacteria (sepsis, infected burns, pneumonia), urinary infections
griseofulvin	fungus infections
kanamycin	intestinal infections, urinary infections
penicillin	gonococcus, pneumococcus, streptococcus, meningococcus, staphylococcus, and other gram-positive bacteria
polymyxin	gram-negative bacteria in wounds, burns, and intestinal infections
streptomycin	tuberculosis, urinary infections
tetracycline	broad spectrum
Vancomycin	staphylococcus infections

Other antibiotics also act by inhibiting bacterial protein synthesis. *Strepto-mycin* is one example; it is used principally to treat tuberculosis (TB). *Chlor-amphenicol,* another such antibiotic, is effective against typhoid and other salmonella infections, Rocky Mountain spotted fever, and certain severe infections. Its use is limited because of the possible side effect of aplastic anemia, an often fatal blood disorder caused by the drug's action on bone marrow. These and other antibiotics are listed in Table 23-2.

23.3 Aspirin and analgesics

An **analgesic** is a drug that relieves pain. It does not impair the patient's consciousness, but decreases the awareness of sensitivity to pain. *Aspirin* is one of the most widely used analgesics available without a prescription.

Aspirin, or acetylsalicylic acid, is most effective for mild to moderate pain in headache, neuralgia (pain from a nerve), muscle pain, pain from joints, and

aspirin
(acetylsalicylic acid)

toothache. It also acts as an antipyretic, lowering an elevated body temperature, and as an antiinflammatory agent, reducing the redness, heat, and swelling that accompany inflammation, which is the local tissue response to injury. Aspirin has multiple effects on the body; its pain-relieving action probably results from dilation of blood vessels at the site from which the pain comes. It may also inhibit the synthesis of prostaglandins (see Section 14.1), which appear to be involved in the inflammation process.

In the digestive tract, aspirin is hydrolyzed to acetate and salicylate, which is the physiologically active species. Salicylate can be administered directly

salicylate
ion

but is too irritating. Aspirin itself in a normal dose of 5 grains (324 mg) causes loss of 2 to 6 ml of blood from irritation of the mucous membranes of the stomach. In addition, certain individuals are allergic to aspirin; their reaction may be asthma or hives (uticaria) in acute forms.

Acetaminophen is the most effective and safest aspirin substitute. This com-

$$HO-\langle\bigcirc\rangle-NH-\overset{\overset{\displaystyle O}{\|}}{C}-CH_3$$

acetaminophen

pound, however, lacks the antiinflammatory action of aspirin. The related molecule, phenacetin (an ethoxy group, —OCH_2CH_3, replaces the hydroxyl group), has also been used as an analgesic. Its use, however, can cause a hemoglobin disorder, methemoglobinemia, producing anemia. Phenacetin may also result in hemolytic anemia, particularly in those individuals who genetically lack a particular enzyme (glucose-6-phosphate dehydrogenase), as do 10 to 15% of blacks in the United States.

Other compounds are available as pain relievers only. Darvon, propoxyphene, is one of the most frequently prescribed, but its effectiveness has been questioned.

$$CH_3-CH_2-\overset{\overset{\displaystyle O}{\|}}{C}-O-\overset{\overset{\displaystyle \bigcirc}{|}}{\underset{\overset{\displaystyle |}{CH_2}}{C}}-\overset{\overset{\displaystyle CH_3}{|}}{CH}-CH_2-N(CH_3)_2$$

propoxyphene
(Darvon)

Table 23-3 lists the composition of commercially available analgesics.

23.4 Narcotic analgesics

Narcotic analgesics are the most powerful pain relieving drugs available. Frequent use, however, leads to a state of **drug dependency** or addiction. The body of an addicted individual adapts to the presence of the narcotic. If the drug is no longer administered, physiological disturbances characterized by "withdrawal symptoms" take place. Therefore, these drugs are medically prescribed only in cases of severe pain, such as that resulting from injury,

Table 23-3 **Composition of Commercial Analgesics**

Name	Analgesic ingredient[a]	Other ingredients
Anacin	aspirin (6.17)	caffeine
Bufferin	aspirin (5)	buffer
Cope	aspirin (6.5)	antihistamine, caffeine, buffer
Empirin	phenacetin (2.5), aspirin (3.5)	caffeine
Excedrin	acetaminophen (1.5), salicylamide (2), aspirin (3)	caffeine
Tylenol	acetaminophen (5)	—
Vanquish	aspirin (3.5), acetaminophen (3)	caffeine, buffer

[a] Numbers in parentheses indicate the amount, in grains, of analgesic ingredient; 1 grain = 64.8 mg.

surgery, or a heart attack, when they are taken for relatively short periods of time.

Opium, the most abundant natural source of narcotic analgesics, comes from a poppy plant (Figure 23-3). It contains a mixture of about 25 alkaloids, the largest fraction (about 10%) of which is morphine. Paregoric, a tincture of opium, contains morphine (0.4 mg/ml) as its active ingredient. *Morphine* is the most effective drug against severe pain. It acts by modifying awareness of

morphine

the pain, regardless of its origin or intensity. *Codeine*, or methylmorphine, is also found in opium but is a less potent pain killer. Because addiction to codeine is much less likely, however, this drug is the most commonly used narcotic analgesic. (It is also found in cough preparations because it inhibits the

Figure 23-3. Opium poppies. (Photo courtesy of U.S. Drug Enforcement Administration.)

cough center of the brain.) Heroin, or diacetylmorphine, is not present in opium, but is made from morphine by a simple chemical process. This narcotic, which is outlawed as a medical drug, is denser than morphine, making it easier to smuggle and to "cut" or dilute with another substance like lactose before selling on the street.

Meperidine (Demerol) is a major synthetic narcotic analgesic. Its potency

meperidine
(Demerol)

lies between that of morphine and that of codeine. Its effect is shorter than morphine (about 2 hours instead of 4), an advantage in short procedures or preparation for delivery. Meperidine, an addicting drug, is the most common narcotic abused by physicians and nurses.

Methadone (shown in Section 12.3), a synthetic narcotic, is not used as an analgesic but as a substitute for heroin. In methadone maintenance programs, heroin addicts receive oral doses of this drug, allowing them to function without having to find a "fix" each day. Methadone itself is addicting but has less severe withdrawal symptoms than heroin. Heroin withdrawal symptoms may include anxiety, perspiration, restlessness, tremors, muscle, joint and abdominal pain, fever, and possible convulsions and mental disorders. Methadone blocks the effects of an injection of heroin. Other drugs such as Naloxone and Nalorphine are narcotic antagonists; they are used as "antidotes" to terminate the effects of a narcotic in the patient's system.

23.5 Anesthetics

Like the analgesics, **anesthetics** decrease a patient's reaction to pain. General anesthetics, however, also cause loss of consciousness, diminished reflexes, and relaxation of muscles in addition to analgesia. They act by depressing the central nervous system, with minimal effect on the vital signs. Anesthetics are administered during surgery and childbirth.

The *anesthetic ethers* are described in Section 11.5. Diethyl ether ($CH_3CH_2OCH_2CH_3$), widely used for many years, is safe and easy to administer, but it is explosive and has a long recovery time accompanied by nausea. Divinyl ether ($CH_2{=}CH-O-CH{=}CH_2$) is toxic to the liver and kidney; its use is restricted to brief procedures. Fluoroxene ($CF_3CH_2OCH{=}CH_2$), also employed for short periods, is limited by its flammability. Methoxyflurane ($Cl_2CHCF_2OCH_3$), administered for analgesia in obstetrics and minor operations, has a slow induction and recovery period.

Halogenated hydrocarbons also serve as inhalation anesthetics. Chloroform ($CHCl_3$) is rarely used because of its possible toxicity to the heart and liver. Trichloroethylene [$HC(Cl){=}CCl_2$] produces rapid analgesia for delivery but causes rapid respiration and affects the heart. Halothane is the most popular anesthetic. It causes little respiratory irritation or postanesthetic nausea and is

$$\begin{array}{c} \quad\;\; F \quad\; H \\ \quad\;\; | \quad\;\; | \\ F-C-C-Br \\ \quad\;\; | \quad\;\; | \\ \quad\;\; F \quad\; Cl \end{array}$$

halothane

not explosive. Halothane, however, relaxes muscles poorly and also may result in respiratory or cardiac depression.

Gases employed as anesthetics include the widely used cyclopropane, which has a rapid induction period and little irritation or aftereffect. Its major disadvantage is the possibility of explosion. Nitrous oxide, or "laughing gas," N_2O, is administered as a mixture (80%/20%) with oxygen. The gas serves mainly as an analgesic in combination with other anesthetics.

Intravenous anesthetics act for very short periods of time. Thiopental (pentothal) is the one used most often. The related molecule, methohexital (Brevital), is more potent and permits faster recovery.

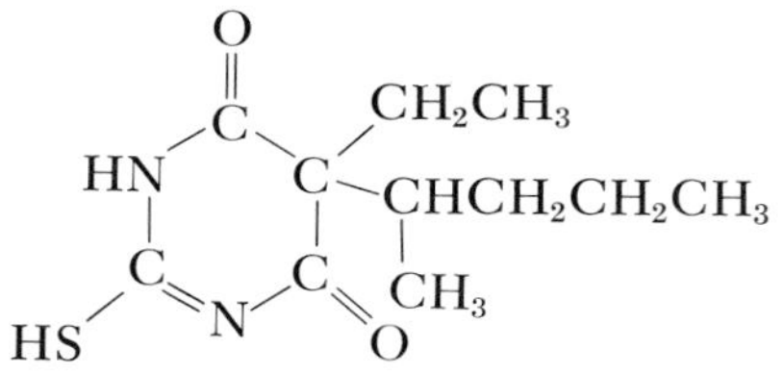

thiopental

Local anesthetics block the transmission of nerve impulses from the site of application without causing the patient to lose consciousness. Procaine (Novocain), lidocaine (Xylocaine), and benzocaine are common examples. Local

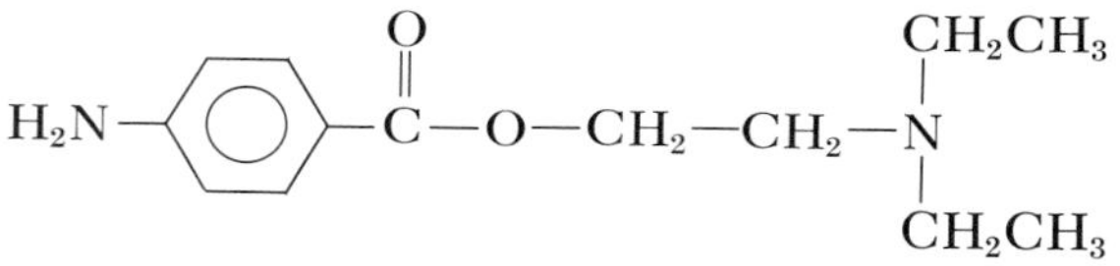

procaine

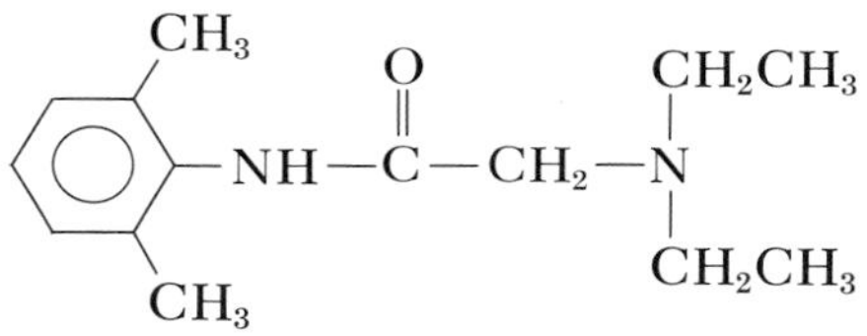

lidocaine

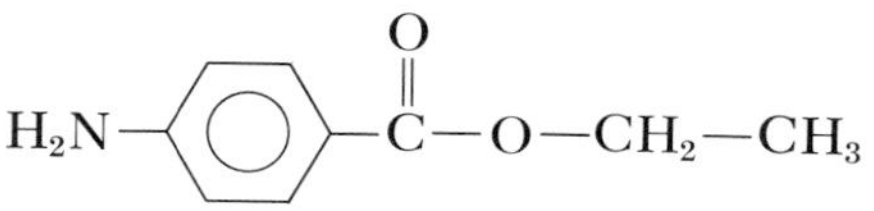

benzocaine

anesthetics are applied either topically or by injection (see Figure 12-1).

23.6 Sedative–hypnotics

Like general anesthetics, the **sedative–hypnotic drugs** can act by depressing the central nervous system. Their effect is longer lasting, however, and they can be administered orally. In small doses these drugs cause sedation—a calming effect. Larger doses result in a state similar to natural sleep (hence the name hypnotic). Still bigger doses first cause intoxication, then anesthesia, followed finally by respiratory depression and death. Continued administration leads to physical dependence and withdrawal symptoms if the drug is withheld.

The *barbiturates* are an important class of sedative–hypnotic drugs. The structures of several examples are as follows:

phenobarbital

amobarbital

pentobarbital

They are divided into groups according to the length of their effect. Long-acting barbiturates, such as phenobarbital, last for 6 to 12 hours and are used to provide continual sedation and treat anxiety, hypertension, and epilepsy. Intermediate-acting barbiturates, like amobarbital (Amytal), are effective for 4 to 6 hours; they are administered in cases of insomnia. Used for preoperative sedation and to treat insomnia, the short-acting barbiturates, which include pentobarbital (Nembutal) and secobarbital (Seconal), wear off after 3 to 4 hours. The ultrashort-acting barbiturates, such as thiopental, are effective for only minutes; as mentioned in the last section, they serve as intravenous anesthetics. Side effects of the barbiturates may include a "hangover" the morning after, listlessness, depression, nausea, and emotional disturbances.

Barbiturates, known as "downs" or "goofballs," are widely abused. (Pentobarbital capsules are called "yellow jackets" and secobarbital capsules are called "red devils" because of their color.) These drugs are taken in high doses to create effects similar to alcohol intoxication, which include euphoria but also sluggishness and impairment of memory, judgment, attention span, and motor coordination. Physical dependence results from above normal doses taken daily for several months. Withdrawal from barbiturates is more dangerous than withdrawal from narcotic analgesics and should be carried out in a hospital.

Chloral hydrate $[Cl_3CCH(OH)_2]$ is a rapidly acting nonbarbiturate hypnotic. This substance is known as "knockout drops" and a "Mickey Finn" when dissolved in alcohol. Paraldehyde (see Section 11.7), another example, is safe but irritating and has a disagreeable taste.

23.7 Antianxiety and antipsychotic drugs

Certain newer drugs are used to treat anxiety and tension, like the barbiturates in low doses. *Meprobamate* (Miltown, Equanil) is similar in action to amobarbital. *Chlordiazepoxide* (Librium), a more popular example, is a longer-acting drug. *Diazepam* (Valium), one of the most widely prescribed

meprobamate
(Miltown, Equanil)

chlordiazepoxide (Librium)   diazepam (Valium)

drugs, has a similar structure. These two antianxiety agents are no more effective than barbiturates as sedatives but have a higher margin of safety and less potential for abuse. They produce less sleepiness or interference with motor activities.

Many nonprescription drugs (such as Compoz, Sleep-eze, Nytol, and Sominex) are sold as sedative–hypnotics. They contain antihistamines, aspirin-like compounds, and other substances which may produce some sedation as secondary effects. Little evidence, however, supports their claims; these products are questionable in terms of both safety and effectiveness.

Antipsychotic agents are a recent class of drugs able to reduce the most severe symptoms in a high percentage of psychotic patients, those such as schizophrenics with major psychological disturbances. The effect of these drugs is very different from that of sedative–hypnotics. Compounds like *chlorpromazine* (Thorazine) alter mood and behavior—reducing hallucinations and illusions, as well as calming the patient. By alleviating serious symptoms without

chlorpromazine
(Thorazine)

causing loss of consciousness, the antipsychotic drugs are responsible in part for reducing the period of hospitalization for mentally ill patients.

23.8 Alcohol

Ethyl alcohol or ethanol (CH_3CH_2OH) is a central nervous system depressant used more often socially than medically. In the United States about 80% of the men and 67% of the women drink at least occasionally; the total consumption per person is approximately 10 gallons of alcohol each year. But some people drink more than others—about 9,000,000 people in this country are either chronic alcohol abusers or alcoholics, physically dependent on the drug. Large quantities of alcohol cause irreversible damage to the brain, nervous system, and heart in addition to the disruption of family life and job. Alcoholic cirrhosis of the liver is the fourth leading cause of death between the ages of 25 and 45 in large U.S. cities.

Alcohol is rapidly absorbed, particularly when the stomach is empty—in this case, absorption is nearly complete in an hour. It is metabolized in the liver, first by the enzyme alcohol dehydrogenase. This reaction, the conversion of ethanol to acetaldehyde, is the slow step in the breakdown of alcohol. It is not affected by black coffee, cold air, a cold shower, or any of the traditional means of "sobering up" someone who has drunk too much. The acetaldehyde is converted to acetate and then to acetyl coenzyme A before entering the citric acid cycle, producing 7 kcal/g of alcohol.

The behavioral effects of various doses of alcohol are listed in Table 23-4.

Table 23-4   **Behavioral Effects of Alcohol**

Volume of 40–50% (80–100 proof) alcohol (ounces)	Blood alcohol level (g/100 ml)	Effect on behavior
2 (60)[a]	0.05	false sense of well-being, impaired vision, querulous
4 (120)	0.10	poor reaction time and coordination, confusion
6 (180)	0.15	very poor reaction time, definite intoxication
8 (240)	0.20	physical and mental depression
8–14 (240–420)	0.20–0.35	confusion, slurred speech
12–16 (360–480)	0.30–0.40	stupor
14–18 (420–540)	0.35–0.45	coma
over 18 (over 540)	more than 0.45	death

[a] Numbers in parentheses indicate the volume of alcohol expressed in milliliters.

This compound always causes depression of the central nervous system—the apparent initial stimulation results from depression of the inhibitory centers of the brain. In an average person, the **blood alcohol concentration** (BAC) may reach about 0.1% 1 hour after drinking 4 ounces of 100 proof (50%) alcohol. Driving with an alcohol level above this value is illegal in all states because of the greatly increased chance of an accident. The blood alcohol level returns to zero after about 6 hours (2/3 ounce of 50% alcohol is metabolized each hour). The National Institute on Alcohol Abuse and Alcoholism recommends that you drink no more than 1½ ounces of alcohol per day, only with food and only in a dilute form. This quantity represents 3 ounces of 100 proof whiskey, or 12 ounces (one-half of a bottle) of wine, or three 12-ounce containers of beer (Figure 23-4).

Alcohol also acts as a diuretic and widens blood vessels; it is sometimes used medically for this purpose. Other therapeutic uses include skin massage, reflex stimulation in cases of collapse, postoperative analgesia, and as a solvent for drugs.

A "hangover" results in part from mild withdrawal symptoms from alcohol, which may include tremors, fatigue, vertigo, headache, gastritis, acidosis,

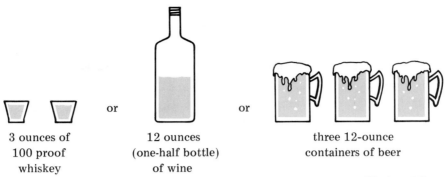

3 ounces of
100 proof
whiskey

or

12 ounces
(one-half bottle)
of wine

or

three 12-ounce
containers of beer

Figure 23-4. The maximum daily limit of alcohol consumption (National Council on Alcohol Abuse and Alcoholism).

weakness, and dehydration. In the case of an alcoholic, withdrawal is dangerous, causing delusion and hallucinations, a condition known as delerium tremens ("DT's"). Drugs used to treat alcoholics, like disulfiram (Antabuse), prevent oxidation of acetaldehyde, causing dizziness, headache, nausea, vomiting, and difficulty in breathing if alcohol is ingested.

23.9 Stimulants

Stimulants are a class of drugs that increase activity of the central nervous system in a manner similar to the hormone epinephrine (adrenaline). *Amphetamine* is a typical stimulant. It elevates the mood, reduces feelings of fatigue and hunger, increases reaction time, concentration, and the capacity to do

$$CH_2-CH-NH_2$$
$$|$$
$$CH_3$$

amphetamine

work. The *d* isomer, dextroamphetamine (Dexedrine), and the racemic mixture of *d* and *l* forms, known as Benzedrine, are both used medically, as is the more potent methyl derivative, methamphetamine (Methedrine). These drugs are administered to treat depression, curb appetite in cases of obesity, improve performance, and treat hyperkinetic (overactive) children.

They create a state of drug dependence, particularly when injected directly into the bloodstream ("mainlining"). A typical "speed freak" (Figure 23-5) continues injection of methamphetamine for 3 to 6 days followed by 12 to 18 hours of sleep before the next round. In addition to weight loss and susceptibility to infection, this form of drug abuse may cause an amphetamine-

Figure 23-5. Injecting "speed." (WHO photo by E. Mandelmann.)

induced psychosis with hallucinations and delusions as found in schizo-phrenia. Cocaine ("coke," "snow," "charlie") is an alkaloid that also acts as a stimulant and has side effects similar to those of the amphetamines.

Caffeine is the most widely used nonmedical central nervous system stimu-

caffeine

lant. Table 23-5 lists the caffeine content of common beverages. It is also available "over the counter" in the form of tablets, such as No-Doz. A dose of

Table 23-5 **Caffeine Content of Beverages**

Beverage	Caffeine (mg/cup)
coffee (regular)	100–150
coffee (instant)	60–80
coffee (decaffeinated)	2–4
tea	25–75
cocoa	under 50
cola	30–50[a]

[a] In 12 ounces.

50 to 200 mg of caffeine increases mental alertness and reduces drowsiness and fatigue. More than this amount causes nervousness, irritability, tremors, and headache. Heavy use of this drug may lead to a craving and physical dependence.

23.10 Hallucinogens

Hallucinogenic agents, also known as psychedelic or psychotomimetic agents, alter thought processes. They affect perception of space and time, mood, and memory without disturbing the central nervous system. These substances are rarely administered medically but a user who reacts poorly (a "bad trip") may seek medical assistance.

Lysergic acid diethylamide, LSD or "acid" (shown in Section 12.7 and Figure 23-6) is the most potent hallucinogen. Effects such as mood fluctuation, intense color perception, and loss of personal identity occur with oral doses of less than 0.05 mg. Common side effects involve feelings of panic resulting from anxiety.

Marijuana contains substances chemically known as tetrahydrocannabinols, one of which has the following structure:

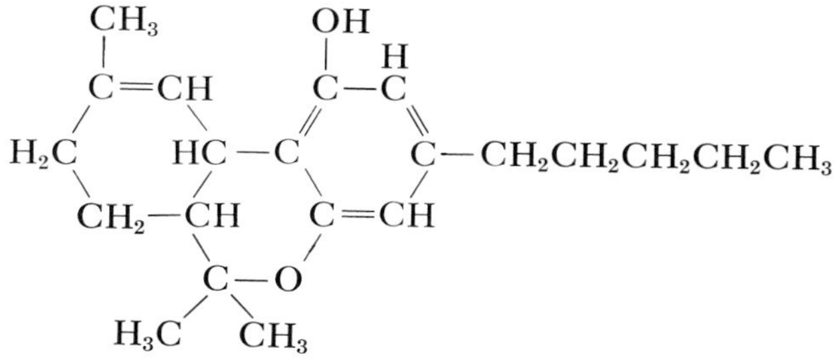

a tetrahydrocannabinol

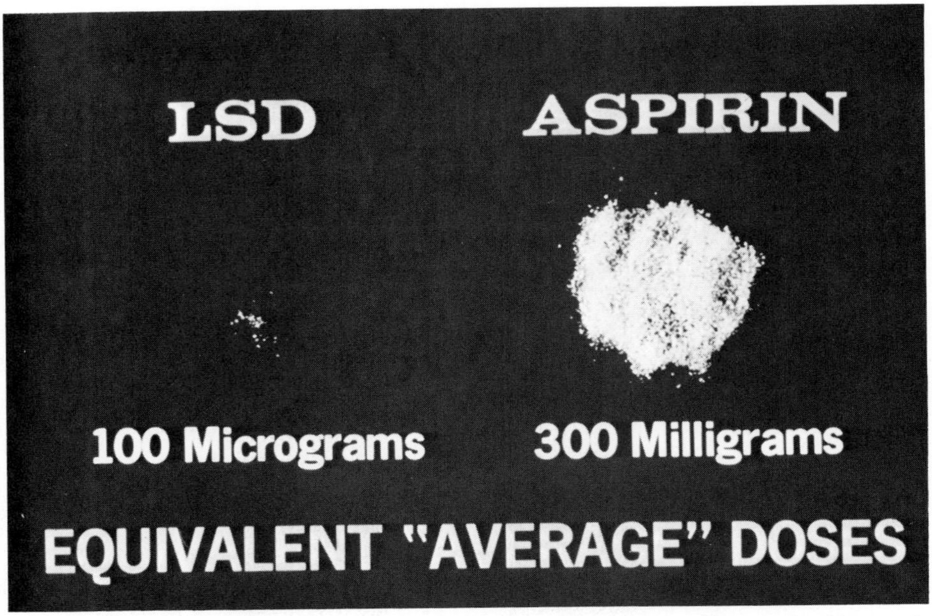

Figure 23-6. LSD. Only a very small amount of this compound is required to produce a physiological effect. (Photo courtesy of Pharmaceutical Manufacturers Association.)

These molecules have effects similar to those of the sedative–hypnotics. They produce initial euphoria and enhanced sense perception, followed by a sedative dreamlike phase. The drug shortens memory span and distorts time and space. Marijuana, which has been tried at least once by 25 million people in the United States, creates no physical dependence. Figure 23-7 is a photograph of marijuana leaves.

Mescaline is a hallucinogen found in peyote cactus that is similar in effect to LSD. Another example, psilocybin, is present in certain mushrooms. Possible medical problems result from the presence of harmful impurities in synthetic hallucinogens sold on the "street."

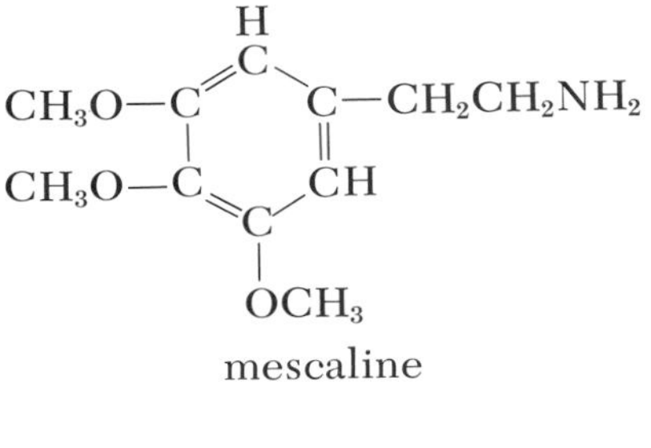

mescaline

Figure 23-7. Marijuana leaves. (Photo courtesy of U.S. Drug Enforcement Administration.)

23.11 Drugs and the autonomic nervous system

In addition to a central nervous system, which produces controlled responses to changes in your surroundings, your body has an autonomic nervous system, which regulates the muscles and glands of internal organs. As described in a previous chapter (Section 16.7), chemical substances known as neurotransmitters are involved in bringing "messages" from the nervous system to other cells of the body. The neurotransmitters of the autonomic nervous system are norepinephrine, epinephrine, and acetylcholine.

Drugs that produce effects similar to those of epinephrine or norepinephrine are called **sympathomimetic (adrenergic) drugs**. The physiological effects of these two catecholamines have been discussed in the chapter on hormones (Section 21.25). A major clinical use of this class of drugs, which includes phenylephrine (Neo-Synephrine), isoproterenol (Isuprel), and ephedrine, in addition to epinephrine itself, is to open the bronchial passages in a person with asthma and to relieve nasal congestion (by acting as a vasoconstrictor). Drugs that block the effect of these amines are known as sympathoplegic drugs or adrenergic blocking agents.

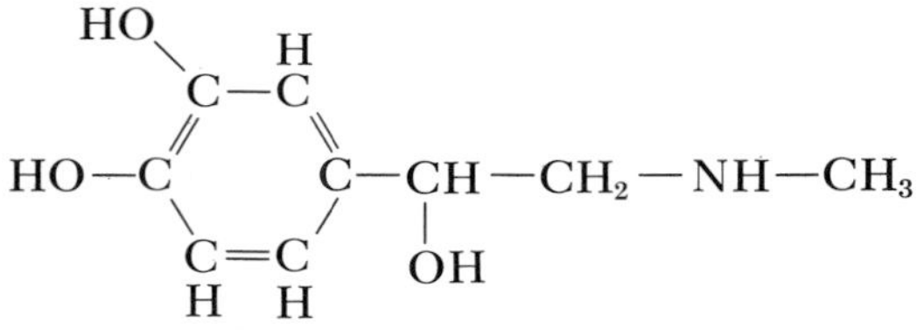

epinephrine

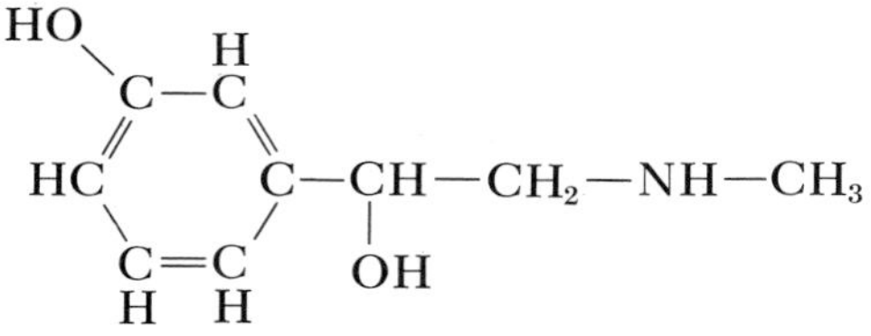

phenylephrine

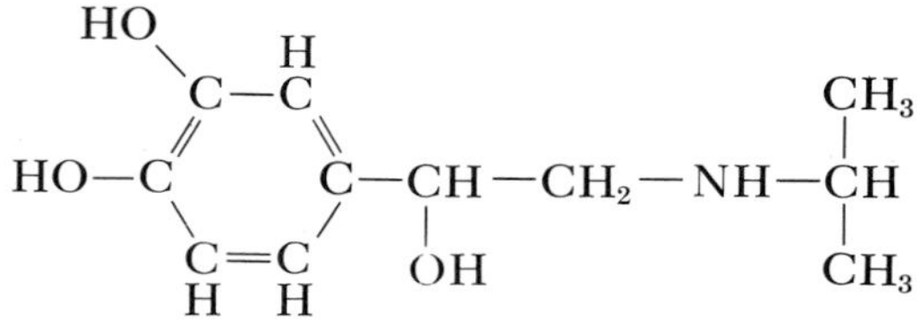

isoproterenol

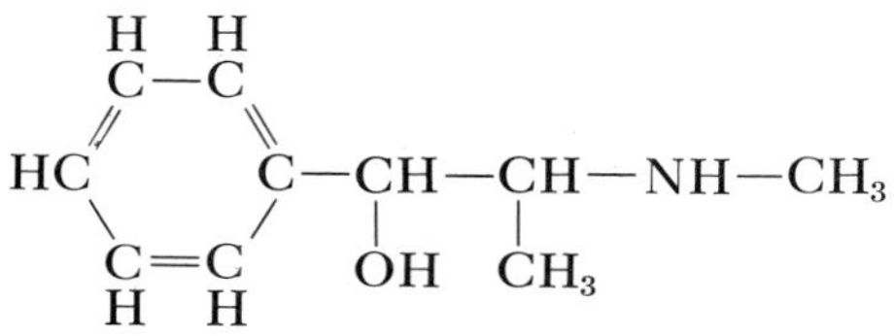

ephedrine

Parasympathomimetic (cholinergic) drugs have effects similar to those of acetylcholine. Some members of this class, such as physostigmine, produce

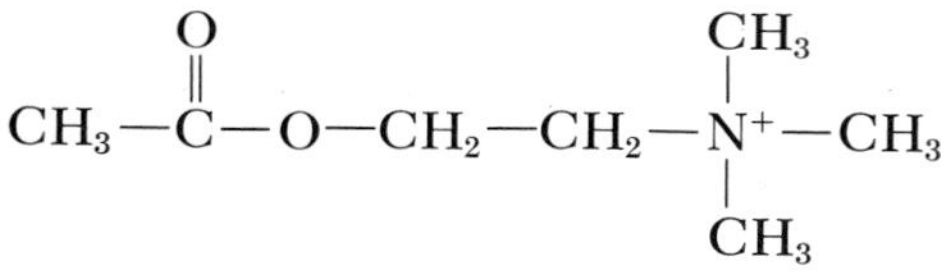

$$CH_3-\overset{\overset{\displaystyle O}{\|}}{C}-O-CH_2-CH_2-\overset{\overset{\displaystyle CH_3}{|}}{\underset{\underset{\displaystyle CH_3}{|}}{N^+}}-CH_3$$

acetylcholine

their effect by inhibiting the enzyme acetylcholinesterase, which inactivates acetylcholine. These drugs are used to treat the eye disorder glaucoma and

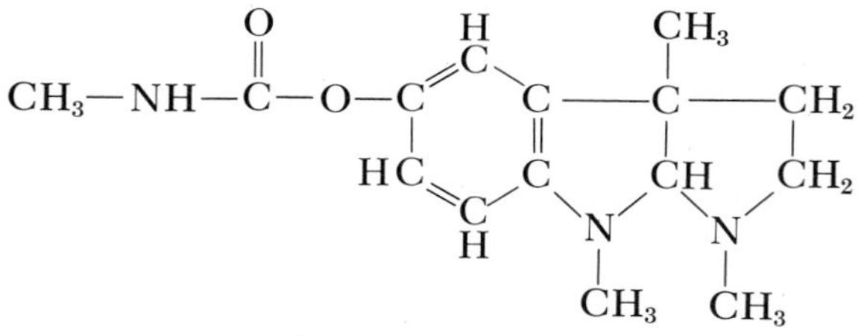

physostigmine

the muscle disease myasthenia gravis. Although not used medically, nicotine, found in tobacco leaves, produces stimulatory effects like acetylcholine.

$$CH_2-CH_2$$

nicotine

Compounds that block some of the effects of acetylcholine are known as **parasympatholytic (anticholinergic) drugs**. Atropine, an alkaloid from the belladonna shrub, is one example. It is used as a preanesthetic medication to pre-

atropine

vent obstruction of the airway by reducing secretions. Atropine relaxes smooth muscle, such as in the bladder and ureter, and suppresses secretion in cases of peptic ulcer. Scopolamine is a drug similar in structure and effect to atropine, decreasing the activity of smooth muscle and glands that produce external secretions.

23.12 Poisons

Most substances taken in excess can act as **poisons**, causing injury or death. Of course, the most dangerous substances are those that are harmful in small amounts. Of the over 1 million cases of poisoning each year in the United States, *the largest cause is medicine designed for internal use.* Most poisonings are accidental and involve such drugs as aspirin, barbiturates, and antianxiety agents. The effects produced by overdoses of the types of drugs described in this chapter are listed in Table 23-6. Children under 5 years old make up about one-third of the accidental poisonings (Figure 23-8).

Poisons can result in damage by local irritation but cause the most serious effects internally when carried by the blood. They can destroy the tissues of an organ, inactivate enzymes, combine with cell components, or cause release of harmful substances. Because they are organs of excretion, the kidneys are especially vulnerable to toxic agents. The liver is the site of detoxification, the attempt to chemically convert poisons into less toxic molecules.

Some poisons have acute effects—their presence is detected shortly after exposure, requiring immediate assistance. Others have chronic effects, which

Table 23-6 **Effects of Drug Poisoning**

Class of drug	Effects
anesthetic	respiratory depression
antibiotic	depends on specific drug
aspirin and salicylates	fever, convulsions, acidosis, stupor
narcotic analgesics	respiratory depression, coma
sedative–hypnotics and antianxiety drugs	anesthesia, circulatory shock, respiratory depression
stimulants	hypertension, toxic psychosis

Figure 23-8. A warning on the dangers of poison. (Courtesy of Children's Memorial Hospital, Nebraska.)

develop after long-term exposure to low concentrations of the poison. In either case, the main concern is removing the source of the poison and keeping the patient alive. Few poisons have specific antidotes that can be administered to counter the effects.

23.13 Environmental pollutants

Many toxic substances are released into the environment either by accident or as waste products. Some of these, the **air pollutants**, have already been briefly mentioned (Section 5.13). *Carbon monoxide,* CO, the major pollutant, is produced by incomplete combustion of gasoline in automobile engines.

$$2C + O_2 \longrightarrow 2CO$$

Its toxic effect results from the formation of carboxyhemoglobin, which prevents oxygen from reaching the tissues and causes death by asphyxiation. The normal "background" blood level is 0.5%; higher concentrations produce the following effects: 1 to 2%—some behavioral changes; 2 to 5%—central nervous system effects; 5 to 10%—cardiac and pulmonary effects; 10 to 80%—headache, fatigue, drowsiness, coma, respiratory failure, and death. A "moderate" cigarette smoker (one pack per day) inhales enough carbon monoxide to reach a blood level of about 5%.

Nitrogen oxides, largely nitric oxide, NO, and some nitrogen dioxide, NO_2, are formed during high-temperature combustion of gasoline, coal, and gas:

$$N_2 + O_2 \longrightarrow 2NO$$
$$2NO + O_2 \longrightarrow 2NO_2$$

Hydrocarbons, of which the most abundant is methane, are released into the atmosphere from the gasoline in automobiles. They produce other pollutants, such as peroxyacetyl nitrates (PAN), by reactions in light known as photochemical oxidation. Ozone, O_3, formation is also related to the presence of hydro-

$$CH_3-\overset{\overset{\displaystyle O}{\|}}{C}-O-O-NO_2$$

a peroxyacetyl nitrate
(PAN)

carbons; high concentrations of this gas (9 ppm) result in severe pulmonary edema (presence of fluid in the lungs). Photochemical smog ("smoke-fog"), which causes eye irritation and interferes with normal respiration, is formed through the interaction of hydrocarbons and nitrogen dioxide; it contains O_3, CO, PAN, and organic compounds such as aldehydes and ketones. *Sulfur oxides*, largely sulfur dioxide, SO_2, with some sulfur trioxide, SO_3, result from coal and fuel oil combustion, as well as from volcanoes:

$$S + O_2 \longrightarrow SO_2$$
$$2SO_2 + O_2 \longrightarrow 2SO_3$$

These pollutants irritate the respiratory system and can cause constriction of the air pathways. They are most dangerous to those people, such as the elderly, who may suffer from chronic respiratory or cardiovascular disease.

Solids, known as *particulates*, consist of very small particles (0.0002 to 500 μm in diameter), which are produced by coal combustion, industrial processes, and forest fires. They may consist of carbon and the oxides of many elements, including Fe_2O_3 (Fe_3O_4), SiO_2, Al_2O_3, K_2O, P_2O_5, CaO, MgO, TiO_2, and Na_2O. The penetrating ability of these particles depends on their size; those smaller than 0.5 μm in diameter may reach the alveoli. Other solids, such as asbestos fibers, are released into the air from building insulation and fireproofing materials. The presence of this mineral in the lungs may lead to asbestosis, a respiratory disease, and lung cancer. Silicosis, a pulmonary disease, is caused by the accumulation of particles of silicon dioxide, SiO_2, found in sand and rocks. Coal dust produces the miner's "black lung" disease (Figure 23-9). The inhalation of beryllium salts may result in berylliosis, an acute pneumonia, or chronic lung disease.

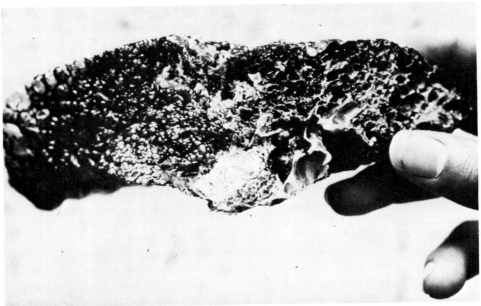

Figure 23-9. A lung taken from a patient suffering from "black lung" disease. (EPA Documerica photo by Leroy Woodson.)

Heavy metals, those with high atomic weights, may be systematic poisons, affecting numerous organs, as shown in Table 23-7. Mercury, for example, acts by binding strongly to sulfur-containing groups in the body, inhibiting enzyme action and causing cellular damage. The vapor of this liquid metal can pass from the lungs into the blood, leading to damage of the brain and central nervous system. Inorganic mercury salts affect the liver and kidney tissue; organic mercury compounds, the alkyl mercurials, pose a greater hazard because they accumulate in brain tissue and are not readily eliminated. Lead can bind to sulfur atoms like mercury but can also gather in the skeleton like calcium. Because of its former use in wall paints, ingestion of lead is common in older buildings, where children may eat pieces of peeling paint. (Its use is

Table 23-7   Heavy-Metal Poisons

Metal	Sources	Symptoms	
		Acute poisoning	Chronic poisoning
arsenic (As)	pesticides, metal industry	gastrointestinal disturbances	skin pigmentation, edema, kidney and liver damage
antimony (Sb)	alloys, batteries, ceramics	nausea, hepatitis	anemia, weight loss, skin diseases
cadmium (Cd)	metal industry	fever, chest pain	respiratory and kidney disease
chromium (Cr)	industry	dermatitis	gastrointestinal, liver and kidney damage, cancer
lead (Pb)	flaking paint (auto emissions)	abdominal pain, irritability	anemia, cramps, "lead line," palsy
mercury (Hg)	fish, shellfish, antiseptics	gastrointestinal disturbances	kidney and central nervous system damage
selenium (Se)	paper, coal, petroleum	gastrointestinal disturbances	kidney and liver disease

now restricted to the outsides of buildings.) Figure 23-10 shows workers in a lead foundry wearing special face masks to protect them from the metal fumes. Certain molecules [such as dimercaprol, penicillamine, and ethylenediaminetetraacetic acid (EDTA)] bind strongly to heavy metals and are used to remove them from the body.

Many organic *pesticides*, substances used to destroy biological pests, are toxic. The most common types are organochlorine compounds (DDT, methoxychlor, aldrin, dieldrin, lindane), organophosphorus compounds (parathion, malathion), and carbamate compounds (carbaryl, baygon); examples are shown in Figure 23-11 (p. 553). The organophosphorus and carbamate pesticides deactivate the enzyme acetylcholinesterase, causing acetylcholine to continue to trigger nerve impulses, which results in tremors, convulsions, and death. Organochlorine pesticides dissolve in the fatty membrane surrounding the nerve fibers, interfering with the transport of sodium and potassium ions and causing effects similar to those of the two other types of pesticides.

Industrial and household solvents are also potential poisons. These include products of petroleum distillation (kerosene, gasoline), aromatic hydrocarbons (benzene), and halogenated hydrocarbons (carbon tetrachloride). Re-

Figure 23-10. Workers in a lead foundry. (WHO photo.)

cently, polychorinated biphenyls (PCB's), have been added to the list of environmental poisons. These compounds affect liver tissue and enzyme systems.

a polychlorinated
biphenyl (PCB)

Several types of "pollutants" may be **carcinogens**, cancer-causing agents, as shown in Table 23-8 (p. 554). A wide variety of compounds having very different chemical structures can act as carcinogens. Aromatic benzypyrene forms when meats are charcoal-broiled. Nitrosamines are believed to form through chemical reactions in the stomach between nitrites (used as meat preservatives) and amines from your diet. Vinyl chloride is the basis for the common plastic material polyvinyl chloride (PVC). Note that naturally occurring sub-

organochlorine

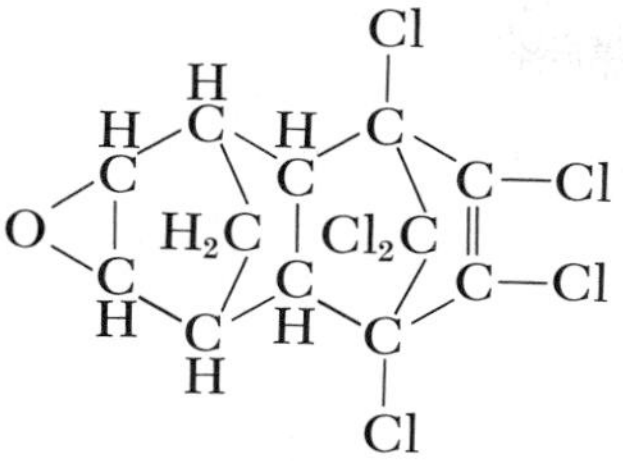

dieldrin

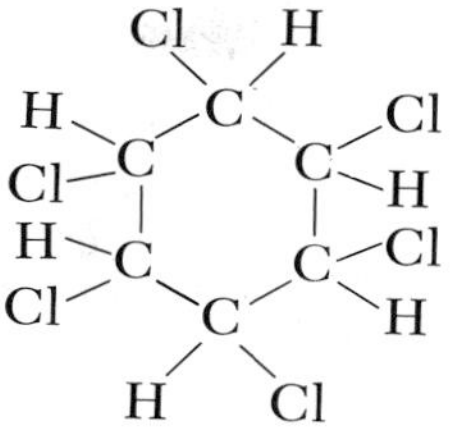

lindane

organophosphorus

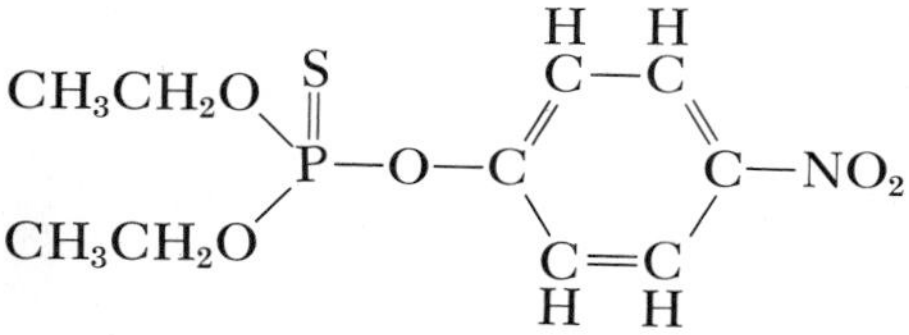

parathion

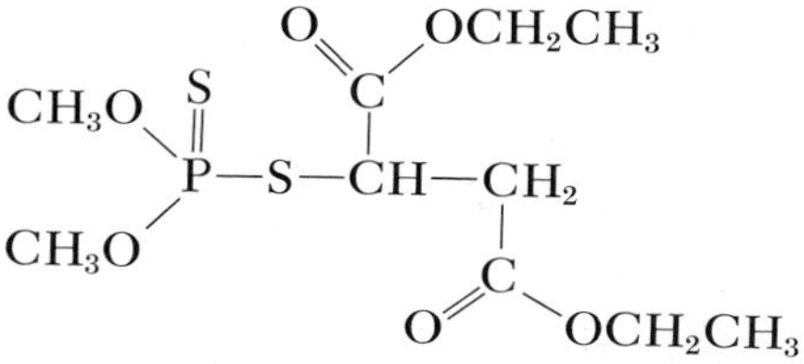

malathion

carbamate

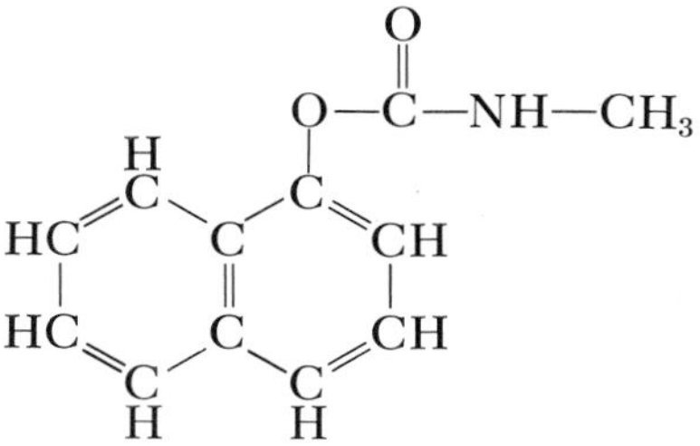

carbaryl

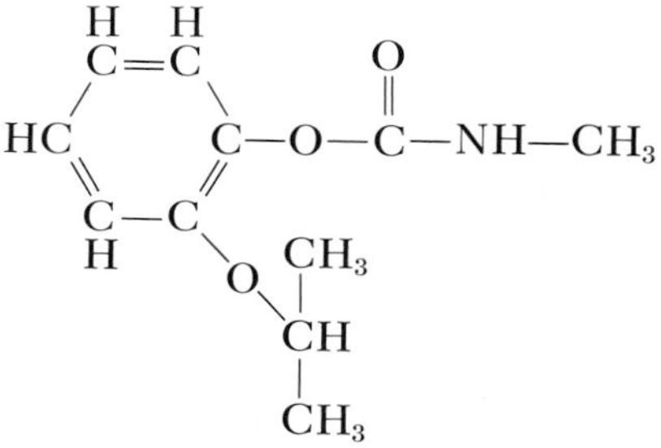

baygon

Figure 23-11. Examples of insecticides.

stances like aflatoxins from fungus can also seve as carcinogens. A large number of cancers appear to be triggered in some way by chemical substances in the environment.

Many harmful pollutants are difficult to trace because of the long waiting period, up to 20 to 30 years, before the effects, such as a type of cancer, become noticeable. Health workers known as epidemiologists must compare the

Table 23-8 Carcinogens

Class	Example
polynuclear aromatic hydrocarbon	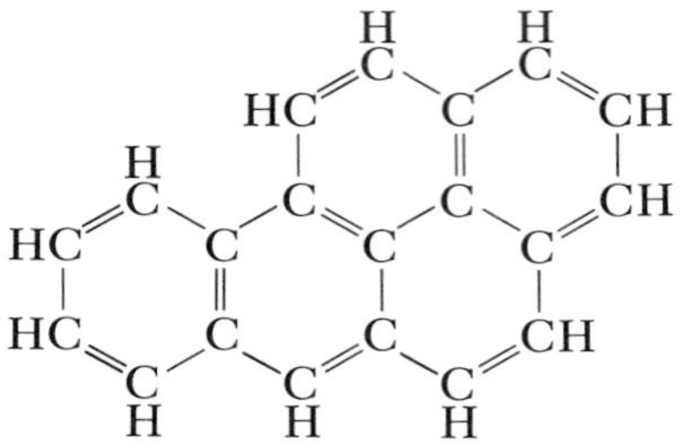 benzpyrene
aromatic amine	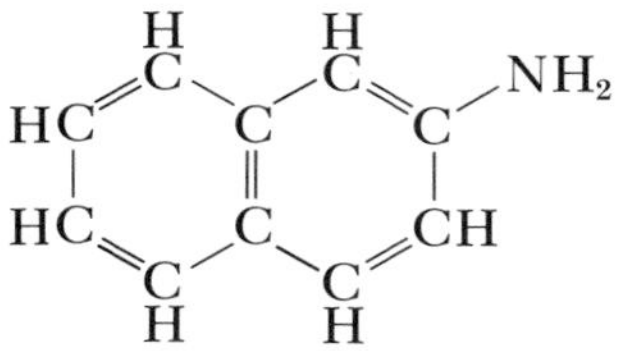 2-naphthylamine
chlorinated hydrocarbon	$CH_2{=}CHCl$ vinyl chloride
N-nitroso compound	$(CH_3)_2N{-}N{=}O$ dimethylnitrosamine
inorganic substance	nickel (Ni)
natural product	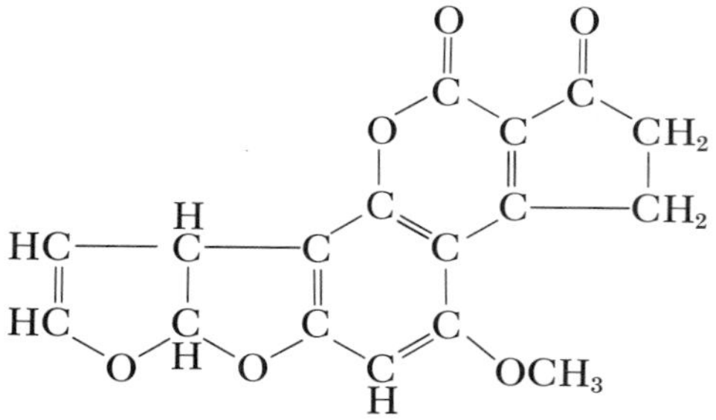 aflatoxin B
alkylating agent	$CH_2Cl{-}CH_2{-}S{-}CH_2{-}CH_2Cl$ bis(2-chloroethyl)sulfide (mustard gas)

benzpyrene

2-naphthylamine

aflatoxin B

distribution of disease within specific populations to identify possible causes in the environment.

SUMMARY

Drugs are chemical substances that affect your body. They are used for medical diagnosis, preventing or curing disease, treating symptoms, or birth control. Most drugs act by stimulating or depressing certain activities in the cell, killing or weakening a foreign organism, or causing irritation.

Antibiotics are chemical substances used to treat infectious disease, an application known as chemotherapy. They are produced from living cells and act by killing or inhibiting the growth of a microorganism. Penicillins are the most potent antibacterial drugs; they inhibit cell wall synthesis in gram-positive bacteria. Tetracyclines, a widely used class of antibiotics, have a broad spectrum of activity.

Analgesics relieve pain by decreasing the patient's sensitivity without loss of consciousness. Aspirin, acetylsalicylic acid, is a nonprescription analgesic effective against mild to moderate pain in headache, neuralgia, muscle pain, joint pain, and toothache. It also acts as an antipyretic (lowering fever) and as an antiinflammatory agent. Acetaminophen is the most effective and safest aspirin substitute.

Narcotic analgesics are the most powerful pain-relieving drugs available. Frequent use, however, leads to a state of drug dependence, or addiction, characterized by withdrawal symptoms if the drug is discontinued. They are medically prescribed only in cases of severe pain for relatively short periods of time. Morphine, an alkaloid from opium poppy, is the most effective narcotic analgesic. Meperidine (Demerol) is a synthetic narcotic analgesic sometimes abused by physicians and nurses.

General anesthetics decrease reaction to pain as well as causing loss of consciousness, diminished reflexes, and muscle relaxation. Anesthetic ethers include diethyl ether, divinyl ether, fluoroxene, and methoxyflurane. Chloroform, trichloroethylene, and popular halothane also serve as inhalation anesthetics. Gases employed are cyclopropane and nitrous oxide, or "laughing gas." Intravenous anesthetics like thiopental last for only short periods. Local anesthetics such as procaine act only at the site of application.

Sedative–hypnotics depress the central nervous system. They are used to cause sedation, a calming effect, or in larger doses to produce a state similar to natural sleep. Barbiturates, a major class of sedative–hypnotics, include long-acting phenobarbital, intermediate-acting amobarbital, and short-acting pentobarbital and secobarbital.

Newer antianxiety drugs include meprobamate, chlordiazepoxide, and diazepam. The latter two, sold as Librium and Valium, respectively, have a higher safety margin and less potential for abuse than barbiturates. Antipsychotic drugs, such as chlorpromazine, reduce the most severe symptoms, such as hallucinations and illusions, in a high percentage of psychotic patients like schizophrenics.

Ethyl alcohol, or ethanol, is a central nervous system depressant used more socially than medically. Continued use in large quantities produces irreversible brain, nervous system, and heart damage as well as cirrhosis of the liver.

Stimulants increase the activity of the central nervous system. Amphetamine, a typical example, elevates the mood, reduces feelings of fatigue and hunger, and increases reaction time, concentration, and the capacity to do work. Caffeine is the most widely used nonmedical stimulant.

Hallucinogens, known as psychedelic or psychotomimetic agents, alter perception of space and time, mood, and memory without disturbing the central nervous system. Lysergic acid diethylamide, LSD, is the most potent hallucinogen.

Drugs that produce effects similar to those of epinephrine or norepinephrine, neurotransmitters of the autonomic nervous system, are called sympathomimetic drugs; examples are phenylephrine and isoproterenol. Parasympathomimetic drugs, like physostigmine, are similar in action to acetylcholine. Parasympatholytic drugs, such as atropine, block the effects of acetylcholine.

Most substances taken in excess can act as poisons, causing injury or death. The largest cause is medicine designed for internal use, such as aspirin, barbiturates, or antianxiety agents. Poisons act by destroying tissue, inactivating enzymes, combining with cell components, or causing release of harmful substances.

Many types of toxic substances are released into the environment. These include gases, heavy metals, pesticides, industrial and household solvents, and particulates. Some pollutants cause immediate or acute effects; others result in chronic diseases, developing over many years from low exposures. Compounds of widely varying structure, such as aromatic hydrocarbons, nitroso compounds, and inorganic substances, can act as carcinogens, cancer-causing agents.

Exercises

1. (Intro.) What is a drug? What are the general uses of drugs?
2. (23.1) In what ways do drugs act?
3. (23.1) What series of events takes place in the body after a drug is administered?
4. (23.1) What factors determine a patient's response to a drug?
5. (23.2) Define antibiotic, chemotherapy, selective toxicity.
6. (23.2) Describe the advantages and disadvantages of the penicillins.

7. (23.2) Why are tetracyclines known as "broad spectrum" antibiotics?

8. (23.3) What is an analgesic?

9. (23.3) Draw the structure and write the chemical name of aspirin. Describe its functions in the body.

10. (23.4) How do narcotic analgesics differ from aspirin?

11. (23.4) Describe the uses of two natural and two synthetic narcotics.

12. (23.5) How does a general anesthetic differ from a local anesthetic?

13. (23.5) Give an example of each of the following types of anesthetic and state its properties: ether, halogenated hydrocarbon, gas, intravenous anesthetic, local anesthetic.

14. (23.6) How is a sedative–hypnotic different from an anesthetic?

15. (23.6) What are the different categories of barbiturates and their uses?

16. (23.7) Describe how the newer antianxiety drugs differ from barbiturates.

17. (23.7) What are the functions of antipsychotic drugs?

18. (23.8) Summarize the physiological and behavioral effects of ethyl alcohol.

19. (23.9) Describe the activity of a stimulant such as amphetamine.

20. (23.10) What are the effects of a hallucinogen?

21. (23.11) Define sympathomimetic drug, parasympathomimetic drug, parasympath-olytic drug. Give one example of each type.

22. (23.12) What kinds of substances can act as poisons? What are the most common poisons?

23. (23.12) Describe the types of effects a poison can have on the body.

24. (23.13) Give an example of each of the following types of environmental pollutant: gas, heavy metal, insecticide, solvent, dust.

25. (23.13) What is the difference between an acute effect and a chronic effect of a pol-lutant?

Basic mathematics
for chemistry

A.1   Fractions, decimals, and percent

A **fraction** represents part of a whole number. It consists of two parts, one number (the numerator) written above another number (the denominator), the two being separated by a bar. For example, the fraction $\frac{1}{2}$, one-half, means 1 part out of a total of 2 equal parts. The fraction represents a ratio, or relationship, between two numbers, in this case 1 to 2. You can also think of a fraction as a division, 1 divided by 2.

If you carry out the division indicated by the fraction $\frac{1}{2}$, you obtain a decimal, 0.5. The **decimal system** is simply a way of writing numbers based on 10, as illustrated in Figure A-1. The numbers in front of the decimal point are larger than 1, while those to the right of it represent fractions. As shown, the position of a number with respect to the decimal point determines its value. Thus, the number 0.5 represents $5 \times 1/10$, or five-tenths. Its meaning is exactly the same as one-half, since the ratio 1/2 is identical to the ratio 5/10. When adding or subtracting decimal numbers, you must line up the numbers in a column, placing the decimal points directly under each other:

$$
\begin{array}{cc}
\textit{addition} & \textit{subtraction} \\
3.23 & 376.94 \\
152. & -\ \ 98.51 \\
\underline{+\ 0.378} & \overline{278.43} \\
155.608 &
\end{array}
$$

In multiplication, the number of decimal places, the numbers after the decimal point, is equal to the sum of the decimal places in the numbers being multiplied. In the following example, 3.21 has two decimal places and 0.95 has two, so the product must have $2 + 2$, or 4:

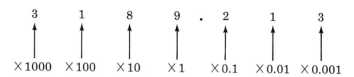

Figure A-1. A number in the decimal system.

3000.	thousands
100.	hundreds
80.	tens
9.	units
0.2	tenths
0.01	hundredths
0.003	thousandths
3189.213	

multiplication

$$
\begin{array}{r}
3.21 \\
\times 0.95 \\
\hline
1605 \\
2889 \\
\hline
3.0495
\end{array}
$$

division

$$
\begin{array}{r}
13.1 \\
1.5\overline{)19.65} \\
\underline{15} \\
4\ 65 \\
\underline{4\ 50} \\
15 \\
\underline{15} \\
0
\end{array}
$$

For division, when the number with which you are dividing, such as 1.5, has a decimal point, you must move it all the way to the right, 15., and then shift the decimal point in the number being divided, 19.65, by exactly the same amount, 196.5.

The term **percent** can represent a special type of fraction—parts per hundred. Fifty percent, 50%, means 50 parts out of 100 equal parts, or 50/100. You can also write 50% as a decimal, 0.50, fifty one-hundredths, by moving the decimal point two places to the left.

A.2 Scientific (exponential) notation

Scientific or **exponential notation** is a shorthand way of writing both large and small numbers. It is based on expressing numbers as *powers of 10*, that is, 10 multiplied by itself, as shown in Table A-1. The raised number, or exponent, indicates how many times 10 must be multiplied by itself; a negative exponent represents a fraction, 1 divided by 10 raised to a power. As you can see, a change of 1 in the exponent means a change of 10 in the size of the number.

Any number can be expressed in scientific notation. For example, 365 can be written as 3.65×10^2. The number in front of the multiplication sign, the

Table A-1 **Powers of 10**

Power of 10	Meaning
10^3	$10 \times 10 \times 10 = 1000$
10^2	$10 \times 10 = 100$
10^1	10
10^0	1
10^{-1}	$1/10^1 = 1/10 = 0.1$
10^{-2}	$1/10^2 = 1/100 = 0.01$
10^{-3}	$1/10^3 = 1/1000 = 0.001$

coefficient, has a value between 1 and 10. The power of 10, the exponential, which in this case is 10^2, determines the location of the decimal point. Since 10^2 means 100, the number 3.65×10^2 is 3.65×100, or 365. Notice that the exponent 2 indicates that the decimal point moves two places. If the number were written 3.65×10^{-2}, the decimal point would be moved in the opposite direction two places, 0.0365. Other examples are presented in Table A-2. By simply counting the number of places you must shift the decimal point to make a number between 1 and 10, you can determine the proper exponent. Just remember that a number larger than 1 has a positive exponent and a fraction has a negative exponent.

Table A-2 **Examples of Numbers in Scientific (Exponential) Notation**

Number	Notation
5,300,000	5.3×10^6
39,800	3.98×10^4
1,480	1.48×10^3
56	5.6×10^1
0.33	3.3×10^{-1}
0.0067	6.7×10^{-3}
0.0000051	5.1×10^{-6}

To add or subtract numbers written in scientific notation, you must express them in terms of the *same* power of 10. For example, the sum $(3 \times 10^4) + (2 \times 10^5)$ must be rewritten as $(0.3 \times 10^5) + (2 \times 10^5)$ or as $(3 \times 10^4) + (20 \times 10^4)$. The sum is then simply equal to the sum of the coefficients expressed to the same power of 10 as the numbers being added. In this case, it is equal to 2.3×10^5. You handle subtraction in a similar way, finding the difference between the coefficients after the numbers have been written in the same power of 10.

A product of numbers in scientific notation is found by *multiplying* the coefficients and *adding* (algebraically) the exponents. Thus, $(3 \times 10^4) \times (2 \times 10^5)$ is equal to $(3 \times 2) \times 10^{(4+5)}$, or 6×10^9. You divide two numbers by *dividing* their coefficients and *subtracting* their exponents. The quotient $(3 \times 10^4)/(2 \times 10^5)$ becomes $(3/2) \times 10^{(4-5)}$, or 1.5×10^{-1}.

A.3   Proportions and algebra

A **proportion** represents two ratios that have the same value and are therefore separated by an equals sign:

$$\frac{1}{2} = \frac{50}{100} \qquad \text{a proportion}$$

An important property of any proportion is that *the products of the diagonal terms are equal;* 1×100 is the same as 2×50. This relationship proves useful for solving problems when one of the four terms in a proportion must be determined from the other three.

An unknown quantity is given the symbol x. It is generally found by solving an **equation**, a mathematical statement that two quantities are equal, such as the following proportion:

$$\frac{1}{2} = \frac{x}{100}$$

To find the value of x, you first cross-multiply, setting the two products equal to each other:

$$1 \times 100 = 2 \times x$$

You must divide both sides of the equation by 2, since you are not interested in the value of $2x$, but of x by itself.

$$\frac{1 \times 100}{2} = \frac{2 \times x}{2}$$

You cancel the 2's on the right side of the equation since any number divided by itself has a value of 1, and multiplication by 1 does not change a number:

$$\frac{1 \times 100}{2} = x$$

$$\frac{100}{2} = x$$

$$50 = x$$

The answer can be checked by substituting 50 for x in the original proportion and cross-multiplying to make sure that the two products are equal:

$$\frac{1}{2} = \frac{50}{100}$$
$$1 \times 100 = 2 \times 50$$
$$100 = 100$$

A.4 The unit-factor method

Many problems can be solved using an approach known as the **unit-factor method,** or **dimensional analysis.** This technique is based on the units or dimensions of the quantities involved in a calculation, such as inches, quarts, pounds, or seconds. Ratios known as *conversion factors* are used; they consist of two different ways of expressing the same thing, like 60 seconds and 1 minute. The conversion factor relating seconds and minutes can be written as follows:

$$\frac{60 \text{ seconds}}{1 \text{ minute}} \quad \text{or} \quad \frac{1 \text{ minute}}{60 \text{ seconds}}$$

Since the quantity on top of each fraction is equal to that on the bottom, conversion factors have the value 1. Thus, multiplying a conversion factor by another quantity will not change its size, only its units.

The unit-factor method is used to convert from one set of units to another. For example, you might want to convert 10 minutes to seconds. To solve this problem, you first identify the desired units of the answer, which is seconds. Then, you set this unit equal to the given quantity, 10 minutes, multiplied by some conversion factor:

$$\text{seconds} = 10 \text{ minutes} \times \text{conversion factor}$$

The conversion factor is the ratio that relates these two units; you must write it so that seconds appear on top and minutes on the bottom:

$$\text{seconds} = 10 \text{ minutes} \times \frac{60 \text{ seconds}}{1 \text{ minute}}$$

$$\left(\frac{\text{unknown}}{\text{quantity}} \right) = \left(\frac{\text{known}}{\text{quantity}} \right) \times \left(\frac{\text{conversion}}{\text{factor}} \right)$$

By expressing the conversion factor this way, you make sure that minutes will cancel out, leaving seconds as the desired answer:

$$\text{seconds} = 10 \; \text{\sout{minutes}} \times \frac{60 \; \text{seconds}}{1 \; \text{\sout{minute}}}$$

$$= 600 \; \text{seconds}$$

Units cancel only when they appear on the top in one fraction and on the bottom in another. If you had used the other form of the conversion factor, 1 minute/60 seconds, you would not have ended up with seconds as the answer, but a meaningless unit, minute²/second. More examples are worked out in Appendix B.

Metric system conversions

B.1   Conversions within the metric system

You can easily convert from one metric unit to another using the relationships listed in Table B-1. To carry out these conversions, you can use either the method of proportions (Appendix A.3) or the unit-factor method (Appendix A.4). Both techniques will be illustrated.

For example, you may need to convert a drug dosage, such as 0.05 gram, into milligrams. To use a proportion, you first write down the known relationship between grams and milligrams:

$$1 \text{ gram} = 1000 \text{ milligrams}$$

Then you draw a bar and underneath it write the given information, 0.05 gram, and the unknown quantity, x:

$$\frac{1 \text{ gram}}{0.05 \text{ gram}} = \frac{1000 \text{ milligrams}}{x}$$

Make sure that grams are on one side of the equation and milligrams are on the other side. Now cross-multiply and solve for x.

$$1 \text{ gram} \times x = 0.05 \text{ gram} \times 1000 \text{ milligrams}$$

$$x = \frac{0.05 \text{ gram} \times 1000 \text{ milligrams}}{1 \text{ gram}}$$

$$x = 50 \text{ milligrams}$$

Therefore, 0.05 gram is equal to 50 milligrams.

Exactly the same problem can be solved by the unit-factor method as follows. The desired answer, milligrams, must be equal to the given quantity, 0.05 gram, multiplied by a conversion factor:

$$\text{milligrams} = 0.05 \text{ gram} \times \text{conversion factor}$$

The conversion factor is the ratio 1000 milligrams/1 gram; you must state the

Table B-1 **Metric System Conversions**

mass	1 kilogram = 1000 grams
	1 gram = 1000 milligrams
	1 milligram = 1000 micrograms
length	1 kilometer = 1000 meters
	1 meter = 100 centimeters
	1 centimeter = 10 millimeters
volume	1 liter = 1000 milliliters
	1 milliliter = 1 cubic centimeter

fraction in this way so that grams will cancel out:

$$\text{milligrams} = 0.05 \, \text{gram} \times \frac{1000 \, \text{milligrams}}{1 \, \text{gram}}$$

$$= 50 \, \text{milligrams}$$

The answer, of course, is the same, 50 milligrams. The unit-factor method is generally faster, but you must be careful to express the conversion factors properly and cancel out the units carefully.

In the following example, another problem is worked out in detail using both techniques.

Example: Convert 600 milliliters to liters

(a) Proportion method

$$\frac{1 \, \text{liter}}{x} = \frac{1000 \, \text{milliliters}}{600 \, \text{milliliters}}$$

$$1 \, \text{liter} \times 600 \, \text{milliliters} = x \times 1000 \, \text{milliliters}$$

$$\frac{1 \, \text{liter} \times 600 \, \text{milliliters}}{1000 \, \text{milliliters}} = x$$

$$0.6 \, \text{liter} = x$$

(b) Unit-factor method

$$\text{liters} = 600 \, \text{milliliters} \times \frac{1 \, \text{liter}}{1000 \, \text{milliliters}}$$

$$= 0.6 \, \text{liter}$$

B.2 Conversions between the English and metric systems

Table B-2 lists the most important conversions between the units of the English system and metric system for mass, length, and volume. Problems

Table B-2   **Conversion between English Units and Metric Units**

Quantity	English system	Metric system
mass	1 pound[a]	454 grams
length	1 inch	2.54 centimeters
volume	1 quart	946 milliliters

[a] The pound is technically a unit of force, measuring the pull of the earth on an object from gravity.

that involve changing English units to metric units, or the other way around, are carried out in exactly the same manner as described in the previous section.

Example: Convert 12 inches to centimeters

(a) Proportion method

$$\frac{1 \text{ inch}}{12 \text{ inches}} = \frac{2.54 \text{ centimeters}}{x}$$

$$1 \text{ inch} \times x = 12 \text{ inches} \times 2.54 \text{ centimeters}$$

$$x = \frac{12 \text{ inches} \times 2.54 \text{ centimeters}}{1 \text{ inch}}$$

$$x = 30.5 \text{ centimeters}$$

(b) Unit-factor method

$$\text{centimeters} = 12 \text{ inches} \times \frac{2.54 \text{ centimeters}}{1 \text{ inch}}$$

$$= 30.5 \text{ centimeters}$$

Some problems may involve more than one conversion. In these cases, you carry out each conversion as a separate step in the proportion method or multiply by several conversion factors in the unit-factor method.

Example: Convert 50 kilograms to pounds

(a) Proportion method

$$\frac{1 \text{ kilogram}}{50 \text{ kilograms}} = \frac{1000 \text{ grams}}{x}$$

$$1 \text{ kilogram} \times x = 50 \text{ kilograms} \times 1000 \text{ grams}$$

$$x = \frac{50 \text{ kilograms} \times 1000 \text{ grams}}{1 \text{ kilogram}}$$

$$x = 50,000 \text{ grams}$$

$$\frac{1 \text{ pound}}{x} = \frac{454 \text{ grams}}{50{,}000 \text{ grams}}$$

$$1 \text{ pound} \times 50{,}000 \text{ grams} = x \times 454 \text{ grams}$$

$$\frac{1 \text{ pound} \times 50{,}000 \text{ grams}}{454 \text{ grams}} = x$$

$$110 \text{ pounds} = x$$

(b) Unit-factor method

$$\text{pounds} = 50 \text{ kilograms} \times \frac{1000 \text{ grams}}{\text{kilogram}} \times \frac{1 \text{ pound}}{454 \text{ grams}}$$

$$= 110 \text{ pounds}$$

In problems of this type, the unit-factor method is much more efficient, but care is needed to prevent mistakes in arranging the conversion factors. An alternative method involves finding a conversion that directly relates the units; in this case, you could use the relationship 1 kilogram = 2.2 pounds.

B.3 Temperature conversions

Conversion between Celsius and Fahrenheit temperatures can be performed using one of the two following formulas:

$$^\circ\text{F} = \frac{9}{5} \, (^\circ\text{C}) + 32 \tag{1}$$

$$^\circ\text{C} = \frac{5}{9} \, (^\circ\text{F} - 32) \tag{2}$$

You use equation (1) when you want to find the Fahrenheit temperature and equation (2) when you want to find the Celsius temperature.

Example: Convert 68°F to Celsius

$$^\circ\text{C} = \frac{5}{9} \, (^\circ\text{F} - 32)$$

$$= \frac{5}{9} \, (68 - 32)$$

$$= \frac{5}{9} \, (36)$$

$$= 20^\circ\text{C}$$

Example: Convert 50°C to Fahrenheit

$$^\circ\text{F} = \frac{9}{5} \, (^\circ\text{C}) + 32$$

$$= \frac{9}{5}(50) + 32$$

$$= 90 + 32$$

$$= 122°F$$

You can also carry out these temperature conversions with the following proportion:

$$\frac{°C}{°F - 32} = \frac{5}{9} \tag{3}$$

By rearranging this proportion, you can obtain formulas (1) and (2). Or the equation can be used directly, by substituting the given information and solving for the unknown.

The Kelvin temperature is derived from the Celsius temperature by simply adding 273:

$$K = °C + 273$$

Thus, if the temperature is 25°C, the Kelvin equivalent is 25 + 273, or 298 K.

Exercises*

1. A patient receives 250 milligrams of a drug. What is this dose in grams?
2. How many micrograms is the dose in problem 1?
3. An infant weighs 6 pounds. What is the weight in grams? in kilograms?
4. A package contains 180 grams of food. How many pounds does it weigh?
5. Convert 50 centimeters to millimeters.
6. A person runs 2.5 kilometers. What is this distance in meters?
7. What is a waistline in millimeters if its value is 32 inches?
8. A woman's height is 130 centimeters. What is her height in inches?
9. If you drink 0.4 liter of water, what is the volume in milliliters?
10. An injection contains 5 cubic centimeters of fluid. What is the volume in milliliters? in liters?
11. How many quarts are there in a liter of milk?
12. A gasoline tank holds 15 gallons. How many liters does it hold? (Note: 4 quarts equal 1 gallon.)
13. If you have a temperature of 104°F, what is its value in the Celsius scale?
14. If a package says "Store at 10°C," do you have to put it in the freezer, refrigerate, or leave at room temperature?
15. Liquid nitrogen has a temperature of −78°C. What is its Kelvin temperature?
16. Convert 296 K to Celsius and to Fahrenheit temperatures.

* Answers may be found in Appendix F.

Orbitals

C.1 Electron configurations of the atoms

The modern "picture" of the atom is more complicated than described in Chapter 2. Not only does an atom have energy levels, but these are divided into **sublevels**. The further an evergy level from the nucleus (and the greater its energy), the more sublevels exist. The first energy level (K) has one sublevel, labeled s; the second level (L) has two sublevels, an s and a p; the third (M) has three, an s, p, and d.

Each sublevel consists of a certain number of **orbitals**, as illustrated in Figure C-1. Orbitals represent regions of space around the atom in which there is the greatest chance of finding an electron. Each orbital can hold a maximum of *two* electrons. There is only one orbital corresponding to the s sublevel, the s orbital, which has a spherical shape. The three possible orbitals of a p sublevel, p orbitals, have shapes similar to dumbbells. An atom can have a total of six electrons in each p sublevel, two in each of the three orbitals.

Table C-1 presents the **electron configurations**, the arrangement of electrons, for the first 20 elements. Hydrogen has only one electron; it is in the s sublevel of the first energy level, identified $1s^1$. The number in front is the major energy level, and the superscript is the number of electrons in the sublevel. With helium, the first energy level is filled; two electrons are in the s sublevel: $1s^2$. In lithium, $1s^22s^1$, and beryllium, $1s^22s^2$, the s orbital of the second energy level fills up. Since the second level has a p sublevel, the next six electrons are added to p orpitals, until this level is filled in the noble gas neon, $1s^22s^22p^6$. This buildup (*aufbau*) process continues for all the atoms, one electron being added each time to the next available (lowest energy) orbital. The order in which the levels are filled is shown in Figure C-2.

You can think of the periodic table as being organized into "blocks," each of which corresponds to the filling of one type of sublevel, as shown in Figure C-3. The s subshell is being filled in the elements of Groups Ia and IIa (as well as helium). Electrons are being added to the p subshell for the elements from Group IIIa up to the noble gases. Notice that in each row (except the

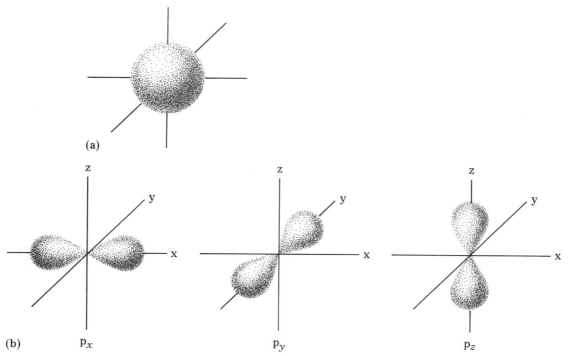

(a)

(b) p_x p_y p_z

Figure C-1. Atomic orbitals: (a) s orbital; (b) p orbitals.

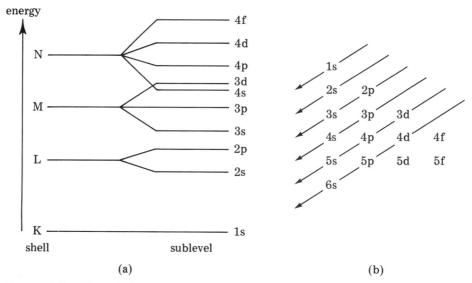

energy

N — 4f
    4d
    4p
M — 3d
    4s
    3p
    3s
L — 2p
    2s

K —————— 1s

shell   sublevel

(a)

1s
2s   2p
3s   3p   3d
4s   4p   4d   4f
5s   5p   5d   5f
6s

(b)

Figure C-2. The relative energies of the atomic sublevels. (a) Electrons are added beginning at the level of lowest energy, at the bottom of the diagram. (b) A way of remembering the order in which the levels are filled.

Table C-1 Electron Configurations

Element	Electron configuration			
H	$1s^1$			
He	$1s^2$			
Li	$1s^2$	$2s^1$		
Be	$1s^2$	$2s^2$		
B	$1s^2$	$2s^22p^1$		
C	$1s^2$	$2s^22p^2$		
N	$1s^2$	$2s^22p^3$		
O	$1s^2$	$2s^22p^4$		
F	$1s^2$	$2s^22p^5$		
Ne	$1s^2$	$2s^22p^6$		
Na	$1s^2$	$2s^22p^6$	$3s^1$	
Mg	$1s^2$	$2s^22p^6$	$3s^2$	
Al	$1s^2$	$2s^22p^6$	$3s^23p^1$	
Si	$1s^2$	$2s^22p^6$	$3s^23p^2$	
P	$1s^2$	$2s^22p^6$	$3s^23p^3$	
S	$1s^2$	$2s^22p^6$	$3s^23p^4$	
Cl	$1s^2$	$2s^22p^6$	$3s^23p^5$	
Ar	$1s^2$	$2s^22p^6$	$3s^23p^6$	
K	$1s^2$	$2s^22p^6$	$3s^23p^6$	$4s^1$
Ca	$1s^2$	$2s^22p^6$	$3s^23p^6$	$4s^2$

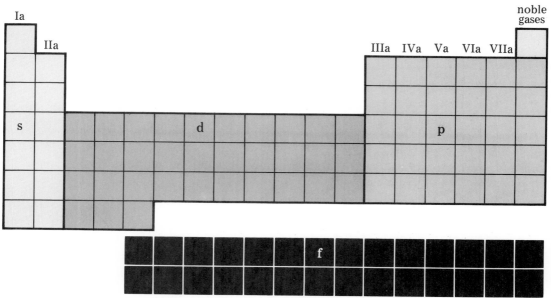

Figure C-3. Arrangement of the periodic table into blocks. Each block corresponds to the filling of one type of sublevel.

first), two elements are in the "s block" and six elements are in the "p block," since the s orbital is completely filled with two electrons and the three p orbitals are filled with a total of six electrons. Those elements in the "d block" are transition metals, while those in the "f block" are lanthanides and actinides. As you can see, the d orbitals can hold a maximum of 10 electrons and the f orbitals can hold a total of 14 electrons.

C.2 Hybrid orbitals

As you learned in Section 10.2, the simplest hydrocarbon, methane, CH_4, has four covalent bonds that point toward the corners of a tetrahedron. Carbon has the electronic configuration $1s^2 2s^2 2p^2$ (see previous section). You would expect the 2s and 2p electrons to form different kinds of bonds, yet all four bonds in the methane molecule are alike. To explain this fact, you can think of the 2s and the 2p orbitals of carbon combining to create a new kind of mixed or **hybrid orbital**, called an sp^3 (pronounced s-p-three) or tetrahedral bond orbital, as shown in Figure C-4. Each of the four resulting hybrid orbitals contains one of the four electrons of the L shell of carbon.

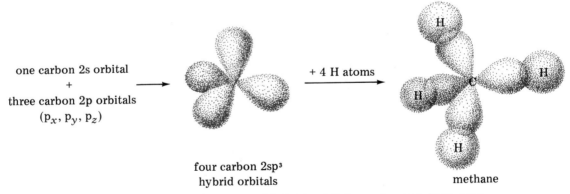

one carbon 2s orbital
+
three carbon 2p orbitals
(p_x, p_y, p_z)

+ 4 H atoms

four carbon $2sp^3$
hybrid orbitals

methane

Figure C-4. Formation of methane. The sp^3 hybrid orbitals are formed from combining the carbon 2s and 2p orbitals.

The methane molecule forms when each of the sp^3 orbitals of carbon overlaps with the 1s orbitals of four hydrogen atoms, as shown in Figure C-4. The covalent bonds that result from this overlap contain shared electrons located directly between the centers of the atomic nuclei; they are called **sigma** (σ) **bonds**. The ethane molecule can be pictured in a similar way. As illustrated in Figure C-5, the single bond between the two carbons results from the overlap of one sp^3 orbital from each atom.

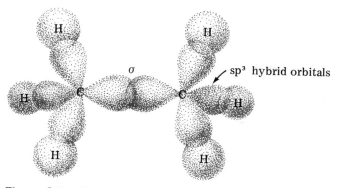

Figure C-5. The bonding in ethane. A sigma (σ) bond is formed by overlap of carbon sp^3 orbitals.

Ethene (ethylene), C_2H_4, has a very different shape from ethane and therefore has different bonding. The 2s orbital on each carbon mixes with two of the 2p orbitals to form three sp^2 (pronounced s-p-two) hybrid orbitals. As shown in Figure C-6, each of the sp^2 orbitals, which are 120° apart, is involved in the formation of a sigma bond—two on each carbon atom bond to hydrogen 1s orbitals, and one bonds to the other carbon. The p orbitals remaining on each carbon atom overlap to form a **pi (π) bond**, which is located above and below the line connecting the centers of the carbon nuclei. The double bond of this molecule consists of the sigma bond formed by overlap of the sp^2 orbitals and the pi bond from overlap of the p orbitals.

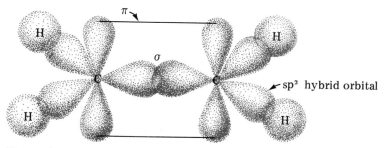

Figure C-6. The bonding in ethene (ethylene). The double bond consists of a sigma (σ) bond from overlap of carbon sp^2 orbitals and a pi (π) bond from overlap of p orbitals.

Ethyne (acetylene), C_2H_2, contains two carbon atoms, each of which has two sp hybrid orbitals, formed from combining one 2s and one 2p orbital. The two remaining 2p orbitals on one carbon atom form two sets of pi bonds with the 2p orbitals on the other carbon atom, as shown in Figure C-7. The triple bond therefore consists of the one sigma bond formed by overlap of the sp orbitals and the two pi bonds from overlap of the p orbitals.

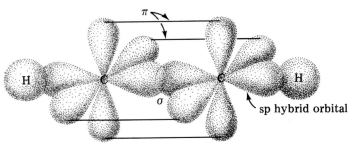

Figure C-7.   The bonding in ethyne (acetylene). The triple bond consists of a sigma (σ) bond from overlap of sp hybrid orbitals and two pi (π) bonds from overlap of two sets of p orbitals.

Details of glucose catabolism

D.1 Glycolysis

Starting with glucose, the first step of glycolysis involves adding a phosphate group or (phosphorylation) using ATP (refer to Figure 17-7):

1. glucose + ATP $\xrightarrow{\text{hexokinase}}$ glucose 6-phosphate + ADP

The second step consists of isomerization of the resulting glucose 6-phosphate to fructose 6-phosphate:

2. glucose 6-phosphate $\xrightarrow{\substack{\text{phospho-}\\\text{glucoisomerase}}}$ fructose 6-phosphate

A second molecule of ATP is added in the third step to form fructose 1,6-diphosphate:

3. fructose 6-phosphate + ATP $\xrightarrow{\substack{\text{phospho-}\\\text{fructokinase}}}$ fructose 1,6-diphosphate + ADP

Step 4 involves cleavage of the six-carbon sugar to 2 three-carbon pieces, glyceraldehyde 3-phosphate, as shown in Figure 17-7, and dihydroxyacetone phosphate:

$$
\begin{array}{l}
CH_2OPO_3{}^{2-} \\
\mid \\
C{=}O \\
\mid \\
CH_2OH
\end{array}
$$

4a. fructose 1,6-diphosphate $\xrightarrow{\text{aldolase}}$

dihydroxyacetone phosphate + glyceraldehyde 3-phosphate

This second molecule is then also converted to glyceraldehyde 3-phosphate since that is the form of triose required for the next phase of glycolysis:

4b. dihydroxyacetone phosphate $\xrightarrow{\substack{\text{triosephosphate} \\ \text{isomerase}}}$ glyceraldehyde 3-phosphate

So far, the glucose molecule has been cleaved into two parts and two molecules of ATP have been used up.

The second part of glycolysis begins with the phosphorylation of glyceraldehyde 3-phosphate to form 1,3-diphosphoglyceric acid:

5. 2[glyceraldehyde 3-phosphate] + 2NAD$^+$ + 2P$_i$ $\xrightarrow{\substack{\text{glyceraldehyde-} \\ \text{3-phosphate dehydrogenase}}}$

2[1,3-diphosphoglyceric acid] + 2NADH + 2H$^+$

In this important step the aldehyde group is oxidized to a carboxylic acid group, releasing two electrons and a hydrogen ion to NAD, which acts as a coenzyme for the dehydrogenase that catalyzes this reaction. In this process, NAD$^+$ is reduced to NADH. The resulting energy is then used to add a phosphate group, forming a high-energy molecule. The 1,3-diphosphoglyceric acid then transfers the phosphate group to a molecule of ADP, producing ATP:

6. 2[1,3-diphosphoglyceric acid] + 2ADP $\xrightarrow{\substack{\text{phosphoglycerate} \\ \text{kinase}}}$

2[3-phosphoglyceric acid] + 2ATP

Thus, the energy of the oxidation of the aldehyde group has been stored in the bond energy of ATP. The 3-phosphoglyceric acid is converted to 2-phosphoglyceric acid by transfer of the remaining phosphate group:

7. 2[3-phosphoglyceric acid] $\xrightarrow{\text{phosphoglyceromutase}}$ 2[2-phosphoglyceric acid]

The resulting molecule is then dehydrated in step 8 to form phosphoenolpyruvic acid:

8. 2[2-phosphoglyceric acid] $\xrightarrow{\text{enolase}}$ 2phosphoenolpyruvic acid + 2H$_2$O

Notice that removal of water, dehydration, results in the formation of a double bond. In the next reaction, the phosphate group is transferred to ADP, again forming a molecule of ATP:

9. 2phosphoenolypyruvic acid + 2ADP $\xrightarrow{\substack{\text{pyruvate} \\ \text{kinase}}}$ 2pyruvic acid + 2ATP

In step 10, pyruvic acid, which is left when the phosphate is lost, becomes reduced:

10. 2pyruvic acid + 2NADH + 2H$^+$ $\xrightarrow{\substack{\text{lactate} \\ \text{dehydrogenase}}}$ 2lactic acid + 2NAD$^+$

NADH, formed in step 5, now transfers its electron and hydrogen to convert the carbonyl group to an alcohol group, resulting in the final product of glycolysis, lactic acid.

D.2   The citric acid cycle

The first step of the citric acid cycle consists of the addition of the two-carbon acetyl group from acetyl-CoA to the four-carbon oxaloacetic acid to form six-carbon citric acid (refer to Figure 17-8):

1. acetyl-CoA + oxaloacetic acid $\xrightarrow{\text{citrate synthase}}$ citric acid + CoA

In the next two steps, *cis*-aconitic acid is produced by dehydration, and isocitric acid is then formed by adding the water again, making this isomer of citric acid:

2a. citric acid $\xrightarrow{\text{aconitase}}$ *cis*-aconitic acid

2b. *cis*-aconitic acid $\xrightarrow{\text{aconitase}}$ isocitric acid

Step 3 involves removing a carboxyl group from isocitric acid as carbon dioxide, in the process reducing NAD^+ to NADH:

3. isocitric acid + NAD^+ $\xrightarrow{\substack{\text{isocitrate} \\ \text{dehydrogenase}}}$

$$\text{alpha-ketoglutaric acid} + CO_2 + \text{NADH}$$

The alpha-ketoglutaric acid thus formed is then oxidized to succinic acid by producing another molecule of carbon dioxide:

4a. alpha-ketoglutaric acid + NAD^+ + GDP + P_i $\xrightarrow{\substack{\text{CoA, succinyl} \\ \text{thiokinase}}}$

$$\text{succinic acid} + CO_2 + \text{NADH} + H^+ + \text{GTP}$$

In this reaction, guanosine triphosphate (GTP), a high-energy compound like ATP, results from the addition of inorganic phosphate to guanosine diphosphate (GDP), and NAD^+ is converted to NADH. The GTP then converts ADP to ATP in a subsequent reaction:

4b. GTP + ADP $\longrightarrow$ GDP + ATP

Step 5 entails oxidation by removal of two hydrogen atoms to flavin adenine dinucleotide (FAD), generating fumaric acid and $FADH_2$:

5. succinic acid + FAD $\xrightarrow{\substack{\text{succinate} \\ \text{dehydrogenase}}}$ fumaric acid + $FADH_2$

Hydration then occurs to form malic acid:

6. fumaric acid + H_2O $\xrightarrow{\text{fumarase}}$ malic acid

In the last step, malic acid is oxidized to oxaloacetic acid again, converting NAD^+ to NADH:

7. malic acid + NAD^+ $\xrightarrow{\substack{\text{malate} \\ \text{dehydrogenase}}}$ oxaloacetic acid + NADH + H^+

The cycle is now ready to start over again.

The cell

The cell is the basic structural unit of all living matter. A "typical" animal cell is shown in Figure E-1. Its components and their functions are defined in the following list:

cell membrane   surrounds cell, separating it from the environment
cytoplasm   aqueous phase in which the cell components are suspended; includes all of the cell except the nucleus
endoplasmic reticulum   system of membrane-covered channels in the cytoplasm; passageway for internal transport; associated with ribosomes in protein synthesis

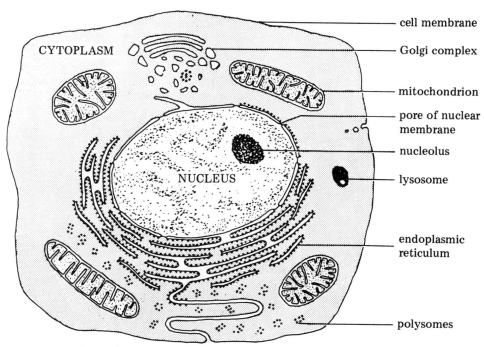

Figure E-1.   A "typical" animal cell.

Golgi complex involved in storage and modification of products of cell secretion

lysosome contains digestive enzymes that break down ingested food

mitochondrion location of aerobic process of respiration; contains many enzyme systems; center for ATP generation

nucleolus involved in rRNA synthesis and ribosome formation; associated with chromosomes

nucleus part of cell containing chromosomes; site of DNA and RNA synthesis; surrounded by its own membrane

polysomes complex of mRNA and ribosomes involved in polypeptide synthesis

Answers to numerical problems

Chapter 2

10. (a) 15, 31, 15, 16, 15
(b) 9, 19, 9, 10, 9
(c) 21, 45, 21, 24, 21

16. 16 amu

Chapter 4

1. (a) 98 amu; (b) 170 amu; (c) 58 amu; (d) 100 amu; (e) 78 amu

2. (a) 2.0% H, 32.7% S, 65.3% O
(b) 63.5% Ag, 8.2% N, 28.3% O
(c) 41.7% Mg, 54.9% O, 3.4% H
(d) 39.1% K, 1.0% H, 12.0% C, 47.9% O
(e) 92.3% C, 7.7% H

5. (a) 98 g; (b) 170 g; (c) 58 g; (d) 100 g; (e) 78 g

6. (a) 164 g; (b) 49 g; (c) 440 g; (d) 6.2 g

15. 46 g

16. 900 g

Chapter 5

8. 5 liters

12. 2 atm

15. 10 liters

19. 900 torr

Chapter 7

13. (a) 2.0 g; (b) 2.25 g; (c) 2.5 g
14. 1.3% (w/v)
16. (a) 360 g; (b) 90 g; (c) 36 g
17. 0.25 M
18. 1.7 ml stock solution diluted to 10 ml
19. 0.5%
20. 10 ml

Chapter 8

14. $[H^+] = 10^{-5}$; $[OH^-] = 10^{-9}$ mol/liter
18. (a) 31 g; (b) 29 g; (c) 33 g
19. 98 g
20. (a) 6 N; (b) 0.5 N; (c) 2 N; (d) 0.1 N
21. 0.25 N

Chapter 9

8. 3.1 g

Appendix B

1. 0.25 g
2. 250,000 g
3. 2724 g; 2.724 kg
4. 0.40 pound
5. 500 mm
6. 2500 m
7. 81.3 mm
8. 51.2 inches

9. 400 ml
10. 5 ml; 0.005 liter
11. 1.06 quart
12. 56.8 liters
13. 40°C
14. 50°F (refrigerate)
15. 195 K
16. 23°C; 99°F

INDEX